With be

from   Or

DR. J. SCAIFE

Anticancer Therapeutics

Anticancer Therapeutics

Edited by

Sotiris Missailidis

The Open University, UK

⊛WILEY-BLACKWELL

A John Wiley & Sons, Ltd., Publication

This edition first published 2008
© 2008 by John Wiley & Sons, Ltd

Wiley-Blackwell is an imprint of John Wiley & Sons, formed by the merger of Wiley's global Scientific, Technical and Medical business with Blackwell Publishing.

Registered office: John Wiley & Sons Ltd, The Atrium, Southern Gate, Chichester, West Sussex, PO19 8SQ, UK

Other Editorial Offices:
9600 Garsington Road, Oxford, OX4 2DQ, UK
111 River Street, Hoboken, NJ 07030-5774, USA

For details of our global editorial offices, for customer services and for information about how to apply for permission to reuse the copyright material in this book please see our website at www.wiley.com/wiley-blackwell

The right of the author to be identified as the author of this work has been asserted in accordance with the Copyright, Designs and Patents Act 1988.

Library of Congress Cataloging-in-Publication Data

Anticancer therapeutics / edited by Sotiris Missailidis.
    p. ; cm.
    Includes bibliographical references and index.
    ISBN 978-0-470-72303-6 (cloth : alk. paper) 1. Cancer – Chemotherapy. 2. Antineoplastic agents. I.
Missailidis, Sotiris.
    [DNLM: 1. Antineoplastic Agents – therapeutic use. 2. Drug Design. 3.
Neoplasms – drug therapy. QZ 267 A6287 2008]
    RC271.C5.A6728 2008
    616.99′4061 – dc22

                                                            200802409

ISBN 978-0-470-72303-6

A catalogue record for this book is available from the British Library.

Typeset in 10/12 Times by Laserwords Private Limited, Chennai, India
Printed and bound in Singapore by Markono Pte Ltd

First printing   2008

To Mary

and all those who may have benefited by

these new drug developments

Contents

Forward   xiii
Acknowledgements   xv
List of contributors   xvii

SECTION I: Development of Anticancer Therapeutics   1

1 Exploring the Potential of Natural Products in Cancer Treatment   3
Fotini N. Lamari and Paul Cordopatis

   1.1   Introduction   3
   1.2   Sources   4
   1.3   Different Approaches to the Search for Bioactive Natural Products   6
   1.4   Methodologies of Lead Compound or New Drug Identification   10
   1.5   Chemoprevention – A New Area for Natural Product Research   13
   1.6   Concluding Remarks   13

2 Combinatorial Approaches to Anticancer Drug Design   17
Sotiris Missailidis

   2.1   Introduction   17
   2.2   Combinatorial Approaches for Small Molecule Drug Design   17
   2.3   Display Technologies   21
   2.4   Aptamer Selection   23
   2.5   Conclusions   27

3 Rational Approaches to Anticancer Drug Design/*in silico* Drug Development   29
Stefano Alcaro, Anna Artese and Francesco Ortuso

   3.1   Introduction   29
   3.2   Approaches to the Drug Discovery Process in Anticancer Research   31

3.3   Ligand-based Examples   32
3.4   Structure-based Examples   36
3.5   Conclusions   44

SECTION II: Anticancer Therapeutics   **47**

4 Introduction to Anticancer Therapeutics   **49**

Teni Boulikas

4.1   Problems in cancer   49
4.2   Cancer treatments   51
4.3   Classification of chemotherapy drugs   53

5 Platinum Drugs   **55**

Teni Boulikas, Alexandros Pantos, Evagelos Bellis and Petros Christofis

5.1   Cisplatin   55
5.2   Lipoplatin   57
5.3   Carboplatin   61
5.4   Oxaliplatin   62
5.5   Lipoxal   62
5.6   New Platinum Compounds   64
5.7   Cisplatin Resistance and Chemotherapy   70

6 Antimicrotubule Agents   **79**

Iain Brown, Jay N Sangrithi-Wallace and Andrew C Schofield

6.1   Taxanes   79
6.2   Vinca Alkaloids   82
6.3   Mechanisms of Resistance to Antimicrotubule Agents   86

7 Antimetabolites in Cancer Therapy   **91**

Jessica Scaife and David Kerr

7.1   Introduction   91
7.2   Folate Antagonists   92
7.3   Pyrimidine Antagonists   96
7.4   Purine Antagonists   104
7.5   Summary   109

8 Antitumour Antibiotics   **111**

Manuel M. Paz

8.1   Introduction   111

8.2   Actinomycin   111
8.3   Mitomycin C   112
8.4   Bleomycin   115
8.5   Anthracyclines   118
8.6   Trabectedin (Ecteinascidin, ET-743)   121
8.7   Camptothecins   123
8.8   Podophyllotoxins   124

9 Alkylating Agents   133
   Ana Paula Francisco, Maria de Jesus Perry, Rui Moreira and Eduarda Mendes
   9.1   Introduction   133
   9.2   Nitrogen Mustards   133
   9.3   Methylmelamines and Ethylenimines   140
   9.4   Methylhydrazine Derivatives   141
   9.5   Alkylsulfonates   143
   9.6   Nitrosoureas   144
   9.7   Triazenes   149

10 Hormone Therapies   159
    George C. Zografos, Nikolaos V. Michalopoulos and Flora Zagouri
    10.1   Introduction   159
    10.2   Oestrogen Receptor Targeted Therapeutics   160
    10.3   Progesterone-Targeted Therapy   174
    10.4   Neuroendocrine Tumours   176

11 Photodynamic Therapy of Cancer   187
    K. Eszter Borbas and Dorothée Lahaye
    11.1   Introduction   187
    11.2   Photosensitizers   196
    11.3   Outlook   212
    11.4   Acknowledgement   212

12 Target-directed Drug Discovery   223
    Tracey D. Bradshaw
    12.1   Introduction   223
    12.2   Tyrosine Kinases – Role and Significance in Cancer   226
    12.3   Targeted Therapy for the Treatment of Non-small Cell Lung Cancer (NSCLC)   226
    12.4   Targeted Therapy for the Treatment of Chronic Myeloid Leukaemia   229
    12.5   Targeted Therapy for the Treatment of Breast Cancer   230
    12.6   Angiogenesis   231

12.7   Targeting Cell Cycling   235
12.8   Targeting Apoptosis   237
12.9   Targeting mTOR   237
12.10 The Future of Molecularly Targeted Therapy   238

13 Tumour Hypoxia: Malignant Mediator   245
Jill L. O'Donnell, Aoife M. Shannon, David Bouchier-Hayes
13.1   Introduction   245
13.2   Hypoxia Inducible Factor-1 and Hypoxia   246
13.3   HIF-1α Post-translational Changes   247
13.4   How Genetics Can Modify HIF   248
13.5   How Tumours Overcome Hypoxia with HIF-1   249
13.6   HIF-1 Therapeutics   252
13.7   Conclusion   255

14 Resistance to Chemotherapy Drugs   263
Robert O'Connor and Laura Breen
14.1   Introduction   263
14.2   What are the Factors Limiting the Efficacy of Cancer Chemotherapy
       Treatment?   263
14.3   A Classification of the Important Chemotherapy Resistance Mechanisms   265
14.4   Illustrative Mechanisms of Pharmacokinetic Resistance   267
14.5   Illustrative Mechanisms of Pharmacodynamic Resistance   273
14.6   Conclusion   277

15 Cancer Immunotherapy   283
Maria Belimezi
15.1   The Molecular Basis of Cancer Immunotherapy and Gene Immunotherapy
       of Cancer   283
15.2   Recombinant Monoclonal Antibodies   284
15.3   Cell Immunotherapy   292
15.4   Cancer Vaccines   296

16 Gene Therapy   305
Maria Belimezi, Teni Boulikas and Michael L. Roberts
16.1   The Concept of Gene Therapy   305
16.2   Steps for Successful Gene Therapy   306
16.3   Retroviruses in Cancer Gene Therapy   307
16.4   Adenoviruses in Cancer Gene Therapy   308
16.5   Gene Therapy of Cancer   310

16.6   Cancer Immunotherapy with Cytokine Genes   311
16.7   IL-12 in Cancer Immunotherapy   311
16.8   Viruses able to Kill Cancer Cells   314

17 Antisense Agents   317
Huma Khan and Sotiris Missailidis

17.1   Introduction   317
17.2   Traditional Antisense Oligonucleotides (ASOs)   318
17.3   Ribozymes and DNAzymes   320
17.4   RNA Interference and siRNAs   321
17.5   Shortcomings of Antisense Therapeutics   322
17.6   Antisense Agents in Clinical Trials   324
17.7   Concluding Remarks   329

18 Aptamers as Anticancer Agents   331
Vaidehi Makwana, Suzanne Simmons and Sotiris Missailidis

18.1   Introduction   331
18.2   Aptamers in Cancer   332
18.3   Final comments   341

**SECTION III: Other Aspects in Anticancer Therapeutic
Development   347**

19 Treatment of Cancer in Conjunction with Other Agents   349
Gary Robert Smith

19.1   Introduction   349
19.2   Non-steroidal Anti-inflammatory Drugs   353
19.3   Angiotensin-converting Enzyme (ACE) Inhibitors and Angiotensin
        Receptor Blockade   358
19.4   Partners in Crime – Dealing with Co-infections   362
19.5   Discussion   363

20 Clinical Trials in Oncology   365
Tim Friede, Janet Dunn and Nigel Stallard

20.1   Clinical Trials   365
20.2   Early-Phase (Phase I and Phase II) Clinical Trials in Oncology 368
20.3   Confirmatory (Phase III) Trials in Oncology   371
20.4   Further Issues in Clinical Trials in Oncology   374

21 Representative Cancers, Treatment and Market   377

Teni Boulikas and Nassos Alevizopoulos

21.1   Lung Cancer   377
21.2   Breast Cancer   378
21.3   Prostate Cancer   378
21.4   Colorectal Cancer   379
21.5   Ovarian Cancer   380
21.6   Pancreatic Cancer   380
21.7   Gastric Cancer   381
21.8   Combination Chemotherapy   382
21.9   The Pharmaceutical World of Anticancer Drugs   383

22 Future Trends in Cancer Therapeutics   387

Sotiris Missailidis

22.1   Introduction   387
22.2   Personalized Medicines   388
22.3   Delivery Systems   390
22.4   Closing Remarks   391

Index   393

Foreword

There remains an urgent unmet clinical need to augment the efforts of surgeons and radiotherapists in combating localized malignant disease. Moreover, better strategies are required to deal with the all-too-common clinical scenario of patients presenting with disseminated metastatic disease, using systemic chemotherapy. In the 60-year history of cancer chemotherapy much has been achieved, particularly through the development of cytotoxic drugs in the 1950s and 1960s.

These are truly exciting times to be involved in anticancer therapeutics. Many would consider that we are now firmly in the second 'golden age' of discovering and developing new treatments for cancer. Such therapies represent the fruits arising from the elucidation of the cancer genotype and phenotype at the molecular and cellular level. This has led to the identification of a plethora of new cancer targets involved in growth signaling, angiogenesis and apoptosis. Notable recent success stories include chimeric (human/mouse) or humanized monoclonal antibodies such as rituximab, trastuzumab and bevacizumab and small molecule kinase inhibitors such as imatinib, erlotinib, sunitinib and sorafenib. These new drugs are making a marked impact in modern chemotherapeutic regimes. However, significant improvements are still required in the process of translating new targets through preclinical discovery and development and into a successful clinical outcome. Evidence suggests that many drugs entering the clinic fail to reach approval for reasons of poor pharmacology, lack of efficacy or unacceptable adverse effects.

The following chapters provide an insight into state-of-the art methodologies to discovering new drugs including combinatorial approaches, natural product libraries and *in silico* techniques. Thereafter, there is a comprehensive guide through traditional cytotoxic drugs (platinum drugs, antimetabolites, tubulin interactive agents, alkylating agents, antitumour antibiotics), hormone and photodynamic therapies and recently approved molecularly-targeted drugs, both small molecules and monoclonal antibody-based. There are also reviews of the important areas of tumour drug resistance and tumour hypoxia. In addition, a glimpse of the future is provided through considerations of new approaches currently under study at the preclinical and early clinical level; these include cancer vaccines, gene therapy, antisense agents and aptamers. Finally, clinical practice issues are described; combination treatments, trials and a focus on particular major cancers.

The current excitement and optimism of improvements in patient benefit arising from modern cancer chemotherapy are illustrated herein. The present trend continues to build on exploiting the molecular hallmarks of cancer; three of the four drugs approved by the US FDA in 2007 were the kinase inhibitors nilotinib and lapatinib and the mTOR inhibitor temsirolimus. In addition, evidence to support a view that new cytotoxics also have a place in contemporary cancer drug development is provided by the 2007 FDA approval of the anti-tubulin epothilone, ixabepilone. There is also considerable attention being paid to the effective combination usage of molecularly-targeted drugs with traditional cytotoxics (for example, with the approval of the angiogenesis inhibitor antibody bevacizumab, with carboplatin and paclitaxel in patients with non-small cell lung cancer). In parallel, there is increasing emphasis being placed on the individualization of treatment options with the identification and use of prognostic and treatment response biomarkers.

There has never been a better era to be involved in discovering and developing anticancer drugs.

Lloyd R. Kelland
Head of Biology
Cancer Research Technology
University College London
London, UK

Acknowledgements

There are a number of people that I would like to acknowledge for their contribution in the completion of this volume. First of all, I would like to acknowledge my wife Dr Giselle Martins dos Santos Ferreira and our daughter Ananda for their patience and support during the frantic time of putting this volume together. I could not have done it without their support and encouragement.

For this volume to be compiled, it took a great many authors, specialists in their field, who volunteered their time when all are busy with their jobs. I would like to thank them for preparing very interesting and comprehensive pieces of work, working meticulously and with good timing to see this effort completed within the time plan that was originally agreed and at the highest standard. I thank all the authors of their chapter contributions. They made the volume what I hope to be a very interesting, informative and authoritative work.

Finally, I would like to thank the team at Wiley, Rachael Ballard, Liz Renwick, Fiona Woods and Robert Hambrook, as well as our editor, Alison Woodhouse, who have guided me through the various stages of this process and with whom I have worked so well in the preparation of this volume and are already preparing the next.

Sotiris Missailidis
The Open university, UK

List of Contributors

Stefano Alcaro
Laboratorio di Chimica Farmaceutica
Dipartimento di Scienze Farmacobiologiche
Università di Catanzaro *'Magna Græcia'*
Roccelletta di Borgia (CZ), Italy

Anna Artese
Laboratorio di Chimica Farmaceutica
Dipartimento di Scienze Farmacobiologiche
Università di Catanzaro *'Magna Græcia'*
Roccelletta di Borgia (CZ), Italy

Maria Belimezi
Regulon Inc. (USA)
Afxentiou 7, Alimos
Athens, Greece

Evagelos Bellis
Regulon Inc. (USA)
Afxentiou 7, Alimos
Athens, Greece

K. Eszter Borbas
Department of Chemistry
North Carolina State University
Raleigh, NC, USA

David Bouchier-Hayes
RCSI Education and Research Unit
Beaumont Hospital
Dublin, Ireland

Teni Boulicas
Regulon Inc. (USA)
Afxentiou 7, Alimos
Athens, Greece

Tracey Bradshaw
Centre for Biomolecular Sciences
School of Pharmacy
University of Nottingham
Nottingham, UK

Laura Breen
National Institute for Cellular Biotechnology
    (NICB)
Dublin City University
Dublin, Ireland

Petros Christofis
Regulon Inc. (USA)
Afxentiou 7, Alimos
Athens, Greece

Paul Cordopatis
Laboratory of Pharmacognosy and Chemistry
    of Natural Products
Department of Pharmacy
University of Patras
Patras, Greece

Janet Dunn
Warwick Medical School
The University of Warwick
Warwick, UK

Ana Paula Francisco
CECF, Faculty of Pharmacy
University of Lisbon
Lisboa, Portugal

Tim Friede
Warwick Medical School
The University of Warwick
Warwick, UK

David Kerr
Department of Clinical Pharmacology
University of Oxford
Old Road Campus Research Building
Oxford, UK

Huma Khan
Department of Chemistry and Analytical
  Sciences
The Open University
Milton Keynes, UK

Dorothée Lahaye
Department of Chemistry
North Carolina State University
Raleigh, NC, USA

Fotini N. Lamari
Laboratory of Pharmacognosy and Chemistry
  of Natural Products
Department of Pharmacy
University of Patras
Patras, Greece

Vaidehi Makwana
Department of Chemistry and Analytical
  Sciences
The Open University
Milton Keynes, UK

Eduarda Mendes
CECF, Faculty of Pharmacy
University of Lisbon
Lisboa, Portugal

Nikolaos V. Michalopoulos
Medical School
University of Athens,
Athens, Greece

Sotiris Missailidis
Department of Chemistry and Analytical
  Sciences
The Open University
Milton Keynes, UK

Rui Moreira
CECF, Faculty of Pharmacy
University of Lisbon
Lisboa, Portugal

Robert O'Connor
National Institute for Cellular Biotechnology
  (NICB)
Dublin City University
Dublin, Ireland

Jill L. O'Donnell
RCSI Education and Research Unit
Beaumont Hospital
Dublin, Ireland

Francesco Ortuso
Laboratorio di Chimica Farmaceutica
Dipartimento di Scienze
  Farmacobiologiche
Università di Catanzaro *'Magna Græcia'*
Roccelletta di Borgia (CZ), Italy

Alexandros Pantos
Regulon Inc. (USA)
Afxentiou 7, Alimos
Athens, Greece

Manuel M. Paz
Department of Organic Chemistry
Faculty of Science
University of Santiago de Compostela
Lugo Campus
Lugo, Spain

Maria de Jesus Perry
CECF, Faculty of Pharmacy
University of Lisbon
Lisboa, Portugal

Michael L. Roberts
Regulon Inc. (USA)
Afxentiou 7, Alimos
Athens, Greece

Jessica Scaiffe
ST2 Core Medical Training
Department of Oncology
Cheltenham General Hospital
Cheltenham, UK

Andrew Schofield
School of Medicine
College of Life Sciences and Medicine
University of Aberdeen
Aberdeen, UK

Aoife M. Shannon
RCSI Education and Research Unit
Beaumont Hospital
Dublin, Ireland

Suzanne Simmons
Department of Chemistry and Analytical
    Sciences,
The Open University,
Walton Hall,
Milton Keynes, UK

Gary Robert Smith
Perses Biosystems Ltd
Warwick, UK

Nigel Stallard
Warwick Medical School
The University of Warwick, UK

Flora Zagouri
Medical School
University of Athens
Athens, Greece

George C. Zografos
Medical School
University of Athens,
Athens, Greece

SECTION I

Development of Anticancer Therapeutics

1
Exploring the Potential of Natural Products in Cancer Treatment

Fotini N. Lamari and Paul Cordopatis

1.1 Introduction

Cancer remains a major cause of mortality worldwide. In 2006 in Europe there were an estimated 3.2 million cancer cases diagnosed (excluding non-melanoma skin cancers) and 1.7 million deaths from cancer (Ferlay *et al.*, 2007). According to the World Health Organization (www.who.int), from a total of 58 million deaths worldwide in 2005, cancer accounts for 7.6 million (or 13 %) of all deaths. Cancer rates are predicted to further increase if nothing changes, mainly due to steadily ageing populations in both developed and developing countries and current trends in smoking prevalence and the growing adoption of unhealthy lifestyles. It is estimated that almost half of cancer cases can be prevented by infection control, adoption of a healthy lifestyle (diet and exercise) and tobacco abstinence. Early diagnosis and effective treatment can further dramatically decrease cancer mortality. In the therapeutic area, the hope is to turn many cases of fatal cancer into 'manageable' chronic illness, as it has happened with other disease entities.

When it comes to treatment, humans have always turned to nature. The medicinal value of plants has been recognized by almost every society on this planet. Up to the nineteenth century, herbal extracts containing mixtures of natural products provided the main source of folk medicines. The first synthetic pharmaceutical drug, aspirin, was developed in the latter half of the nineteenth century and its potency as pain reliever spawned the era of synthetic therapeutic agents. Since the 1940s, 175 anticancer drugs have been developed and are commercially available in the United States, Europe and Japan; 65 % of these were inspired from natural products, i.e. pure natural products (14 % of total), semisynthetic modifications of

Anticancer Therapeutics   Edited by Sotiris Missailidis
© 2008 John Wiley & Sons, Ltd

natural products, natural product mimics or synthetic molecules with pharmacophores from natural products (Newman and Cragg, 2007). The sheer numbers prove the importance and contribution of Nature's biodiversity to the development of efficient therapies.

At present, the role of natural products in drug discovery programmes of large pharmaceutical companies has been de-emphasized. The reasons for this situation can be traced back to the advent of high-throughput screening assays in the drug discovery process that favour single molecules and not mixtures of these (non-specific interferences, presence of fluorescent compounds and insoluble materials), the better suitability of combinatorial chemistry in this set-up and its 'overhyped' potential to deliver new lead compounds, the inherent difficulties of natural product research and the concerns about ownership. The Convention on Biological Diversity (www.cbd.int) in 1992 brought into the agenda the access of developing countries 'on a fair and equitable basis to the results and benefits arising from biotechnologies based upon genetic resources provided by these'. Therefore, apart from the risk of being called 'biopirates', pharmaceutical companies opted not to be engaged in natural product research, which suffers from lack of reproducibility of extracts, the inaccessibility of collection sites, laborious procedures to isolate and purify bioactive chemical compounds (often present in trace amounts), and the rediscovery of known compounds.

The impact of this shift has started to become obvious. Although molecular and cellular biology brought about a surge of molecular targets for therapy, this has not been accompanied by a rise in the number of anticancer drugs developed. Anticancer research has mainly focused on the cancer cells and the development of cytotoxic drugs for efficient and selective chemotherapy. Despite the identification of more than 100 distinct types of cancer and the puzzling findings of molecular and cellular biology, it has been suggested that there are six essential alterations in cell physiology that collectively dictate malignant growth: self-sufficiency in growth signals, insensitivity to growth-inhibitory signals, evasion of apoptosis, limitless replicative potential, sustained angiogenesis, and tissue invasion and metastasis (Hanahan and Weinber, 2000). Thus, the probability of finding one 'magic bullet' drug to cure cancer seems to be nil. Indeed, in chemotherapy, combinations of chemical compounds are used. In recent years, other non-cytotoxic therapeutic agents such as hormonal and biological response modalities are also under study. This is due to the realization that tumours are complex tissues in which mutant cancer cells have conscripted and subverted normal cell types to serve as active collaborators, and, furthermore, the realization of the potential of the immune system. This different approach (to restore balance) is common in many traditional holistic medicinal systems incorporating plants that boost the immune response, like *Withania somnifera* in ayurvedic medicine (Balachandran and Govindarajan, 2005).

The scope of this chapter is to illustrate the potential of natural products as sources of traditional and novel anticancer drugs, with emphasis on the philosophy and rationale directing source screening, lead compounds or new drug identification and subsequent drug development.

1.2 Sources

Plants and microorganisms have been the major sources of natural products throughout the centuries (Balunas and Kinghorn, 2005), but after one century of isolation and structural characterization of their natural products we realize that we have achieved only a glimpse into this vast reservoir of biodiversity. From the approximately 250 000–300 000 plants all over the world, about 10 % have been systematically investigated for the presence of bioactive

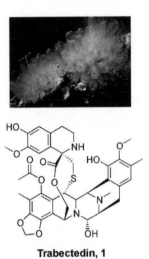

Trabectedin, 1

Figure 1.1 Ecteinascidin-743 (trabectedin or Yondelis) from *Ecteinascidia turbinate* is the first marine organism-derived anticancer drug. Photo Pharma Mar copyright

phytochemicals (McChesney *et al.*, 2007). However, these numbers do not reveal the wealth of natural products therein, since more than one bioactive natural product may be in one plant, either all through the plant or in a particular part (e.g. roots). Their concentration varies greatly according to developmental age, location, time, climate or environment; they may be produced only as a response to an elicitor (chemical communication, physical stimuli). Similarly, out of the million microorganisms, it is estimated that more than 99 % of these remain to be studied. The diversity of marine microorganisms is estimated to be more than 10 million species, more than 60 % of which are unknown (Jensen and Fenical, 1994). In addition to the open ocean, there are diverse and dynamic areas such as mangrove swamps, coral reefs, hydrothermal vents, and deep-sea sediments in which to search for microbes. Other marine organisms (e.g. sponges, tunicates) are an untapped source of novel natural products, which we have just started to explore. Marine organisms produce really novel and complex secondary metabolites. The richest sources of anticancer marine natural products have been soft-bodied and mainly sessile organisms, such as sponges, sea slugs and tunicates, which lack physical defence against their predators, and hence rely on chemical warfare using cytotoxic secondary metabolites. The first drug (ziconotide) isolated from marine organisms was launched in 2006 for chronic pain (Newman and Cragg, 2007), whereas several marine natural products are now tested for their anticancer efficacy in clinical trials (Simmons *et al.*, 2005). Yondelis (ecteinascidin-743, **1**) is a *tris*-tetrahydroisoquiloline isolated from the sea squirt, *Ecteinascidia turbinate* which received authorization in 2007 for commercialization from the European Commission for advanced soft tissue sarcoma (Figure 1.1). Ecteinascidin-743 is under study in two phase III clinical trials for the treatment of sarcoma and ovarian cancer and in other 16 phase I or phase II trials (www.clinicaltrials.gov). It interacts with the minor groove of DNA and alkylates guanine at the N2 position, which bends towards the major groove. Thus, the drug affects various transcription factors involved in cell proliferation, particularly *via* the transcription-coupled nucleotide excision repair system (von Mehren, 2007).

The list of potential sources of natural products can be extended to alcoholic and non-alcoholic beverages, processed food, animals, animal food and excreta, etc. (Tulp and Bohlin, 2004).

1.3 Different Approaches to the Search for Bioactive Natural Products

Given the wealth of sources, searching for a potent anticancer natural product does seem analogous to looking for a needle in a haystack. There are two main approaches to the study of those, the random screening and the rational selection. Random screening is nowadays possible, mainly in the pharmaceutical companies, due to the automated, high-throughput screening assays and the progress in isolation and fractionation techniques. Camptothecin (**2**), a potent cytotoxic alkaloid, was first extracted from the stem wood of the Chinese ornamental tree *Camptotheca acuminata* during the screening of thousands of plants in 1958 (Wall and Wani, 1995) (Figure 1.2). Monroe Wall and Mansukh Wani, while screening thousands of plant extracts as a possible source of steroidal precursors for cortisone, sent 1000 of the extracts to Jonathan Hartwell of the National Cancer Institute (NCI) Cancer Chemotherapy National Service Center (CCNSC) for investigation of their potential antitumour activity.

The extracts of *C. acuminata* were identified as the only ones showing significant activity in an adenocarcinoma assay. Camptothecin inhibits DNA topoisomerase I and the first

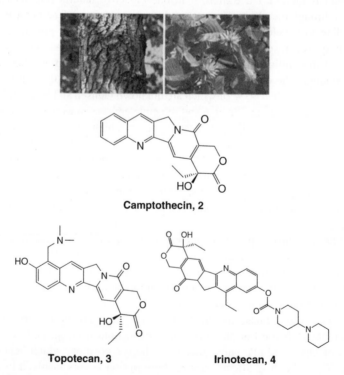

Camptothecin, 2

Topotecan, 3   **Irinotecan, 4**

Figure 1.2 Camptothecin was originally isolated from *Camptotheca acuminata*. The analogues, topotecan and irinotecan are used in the clinical practice. Photos by Shu Suehiro

generation analogues, hycamtin (topotecan, **3**) and camptosar (irinotecan, CPT-11, **4**), marketed by Glaxo-SmithKline and Pfizer, respectively, are used for the treatment of ovarian and colon cancers (Figure 1.2). Over a dozen more camptothecin analogues are currently at various stages of clinical development (Sriram *et al.*, 2005; Srivastava *et al.*, 2005). The same research team discovered taxol (**5**) from the bark of the Pacific Yew tree (*Taxus brevifolia* Nutt.) as result of an exploratory plant screening programme sponsored by the USA National Cancer Institute (Figure 1.3). The discovery of taxol assumed an added measure of importance through the ground-breaking discovery by Horwitz and colleagues (Horwitz, 2004) of its unique mechanism of action, namely, 'its ability to induce the formation of characteristic microtubule bundles in cells'. In 1992, taxol was marketed for the treatment of refractory ovarian cancer and in 1994 for metastatic breast cancer. Taxotere (**6**), one of its semisynthetic derivatives was approved in 1996 for breast cancer and is now known as a better anticancer drug than taxol (Srivastava *et al.*, 2005) (Figure 1.3).

However, it is believed that the effectiveness of random screening in delivering new anticancer drugs or novel lead compounds is surpassed by that of the rational approach. With regards to plants, the most efficient approach has been the recording of the therapeutic uses of the plants in traditional or folkloric medicine systems. This field of research, ethnopharmacology, is part of ethnomedicine and ethnobotany. Information on the alleged medicinal uses of plants comes either from anthropological field studies or from study of ancient literature. There are many local communities in which healing traditions have only been transferred from one generation to the next orally and are now at a risk or are lost since those are displaced by modern western therapeutic systems. The fact that these have been refined through centuries

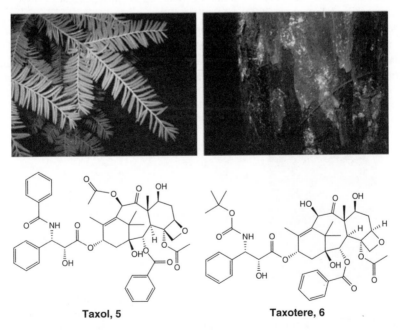

Taxol, 5   **Taxotere, 6**

Figure 1.3 *Taxus brevifolia* was the original source of the anticancer drug taxol. The semisynthetic analogue taxotere is used for breast cancer. Photos by Walter Siegmund, retrieved from http://en.wikipedia.org on December 1 2007 and under GNU Free Documentation license

via trial and error and they have produced important anticancer drugs, such as podophyllotoxin, shows the necessity for taking actions for their preservation. The limited number of field ethnopharmacological studies at present can be attributed to the difficulties and the necessary expertise for the translation of indigenous diseases or concepts of illness into their modern counterparts and vice versa. Anticancer ethnopharmacological research is particularly problematic in that aspect, since cancer is not one disease entity and has multiple signs and symptoms. It has been suggested that the plants used for the following ethnomedical claims should be investigated; cancer treatment, treatment of 'hard swellings', abscesses, calluses, corns, warts, polyps, or tumours, immune disorders, infectious diseases, parasitic and viral diseases, abortive (Mans *et al.*, 2000; Hartwell, 1971). Another practice, though, is to give priority for anticancer research to plants with any alleged medicinal effect. The plants and their medicinal uses by indigenous healers must be recorded, specimens need to be taken back in the laboratories and be correctly identified by botanists, the herbal recipes, the dosage schedules, the total duration of therapy and the side effects must be observed and written down (Raza, 2006).

The species *Podophyllum peltatum* Linnaeus and *Podophyllum emodii* (Podophyllaceae) are the most representative example of plants with a long history of medicinal use for the treatment of cancer, i.e. treatment of skin cancers and warts (Figure 1.4). Two anticancer drugs, teniposide (**7**) and etoposide (**8**), which are semisynthetic derivatives of epipodophyllotoxin, show good clinical effects against several types of neoplasms including testicular and small-cell lung cancers, lymphoma, leukaemia, Kaposi's sarcoma, etc. (Figure 1.4). The major active constituent, podophyllotoxin, was first isolated in 1880, but its correct structure was only reported in the 1950s and it has not been used due to unacceptable toxicity. Podophyllotoxin inhibits the polymerization of tubulin and causes arrest of the cell cycle in the metaphase (Gordaliza *et al.*, 2004).

The development of well-organized electronic databases of relevant research work greatly helps to restrict the initial target group. In 2000 Graham *et al.* provided a list of over 350 plants that are used ethnomedically against cancer after a query in the NAPRALERT database. Apart from the useful input of ethnopharmacological research to modern drug discovery it also

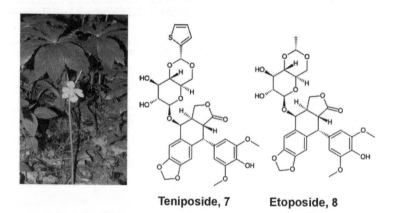

Teniposide, 7   **Etoposide, 8**

Figure 1.4   Teniposide and etoposide are anticancer drugs, which are semisynthetic analogues of epipodophyllotoxin. Epipodophyllotoxin is a natural product present in the rhizome of *Podophyllum peltatum*. Photo by the United States Department of Agriculture, retrieved from http://en.wikipedia.org on December 1 2007 and in the public domain

offers improved therapeutic options in the regions of study through interaction of indigenous healers and physicians trained in western medicine. Furthermore, conservation and refinement of ethnomedicine will likely have a significant positive impact on global health, since more than 80 % of the population still rely on their traditional *materia medica* for their everyday health care needs.

Other rational approaches to the selection of plant or other sources for screening are the taxonomic, the phytochemical and that of chemical ecology. Screening within a given plant genus or family of particular interest (*taxonomic* approach), e.g. the genus *Taxus*, is a strategy which will likely reveal bioactive phytochemicals, especially of the same structural group, and thus promote the structure-activity studies. The high likelihood that a single compound may be present in several taxa of the same genus, and thus be responsible for the bioactivity of these, decreases the possibilities of finding a new lead compound. About 400 taxane-type diterpenoids have been isolated from various *Taxus* plants so far, and some of them possess interesting anticancer activities (Shi and Kiyota, 2005; Kingston and Newman, 2007). Furthermore, this screening introduced a solution to the problem of the supply of taxol. Taxol was isolated in a yield of 0.014 % from the extract of the bark of *Taxus brevifolia* in 1967; this low yield and the high quantities of taxol necessary drove *Taxus brevifolia* close to extinction. A biosynthetic precursor, 10-deacetylbaccatin III, was isolated from the needles of several yew trees, which are an abundant and renewable source, and served as a starting material for the semisynthesis of taxol or taxotere. Total synthesis of taxol, a great challenge to organic chemists, was reported in 1994 by Nicolaou *et al.* giving another solution to the production of taxol and other designed taxoids.

Focusing on a particular group of natural products, e.g. indole alkaloids, and screening all plants likely to contain those, constitutes the *phytochemical* approach, which is of particular significance to systematic botany. However, every group of natural products has a representative with anticancer activity. This has been traditionally performed *via* isolation and structural characterization of the natural products. The advances in molecular biology and genomics change the methodology in this traditional field; Zazopoulos *et al.* (2003) developed and used a high-throughput genome scanning method to detect and analyse gene clusters involved in natural-product biosynthesis. This method was applied to uncover biosynthetic pathways encoding enediyne antitumour antibiotics in a variety of actinomycetes and showed that a conserved cassette of five genes is widely dispersed among actinomycetes and that selective growth conditions can induce the expression of these loci, suggesting that the range of enediyne natural products may be much greater than previously thought. As discussed also in the taxonomic approach, it might be argued that screening in this way limits the possibility of finding novel lead compounds.

Microorganisms have produced in the past a significant number of FDA-approved anticancer drugs, such as bleomycin and doxorubicin, but present a particular problem as a source of natural products. The identity of most of them is not known and less than 1 % of these can be cultivated. A novel way of searching for natural products in these is the construction of environmental DNA libraries (Martinez *et al.*, 2000; Pettit, 2004). This method involves isolating large DNA fragments (100–300 kb) from microorganisms derived from soil, marine or other environments, and inserting these fragments into bacterial vectors, thus generating recombinant DNA libraries. Such libraries are then used to identify novel natural products by various means, including expression of the DNA in a heterologous host strain and screening for activities, or by directly analysing the DNA for genes of interest. Furthermore, the possibility of regulating the expression of isolated environmental gene clusters or combining them with genes of other pathways to obtain new compounds (combinatorial biosynthesis) could be a further advantage over traditional natural product discovery methodologies.

Study of ecological interactions among organisms and between them and their environment, mediated by secondary metabolites that organisms produce (chemical ecology), e.g. plants poisonous to animals, has long been a successful strategy in discovering bioactive natural products (Harborne, 2001). One example of how this field changes under the light of findings from cellular and molecular biology is the search not for general cytotoxic compounds, but for effective nuclear factor (NF)-κB inhibitors in marine organisms (Folmer *et al.*, 2007). Several bacteria and viruses have been reported to modulate NF-κB activity in host cells in order to increase their chances to survive as parasites within the host. Pancer *et al.*, (1999) have shown that sea urchins use a NF-κB analogue to protect themselves against apoptosis-inducing compounds released by the diatoms on which they graze, and to respond to bacterial infection and other pathogens. From the evolutionary point of view, one interesting potential explanation for the finding of NF-κB inhibitors in marine organisms is the fact that marine invertebrates and fish, no matter how distantly related they appear to be, possess, in many cases, NF-κB or closely related analogues (Folmer *et al.*, 2007).

1.4 Methodologies of Lead Compound or New Drug Identification

The tools for evaluating the 'anticancer' potential of natural products are rapidly increasing. Preclinical evaluation involves *in vitro* assays of the effect of natural products on specific molecular targets involved in apoptosis, mitosis, cell cycle control or signal transduction, *in vitro* evaluation of cytotoxicity or other mechanisms of action in cultured cancer cells and other normal cells, and evaluation of the antitumour activity in animal models (Mishra *et al.*, 2007). Bioactivity evaluation should also incorporate methods for evaluating the immunomodulating, anti-invasion or angiosuppressive potential of natural products. This vast array of available bioassays necessitates a strategic decision of which will be the first-line assays, which will determine the natural products that are candidates for the next round of bioactivity evaluation; it is obvious that the complete preclinical evaluation of all natural products is not possible not only because of incredibly high cost but also of ethical considerations. The recent overemphasis on the molecular targets is criticized as simplistic and reductionist, and the study of the effect at the cellular level is reappraised (Houghton *et al.*, 2007; Subramanian *et al.*, 2006). The case of the discovery of vincristine (**9**) and vinblastine (**10**), anticancer drugs approved in the early 1960s, which led to the semisynthesis of vinorelbine and vindesine, poses other interesting implications. *Catharanthus roseus* (former *Vinca rosa*) is used in ayurvedic medicine for the treatment of diabetes mellitus (Figure 1.5). In search for effective hypoglycaemic agents, Robert Noble and Charles Beer were surprised to observe that intravenous administration of the *C. roseus* extract to experimental mice resulted in a rapidly falling white blood count, granulocytopenia, and profoundly depressed bone marrow (Duffin, 2002). This chance observation led to the isolation and identification of vinblastine and vincristine as potent therapeutic agents and novel lead compounds.

Bioactivity evaluation is performed on isolated natural products and/or on extracts and/or purified fractions of those. The classic phytochemical approach has the risk of missing natural products that are in trace amounts, and, thus, rediscover known compounds,

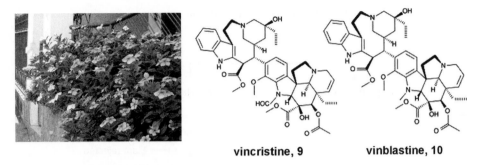

vincristine, 9   vinblastine, 10

Figure 1.5 *Catharanthus roseus* was traditionally used for treatment of diabetes but by chance discovery it was shown to possess anticancer activity. The useful anticancer agents, vincristine and vinblastine, were isolated from *C. roseus*. Photo by Conrado, retrieved from http://en.wikipedia.org on December 1 2007 and under GNU Free Documentation license

e.g. polyphenolics, in high abundance. Bioassay-guided fractionation is the most commonly used strategy for the identification of the bioactive lead compounds. Fractionation reduces complexity, increases the titer of low abundance components and removes 'nuisance' substances. The strategy followed for the isolation of camptothecin is shown in Figure 1.6.

Bioassay-guided fractionation can lead to 'strange' findings (Pieters and Vlietinck, 2005). The bioactivity of the extract might be higher than that of the isolated compounds or fractions and this may be attributed to synergy of the phytochemicals present or decomposition and/or oxidation of the phytochemicals due to the lower amounts of antioxidants present in the fractions and the materials/solvents used in the fractionation process. The possibility of the presence of not one active compound but of several is very strong. On the other hand, the bioactivity of the isolated compounds might be higher than 100 % of the extract due to competition of the phytochemicals present in the extract. The pharmacokinetics of the fractions may also be different from that of the extract (better or worse) since it has been shown that certain natural products affect absorption, e.g. tannins decrease absorption from intestine.

These procedures may lead to the re-isolation and re-identification of known compounds as the bioactive constituents, which is regarded as a considerable loss of time and funds in the search for novel bioactive lead compounds. Thus, de-replication is necessary at an early stage of the discovery process, preferably in the primary extracts, so as to allow the prioritization of work and concentration on those sources that produce novel compounds. Liquid chromatography–mass spectroscopy (LC-MS)/MS coupled with the on-line acquisition of UV/vis spectra and the construction of libraries is a tool for correct structure identification of phytochemicals in an extract (Fredenhagen *et al.*, 2005). Böröczky *et al.* (2006) suggested a simple gas chromatography-based method using cluster analysis as a data-mining tool to select samples of interest for further analysis of lipophilic extracts. Furthermore, the construction of natural product-focused spectral libraries of nuclear magnetic resonance data of isolated compounds allow for the rapid structural elucidation and thus an early de-replication (Dunkel *et al.*, 2006; López-Pérez *et al.*, 2007).

Once identified and the results of preclinical evaluation are good, the bioactive natural product will be directed to clinical evaluation. Results of clinical phase I and II trials will determine if the compound will be evaluated in phase III trials, will be sent back to the laboratory for optimization, or abandoned (Connors, 1996). On the way to the marketplace, the crucial problems of supply and large-scale production must be solved (McChesney *et al.*,

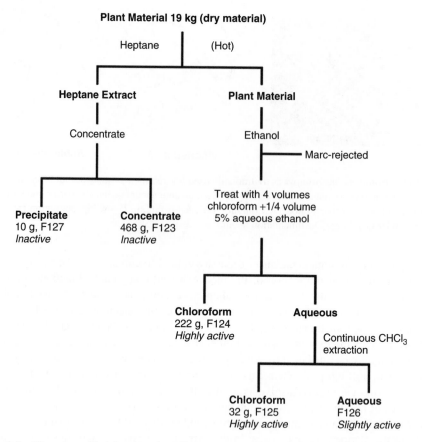

Figure 1.6 Bioassay-guided fractionation of *Camptotheca acuminata* wood plus wood bark. Bioactivity of the fractions was evaluated by the *in vivo* L1210 mouse life prolongation activity. Most of the chloroform phase after concentration was subjected to an 11-stage preparative Craig countercurrent distribution. The bioactive fractions were then combined and further purified by chromatography on a silica gel column and crystallization. The pure bioactive compound was proved to be camptothecin. Reproduced from Wall ME, Wani MC. Camptothecin and taxol: discovery to clinic – thirteenth Bruce F. Cain Memorial Award Lecture. *Cancer Res* 1995, **55**(4), 753–60, with permission from the American Association of Cancer Research

2007). Medicinal chemists play a key role in the generation of structural analogues and introduction of 'drug-like' features. Apart from the traditional chemoenzymatic approach, combinatorial chemistry of a natural product-lead compound is often involved in construction of libraries whereas combinatorial biosynthesis holds a great promise in the field (Boldi, 2004; Harvey, 2007). Combinatorial biosynthesis can be defined as the application of genetic engineering to modify biosynthetic pathways to natural products in order to produce new and altered structures using nature's biosynthetic machinery (Floss, 2006; Julsing *et al.*, 2006). The introduced structural alterations range from simple reactions such as glycosylation, oxidations and reductions, methylations, isoprenylations, halogenations and acylations to the generation of complex hybrid 'unnatural' compounds. The screening of natural product libraries and extracts usually yields a substantially higher percentage of bioactive hits than that of chemical libraries; a recent review (Berdy, 2005) estimates an approximately 100-fold higher hit rate for natural products.

1.5 Chemoprevention – A New Area for Natural Product Research

Epidemiological studies, showing that increased intake of fruits and vegetables is associated with reduced risk of cancer, triggered research on the identification and characterization of the biological properties of the natural products in edible plants and the creation of a new scientific field, that of chemoprevention (Reddy *et al.*, 2003). Chemoprevention, as a scientific field, may be considered still at its infancy, and includes the use of natural or pharmacological agents to suppress, arrest or reverse carcinogenesis, at its early stages. Studies, mainly *in vitro*, have shown that most dietary natural products exhibit pleiotropy; they affect several biological processes (even opposing functions) and act on a multitude of molecular targets (Reddy *et al.*, 2003; Russo, 2007). The 'antioxidant' effect is put forward by most scientists and helps unify the positive effects on different systems, e.g. cardiovascular, neurodegenerative diseases and cancer. Natural products, like genistein, resveratrol, curcumin, retinoic acid and epigallocatechin-3-gallate, became the focus of intense research and public interest. In parallel, a lot of dietary supplements, functional or medical foods, became available to the public and this created a lot of concerns about the safety, the quality, the efficacy and the legislative status of these products.

The field of the study of natural products as chemopreventive agents has to address many problems and challenges. A major problem is the confusion in the literature; from experiments in cell cultures with concentrations of natural products equal to or even higher than those appropriate for pharmacological agents and with no knowledge or study of the absorption and the bioavailability, some scientists jump to conclusions about anticancer or chemopreventive potential (Russo, 2007). Other important questions are 'when' the chemopreventive intervention must take place to show efficacy and what happens if the antioxidant treatment does not occur at the 'appropriate time' and especially what happens when it takes place with standard chemotherapy (Russo, 2007). The clinical evaluation of the chemopreventive properties of a natural product is particularly challenging due to the time involved, the lack of appropriate biomarkers and the fact that it involves healthy people. However, the fact that selective oestrogen receptor modulators, like tamoxifen and raloxifene, do decrease the incidence of breast cancer in post-menopausal women suggests a bright future for chemoprevention; raloxifene especially is a multifunctional medicine that was approved for reducing the risk of invasive breast cancer in postmenopausal women with osteoporosis and in postmenopausal women at high risk for invasive breast cancer (Jordan, 2007). Thus, it is possible that natural products, analogues or combinations of these will be used as chemopreventive agents. Furthermore, the public attention paid to dietary chemoprevention can only be viewed as positive since it increases the awareness of people of the significance of well-balanced diet rich in fruits and vegetables.

1.6 Concluding Remarks

Plants and microorganisms have been sources of a significant percentage of potent anticancer drugs used nowadays, although a small portion of these have been studied. Further exploration of those, of marine organisms and of other novel sources will certainly reveal new drugs, novel lead compounds and new mechanisms of action. However, this process is time-consuming since it involves several steps: selection of the sources, screening and identification of bioactive drugs or lead compounds, *in vitro* and *in vivo* studies of the toxicity and mechanism of action, production or synthesis in high quantities, preclinical and clinical evaluation, approval and development of analogues with better characteristics which enter again the same cycle of drug development. Recognition that the biological diversity of the earth and, thus, the chemical diversity is rapidly diminishing is a very important stimulus for natural products research in the face of irreversible loss of sources of potential drugs. Moreover, in the light of increasing cancer rates, the area of cancer prevention using natural products is very important.

Progress can only be realized with sufficient funds. The immediate co-operation of universities, institutes, big pharmaceutical companies and small biotechnology firms is necessary in order to meet the demands for effective pharmaceuticals. Each sector can contribute in a different way; large-scale random high-throughput screening and clinical development can take place in pharmaceutical companies and in large institutes; universities and institutions can take on research directions that require lengthy procedures and are not so expensive, e.g. screening according to ethnopharmacology. The involvement of scientists from all fields in natural products research has and will further transform the field and the techniques involved in order to meet the demands of modern drug discovery and development.

References

Balunas MJ, Kinghorn AD. Drug discovery from medicinal plants. *Life Sci* 2005, **78**(5), 431–41.

Balachandran P, Govindarajan R. Cancer – an ayurvedic perspective. *Pharmacol Res* 2005, **51**(1), 19–30.

Berdy J. Bioactive microbial metabolites: a personal view. *J Antibiot* 2005, **58**, 1–26.

Boldi AM. Libraries from natural product-like scaffolds. *Curr Opin Chem Biol* 2004, **8**, 281–286.

Böröczky K, Laatsch H, Wagner-Döbler I, Stritzke K, Schulz S. Cluster analysis as selection and dereplication tool for the identification of new natural compounds from large sample sets. *Chem Biodivers* 2006, **3**(6), 622–34.

Connors T. Anticancer drug development: the way forward. *Oncologist* 1996, **1**(3), 180–81.

Duffin J. Poisoning the spindle: serendipity and discovery of the anti-tumour properties of the Vinca alkaloids. *Pharm Hist* 2002, **44**(2), 64–76.

Dunkel M, Fullbeck M, Neumann S, Preissner R. SuperNatural: a searchable database of available natural compounds. *Nucleic Acids Res* 2006, **34**(Database issue): D678–83.

Ferlay J, Autier P, Boniol M, Heanue M, Colombet M, Boyle P. Estimates of the cancer incidence and mortality in Europe in 2006. *Ann Oncol* 2007, **18**(3), 581–92.

Floss HG. Combinatorial biosynthesis – Potential and problems. *J Biotech* 2006, **124**, 242–57.

Folmer F, Jaspars M, Dicato M, Diederich M. Marine natural products as targeted modulators of the transcription factor NF-kappaB. *Biochem Pharmacol* 2008, **75**(3), 603–17.

Fredenhagen A, Derrien C, Gassmann E. An MS/MS library on an ion-trap instrument for efficient dereplication of natural products. Different fragmentation patterns for [M + H]⁺ and [M + Na]⁺ ions. *J Nat Prod* 2005, **68**(3), 385–91.

Gordaliza M, García PA, del Corral JM, Castro MA, Gómez-Zurita MA. Podophyllotoxin: distribution, sources, applications and new cytotoxic derivatives. *Toxicon* 2004, **44**(4), 441–59.

Graham JG, Quinn ML, Fabricant DS, Farnsworth NR. Podophyllotoxin: distribution, sources, applications and new cytotoxic derivatives. *J Ethnopharmacol* 2000, **73**(3), 347–77.

Harvey AL. Natural products as a screening source. *Curr Opin Chem Biol* 2007, **11**, 480–484.

Houghton PJ, Howes M-J, Lee CC, Steventon G. Uses and abuses of in vitso tests in ethnopharmacology: visualizing an elephant. *J Ethnopharmacol* 2007, **110**, 391–400.

Hanahan D, Weinber R. The hallmarks of cancer. *Cell* 2000, **100**(1), 57–70.

Hartwell JL. Plants used against cancer. A survey. *Lloydia* 1971, **34**(2), 204–55.

Harborne JB. Twenty-five years of chemical ecology. *Nat Prod Rep* 2001, **18**(4), 361–79.

Horwitz SB. Personal recollections on the early development of taxol. *J Nat Prod* 2004, **67**(2), 136–8.

Jensen PR, Fenical W. Strategies for the discovery of secondary metabolites from marine bacteria: ecological perspectives. *Annu Rev Microbiol* 1994, **48**, 559–84.

Julsing MK, Koulman A, Woerdenbag HJ, Quax WJ, Kayser O. Combinatorial biosynthesis of medicinal plant secondary metabolites. *Biomol Eng* 2006, **23**(6), 265–79.

Jordan VC. SERMs: meeting the promise of multifunctional medicines. *J Natl Cancer Inst* 2007, **99**(5), 350–6.

Kingston DG, Newman DJ. Taxoids: cancer-fighting compounds from nature. *Curr Opin Drug Discov Devel* 2007, **10**(2), 130–44.

López-Pérez JL, Therón R, del Olmo E, Díaz D. NAPROC-13: a database for the dereplication of natural product mixtures in bioassay-guided protocols. *Bioinformatics* 2007, **23**(23), 3256–7.

Mans DR, da Rocha AB, Schwartsmann G. Anti-cancer drug discovery and development in Brazil: targeted plant collection as a rational strategy to acquire candidate anti-cancer compounds. *Oncologist* 2000, **5**(3), 185–98.

Martinez A, Hopke J, MacNeil IA, Osburne MS. Accessing the genomes of uncultivated microbes for novel natural products. In: *Natural Products: Drug Discovery and Therapeutic Medicine*, edited by L Zhang and AL Demain. Humana Press Inc., Totowa, NJ, 2000.

McChesney JD, Venkataramans SK, Henri JT. Plant natural products: back to the future or into extinction? *Phytochemistry* 2007, **68**, 2015–22.

Mishra KP, Ganju L, Sairam M, Banerjee PK, Sawhney RC. A review of high throughput technology for the screening of natural products. *Biomed Pharmacother* 2008, **62**(2), 94–8.

Newman DJ, Cragg GM. Natural products as sources of new drugs over the last 25 years. *J Nat Prod* 2007, **70**(3), 461–77.

Nicolaou KC, Yang Z, Liu JJ, *et al.* Total synthesis of taxol. *Nature* 1994, **367**(6464), 630–4.

Pancer Z, Rast JP, Davidson EH. Origins of immunity: transcription factors and homologues of effector genes of the vertebrate immune system expressed in sea urchin coelomocytes. *Immunogenetics* 1999, **49**, 773–86.

Pettit RK. Soil DNA libraries for anticancer drug discovery. *Cancer Chemother Pharmacol* 2004, **54**, 1–6

Pieters LA, Vlietinck AJ. Bioguided isolation of pharmacologically active plant components, still a valuable strategy for the finding of new lead compounds? *J Ethnopharmacol* 2005, **100**, 57–60

Raza M. A role for physicians in ethnopharmacology and drug discovery. *J Ethnopharm* 2006, **104**, 297–301

Reddy L, Odhav B, Bhoola KD. Natural products for cancer prevention: a global perspective. *Pharmacol Ther* 2003, **99**(1), 1–13.

Russo GL. Ins and outs of dietary phytochemicals in cancer chemoprevention. *Biochem Pharmacol* 2007, **74**(4), 533–44.

Shi QW, Kiyota H. New natural taxane diterpenoids from *Taxus* species since 1999. *Chem Biodivers* 2005, **2**(12), 1597–623.

Simmons TL, Andrianasolo E, McPhail K, Flatt P, Gerwick WH. Marine natural products as anticancer drugs. *Mol Cancer Ther* 2005, **4**(2), 333–42.

Sriram D, Yogeeswari P, Thirumurugan R, Bal TR. Camptothecin and its analogues: a review on their chemotherapeutic potential. *Nat Prod Res* 2005, **19**(4), 393–412.

Srivastava V, Negi AS, Kumar JK, Gupta MM, Khanuja SP. Plant-based anticancer molecules: a chemical and biological profile of some important leads. *Bioorg Med Chem* 2005, **13**(21), 5892–908.

Subramanian B, Nakeff A, Tenney K, Crews P, Gunatilaka L, Valeriote F. A new paradigm for the development of anticancer agents from natural products. *J Exp Ther Oncol* 2006, **5**(3), 195–204.

Tulp M, Bohlin L. Unconventional natural sources for future drug discovery. *Drug Discov Today* 2004, **9**(10), 450–8.

von Mehren M. Trabectedin–a targeted chemotherapy? *Lancet Oncol* 2007, **8**(7), 565–7.

Wall ME, Wani MC. Camptothecin and taxol: discovery to clinic – thirteenth Bruce F. Cain Memorial Award Lecture. *Cancer Res* 1995, **55**(4), 753–60.

Zazopoulos E, Huang K, Staffa A, *et al.* A genomics-guided approach for discovering and expressing cryptic metabolic pathways. *Nat Biotechnol* 2003, **21**(2), 187–90.

2

Combinatorial Approaches to Anticancer Drug Design

Sotiris Missailidis

2.1 Introduction

In the first instance, the primary sources of anticancer agents were natural sources, plant extracts, sea molluscs and other such life forms or parts thereof, which were on some occasions initially used in traditional medicines and preparations, or were discovered though extensive screening of materials from such sources on cancer cell lines for the identification of active ingredients as a first generation agents. These were subsequently chemically manipulated to improve their anticancer or pharmaceutical properties. However, as chemical and technological developments were progressing, it became possible to synthesize, using automated solid-phase or solution-phase synthetic techniques, vast libraries of *de novo* compounds that could be tested against cancer cell lines or particular targets of interest to generate high affinity and selectivity agents with appropriate desirable properties. A number of combinatorial techniques are currently available, ranging in application from the preparation of small organic compounds to the synthesis of the latest biologics that seem to currently dominate the market in terms of FDA approvals and sales in the past few years. Some of the combinatorial techniques currently available and widely used in anticancer drug design will be presented in this chapter.

2.2 Combinatorial Approaches for Small Molecule Drug Design

The increasing demand for new therapeutics, or compounds that could be screened in biological assays for the identification of novel leads, has pushed development of technological and chemical methodologies in drug design. Combinatorial chemistry or automated medicinal chemistry approaches for the development of novel inhibitors from libraries of

Anticancer Therapeutics   Edited by Sotiris Missailidis
© 2008 John Wiley & Sons, Ltd

organic molecules has been the focus of large pharmaceutical industries perhaps for the past 10–20 years, but a decade ago it remained the source of frustration, originally failing to fulfil its promise. This was partly due to the limitations of the robust chemical protocols and synthetic reactions available for automation; the fact that methodologies available often resulted in mixtures of compounds, for which few or no purification methods were available, resulting in poor quality control and thus frustrating screening results and difficulties in identifying the active ingredients. However, as synthetic approaches and technological developments have been progressing, combinatorial chemistry, coupled with traditional chemistry and *in silico* drug design, is expected to lead to a number of new drug molecules in the future.

Combinatorial chemistry for the synthesis of small organic compounds is based on the use of libraries of building blocks, scaffolds, and standard chemical reactions that can be achieved automatically, through simple addition and incubation steps, either in solid or solution phase.

Such building blocks could include heterocyclic scaffolds, which are a popular choice due to the fact that the majority of drugs are heterocycles, or acyclic compounds, which are also important in drug discovery as they contain groups such as the guanidine functional group present in numerous biologically active molecules (Figure 2.1) (Ganesan, 2002). Finally, complex combinatorial scaffolds are produced based on the use of natural products, such as alkaloids, flavonoids, terpenoids, etc.

The selection of the scaffolds may also be determined not only by the nature of the molecules used, but by the application that they are aimed for. Thus, the aim is not only to produce large numbers of random diverse compounds, but the generation of focused libraries for specific targets. Focused libraries include target-family-oriented libraries, defined by the use of compounds specifically designed to fit certain target-family proteins, such as kinases, G-coupled receptors, etc. (Koppitz and Eis, 2006). These libraries usually contain 10 000 to 50 000 compounds, derived from several dozen scaffolds and chemistries and they are preferable for particular applications to corporate collections of compounds that are often in the range of millions of compounds. Synthesized starting libraries in a target-family-oriented library setting typically consist of 2000 compounds based on a common scaffold. Other types

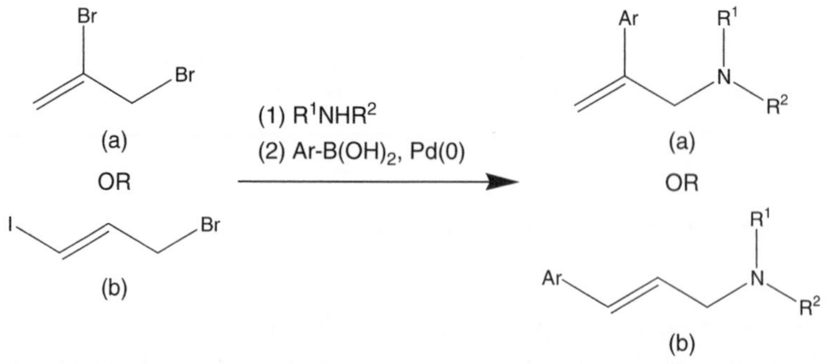

Figure 2.1 An example of acyclic template used in combinatorial synthesis. The synthesis above results in a library of 1344 compounds, with an average yield of 35 % and an average purity of >80 %. The reaction (a) is also an example of the Suzuki cross-coupling reactions often used in combinatorial approaches (adapted from Ganesan, 2002)

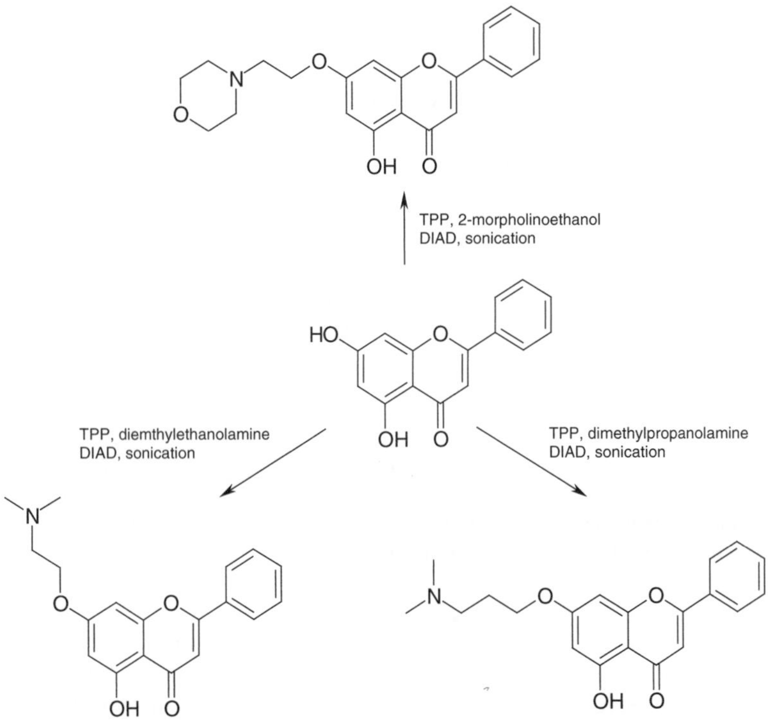

Figure 2.2 Synthesis of tertiary amines by Mitsunobu organic reaction

of libraries used in the automated medicinal chemistry setting include diversity-oriented synthesis-related libraries and dynamic combinatorial libraries (Koppitz and Eis, 2006).

A variety of reactions are now available for automation in an automated medicinal chemistry setting. Such reactions include the Mitsunobu reaction (Figure 2.2), the Suzuki cross-coupling reaction (Figure 2.1a), cyclization reactions or ring forming Diels–Alder cycloaddition reactions (Figure 2.3a and b, respectively).

The libraries are now carefully designed prior to synthesis. Original libraries aimed at the synthesis of large number of compounds at random for identification of hit compounds. However, such libraries offered little in terms of selection of compounds with the desired properties. More recently, the libraries are carefully designed, using *in silico* techniques, molecular modelling, to identify structural features necessary to fill voids, or neural networks that provide scores of drug- or lead-likenesses that may be used in designing the library.

Once the selection of the building blocks has been completed and the chemistry used has been identified, the actual library synthesis occurs, using automated dedicated equipment. Various types of automated equipment are available in the market, ranging from fully automated instruments, where there is no need for manual interference, to independent modular equipment that is separating the synthesis in distinct sub-procedures. Each of these instruments has their advantages and disadvantages, which include costs and flexibility of use. Microwave synthesizers have offered additional advantages over traditional approaches and are now well established in Medicinal Chemistry laboratories (Koppitz and Eis, 2006).

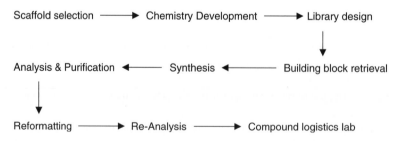

Figure 2.3　Further examples of reactions used in combinatorial chemistry approaches: (a) a cycliza-tion reaction and (b) a Diels–Alder cycloaddition reaction

Following the library synthesis, the crude products are collected and used for high-throughput purification and further analysis. The method of choice for such analysis is based on liquid chromatography–mass spectrometry (LC-MS), using mostly reverse phase chromatography and UV detection. Other detectors can include evaporative light scattering detection (ELSD), chemiluminescent nitrogen detectors (CLND) (for compounds that do not present a UV spectrum) or charged aerosol detectors (CAD). An alternative to high pressure liquid chromatography, also used in many laboratories, is high-throughput nuclear magnetic resonance (NMR), but this suffers from lack of fully automated interpretation of the NMR spectra, thus limiting the technique to library subsets.

The complete process of an automated medicinal chemistry process is shown schematically below (Koppitz and Eis, 2006):

Scaffold selection ⟶ Chemistry Development ⟶ Library design

Analysis & Purification ⟵ Synthesis ⟵ Building block retrieval

Reformatting ⟶ Re-Analysis ⟶ Compound logistics lab

Somewhere between the automated medicinal chemistry and the more biological combina-torial chemistry approaches stand the synthesis of peptides. Peptides have provided candidate therapeutics for a number of applications and have been a useful entity in drug design. How-ever, they can be considered as natural product libraries, synthesized either using solid phase (or even solution phase) combinatorial chemistry approaches, where the procedure described above can be utilized with amino acids used as scaffolds, or, because of their biological nature, they can be used in phage display technologies described in the next section of this chapter.

Solid phase peptide synthesis was introduced by Merifield in 1963 and has been the basis of peptide-based solid-phase combinatorial approaches. The original approaches have been improved by the introduction of various synthetic methodologies, such as the pin-shaped solid support synthesis introduced by Geysen, the 'teabag' approach introduced by Houghten for epitope mapping, or the split-and-mix solid phase synthesis introduced by Furka. These solid-phase approaches have offered various advantages over other methodologies as they allow the compounds to be isolated once they are attached to solid supports, the isolation of the immobilized products by filtration allows the use of large reagent excesses that offer high-yield at each step, whereas the use of split-and-mix synthesis simplifies the preparation of large libraries (Vicent *et al.*, 2007).

Most libraries synthesized by solid phase synthesis, particularly using the pin approach, offer limited libraries in terms of number of compounds synthesized. Thus, they mostly favour the development of short peptides. The limited number of positions would then allow the potential exploration of all possible amino acid combinations for each particular position. However, this is often not necessary and amino acids with similar properties are used, thus simplifying the number of possible combinations.

In addition to solid phase peptide synthesis, solution-phase synthetic approaches have also been developed by a number of groups. Such approaches offer the possibility of expanding the repertoire of chemical reactions and allow the application of convergent synthetic strategies, the synthesis of mixture libraries or the use of dynamic libraries. The use of dynamic combinatorial chemistry is driven by the interactions of the library constituents with the target site, thus allowing a target-driven generation or amplification of active constituents (Vicent *et al.*, 2007). This approach is similar in concept with the Phage displayed libraries, which are biologically-displayed libraries, examined in the next section of this chapter.

2.3 Display Technologies

2.3.1 Phage display technology

As mentioned before, peptides have been at the intersection between small molecule therapeutics and their larger proteinic counterparts, antibodies. Furthermore, peptides have some advantages over antibodies, in that they have lower manufacturing costs, higher activity per mass, better tumour penetration, increased stability and reduced immunogenicity. Furthermore, they have the advantage of increased specificity over small molecule therapeutics, which makes them interesting modalities for the pharmaceutical industry (Ladner *et al.*, 2004).

Although synthetic approaches to peptide libraries have been described for over 45 years, the greatest boost of combinatorial approaches to selection of peptides from vast libraries came with the development of the phage display technology by Smith in 1985. Subsequently, this was improved upon and it was reported that a filamentous phage was used to display a random oligopeptide on the N-terminus of the viral PIII coat protein, by inserting a stretch of random deoxyoligonucleotide into the pIII gene of filamentous phage (Smith, 1985; Parmley and Smith, 1988), thus initiating a process that has resulted in a number of modern therapeutics. The phage display technique is very effective in producing large number (up to 10^{10}) of diverse peptides and proteins and isolating molecules that perform particular functions. The technique involves the expression of proteins, such as peptides or antibodies, on the surface of filamentous phage. The DNA sequences of interest are inserted into a location in the genome of the phage, so that the encoded protein is expressed or 'displayed' on the surface of the phage as a fusion product to one of the phage coat proteins. It is now known that insertion of foreign DNA into the filamentous phase gene III is expressed as a fusion protein and displayed on the surface of the phage, resulting in phage display libraries containing billions of variants that can then be used to select and purify the specific phage particles with sequences that offer the desired binding specificities (Azzazy and Highsmith, 2002).

The phage display technique can select peptides that bind protein targets with high affinity and specificity. Such peptides have been selected for oncological targets and include the ErbB-2 tyrosine kinase receptor, which has been implicated in many human malignancies, with a dissociation constant in the micromolar range and significant specificity. Other peptides that have resulted from peptide-based phage display technology include peptides against the vascular endothelial growth factor (VEGF) and the MUC1 glycoprotein, whereas peptide therapeutics that have reached the market include representatives such as Lupron (leuprolide), for the treatment of prostate cancer, endometriosis, fibrosis or precocious puberty, and Zoladex (goserelin) for the treatment of breast cancer, prostate cancer and endometriosis (Ladner et al., 2004). A peptide, after selection, can be modified appropriately to confer to it higher resistance to peptidase degradation and through coupling to polyethylene glycol (PEG) improve its pharmacokinetic properties.

As mentioned above, phage display technology is not only used for the selection of peptides, but can be used for the 'display' of larger protein molecules in the surface of the phage, including antibodies and antibody fragments. Thus, phage display offers a powerful alternative to the standard hybridoma technology for the development and isolation of recombinant antibodies with unique specificities. In this regards, phage display technology can be used to either generate human monoclonal antibodies and antibody fragments such as single chain Fvs, or humanized mouse antibodies, significant for cancer immunotherapy. A scFv library can be prepared using phage display by preparation of the genes of the variable heavy and light chains of the antibodies through reverse transcription or the mRNA obtained from B-lymphocytes, the amplification of the gene products and their assembly into a single gene, the incorporation of the assembled scFv DNA fragment into the phage vector, resulting into the expression of the protein on the surface of the phage. This is followed by selection through interrogation for binding to the specific target of interest and the genotype of selected variants is recovered and amplified by infecting and subsequently culturing fresh *Escherichia coli* cells (Azzay and Highsmith, 2002; Dufner *et al.*, 2006).

In addition to phage display libraries, other 'display' systems can and have been used successfully for combinatorial approaches, including bacterial display, yeast display and mammalian cell display (reviewed by Sergeeva *et al.*, 2006).

2.3.2 Ribosomal display technologies

Another particularly interesting technology is the ribosomal display, which has come to alleviate several problems associated with the phage display, mainly the limiting step of constructing large-size phage libraries due to transformation efficiency. Other such limitations in the phage display technology include gene deletion, plasmid instability, gross phage cross-contamination and limited library size and diversity. Ribosomal display, originally described by Mattheakis *et al.* (1994), aims to circumvent these limitations. In ribosomal display, the DNA library that encodes for the peptide or protein of interest is transcribed and translated *in vitro*, using prokaryotic or eukaryotic cell-free expression systems. Thus, the DNA that éncodes the protein library is first transcribed to mRNA that is then purified and used for *in vitro* translation. This DNA library contains all the signals required for the cell-free *in vitro* transcription and translation, and in it its most widely used version is designed with an absence of a stop codon at the end of the coding sequence, so as to prevent the release of the mRNA and the nascent polypeptide from the ribosomes. The *in vitro* translation of mRNA is, as described above, designed so as to prevent dissociation of mRNA, ribosomes, and the translated peptide, but this complex is further stabilized by use of low temperature or elevated magnesium ions. The mRNA–ribosome–peptide complexes are subsequently used for affinity selection on an immobilized target, where only the binding complexes that specifically recognize the target are bound and are subsequently removed through elution steps and reverse transcribed into cDNA. The cDNA is then amplified by polymerase chain reaction (PCR) and used for the next cycle of enrichment and PCR and is finally analysed by sequencing (Azzazy and Highsmith, 2002; Dufner *et al.*, 2006). As there is no transformation step, large libraries can be constructed and used for selection. Furthermore, further diversification of the library can be introduced wither before or in-between selection cycles *via* DNA shuffling or error-prone PCR (Azzazy and Highsmith, 2002). Thus, a number of advantages have been described for this *in vitro* display system of combinatorial chemistry, over their *in vivo* counterparts. Such advantages include increased library size and diversion, minimal library bias by improved *in vivo* expression and folding, simple affinity maturation cycles, rapid library construction and selection, ease of recovery of genotype by reverse transcription (RT)–PCR and ability to display and select cytotoxic molecules. One of the principal applications of this technology in oncology has been for the development of human antibodies against various tumour markers that have seen a tremendous rise in the anticancer therapeutic market (Irving *et al.*, 2001), whereas the peptide therapeutics developed using this technology include the peptides against the prostate-specific antigen (PSA) (Gersuk *et al.*, 1997).

Another *in vitro* selection technology for the identification of ligands, this time of oligonucleotide, rather than peptidic origin, is the technology described in the next and final section of this chapter.

2.4 Aptamer Selection

2.4.1 The SELEX methodology

Another combinatorial methodology employed in drug design is called Systematic Evolution of Ligands by Exponential Enrichment (SELEX) and is focused on the selection of high affinity and specificity oligonucleotide ligands, termed aptamers, from vast combinatorial libraries. This process was first described in the same year (1990) by two independent groups in the USA (Tuerk and Gold, 1990; Ellington and Szostak, 1990).

The SELEX methodology is based on an evolutionary process characterized by rounds of interaction of an oligonucleotide library with a particular target, partition of bound from unbound species, elution of bound species from target and amplification of those using PCR. Amplified aptamers are allowed to interact again with the target for a number of iterative steps, to result in isolation of higher binding species through competitive binding. This was inspired by the evolution of RNA molecules through the ages into binding specifically to particular target proteins to facilitate biological functions. To reproduce this in the laboratory and within a short time interval, a number of competitive steps may be required to separate the highest binding molecules. Similarly, by counter-selecting and excluding molecules that bind to homologous proteins or isoforms as part of the selection methodology, the process can result in very selective and highly specific molecular entities that offer a high degree of differentiation and can thus target specifically mutant proteins over their native counterparts, thus conferring disease-specific targeting properties to these molecules.

In order to start the SELEX process, a single-stranded DNA oligonucleotide pool is chemically synthesized with a random region of nucleotides, varying from 10 to 100 bases, and a fixed primer sequence at either end to allow for amplification by PCR. From this theoretical maximum, which for a 25-base long degenerate region corresponds to 4^{25} or about 10^{15} sequences, the pool has at least one copy of each possible sequence and structure to be selected for optimal affinity towards the target. Following the synthesis of the oligonucleotide pool, a selection method based on an evolutionary approach of repeated steps of interaction, elution and amplification is applied. This methodology includes the exposure of the target molecule to the library in order to allow for interaction of all binding aptamers with this target. The target is exposed to the aptamer library in a number of ways, usually attached to a solid support such as agarose or sepharose in affinity chromatography matrices, magnetic bids or polymers, but often also in solution, separated by filtration and molecular weight differences. The sequences that do not bind to the target are washed away. The bound oligonucleotides are then eluted *via* a chosen elution method, which can be temperature, alterations of pH, salt concentration, use of chaotropic agents or another method of choice. Eluted aptamers are pooled for amplification *via* a PCR (DNA) or RT–PCR (RNA) amplification protocol. The selection, elution and amplification protocols comprise a SELEX round. In order to select for the best binder, a number of rounds ranging from 6 to as many as 15, have been reported to allow, through competitive binding against a limited target, the displacement of all non-specific or weak binders by the best binding aptamer. The method was conceived as an evolutionary process that allows for the selection of better, higher affinity binders. To exclude aptamers that may bind to similar proteins and ensure specificity and selectivity, counter-selection rounds are also included at this stage, allowing aptamers that bound to the target, but not to related molecules, to be amplified for the next selection round. After the aptamer selection process, the PCR products of the last round are double stranded and cloned into a vector and, subsequently, sequenced to allow for identification of the best binding sequence, which can then be chemically synthesized. The SELEX methodology is summarized in Figure 2.4.

There are features of the technique that need to be tailored to suit the requirements of each selection. First consideration is the choice of the nature (RNA, DNA or unnatural bases) and length of the nucleic acid library. The second stage is the definition of the method used for the selection process. The molecules can be selected in a variety of *in vitro* environments/methods that can be chosen and modified to fit the need. Such selection methods include immobilization of the target in affinity chromatography matrices, magnetic beads or other support systems, or the selection in solution and separation of bound complexes from unbound species with the use of nitrocellulose filters of appropriate pore size. Finally, the number of SELEX rounds may need to be optimized, depending on the stringency of the selection required and the amount of target available for competitive binding.

Following the identification and synthesis of the best binding aptamer, further studies are carried out to determine the binding affinity (K_d) and specificity of the selected molecule against the target and, depending on the application, if desired, the molecule can then be 'fashioned' (e.g. 5′ or 3′ modification for chelator bioconjugation or nuclease resistance) for the required end use. Further additional chemical modifications, such as PEGylation (Floege

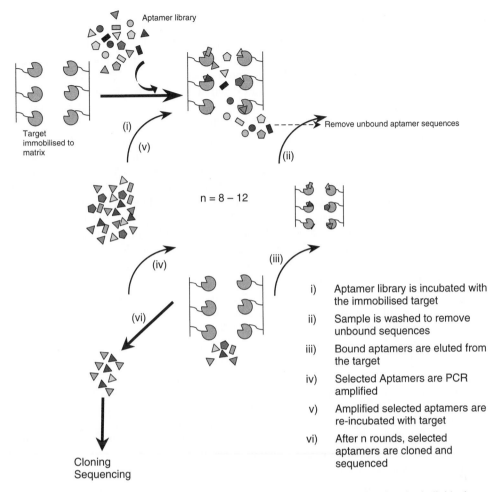

Aptamer library

Target immobilised to matrix

Remove unbound aptamer sequences

(i)

(v)

(ii)

n = 8 – 12

(iv)

(iii)

(vi)

Cloning
Sequencing

i) Aptamer library is incubated with the immobilised target

ii) Sample is washed to remove unbound sequences

iii) Bound aptamers are eluted from the target

iv) Selected Aptamers are PCR amplified

v) Amplified selected aptamers are re-incubated with target

vi) After n rounds, selected aptamers are cloned and sequenced

Figure 2.4 An interpretation of the SELEX methodology is presented, showing the individual steps of selection, partitioning and amplification, which results in the identification of the highest binding molecule that can subsequently form the lead molecule in diagnostic or therapeutic applications

et al., 1999), or liposomes, can be easily made at desired positions of the aptamer to improve its pharmaceutical, therapeutic and/or diagnostic application.

Attachments of such molecules serve to decrease the renal clearance time of aptamers, as the pharmacokinetic properties of these molecules are limited by their small size (5–25 kDa) and hydrophilic nature. Signalling molecules such as fluorophores can also be incorporated into the aptamer for imaging/signalling purposes, while the attachment of drugs can facilitate the therapeutic effect of the aptamer, if required (Floege *et al.*, 1999). Although chemical modifications can be completed relatively easily, it is important that such alterations to the backbone structure of the aptamer do not affect its binding to the target, particularly given that binding is largely governed by shape-shape interactions between the two molecules.

2.4.2 Non-SELEX methods for the selection of aptamers

SELEX has proved to be a robust and powerful methodology for the isolation of many aptamers directed towards a variety of targets. However, a number of attempts have been made to improve, or bypass this technique, both to overcome some of the drawbacks associated with traditional SELEX procedures and to bypass the patent stronghold on SELEX by Archemix and Somalogic in the USA. Thus, 'non-SELEX'-based methods for the selection of aptamers have recently been put into practice. The utilization of capillary electrophoresis has demonstrated to be a highly efficient approach into the partitioning of aptamers with desired properties from a randomized pool (Drabovich *et al.*, 2005; Berezovski *et al.*, 2005). Using this technique, aptamers to h-Ras, a protein involved in the development and progression of cancer, were isolated with predetermined kinetic parameters (Berezovski *et al.*, 2005). The isolation of aptamers with predefined kinetic and thermodynamic properties of their interaction with the target has so far been obstructed by the standard SELEX technology. Furthermore, this method of aptamer selection only employed the partitioning steps of SELEX without the need for amplification between them. Hence, one of the most significant advantages of this non-SELEX method is its application to libraries which are difficult or cannot be amplified, thus overcoming the problems associated with using modified oligonucleotide libraries, as mentioned above. As well as the relative simplicity and easy-to-use nature of this procedure, aptamers are selected within only a few hours, which contrasts the several days or weeks needed for standard SELEX systems.

Exploitation of computational methods has also led the way into the development of non-SELEX methods for aptamer selection. More importantly, computational methods have been powerful in selecting aptamers with inhibitory activities or sequences that undergo ligand dependent conformational changes, a property useful for the design of molecular and aptamer beacons. One of the major drawbacks associated with SELEX is the selection of aptamers that may not have any inhibitory activity towards its target, since the selection of aptamers is based on affinity. Consequently, researchers have used this drawback to drive the engineering of alternative selection methods based on inhibitory activity of aptamers.

Algorithm methods have shown to be sufficiently effective in selecting aptamers with such properties. This computational method has been used to predict the secondary structure of nucleic acids under different conditions e.g. in the presence and absence of a ligand (Hall *et al.*, 2007). In general, algorithmic methods use aptamers with known structures and/or features (such as aptamers that undergo ligand-induced conformation changes) to rapidly select oligonucleotides from virtual pools, which may present similar properties or adopt similar structures. Hence, sequences are selected that match a defined profile. One of the most valuable structures applied to computational selection is that of the G-quartet. Such structures are suggested to have important implications in the biology of cancer and thus aptamers that adopt such configurations are of great interest. An aptamer selected for thrombin has been thoroughly investigated and is known to adopt such a configuration. Consequently, this aptamer has been used as a model to investigate the potential of new selection methods, based on inhibitory activity. By randomizing the sequence of the duplex region of the anti-thrombin aptamer, Ikebukuro and co-workers (2006) selected inhibitory aptamers using genetic algorithm on a library of limited sequences. In another report, evolution mimicking algorithms were used to select aptamers with potent inhibitory activity from a pool that was designed to form G-quartet structures and contain a limited number of sequences (Noma and Ikebukuro, 2006). Computational selection methods of aptamers require detailed information on the prerequisites of the basis of selection, for example the structure of the aptamer that new aptamers are to be modelled on or the structural changes that the profile aptamer undergoes upon ligand binding. Given that this is not always feasible and that gaining this information can sometimes be a lengthy process, computational selection of aptamers may be delayed in materializing as a potential option for ligand identification. Meanwhile, SELEX itself is still widely used for aptamer selection and is constantly advancing to overcome many of its initial challenges.

2.5 Conclusions

Combinatorial chemistry techniques have been a powerful drive for anticancer drug discovery during the last decade or two. Both chemical and biological approaches to drug development have resulted in a number of compounds that are currently in clinical trials or in the market. Though chemical combinatorial approaches originally failed to deliver small molecule therapeutics, despite the vast investment from the pharmaceutical industry on automated procedures and instrumentation, it has progressed through the years to a level where chemical scaffolds, synthetic chemical approaches and reaction mechanisms, purification procedures and technological developments have now allowed the selection and preparation of small molecular therapeutics to a large number of targets. Furthermore, in parallel, a range of *in vitro* and *in vivo* biological combinatorial chemistry approaches have been developed. These approaches have resulted to the major blockbuster drugs of our time, antibody and peptide therapeutics, such as Herceptin and Zevalin, and are also the source of oligonucleotide therapeutics, such as aptamers, that offer a significant promise for the future of anticancer therapies. The products of these methodologies will be analysed in detail in the relevant chapters of immunotherapy and oligonucleotide therapeutics later on in this volume.

References

Azzazy HME, Highsmith WE Jr. Phage display technology: clinical applications and recent innovations, *Clin Biochem* 2002, **35**, 425–45.

Berezovski M, Drabovich A, Krylova SM, *et al.* Nonequilibrium capillary electrophoresis of equilibrium mixtures: a universal tool for development of aptamers. *J Am Chem Soc* 2005, **127**(9), 3165–71.

Drabovich A, Berezovski M, Krylov SN. Selection of smart aptamers by equilibrium capillary electrophoresis of equilibrium mixtures (ECEEM). *J Am Chem Soc* 2005, **127**(32), 11224–5.

Dufner P, Jermutus L, Minter RR. Harnessing phage and ribosome display for antibody optimisation, *Trends Biotechnol* 2006, **24**, 523–9.

Ellington AD, Szostak JW. *In vitro* selection of RNA molecules that bind specific ligands. *Nature* 1990, **346**(6287), 818–22.

Floege J, Ostendorf T, Janssen U, *et al.* Novel approach to specific growth factor inhibition *in vivo*: antagonism of platelet-derived growth factor inglomerulonephritis by aptamers. *Am J Pathol* 1999, **154**(1), 169–79.

Ganesan A. Recent developments in combinatorial organic synthesis. *DDT* 2002, **7**, 47–55.

Gersuk GM, Corey MJ, Corey E, Stray E, Kawasaki GH, Vessela RL. High affinity peptide ligands to prostate-specific antigen identified by polysome selection. *Biochem Biophys Res Commun.* 1997, **232**, 578–82.

Hall B, Hesselberth JR, Ellington AD. Computational selection of nucleic acid biosensors via a slip structure model. *Biosensors and Bioelectronics* 2007, **22**(9–10), 1939–47.

Ikebukuro K, Yoshida W, Noma T, Sode K, Analysis of the evolution of the thrombin-inhibiting DNA aptamers using a genetic algorithm. *Biotechnol Lett* 2006, **28**, 1933–37.

Irving RA, Coia G, Roberts A, Nuttall SD, Hudson PJ. Ribosome display and affinity maturation: from antibodies to single V-domains and steps towards cancer therapeutics. *J Immunol Methods* 2001, **248**, 31–45.

Koppitz M, Eis K. Automated medicinal chemistry. *DDT* 2006, **11**, 561–8.

Ladner RC, Sato AK, Gorzelany J, de Souza M. *DDT* 2004, **12**, 525–9.

Noma T, Ikebukuro K. Aptamer selection based on inhibitory activity using an evolution-mimicking algorithm. *Biochem Biophys Res Commun* 2006, **347**(1), 226–31.

Parmley SF, Smith GP. Antibody-selectable filamentous fd phage vectors: Affinity purification of target genes. *Gene (Amsterdam)* 1988, **73**, 305–18.

Sergeeva A, Kolonin MG, Molldrem JJ, Pasqualini R, Arap W. Display technologies: Application for the discovery of drug and gene delivery agents, *Adv Drug Deliv Rev* 2006, **58**, 1622–54.

Smith GP. Filamentous fusion phage – novel expression vectors that display cloned antigens on the virion surface. *Science* 1985, **228**, 1315–17.

Tuerk C, Gold L. Systematic evolution of ligands by exponential enrichment – RNA ligands to bacteriophage-T4 DNA-polymerase. *Science* 1990, **249**(4968), 505–10.

Vicent MJ, Perez-Paya E, Orzael M. Discovery of inhibitors of protein–protein interactions from combinatorial libraries. *Curr Top Med Chem* 2007, **7**, 83–95.

3

Rational Approaches to Anticancer Drug Design/*in silico* Drug Development

Stefano Alcaro, Anna Artese and Francesco Ortuso

3.1 Introduction

The drug development can be considered like a sort of 'race with obstacles' (Figure 3.1). There are a number of key barriers to be passed before an idea, i.e. a novel compound, can be introduced onto the market.

1. Step I is typically related to the identification of the macromolecule(s) responsible for a pathological phenomenon – the target(s) to be selectively influenced by novel compounds (molecular target selection).

2. Step II is the identification of lead compound(s) usually by computational or high-throughput approaches.

3. Step III is related to the preliminary biological tests and Quantitative Structure-Activity Relationships (QSAR) studies.

4. Steps IV and V are related to the biological activities respectively *in vitro* (cell lines) and *in vivo* (animal models). Steps II to V are connected by a sort of cycle defined as the 'lead optimization process'.

5. Step VI is the creation of a patent, if one or more novel compounds have been identified and optimized.

6. Step VII represents the pharmacokinetic study, regarding the four ADME aspects: adsorption, distribution, metabolism and excretion.

Anticancer Therapeutics   Edited by Sotiris Missailidis
© 2008 John Wiley & Sons, Ltd

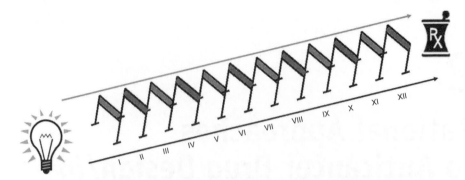

Figure 3.1 The development of a new drug into the market is like a 'race with obstacles'. The most relevant steps are labelled in roman numbers and are briefly discussed in the text

7. Step VIII is the scale-up, generically related to the industrial synthesis procedures.

8. Step IX is the formulation of the drug.

9. Step X regards safety studies, focusing on the toxicological issues.

10. Step XI is the clinical phase in humans, usually divided into two or more sections, as seen later in this book.

11. Step XII is registration and marketing.

A failure in one of the above-mentioned steps results in deleting the entire development of the new drug. This process is usually a long procedure that, in successful cases, takes about 8–12 years and requires a lot of funding resources.

The availability of detailed structural information about the interacting compounds is another central issue to be considered in order to understand the success of the theoretical methods in drug design. The Cambridge Structural Database (CSD) contains more than a quarter of a million small-molecule crystal structures, most of them having biological activity (Allen, 2002). The implications of the CSD in life sciences, especially using theoretical methods related to drug design, have extensively been reported in a recent review (Taylor, 2002). Another popular source of structural information is the Protein Data

Therefore it is quite important to reduce as much as possible any wrong investments on compounds that have no chance of being biologically active; possibly trying to pre-empt and exclude inadequate properties in order to achieve the desired effect. With this aim, in the last decades, drug development has been dramatically influenced by the availability of computational methods, performing what it is now called 'rational drug design'. Several reasons justified this success. First of all the introduction of the molecular recognition concept; Emil Fischer was the first author to introduce the well known paradigm of the 'lock and key' related to enzyme–ligand interaction (Fischer and Thierfelder, 1894). Daniel Koshland modified the original rigid model introducing the issue of flexibility in molecular recognition, proposing the 'induced-fit' theory (Koshland, 1958). Molecular recognition can be defined as the ability of biomolecules to distinguish ligands and selectively interact with them in order to promote pivotal biological events such as transcription, translation, signal transduction, transport, regulation, enzymatic catalysis, viral and bacterial infection and the immune response (Wodak and Janin, 2002).

Bank (PDB) (Berman *et al.*, 2000), that includes more than 45 000 models of proteins, nucleic acids and other polymers. Even if the amount of data is much lower than the CSD, the PDB has the great advantage in that it also implements co-crystallized structures in the presence of ligands and drugs. In these cases, the information is particularly interesting because the three-dimensional structures of the small co-crystallized compounds can be considered as the bioactive conformations. Moreover, the PDB also includes ligand–receptor complexes obtained by nuclear magnetic resonance (NMR) methods, that in few, but quite interesting cases, provide information about the dynamic equilibrium among bioactive conformations.

When drug development is carried out using as a starting structure an experimentally determined complex structure, such as those deposited into the PDB, the term 'structure-based' drug design is adopted (Kroemer, 2007). This term is opposed to a more classical paradigm, still in use especially when no structural information is available for the receptor–ligand complex, and known as 'ligand-based' drug design (Bacilieri and Moro, 2006). These two terms differ in their relationships with the diseases and are respectively related to the target (protein, receptors or genes) and to reference compounds (known drugs or ligands).

3.2 Approaches to the Drug Discovery Process in Anticancer Research

The anticancer drug discovery procedure undergoes the same paradigms applied to other drug design projects. The differences between the structure-based and the ligand-based approaches can be summarized in terms of different links with respect to cancer disease (Figure 3.2).

It is worth noting that despite the fact that knowledge of the target macromolecular structure is often available, ligand-based drug design represents still an important approach for the discovery of new anticancer agents. Moreover, the opportunity to follow both paradigms in the same drug discovery process usually brings a synergy to the entire development, because the structure-based and the ligand-based approaches are totally different theoretical methodologies.

In order to give an idea of the large application of the computational methods available to rational anticancer drug design, we have selected some relevant examples for both paradigms.

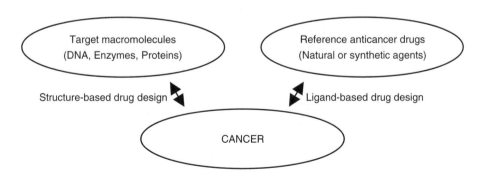

Figure 3.2 The relationships in cancer disease with target macromolecules and anticancer drugs as function of the drug design paradigm

A categorization of them has been applied to those pertinent to the structure-based approach, taking into account the nature of the target involved in the drug discovery process. No categorization has been considered for the ligand-based approach, because the target is not explicitly considered in the computational models.

3.3 Ligand-based Examples

One of the most interesting examples of ligand-based study is related to the inhibitors of the aromatase enzyme. This belongs to the cytochrome P-450 superfamily, whose function is to catalyse the aromatization of androgens, the final step in the biosynthesis of oestrogens (Figure 3.3). Therefore, it is an important factor in sexual development and its inhibition, with the consequent arrest of oestrogen biosynthesis, represents a good therapeutic route for the treatment of oestrogen-dependent breast cancers.

Many research groups have identified some potent inhibitors of aromatase that can be collected into two main classes: steroidal analogues of androst-4-ene-3,17-dione, the natural substrate of aromatase, such as compound **1**, and the non-steroidal inhibitors, such as compound **2** (Figure 3.4).

In absence of any direct information on the tertiary structure of aromatase some molecular modelling studies were carried out with the aim of mapping the active site of the enzyme, performing conformational analysis of the chemical structures of known aromatase inhibitors

CHOLESTEROL

TESTOSTERONE

AROMATASE ⟵ AROMATASE INHIBITORS

ESTROGEN

estrogen receptor ⟷ estrogen receptor

BREAST   UTERUS

PROLIFERATION

CANCER

Figure 3.3 Schematic representation of oestrogen biosynthesis catalysed by the aromatase with the consequent proliferation in breast and uterus cells

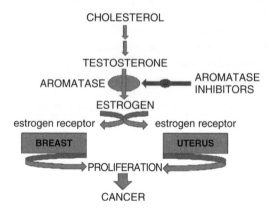

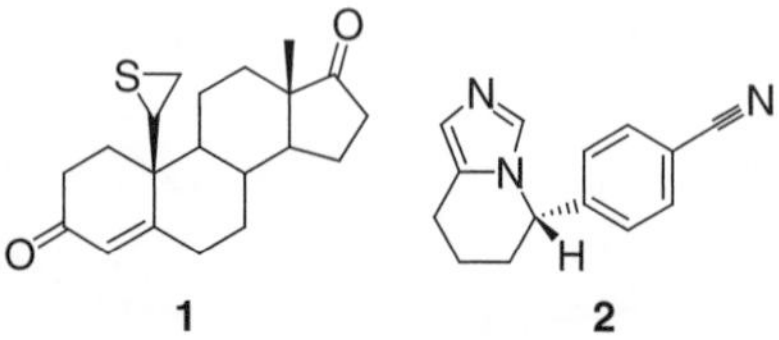

1   **2**

Figure 3.4 Chemical structures of the steroid (19R)-10-thiiranylestr-4-ene-3,17-dione (**1**) and the non-steroid compound CGS 16949A (**2**)

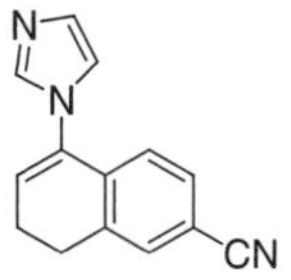

Figure 3.5 Chemical structure of the azole inhibitor **3** obtained by a rational drug design approach

and comparing the azole-type and the most potent steroidal inhibitors, with the final identification of a relative binding model for both inhibitors. The nitrogen of the imidazole ring in compound **2** and the sulfur of the thiirane ring in the steroid (**1**) are assumed to bind the porphyrin of the enzyme by coordinating the iron atom of the heme moiety. Since a cyanophenyl function is an essential structural requirement for high inhibitory activity in compound **2**, it would seem possible that this polar group might mimic one of the carbonyl groups of the steroidal inhibitor. On the basis of these modelling considerations, a new potent azole analogue of compound **2**, compound **3** (Figure 3.5), has been developed by adding a new ring which could mimic the steroid moiety and it has been shown that it exhibits a high level of activity comparable to compound **2**, thus validating the identified binding model.

In summary, all compounds were superimposed computing several descriptors, such as the accessible volumes to the solvent. In summary, all the molecules have been superposed together, computing solvent accessible surface areas. The resulting observations suggest that aromatase's cavity is quite large with a pocket in the region located below the α-face of the steroidal substrates, distal to the enzyme heme group. Moreover, the cyanobenzyl moiety present in the imidazole-type inhibitors partially mimics the steroid backbone of the enzyme's natural substrate (Furet *et al.*, 1993).

Another successful case of ligand-based drug discovery is represented by the poly (ADP-ribose) polymerase-1 (PARP-1) inhibitors. PARP-1 is a nuclear enzyme involved in several cellular processes, such as DNA repair and apoptosis (de Murcia *et al.*, 1994; Dantzer *et al.*, 1999). The enzyme initially recognizes and binds to DNA damage sites and catalyses the synthesis of linear and branched ADP-ribose polymers from the substrate nicotinamide adenine dinucleotide (NAD$^+$), with concomitant release of nicotinamide. Thus, PARP-1 inhibitors can be useful in the repair of DNA damage induced by radiation during cancer therapy. Early studies identified various analogues of 3-aminobenzamide, such as compound **4** (Figure 3.6) as inhibitors of PARP-1; however, these compounds lack potency and specificity, and some are poorly water soluble. Subsequent design of constrained analogues locking the carboxamide group in its probable bioactive conformation (compounds **5, 6** and **7**, Figure 3.6) led to compounds with poor potency; however, compounds **6** and **7** displayed the desired biological effect at high concentrations.

Since compound **7** can form an intramolecular hydrogen bond between the carboxamide and the nitrogen of the heterocycle ring (Figure 3.6), a restricted scaffold such as compound **8** (Figure 3.7) has been suggested.

Derivatives based on compound **8** were confirmed to act as resistance-modifying agents for radiotherapy, but they present potential problems in both synthesis and stability. Another series of tricyclic compounds is based on the cyclo-homologation of the lactam ring corresponding to the 2-phenyl-1H-benzimidazoles (Figure 3.8, **9**).

Derivatives based on **9** appeared to be promising candidates, able to mimic well the reference benzimidazole-4-carboxamide molecule **7**, which has been adopted as a benchmark

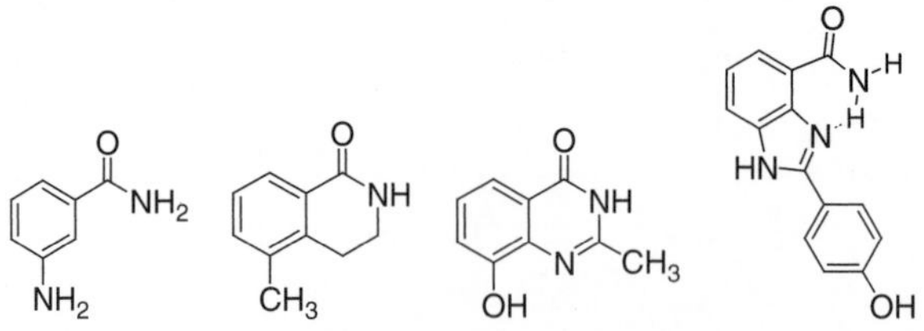

Figure 3.6   Chemical structures of PARP-1inhibitors **4, 5, 6** and **7**

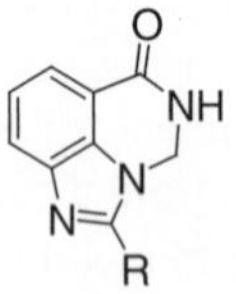

Figure 3.7   Chemical scaffold **8**, as the prototype of **7** restricted analogues

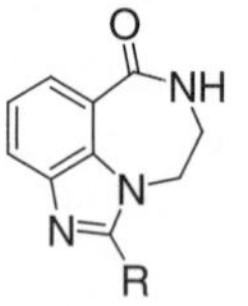

Figure 3.8   Tricyclic 2-phenyl-1H-benzimidazole scaffold **9**

inhibitor of PARP due to its potency, synthetic accessibility, and relatively good water solubility.

Some examples of the derivatives of compound **9** were synthesized and led to the discovery of potent inhibitors compounds **10, 11** and **12** (Figure 3.9). Structure-activity relationships studies indicate that the carboxamide hydrogen is essential; in fact, its replacement by a methyl group causes the complete loss of binding affinity.

Compound **12** shows a 25-fold increase in efficacy when compared to the reference molecule **7** and presents the best biological profile in cellular assays. The design concept was validated by X-ray crystallographic analyses of complexes with the PARP-1 target. In Figure 3.10 the representative inhibitor 2-(3-methoxyphenyl)benzimidazole-4-carboxamide, analogue to compound **7**, bound at the PARP active site is reported (PDB code 1EFY). Its carbonyl oxygen accepts hydrogen bonds from the side chain OH of Ser904 and from the Gly863 backbone amide. The ligand amide donates a hydrogen bond to the Gly863 backbone carbonyl oxygen.

The conformation observed for the carboxamide group of the co-crystallized ligand was also observed in the solid state (i.e. measured for the protein-free compound) of other benzimidazole analogues, confirming the essential role of molecular modelling data in the drug discovery process (White *et al.*, 2000).

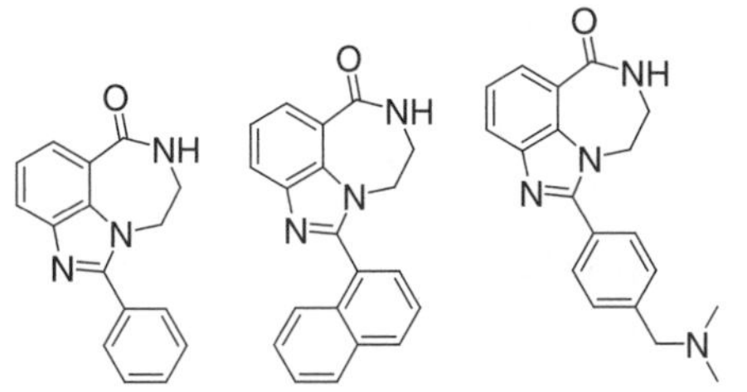

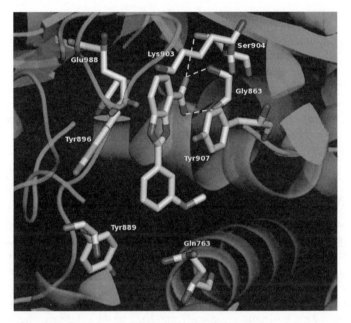

Figure 3.9   Chemical structures of the tricyclic PARP-1 inhibitors **10** (K_i = 4.1 nM), **11** (K_i = 4.9 nM) and **12** (K_i = 5.8 nM)

Figure 3.10   Complex between a representative inhibitor 2-(3-methoxyphenyl)benzimidazole-4-carboxamide and the catalytic domain of chicken PARP. The dashed lines indicate the hydrogen bonds between the inhibitor and some crucial residues of the active site

Another emerging computational ligand-based approach, potentially leading to the identification of new compounds, is the 3D-pharmacophore search. It is based on the fast virtual screening of large chemical databases by pharmacophore models implementing 3D information (Martin, 1992; Sheridan *et al.*, 1989). An interesting example of this technique is related to the discovery of protein kinase C (PK-C) ligands. PK-C comprises a family of phospholipid-dependent enzymes that vary in their activation properties and tissue distribution. The involvement of PK-C isozymes in cellular signal transduction, which results in either differentiation or uncontrolled cellular proliferation, makes them attractive targets for anticancer chemotherapy. Phorbol esters and other chemically diverse tumour promotors are known to activate PK-C. So, since no 3D structural data for PK-C isozymes are available, a

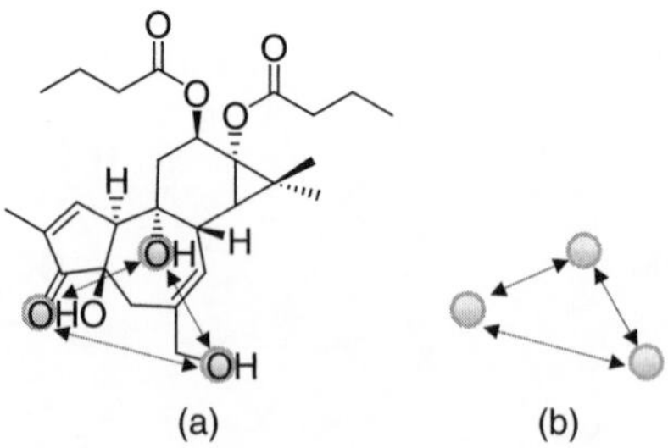

(a)   (b)

Figure 3.11   (a) Chemical structure of the phorbol 12,13-dibutyrate (PDBU) **13**; (b) pharmacophore model deduced from **13**

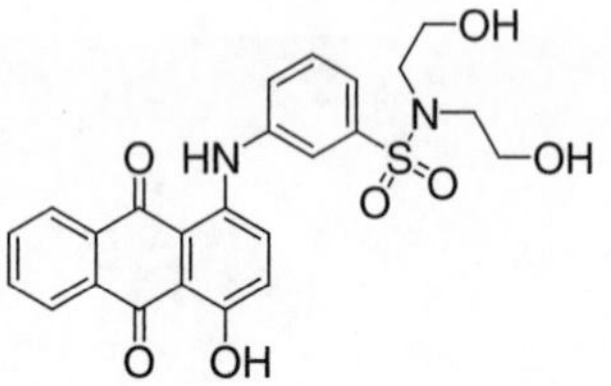

Figure 3.12   Chemical structure of the quinone derivative **14**, identified by the virtual screening and recognized as a potent PK-C agonist

computerized 3D-database search on more than 200 000 structures was carried out using the same pharmacophore model deduced from the active phorbol 12,13-dibutyrate (PDBU) **13** (Figure 3.11a).

The 3D pharmacophore model (Figure 3.11b) was used as a query of virtual screening experiments onto the NCI 3D database containing more than 200 000 anticancer agents. Two hundred and eighty-six hit compounds were found and analysed by considering the presence of hydrophobic substituents, such as a phenyl ring or a butyl group. Compounds lacking this requisite were discarded. The selected 125 compounds were submitted for evaluation in the PK-C binding assay. This analysis led to the discovery of the new potent quinone PK-C agonist **14** (Figure 3.12), a molecule chemically unrelated to the reference phorbol lead (Wang *et al.*, 1994).

The discovery of a number of novel, structurally diverse, PK-C agonists has demonstrated that computer 3D-database pharmacophore searching, combined with molecular modelling studies, is an effective and promising tool to identify novel lead compounds in drug development (Wang *et al.*, 1994).

3.4 Structure-based Examples

This section has been intentionally separated into subsections as the function of the target is considered in each example. Its chemical nature is actually responsible for the different computational approach considered for studying the interaction with anticancer agents. In fact, the nucleic acids are characterized by electrostatic and conformational properties remarkably different with respect to enzymes or proteins. Thus, we have divided into two subgroups our selected examples.

3.4.1 DNA targets

Probably the most relevant macromolecular target for the design of new anticancer compounds is DNA. It is a flexible macromolecule that in biological systems can assume different conformations (Mills *et al.*, 2002). Two conformations have been recently considered for the rational design of anticancer agents specific for certain oligonucleotide sequences and DNA folding structures.

The first one is the classic double helix (B-form), considered as the most probable *in vivo* conformation (Watson and Crick, 1953). This structure is the target of several anticancer drugs currently approved in therapy, such as cisplatin, which covalently binds the DNA preferably at the N7 of purine bases (Boulikas and Vougiouka, 2003). This mechanism of action is, by definition, non-specific toward certain DNA sequences, so severe side effects are caused by the interaction with the DNA of normal cells. One attempt to identify anticancer agents able to distinguish specific DNA sequences resulted in the discovery of the natural antibiotics azinomycin A and B (Nagaoka *et al.*, 1986) (Figure 3.13).

The mechanism of action of the azinomycins is based on a selective bis alkylation of the duplex DNA toward specific triplets satisfying the general formula 5′-d(PuNPy)- 3′ as target receptors (Pu = purine, Py = pyrimidine). It has been proposed that the former alkylation toward the N7 of the first DNA strand occurs with the C10 methylene aziridine moiety fused into the unusual azabicyclohexane ring. The latter alkylation occurs by the C21 methylene of the epoxide ring toward the N7 atom of purine bases, two positions far from the previous one and belonging to the opposite strand. This rare binding fashion implies that the cross-link involves the entire triplet with drug–target interactions within the DNA major groove.

Coleman and co-workers completed the total synthesis of compound **15**, making it possible to access the simplified azinomycin-like compounds (Coleman *et al.*, 2001). Thus, the explanation of the mechanism of the high selectivity toward specific DNA was addressed as a crucial issue for the rational design of new potent and easy synthetically accessible cross-linking agents.

The computational study, originally developed for this class of compounds, was based on a molecular mechanics (MM) approach performed in multiple steps, i.e. looking at the conformational properties of the adduct formed when the aziridine is open and covalently linked to the N7 purine base.

The first step started considering the conformational properties of both natural antibiotics (Alcaro and Coleman, 1998). The presence of unusual chemical moieties and low quality of AMBER* (McDonald & Still, 1992), one of the most trusted force fields in the scientific community by the implementation of a new set of *ab initio* and experimental data. The

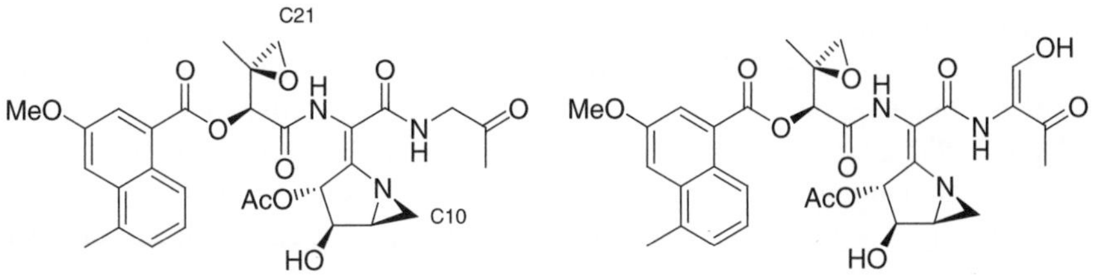

Figure 3.13 Chemical structures of azinomycins A (**15**) and B (**16**)

conformational pre-organization degree of compounds **15** and **16** was computed, considering the sum of the Boltzmann population computed at room temperature of all energy minimized conformers satisfying geometric criteria compatible with a concurrent bis-alkylation of an experimentally known DNA triplet target. The computational results indicated that in both cases there is a good level of pre-organization, possibly enhanced by the rational design of new analogues.

The second step included explicitly in the computational model the DNA in the triplet 5'-d(GCT)-3', experimentally known to be selectively cross-linked by compound **16** (Fujiwara *et al.*, 1999). A hypothesis of the reaction path, compatible with the intercalative recognition of the naphthoate, was also taken into account. The Monte Carlo (MC) simulations carried out with several adducts, built considering the combinatorial product of alkylating moieties and DNA sites of alkylation, led to the explanation of the experimentally known mechanism of cross-linking of azinomycin B (Chang *et al.*, 1989). The development and application of a filtering criterion allowed to uniquely identify the pattern of cross-linking (Alcaro and Coleman, 2000).

The third step extended the study to the selectivity issue. Four sequences, satisfying the general triplet cross-linkable formula, were considered and modelled by MC simulations. A force field correction to the electrostatic treatment, coherent with the net +1 charge adduct obtained in adduct after the first alkylation, allowed the improvement in the quality of the energy calculation. In this condition, the filtering protocol, coupled with the Boltzmann analysis at room temperature, reproduced the experimental selectivity with monoadducts as well as with intercalative models (Alcaro *et al.*, 2002). The difference in the hydrogen bond network between compound **16** and the DNA sequences, within the major groove, clearly explained the details of such selectivity.

In the fourth step, a set of six antitumour drugs, including compound **16**, was considered against six triplex sequences. The cytotoxicity in tumour cell lines, expressed as pIC_{50}, allowed the formulation of a reference classification of the potency of the six drugs. The molecular models of monoadducts were generated according to the previously reported MC search and other protocols, based on the stochastic dynamics (SD) simulations, were implemented and compared. The results were evaluated on the basis of their correlation with known cross-linking experimental and cell proliferation data. In both cases, high correlations were found especially with the stochastic dynamics with energy minimization (SDEM) approach (Alcaro *et al.*, 2005).

In the fifth step, analysis of the DNA structure in the B-form conformation by the well-known approach of the GRID program (Goodford, 1985) allows us to understand the best location of chemical probes around the surface of the target molecules, including nucleic acids. Coupling the GRID information and the model of the bioactive conformation by molecular mechanics methods highlighted the molecule region to be optimized, designing new azinomycin-like compounds. In Figure 3.14, an example of the GRID analysis is reported for the O probe, mimicking the carbonyl oxygen atom, with the putative bioactive conformation of **16** obtained with a target DNA triplet (Fujiwara *et al.*, 1999) in the B-form.

The computational multi-step approach adopted in these studies provides useful information for the design of new anticancer drugs that can exert specific cross-linking activity toward specific triplets.

DNA has been recognized as being far more polymorphic than originally assumed and can form left-handed and parallel stranded duplexes, hairpins with ordered loops, triplexes and quadruplexes, which have been shown to play an important role in the biology of the cell.

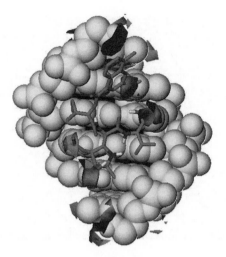

Figure 3.14 Overlap of the monoadduct of 16 (polytube model) with the DNA sequence of 5'-d(TAGCTA)-3'(spacefill model) and the GRID maps obtained with the probe O (solid surface)

Such higher-order DNA structures have provided novel targets for drug design of compounds with improved specificity.

The DNA quadruplex has been extensively characterized by crystallographic methods (Parkinson *et al.*, 2002) giving a great opportunity to carry out structure-based drug design. The presence of two flat interaction regions, denominated as 'bottom' and 'top' sites (Alcaro *et al.*, 2007), makes the drug-interaction studies not so trivial. In this case, an automatic docking procedure is absolutely recommended to properly consider all possible configurations or structures. Since no intercalation seemed possible for the quite compact quadruplex structure (Ou *et al.*, 2007), the end-stacking recognition resulted as the most probable mechanism of flat inhibitors. The in-house docking software MOLINE (Alcaro *et al.*, 2000, 2007) was applied to explain the difference in the binding of two fluorenone derivatives **17** and **18** (Figure 3.15) toward the telomeric DNA sequence d[AG$_3$(T$_2$AG$_3$)$_3$].

The goal of the computational work was the identification of a protocol able to reproduce the different experimental selectivity of compounds **17** and **18** for the telomeric sequence.

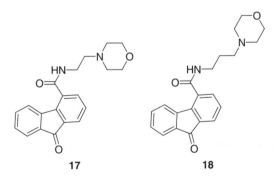

Figure 3.15   Chemical structure of the morpholino fluorenone carboxamides **17** and **18**

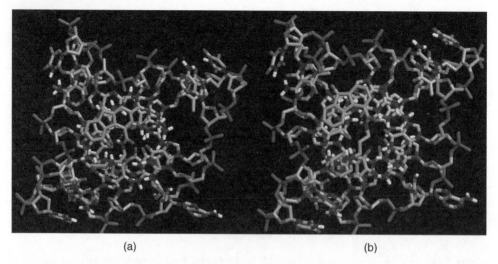

(a)   (b)

Figure 3.16   Top view polytube models of **17** (a) and **18** (b) energy most favored configurations with the d[AG$_3$(T$_2$AG$_3$)$_3$] target obtained with the quasi-flexible automatic docking MOLINE protocol. The telomeric DNA and the ligands are respectively reported in background and foreground views. Intermolecular hydrogen bonds are depicted as dashed lines

Even if only a methylene differs in the structure of these ligands, it is worth noting that competition dialysis experiments reported a stronger affinity of compound **18** versus **17** (Alcaro *et al.*, 2007). The MOLINE calculations were carried out considering as receptor conformations the original (PDB code 1KF1) and the energy-minimized crystallographic coordinates, and as ligands, the lowest energy structures of compounds **17** and **18** obtained after an exhaustive MC search. The thermodynamic module of the MOLINE program allowed the authors to estimate state equations for the recognition mechanism according to Boltzmann population analysis. Since theoretical and experimental data were in agreement, the analysis of the binding modes was performed to verify the most probable mechanism of recognition of both ligands with the telomeric target. The first issue was to address which of the binding sites was occupied in the most stable 1:1 complexes. In both cases it resulted in the 'top' site (Figure 3.16).

Interestingly, the computational estimation of the binding also resulted different for both ligands. **17** and **18** showed respectively a 95 % and 72 % of preference for the 'top' site and the complement to 100 for the 'bottom' one. This scenario is definitively compatible, especially for the compound **18**, with a 1:2 stoichiometry in which two ligand molecules can concomitantly bind the telomeric target, blocking its biological function that is over-expressed in tumour cells. This information is an important issue for the design of a next generation of anticancer agents based on the interaction with DNA structures involved, especially in neoplastic processes.

3.4.2 Enzyme and protein targets

The structure-based drug design paradigm for targets based on peptides, key enzymes or proteins, is generally strictly related to experimental information about their folding properties, usually by crystallographic or NMR methods. When no direct information about the target is

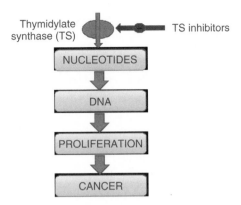

Figure 3.17   Schematic representation of DNA synthesis catalysed by the over-expressed thymidylate synthase with consequent cancerous cellular proliferation

available, but other related proteins have been characterized, it is possible to get theoretical models by homology techniques (Ginalski, 2006). Such an approach is often the unique solution to perform the paradigm but, by definition, it is also fairly accurate. If we consider other approximations, that in a docking or a virtual screening study must be adopted, the homology modelling should be prudently taken into account.

For this subsection we have considered two experimental models involved in anticancer drug design. The former is the enzyme thymidylate synthase (TS) and the latter is a protein, the baculovirus IAP repeat domain (BIR3).

TS is a metabolic enzyme involved in DNA synthesis and cell division. Since TS is over-expressed in tumour cells, many research efforts were developed in order to block or inhibit it (Figure 3.17).

The TS active site consists of a substrate binding site and a cofactor binding site. TS catalyses the reductive methylation of deoxyuridylate (dUMP) to produce deoxythymidylate (dTMP). The methyl group is taken from a cofactor N5–N10 methylene tetrahydrofolate (FH4) that is oxidized to dihydrofolate (FH2), as shown in Figure 3.18.

Due to the crucial role of the folate cofactor in TS catalytic activity, two possible strategies can be adopted to block the enzyme. The first one is to design a ligand for the substrate site, while the second one is to develop a molecule interacting specifically with the folate cofactor site.

A clinically useful TS inhibitor is the anticancer agent 5-fluorouracil (5-FU), which, through its metabolite 5-fluorodeoxyuridine monophosphate (5-F-dUMP), binds into the substrate site (Figure 3.19); however it shows some limitations, such as neurotoxicity, cardiotoxicity and resistance.

TS inhibitors binding in the folate cofactor site are expected to be devoid of such negative properties. One example is represented by compound **19** (Figure 3.20), which shows antitumour activity.

In 1991, the crystal structure of compound **19** complexed to TS was resolved (PDB code 2TSC) and provided important information about the interactions within the active site. The shape of this binding pocket is similar to a cone, whose base is the opening active site where glutamic acid fragments are located. The quinazoline ring of **19** produces hydrogen bonds to Asp169 and Ala263, as shown in Figure 3.21; the carboxylate groups of the inhibitor establish hydrogen bonds with Lys48 and Thr78 through water molecules. The binding pocket is

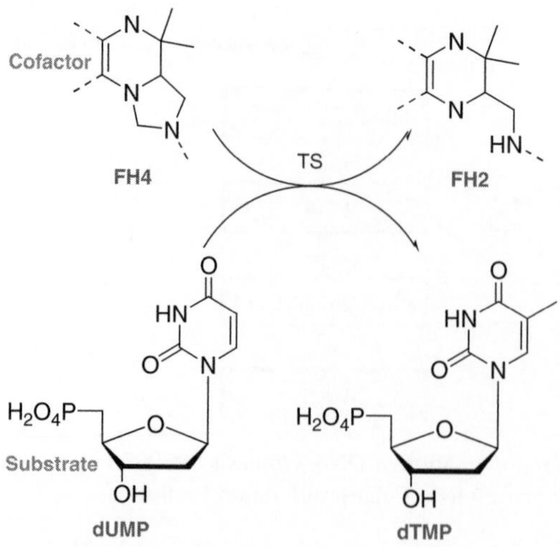

Figure 3.18 Scheme of the reductive methylation of deoxyuridylate (dUMP) to deoxythymidylate (dTMP) catalysed by thymidylate synthase

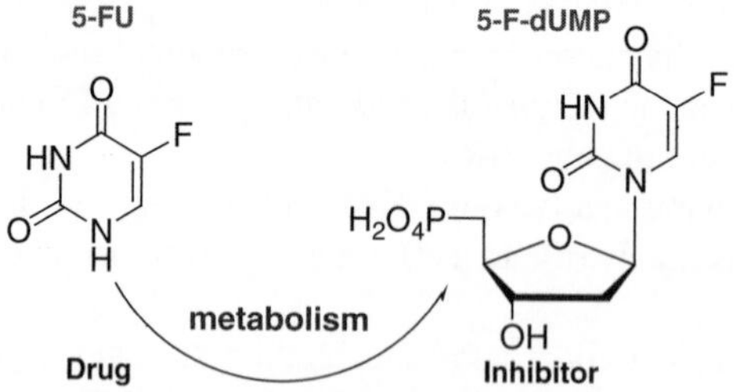

Figure 3.19 Scheme of the metabolic conversion of 5-FU into the active molecule 5-F-dUMP

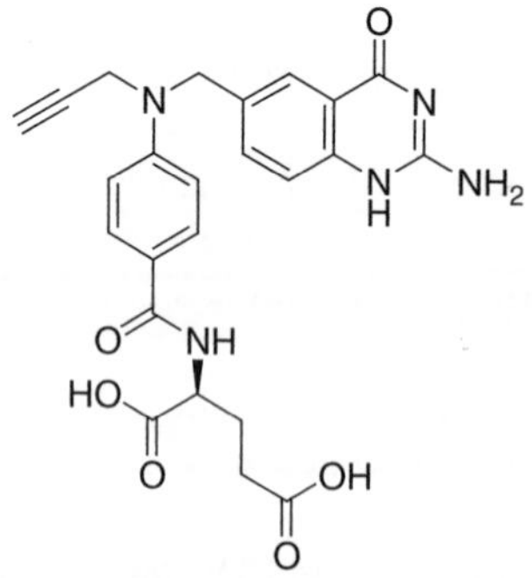

Figure 3.20 Chemical structure of CB3717 10-propargyl-5,8-dideazafolate **19**

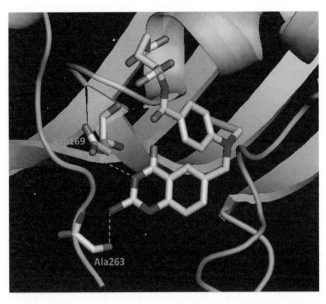

Figure 3.21 Complex between compound 19 and the active site of the thymidylate synthase crystallographic model 2TSC. The dashed lines indicate the hydrogen bonds between the inhibitor and some crucial residues of the binding pocket

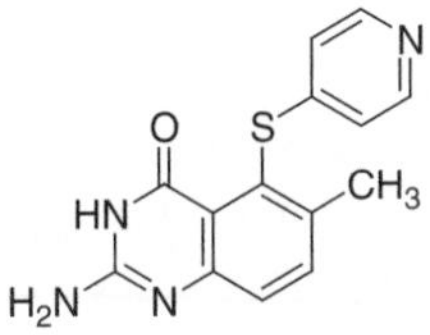

Figure 3.22 Chemical structure of 5-(aryltio)quinazolinone compound **20**

surrounded by hydrophobic chains and the quinazoline ring is also stabilized by a stacking interaction with the pyrimidine nucleus of dUMP.

Upon analysis of this initial X-ray complex, further structural elaborations were made, and, by means of docking experiments, a series of active 5-(arylthio)-quinazolinones was developed. One example of these compounds is shown in Figure 3.22.

Since its importance in interacting with the enzyme, the quinazolinone ring was conserved; examining the [TS·**19**] complex, it was observed that a one-atom linkage from the quinazolinone system was sufficient to place an aromatic ring at a proper position (instead of a two-atom linkage in **19**). Moreover, the glutamic acid component was not incorporated in the new molecules, because it was shown to cause side effects (Costi, 1998; Jones *et al.*, 1996; Webber *et al.*, 1993).

The BIR3 target is the third domain of XIAP, a protein involved in the mechanism of apoptotic cell death. XIAP has recently been considered as an interesting target for the development of new anticancer agents (Dean, 2007). Some complexes of the BIR3 domain with peptides or peptidomimetics, obtained with different spectroscopic methods, have been recently deposited

Figure 3.23 Best fit of the four-featured GBPM pharmacophore model (grid spheres) superimposed with the ligand complexed in the PDB model 1XB1 (polytube model) Two hydrophobic features are centred onto methyl groups of residue side chains (grid spheres on the right position). The hydrogen bond donor feature, corresponding to one amide hydrogen of the backbone, is shown as a grid sphere with projection; the positive ionizable feature, fitting onto the amino terminal nitrogen, is the remaining grid sphere (bottom left position)

into the PDB with codes 1G3F, 1XB1, 1XB0, 1TW6 and 1TFQ. This information is an excellent opportunity to develop common 3D pharmacophore models to be used for the virtual screening of large chemical databases.

In the presence of a multitude of experimental information about ligand–protein interactions, it is possible to apply a new approach that automatically generates pharmacophore models using the GRID program and probes in the 3D characterization of the target. Such an approach is known as GBPM (grid-based pharmacophore model). It has been recently described and tested in two different protein–protein case studies, including the XIAP target (Ortuso F, Langer T, Alcaro S, 2006). The method creates an objective 3D pharmacophore model merging the results of GRID target maps.

In the XIAP case study, the GBPM analysis leads to the identification of a four-featured 3D-pharmacophore model (Figure 3.23).

The GBPM pharmacophore model, validated by the compounds used as a training set, represents an important tool for the drug discovery process and it can be directly used to perform virtual screening experiments to identify new leads as XIAP inhibitors based on chemical scaffolds unrelated to the peptides used in the computational study.

3.5 Conclusions

Ccomputational methods are strongly affecting the drug design process, by adding to the classic routes the term 'rational drug design'. In anticancer drug design, there are multiple mechanisms that can be modulated in order to reduce the hyperproliferation of the tumour cells. The computational approaches can be generally divided into two sections on the basis of the structural knowledge of the target.

If no information about the conformation adopted by the macromolecular target is available, rational drug design will be based on known active compounds and belongs to the 'ligand-based' paradigm. Nowadays, the software and hardware available can allow researchers to carry out the computational work no matter the size and chemical nature of the compounds.

If crystallographic or NMR data are available, or precise structural models of the target can be built by modelling techniques, such as in the case of the DNA B-form conformation, rational drug design will be likely, but not exclusively, performed by the 'structure-based' paradigm. In this case, a differentiation should be made based on the chemical nature of the target. Typically, nucleic acids should be modelled carefully considering the charge and conformational properties that usually are quite different from those of enzymes or proteins.

When both paradigms can be performed, there is a great opportunity to compare totally different approaches in the same case study with consistent synergistic effects and benefits to the entire drug design development. However, the computational tools do not always perfectly fit the special requirements that can often be found in the anticancer drug discovery process. The availability of in-house methods coupled with available approaches seems to be the best compromise for the medicinal chemist involved in these studies.

References

Alcaro S, Coleman RS. Molecular modeling of the antitumour agents azinomycins A and B: force-field parameterization and DNA cross-linking based filtering. *J Org Chem* 1998, **63**, 4620–25.

Alcaro S, Coleman RS. A molecular model for DNA cross-linking by the antitumour agent azinomycin B. *J Med Chem* 2000, **43**, 2783–8.

Alcaro S, Gasparrini F, Incani O, *et al*. A quasi-flexible automatic docking processing for studying stereoselective recognition mechanisms. Part I. Protocol validation. *J Comp Chem* 2000, **21**, 515–30.

Alcaro S, Ortuso F, Coleman RS. DNA Cross-linking by azinomycin B: Monte Carlo simulations in the evaluation of sequence selectivity. *J Med Chem* 2002, **45**, 861–70.

Alcaro S, Ortuso F, Coleman RS. Molecular modeling of DNA cross-linking analogues based on the azinomycin scaffold. *J Chem Inf Model* 2005, **45**, 602–9.

Alcaro S, Gasparrini F, Incani O, Caglioti L, Pierini M, Villani C. A quasi flexible automatic docking processing for studying stereoselective recognition mechanisms. Part II. Prediction of $\Delta\Delta G$ Of complexation and H1-NMR NOE correlation. *J Comput Chem* 2007, **28**, 1119–28.

Allen FH. The Cambridge Structural Database: a quarter of a million crystal structures and rising. *Acta Cryst* 2002, **B58**, 380–8.

Bacilieri M, Moro S. Ligand-based drug design methodologies in drug discovery process: an overview. *Curr Drug Discov Technol* 2006, **3**, 155–65.

Berman HM, Westbrook J, Feng Z, *et al*. The Protein Data Bank. *Nucl Acids Res* 2000, **28**, 235–42.

Boulikas T, Vougiouka M. Cisplatin and platinum drugs at the molecular level (review). *Oncol Rep* 2003, **10**, 1663–82.

Chang G, Guida WC, Still WC. An internal coordinate Monte Carlo method for searching conformational space. *J Am Chem Soc* 1989, **111**, 4379–86.

Coleman RS, Li J, Navarro A. Total synthesis of azinomycin A. *Angew Chem, Int Ed* 2001, **40**, 1736–9.

Costi MP. Thymidylate synthase inhibition: a structure-based rationale for drug design. *Med Res Rev* 1998, **18**, 21–48.

Dantzer F, Schreiber V, Niedergang C, *et al*. Involvement of poly (ADP-ribose) polymerase in base excision repair. *Biochemie* 1999, **81**, 69–75.

Dean EJ, Ranson M, Blackhall F, Dive C. X-linked inhibitor of apoptosis protein as a therapeutic target. *Exp Opin Ther Targets* 2007, **11**, 1459–71.

de Murcia G, Ménissier de Murcia J. Poly (ADP-ribose) polymerase: a molecular nick-sensor. *TIBS* 1994, **19**, 172–6.

Fischer E, Thierfelder H. Verhalten der verschiedenen Zucker gegen reine Hefen. *Ber Dtsch Chem Ges* 1894, **27**, 2031–7.

Fujiwara T, Saito I, Sugiyama H. Highly efficient DNA interstrand cross-linking induced by an antitumour antibiotic, carzinophilin *Tetrahedron Lett* 1999, **40**, 315–18.

Furet P, Batzl C, Bhatnager A, Francotte E, Rihs, G, Lang M. Aromatase inhibitors: synthesis, biological activity and binding mode of azole-type compounds. *J Med Chem* 1993, **36**, 1393–400.

Jones TR, Varney MD, Webber SE, *et al*. Structure-based design of lipophilic quinazoline inhibitors of thymidylate synthase. *J Med Chem* 1996, **39**, 904–17.

Ginalski K. Comparative modeling for protein structure prediction. *Curr Opin Struct Biol* 2006, **16**, 172–7.

Goodford PJ. A computational procedure for determining energetically favorable binding sites on biologically important macromolecules. *J Med Chem* 1985, **28**, 849–57.

Koshland Jr, DE. Application of a theory of enzyme specificity to protein synthesis. *Proc Natl Acad Sci U S A* 1958, **44**, 98–104.

Kroemer RT. Structure-based drug design: docking and scoring. *Curr Protein Pept Sci* 2007, **8**, 312–28.

Martin YC. 3D database searching in drug design. *J Med Chem* 1992, **35**, 2145–54.

Mills M, Lacroix L, Arimondo PB, Leroy JL, Fran̦cois JC, Klump H, Mergny JL. Unusual DNA conformations: implications for telomeres. *Curr Med Chem Anticancer Agents* 2002, **2**, 627–44.

Nagaoka K, Matsumoto M, Oono J, Yokoi K, Ishizeki S, Nakashima T. Azinomycins A and B, new antitumour antibiotics. I. Producing organism, fermentation, isolation and characterization. *J Antibiot* 1986, **39**, 1527–32.

Ortuso F, Langer T, Alcaro S. GBPM: GRID-based pharmacophore model: concept and application studies to protein-protein recognition. *Bioinformatics*, 2006, **22**, 1449–55.

Ou T, Lu YJ, Zhang C, *et al*. Stabilization of G-quadruplex DNA and down-regulation of oncogene c-myc by quindoline derivatives. *J Med Chem* 2007, **50**, 1465–74.

Parkinson GN, Lee MP, Neidle S. Crystal structure of parallel quadruplexes from Human telomeric DNA. *Nature* 2002, **417**, 876–80.

Sheridan RP, Rusinko A, III, Nilakantan R, Venkataraghavan R. Searching for pharmacophores in large coordinate data bases and its use in drug design. *Proc Natl Acad Sci U S A* 1989, **86**, 8165–9.

Taylor R. Life Science applications of the Cambridge Structural Database. *Acta Cryst* 2002, **D58**, 879–88.

Wang S, Zaharevitz DW, Sharma R, *et al*. The discovery of novel, structurally diverse protein kinase C agonists through computer 3D-database pharmacophore search. Molecular modeling studies. *J Med Chem* 1994, **37**, 4479–89.

Watson JD, Crick FHC. Molecular structure of nucleic acids – a structure for deoxyribose nucleic acid. *Nature* 1953, **171**, 737–8.

Webber SE, Bleckman TM, Attard J, *et al*. Design of thymidylate synthase inhibitors using protein crystal structures: the synthesis and biological evaluation of a novel class of 5-substituted quinazolinones. *J Med Chem* 1993, **36**, 733–46.

White AW, Almassy R, Calvert AH, *et al*. Resistance-modifying agents. 9. Synthesis and biological properties of benzimidazole inhibitors of the DNA repair enzyme poly(ADP-ribose) polymerase. *J Med Chem* 2000, **43**, 4084–97.

Wodak SJ, Janin J. Structural basis of macromolecular recognition. *Adv Protein Chem* 2002, **61**, 9–73.

SECTION II

Anticancer Therapeutics

4

Introduction to Anticancer Therapeutics

Teni Boulikas

Approximately 210 cancer types are present in the entries of the National Cancer Institute in USA (http://www.cancer.gov/cancertopics/alphalist).

Overall, cancers of the lung, breast, bowel, stomach and prostate account for almost half of all cancers diagnosed worldwide. However, the types of cancer being diagnosed vary enormously across the world. For example, prostate cancer represents 16 % of all cancers in North America but only 8 % of cancers in southern Europe. Even more strikingly, gastric and liver cancers represent 19 and 14 %, respectively, of all cancer cases in eastern Asia, whereas these two cancer forms have a very low incidence in North America and all four European areas, and are not present in the four most frequent cancers (Table 4.1).

Total new cancer cases per 10 000 of the population per year also vary among geographical regions from the highest 331 in North America to 93 in western Africa (Table 4.2). The cancer death rate per 10 000 of the population per year also varies among geographical regions from 133 in Eastern Europe to 70 in South central Asia (Table 4.2).

4.1 Problems in cancer

Problems in cancer biology that are crucial points for intervention to combat cancer include early detection, genotyping and identification of gene expression profiling that might determine which chemotherapy regime to follow, the efficacy of a chemotherapy treatment and the side effects, killing chemoresistant tumours after failure of front-line chemotherapy, targeting solid tumours and metastases with minimal damage to normal tissue, arresting the process of neoangiogenesis responsible for sprouting of new vessels within growing tumours, preventing metastasis of malignant cells and the management of the cancer patient (Figure 4.1).

Anticancer Therapeutics   Edited by Sotiris Missailidis
© 2008 John Wiley & Sons, Ltd

Table 4.1 The four most frequent cancers (% of all cancers) by geographical region

	Number 1 cancer per region	Number 2 cancer per region	Number 3 cancer per region	Number 4 cancer per region
North America	Prostate (16 %)	Breast (15 %)	Lung (14 %)	Colon and rectum (12 %)
Central America	Cervix uteri (11 %)	Breast (9 %)	Prostate (8 %)	Stomach (8 %)
South America	Breast (12 %)	Prostate (10 %)	Stomach (9 %)	Cervix uteri (8 %)
Northern Europe	Breast (15 %)	Lung (13 %)	Colon and rectum (13 %)	Prostate (11 %)
Southern Europe	Lung (13 %)	Colon and rectum (13 %)	Breast (12 %)	Prostate (8 %)
Western Europe	Breast (14 %)	Colon and rectum (14 %)	Prostate (11 %)	Lung (11 %)
Eastern Europe	Lung (16 %)	Colon and rectum (12 %)	Breast (11 %)	Stomach (10 %)
East Asia	Stomach (19 %)	Lung (17 %)	Liver (14 %)	Oesophagus (9 %)
South-east Asia	Lung (13 %)	Breast (11 %)	Liver (9 %)	Colon and rectum (9 %)
South central Asia	Cervix uteri (13 %)	Breast (11 %)	Oral cavity (9 %)	Lung (6 %)
Western Asia	Lung (13 %)	Breast (13 %)	Colon and rectum (7 %)	Stomach (6 %)
Australia/New Zealand	Colon and rectum (14 %)	Breast (13 %)	Prostate (13 %)	Melanoma of skin (10 %)
Melanesia	Oral cavity (16 %)	Cervix uteri (13 %)	Liver (8 %)	Breast (7 %)
Micronesia	Lung (19 %)	Breast (17 %)	Colon and rectum (9 %)	Prostate (6 %)
Polynesia	Breast (11 %)	Stomach (10 %)	Cervix uteri (9 %)	Colon and rectum (7 %)
East Africa	Kaposi's sarcoma (14 %)	Cervix uteri (14 %)	Liver (8 %)	Oesophagus (7 %)
Central Africa	Kaposi's sarcoma (17 %)	Liver (16 %)	Cervix uteri (11 %)	Stomach (9 %)
North Africa	Breast (14 %)	Bladder (12 %)	Cervix uteri (7 %)	Lung (7 %)
Southern Africa	Cervix uteri (12 %)	Breast (10 %)	Prostate (7 %)	Kaposi sarcoma (7 %)
West Africa	Breast (15 %)	Cervix uteri (15 %)	Liver (11 %)	Prostate (7 %)
Caribbean	Prostate (14 %)	Lung (10 %)	Breast (10 %)	Cervix uteri (10 %)

Source: http://info.cancerresearchuk.org/cancerstats/geographic/world/commoncancers/

Major hurdles arise from toxicity of currently available chemotherapy regimes and inefficiency of cancer treatments, especially for advanced stages of the disease, chemoresistance of tumours and inability to cope with many sites of tumour progression after spreading of the disease. One additional obstacle in oncology is how to implement ingenious discoveries deciphering pathways of molecular carcinogenesis and ways to arrest tumour cell proliferation leading to new experimental molecules targeting a plethora of cancer mechanisms, in the clinic. Although these studies often work in cell culture, the success rate of new drugs from inception to clinical application and marketing is 1 %. Unmet needs across the cancer market remain high, with most therapies conferring low levels of specificity and high toxicity.

Table 4.2 Total cancer cases and total deaths from cancer by geographical region

	Total cancer cases	Cancer rate	Total deaths from cancer	Cancer death rate
North America	1 570 500	331	632 000	125
Central America	153 600	144	91 400	89
South America	612 900	194	345 300	111
Northern Europe	426 400	252	241 000	132
Southern Europe	617 300	237	348 400	122
Western Europe	873 700	267	475 100	131
Eastern Europe	903 400	196	637 000	133
East Asia	2 890 300	169	2 016 300	118
South-east Asia	524 900	119	363 400	84
South central Asia	1 261 500	102	846 200	70
Western Asia	200 200	130	130 000	86
Australia/New Zealand	103 700	299	44 400	120
Melanesia	6 400	147	4 200	99
Micronesia	600	142	400	97
Polynesia	800	155	500	105
East Africa	247 900	152	195 600	123
Central Africa	78 100	125	64 300	105
North Africa	119 600	87	94 000	70
Southern Africa	63 800	174	44 200	122
West Africa	140 400	93	108 000	73
Caribbean	66 500	172	43 300	112

Source: http://info.cancerresearchuk.org/cancerstats/geographic/world/commoncancers/

4.2 Cancer treatments

Although chemotherapy, surgery and radiotherapy continue to be the mainstay treatments of cancer, a number of additional modalities are currently used or are expected to change the landscape of the anticancer drug market and the ways hospitals undertake management of cancer patients. Biological drugs and targeted therapies (Figure 4.2) are those that are aimed at a specific cellular target such as small molecules that inhibit a specific protein molecule that is a key player in signal transduction, in apoptosis, in the cell cycle or in other important cellular pathways. For example, tarceva is a small molecule (see below) that functions by inhibiting the endothelial growth factor receptor (EGFR) whose function is required by many cancer cells and, as a result, EGFR is overexpressed in several cancers. Drug discovery by screening of libraries of 10^{12} or more small molecules for a specific function is an effervescent field in anticancer drug discovery.

Proteins that play a key function in cellular processes of cancer cells might also be targets of gene therapy, of RNAi (inhibitory RNA also called small interfering RNA, siRNA), of antisense, of ribozymes, or aptamers and of triplex-forming oligodeoxyribonucleotides (see below). Cell therapy, still at its infancy in the field of cancer, can become an important future player especially in cancer immunotherapy. Cancer cells can be removed from a patient, transduced *ex vivo* with genes such as IL-2 and reintroduced under the skin of the patient to elicit an immune reaction against their tumour. This approach has been used in clinical trials more than 10 years ago.

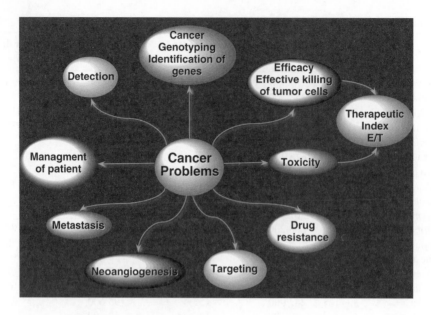

Figure 4.1   Summary of the unmet needs or problems in cancer

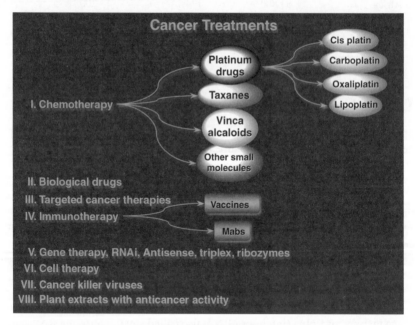

Figure 4.2   A glimpse of cancer treatments currently in use or expected to provide important drugs or approaches in the near future

Use of monoclonal antibodies aimed at a target such as Her2/neu (trastuzumab, commercial name: Herceptin), vascular endothelial growth factor (bevacizumab) or EGFR (cetuximab commercial name: Erbitux) made an important impact recently in the treatment of cancer. Herceptin is the first antibody to receive FDA approval whereas Avastin received FDA approval as first line treatment against non-small cell lung cancer and is expected to become a blockbuster drug. A method older than Hippocrates has been used extensively by modern drug designers and pharmaceutical companies; we know that nature has developed such mechanisms in plants, marine organisms and others to keep proliferating cells under control. Our job then, is to test whole extracts, then fractions and finally individual components for anticancer activities in cell cultures or animals. We then try to come up with the identification of the compounds endowed with anticancer activity, the elucidation of their structure and methods for chemical synthesis to spare depletion or destruction of the plant or marine organism. Such an example of human ingenuity led to the development of taxanes and epothilones, whereas vinca alkaloids also owe their development to plants.

4.3 Classification of chemotherapy drugs

Chemotherapy drugs can be classified into six major groups (L01 class by ATC criteria) (Figure 4.3). These are:

1. Platinum co-ordination complex
   a. Cisplatin
   b. Carboplatin
   c. Oxaliplatin

2. Antimicrotubule agents
   a. Vinca alkaloids (vinblastine, vinorelbine)
   b. Taxanes (paclitaxel, docetaxel)

3. Antimetabolites
   a. Pyrimidine antimetabolites (5-FU, azacitidine, capecitabine, cytarabine, gemcitabine)
   b. Purine antimetabolites (mercaptopurine, fludarabine, pentostatin, cladribine etc.)
   c. DHFR inhibitors (antifolates) (methotrexate, permetrexed, Raltitrexed, etc.)
   d. Thymidilate synthase inhibitors
   e. Adenosine deaminase inhibitors
   f. Ribonucleotide reductase inhibitors

4. Antitumour antibiotics
   a. Actinomycin D
   b. Mitomycin C
   c. Bleomycin
   d. Anthracyclines (doxorubicin, daunorubicin)
   e. Podofyllotoxins (etoposide, teniposide)
   f. Camptothecins (irinotecan, topotecan)

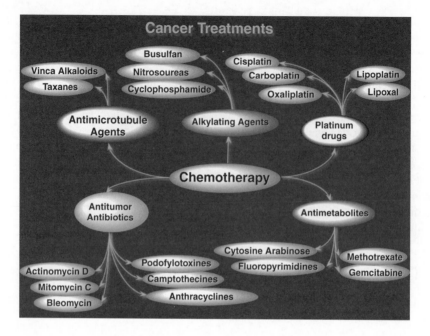

Figure 4.3   Classification of chemotherapy drugs

5. Alkylating agents

   1. Cyclophosphamide

   2. Nitrogen mustard- or L-phenylalanine mustard-based agents (mechorethamine and mephalan, respectively)

   3. Nitrosoureas, e.g. carmustine

   4. Alkane sulfonates, e.g. busulfan

6. Others, includes drugs that do not fall into any of these categories.

Although this may have been the initial classification of anticancer chemotherapy agents, a number of other categories are present, which deserve special mention in this section and this volume. These categories include many blockbuster drugs, like Gleevec, Tarceva and Tamoxifen, corresponding to categories such as kinase inhibitors, signal transduction inhibitors and hormone therapeutics respectively. In addition to those, we will briefly examine some categories of therapeutics that include molecular entities but are not solely based on those, but rather they are based on their physical properties and not their targeting capabilities, such as photodynamic therapy agents.

Since the scope of this volume is to present current and emerging therapeutics in cancer, a number of drugs at the clinical level or preclinical stage will be presented. Furthermore, this volume will examine quite extensively the various existing and emerging biological therapies, whether protein or nucleic acid based, as well as immunotherapies, including gene, cell and viral treatments. Finally, it will offer an overview on other aspects relevant to cancer research, such as clinical trials in oncology and the market for anticancer drugs as well as a brief look into the future.

5

Platinum Drugs

Teni Boulikas, Alexandros Pantos, Evagelos Bellis
and Petros Christofis

5.1 Cisplatin

Cisplatin, since its serendipitous discovery in 1965, its identification in 1969 and its clinical application in the early 1970s continues to be a cornerstone in modern chemotherapy playing an important role among cytotoxic agents in the treatment of epithelial malignancies.

Cisplatin, usually in combination with other drugs, is being used as first-line chemotherapy against cancers of the lung, head-and-neck, oesophagus, stomach, colon, bladder, testis, ovaries, cervix, uterus and as second-line treatment against most other advanced cancers such as cancers of the breast, pancreas, liver, kidney, prostate as well as against glioblastomas, metastatic melanomas, and peritoneal or pleural mesotheliomas (reviewed by Rosenberg, 1977; Hill and Speer, 1982). The clinical use of cisplatin has been impeded by severe adverse reactions including renal toxicity, gastrointestinal toxicity, peripheral neuropathy, asthenia, and ototoxicity.

Cisplatin reacts directly with sulfur groups (such as glutathione) and intracellular levels of glutathione have been linked to cisplatin detoxification. The antitumour properties of cisplatin are attributed to the kinetics of its chloride ligand displacement reactions leading to DNA crosslinking activities (Figure 5.1). DNA crosslinks inhibit replication, transcription and other nuclear functions and arrest cancer cell proliferation and tumour growth. A number of additional properties of cisplatin are now emerging, including activation of signal transduction pathways leading to apoptosis. Firing of such pathways may originate at the level of the cell membrane after damage of receptor or lipid molecules by cisplatin, in the cytoplasm by modulation of proteins *via* interaction of their thiol groups with cisplatin; for example, involving kinases, and other enzymes or finally from DNA damage *via* activation of the DNA repair pathways (reviewed by Boulikas and Vougiouka, 2003, 2004).

Cisplatin induces oxidative stress and is an activator of stress-signaling pathways especially of the mitogen-activated protein (MAP) kinase cascades. Cisplatin adducts are repaired by the nucleotide excision repair (NER) pathway involving, among others, recognition of the

Anticancer Therapeutics   Edited by Sotiris Missailidis
© 2008 John Wiley & Sons, Ltd

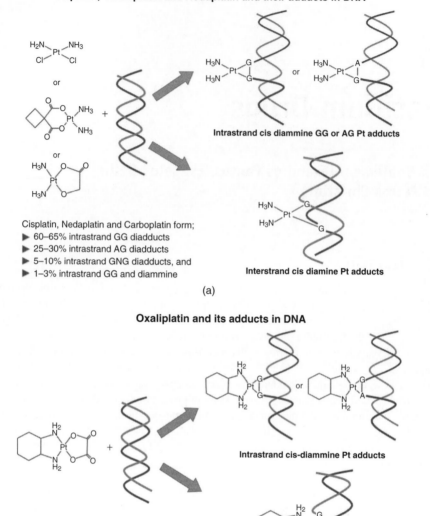

Cisplatin, Carboplatin and Nedaplatin and their adducts in DNA

Intrastrand cis diammine GG or AG Pt adducts

Cisplatin, Nedaplatin and Carboplatin form;
▶ 60–65% intrastrand GG diadducts
▶ 25–30% intrastrand AG diadducts
▶ 5–10% intrastrand GNG diadducts, and
▶ 1–3% intrastrand GG and diammine

Interstrand cis diamine Pt adducts

(a)

Oxaliplatin and its adducts in DNA

Intrastrand cis-diammine Pt adducts

Oxaliplatin form;
▶ 60–65% intrastrand GG diadducts
▶ 25–30% intrastrand AG diadducts
▶ 5–10% intrastrand GNG diadducts, and
▶ 1–3% intrastrand GG

Interstrand cis-diammine Pt adducts

(b)

Figure 5.1 (A and B) Differences in DNA adduct formation between platinum compounds with two amino groups on the Pt atom and those of oxaliplatin

damage by high mobility group (HMG) non-histone proteins and mismatch repair proteins as well as ERCC-1, one of the essential proteins in NER. Defects in DNA mismatch repair produce low-level resistance to cisplatin from the failure to recognize the cisplatin adduct and propagate a signal to the apoptotic machinery. Therapeutic interventions at all these molecular levels, either with gene transfer or with small molecules that interfere with these processes, would greatly affect the ability of cancer cells to cope with cisplatin damage. The discovery of novel platinum molecules could also lead to novel advancements in bypassing cisplatin resistance (McKeage, 2005).

The ototoxicity arises from apoptosis in auditory sensory cells induced by cisplatin (Devarajan *et al.*, 2002). The cumulative dose of cisplatin is a strong risk factor for the development of nephrotoxicity in patients who receive high doses of ifosfamide, cisplatin and etoposide combinations (Caglar *et al.*, 2002). Neutrophil gelatinase-associated lipocalin (NGAL) is a biomarker for renal damage after cisplatin treatment of cancer patients of greater diagnostic value than creatinine (Devarajan *et al.*, 2004). The significant risk of nephrotoxicity caused by cisplatin frequently hinders the use of higher doses to maximize its antineoplastic effects (Humes, 1999; Arany and Safirstein, 2003).

Since the reduction of the renal side effects with hydration and of the gastrointestinal side effects with antiemetics, neurotoxicity is the most important adverse effect associated with cisplatin chemotherapy occurring in $\sim$47 % of treated patients. The symptoms include numbness, tingling, paraesthesiae in the extremities, difficulty in walking, decreased vibration sense in the toes, deep tendon reflexes, loss of ankle jerks, difficulty with manual dexterity, difficulty with ambulation from a deficit in proprioception and gait disturbances. Unfortunately, neuropathy is long term with significant worsening of the symptoms in the first 4 months that may persist for over 52 months after stopping cisplatin treatment. Higher platinum concentrations in tissues from the peripheral nervous system (peripheral nerves and dorsal root ganglia) compared to tissues from the central nervous system (brain, spinal cord) seem to correlate with clinical symptoms of peripheral neuropathy (Al-Sarraf, 1987; Sternberg *et al*, 1989; Hill and Speer, 1982). Cisplatin, when combined with other cytotoxic agents, has shown an improved response rate and survival in a moderate to high number of patients suffering from a number of malignancies.

A number of agents have been shown to ameliorate experimental cisplatin nephrotoxicity; these include antioxidants (e.g. melatonin, vitamin E, selenium, and many others), modulators of nitric oxide (e.g. zinc histidine complex), agents interfering with metabolic pathways of cisplatin (e.g. procaine HCl), diuretics (e.g. furosemide and mannitol), and cytoprotective and antiapoptotic agents (e.g. amifostine and erythropoietin). On the other hand, nitric oxide synthase inhibitors, spironolactone and gemcitabine augment cisplatin nephrotoxicity (reviewed by Ali and Al Moundhri, 2006).

5.2 Lipoplatin

The Lipoplatin formulation is based on the formation of reverse micelles between cisplatin and dipalmitoyl phosphatidyl glycerol (DPPG) under special conditions of pH, ethanol, ionic strength and other parameters. Cisplatin-DPPG reverse micelles are subsequently converted into liposomes by interaction with neutral lipids. Lipoplatin and the platform encapsulation technology applied to its manufacturing procedure adds a strong tool in molecular oncology to wrap up pre-existing anticancer drugs into nanoparticle formulations that alter the biodistribution, lower the side effects, minimize the toxic exposure to normal tissues while

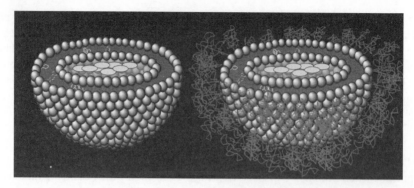

Figure 5.2 Cut through a nanoparticle showing the lipid bilayer and the cisplatin molecules in its lumen (light spheres in the inner core) (left) and that of Lipoplatin nanoparticle (right) showing the PEG molecules on its surface (medium-grey hair-like structures) coating the particle with a hydrophilic inert polymer giving the ability to escape detection from macrophages and evade immune surveillance. From: Boulikas T. Molecular mechanisms of cisplatin and its liposomally encapsulated form, Lipoplatin™. Lipoplatin™ as a chemotherapy and antiangiogenesis drug. *Cancer Therapy*, 2007, **5**, 349–76, reproduced with permission from Gene Therapy Press

maximizing tumour uptake and penetration of the drug. The shell of the liposome in the Lipoplatin formulation has a number of patented features that differentiates it from previous drug formulations. The negatively-charged DPPG molecule on the surface gives to the nanoparticles their fusogenic properties, an important feature for cell entry across the nuclear membrane barrier (Figure 5.2). In addition, their small size results in passive extravasation to tumours whereas a more avid phagocytosis characteristic of tumour cells further enhances the intracellular and nuclear uptake of the drug. A polyethylene glycol (PEG) coating also gives to the particles long circulation properties in body fluids essential for tumour accumulation. For example, phase I studies have shown a half-life of 120 h for Lipoplatin at $100 \, mg/m^2$ compared to 6 h for cisplatin.

One important issue contributing to the therapeutic efficacy of Lipoplatin results from its ability to target primary tumours and metastases and to cause a greater damage to tumour tissue compared to normal tissue. During tumour growth neo-angiogenesis is needed to develop tumour vasculature to enable supply with nutrients for growth and expansion in a process known as neoangiogenesis. The tumour uptake of Lipoplatin results from the preferential extravasation of the 100-nm liposome nanoparticles through the leaky vasculature of tumours. Indeed, the endothelium of the vascular walls during angiogenesis has imperfections that need a certain period for maturation. During angiogenesis, Lipoplatin particles with long circulation properties evade immune surveillance and are able to pass through the leaky vasculature and concentrate in the tumour at about 2- to 40-fold higher concentrations compared to the adjacent normal tissue in human studies. One additional mechanism for the higher accumulation of Lipoplatin in tumour tissue, compared to normal tissue, arises from the higher uptake of Lipoplatin nanoparticles by tumours presumably arising from a more avid phagocytosis by tumour cells. The second mechanism results in an average of 5- to 10-fold higher uptake of Lipoplatin by tumour cells, compared to normal cells in human studies giving an overall 10- to 400-fold higher tumour cell uptake and binding to macromolecules.

Intravenous infusion of Lipoplatin resulted in targeting of primary tumours and metastases in four independent patient cases (one with hepatocellular adenocarcinoma, two with gastric cancer, and one with colon cancer) who underwent Lipoplatin infusion followed by a prescheduled surgery ~20 h later. Direct measurement of platinum levels in specimens from the excised tumours and normal tissues showed that total platinum levels were on the average 10–50 times higher in malignant tissue compared to the adjacent normal tissue specimens (Boulikas *et al.*, 2005). Most effective targeting was observed in colon cancer with an accumulation up to 200-fold higher in colon tumours compared to normal colon tissue. Of the several surgical specimens, gastric tumours displayed the highest levels of total platinum suggesting Lipoplatin as a candidate anticancer agent for gastric tumours; gastric tumour specimens had up to 260 µg platinum/g tissue that was higher than any tissue level in animals treated at much higher doses. Fat tissue displayed a high accumulation of total platinum in surgical specimens in three different patients correlating to the lipid capsule of cisplatin in its Lipoplatin formulation. It was also inferred that normal tissue had more platinum trapped in the tissue but not reacted with macromolecules whereas tumour tissue displayed platinum that reacted with cellular macromolecules; the data were consistent with a model where Lipoplatin damages more tumour compared to normal cells.

Lipoplatin formulation uses several advancements in its liposome encapsulation: (i) the anionic lipid DPPG gives Lipoplatin its fusogenic properties, presumably acting at the level of entry of the drug through the cell membrane after reaching the target tissue; (ii) the total lipid to cisplatin ratio is low (10.24:1 mg lipid/mg cisplatin) in Lipoplatin, which means that less lipid is injected into the patient. For comparison, the ratio of lipids to cisplatin in the liposomal formulation SPI-77 is 71.43:1 (Veal *et al.*, 2001) which is seven-fold higher lipids per mg cisplatin compared to Lipoplatin; and (iii) The PEG polymer coating used on Lipoplatin is meant to give the drug particles the ability to pass undetected by the macrophages and immune cells, to remain in circulation in body fluids and tissues for long periods and to extravasate preferentially and infiltrate solid tumours and metastases through the altered and often compromised tumour vasculature.

The introduction of Lipoplatin has been an advancement in the field of platinum drugs mainly because of its ability to circulate with a half-life of ~100 h compared to 6 h for cisplatin, to have substantially reduced the nephrotoxicity, ototoxicity and neurotoxicity (Stathopoulos *et al.*, 2005) and to concentrate in tumours (Boulikas *et al.*, 2005) (Figure 5.3). Lipoplatin at its recommended dose of 120 mg/m^2, as a 4–6 h intravenous infusion, has also reduced the myelotoxicity and nausea/vomiting of cisplatin. Indeed, Lipoplatin nanovehicles accumulate in cancer tissue with altered vascularization about 40 times more than in normal tissue thereby reducing the potential toxic effects on normal tissue.

Lipoplatin is currently under several phase III evaluations. A phase III multicentre clinical trial uses weekly 120 mg/m^2 Lipoplatin in combination with gemcitabine as first line treatment against non-small cell lung cancer (NSCLC) and is being compared to cisplatin plus gemcitabine. Another phase III study compares weekly Lipoplatin plus 5-fluorouracil (5-FU) versus cisplatin plus 5-FU against head and neck cancers. A third phase III study uses Lipoplatin in combination with paclitaxel as first line treatment against NSCLC and is being compared to cisplatin plus paclitaxel, where the response was found to be similar, but the toxicity, and in particular nephrotoxicity, neurotoxicity and myelotoxicity, was significantly lower with Lipoplatin (Stathopoulos *et al.*, 2007).

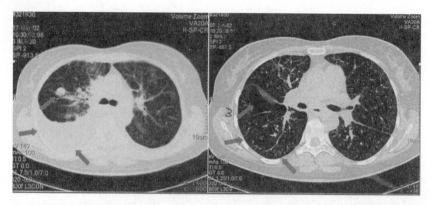

Figure 5.3   Disappearance of a metastatic lesion from pancreatic cancer in the lung lobe with concomitant disappearance of pleural effusion after 5 treatments with Lipoplatin plus gemcitabine. Left: before; right: after treatment. *Cancer Therapy*, 2007, **5**, 537–83, reproduced with permission from Gene Therapy Press

The dosage schemes currently in phase III are Lipoplatin 100–150 mg/m^2, given every 7 days, for 9 weeks in combination with other chemotherapy drugs. Thus, the total dose of cisplatin in Lipoplatin regimens is 1350 mg cisplatin/m^2. This is in contrast to cisplatin that is administered at 100 mg/m^2 every 21 days for four cycles (total 400 mg cisplatin/m^2) because of cumulative toxicity.

Lipoplatin does not show nephrotoxicity, ototoxicity, neurotoxicity, or any other toxicity except a manageable nausea/vomiting (grade I–II in 15 % of patients) and mild myelotoxicity (grade I–II in 57.6 % and grade III in 4 % of patients) in the initial phase I studies at doses of 100 mg/m^2 every 14 days (Stathopoulos *et al.*, 2005). At 100 mg/m^2 Lipoplatin every 14 days no hepatotoxicity, cardiotoxicity, hair loss or allergic reaction was observed but mild asthenia was common. Acute epigastric and back pain at the beginning of Lipoplatin infusion lasting a few minutes was seen in 30 % of patients.

There is no need for pre- or post-hydration of the patient with Lipoplatin. This is in contrast to cisplatin chemotherapy, which requires admittance of the patient the night before infusion for hydration as well as extended stay in the hospital after infusion for post hydration to reduce the nephrotoxicity of the drug.

Myelosuppression is the principal dose-limiting toxicity at a dose of 120 mg/m^2 Lipoplatin every week in combination with standard doses of gemcitabine, where the two drugs have additive myelotoxic effects. The dose limiting myelotoxicity is manifested as neutropenia and thrombocytopenia. Gastrointestinal toxicity, nausea and vomiting are commonly reported (69 % of patients) but are usually of mild to moderate severity. Severe nausea and vomiting (WHO grade 3) occurs in less than 5 % of patients. No grade 4 nausea and vomiting have been reported. Mild diarrhoeas are reported in 20 % of patients.

In conclusion, Lipoplatin has the ability to preferentially concentrate in malignant tissue both of primary and metastatic origin following intravenous infusion to patients. In this respect, Lipoplatin emerges as a very promising drug in the arsenal of chemotherapeutics.

5.3 Carboplatin

Over 20 years of intensive work toward improvement of cisplatin, and with hundreds of platinum drugs tested, has resulted
in the introduction of the widely used carboplatin and of oxaliplatin used only for a very narrow spectrum of cancers. The
inception and promotion to the clinic of platinum drugs have
been milestone achievements in clinical oncology.

Cisplatin analogues have been marketed (carboplatin, oxaliplatin) but none as yet has
achieved a similar broad-spectrum effectiveness. Carboplatin proved markedly less toxic to
the kidneys and nervous system than cisplatin and caused less nausea and vomiting, while
generally (and certainly for ovarian cancer) retaining equivalent antitumour activity. Carboplatin constitutes a reasonable alternative to cisplatin in a combination with gemcitabine, since
it shows synergy with gemcitabine *in vitro*, is easier to use in ambulatory patients, and has a
better non-haematological toxicity profile. The combination of carboplatin with gemcitabine
(Gemzar) initially hampered by unacceptable platelet toxicity, has gained increasing accep-
tance against NSCLC. Combinations of carboplatin with paclitaxel or docetaxel are also used
against NSCLC (Kosmidis *et al.*, 2002).

Paclitaxel plus carboplatin was equally active and well tolerated compared to gemcitabine
plus paclitaxel in a phase III randomized trial and gave response rates of 28.0 % versus
35.0 % (Kosmidis *et al.*, 2002) A phase III randomized trial in 618 patients with advanced
NSCLC gave a response rate of 25 % (70 of 279) in the paclitaxel/carboplatin arm and 28 %
(80 of 284) in the paclitaxel/cisplatin arm (Rosell *et al.*, 2002). The use of vinorelbine, gemcitabine, paclitaxel and docetaxel in combination with cisplatin or carboplatin against NSCLC
has increased by as much as 10 % the overall survival at one year. Carboplatin/paclitaxel-
based combination chemotherapy has become a very popular combination in the US against
advanced NSCLC and has advantages to the older cisplatin-based chemotherapy (reviewed
by Ramalingam and Belani, 2002).

Carboplatin is a safe and effective first-line treatment for women with advanced ovarian cancer as deduced from four large randomized trials of paclitaxel in combination with
platinum against a platinum-based control treatment representing 3588 patients (reviewed
in Sandercock *et al.*, 2002). Carboplatin as single agent has demonstrated a 17 % response
rate against measurable hormone refractory prostate cancer. Hormone refractory prostate cancer has been treated with combination chemotherapy using docetaxel, estramustine phosphate
and carboplatin to an area under the curve of 6 on day 1 of a 4-week cycle (Kikuno *et al.*,
2007). Children with Wilm's tumour (nephroblastoma) were treated with high-dose melphalan, etoposide and carboplatin and autologous peripheral blood stem cell rescue in order to
improve their probability of survival (Pein *et al.*, 1998). Anaplastic astrocytomas and glioblastomas were treated with intravenous administration of carboplatin (300 mg/m^2) on day 1 and
etoposide (60 mg/m^2) on day 1 to 5, repeated every 6 weeks (Yamamoto *et al.*, 2002). A combination of paclitaxel, etoposide and carboplatin is an often used regimen against small-cell
lung cancer (Reck *et al.*, 2006; Baas *et al.*, 2006). Docetaxel combined with cisplatin gives an
overall response rate of 33−46 % compared to 30−48 % of docetaxel combined with carboplatin (reviewed by El Maalouf *et al.*, 2007).

5.4 Oxaliplatin

The alkaline hydrolysis of oxaliplatin produces the oxalato monodentate intermediate complex (pKa 7.23) and the dihydrated oxaliplatin complex in two consecutive steps. The monodentate intermediate is assumed to rapidly react with endogenous compounds (Jerremalm *et al.*, 2003). The crystal structures of oxaliplatin bound to a DNA dodecamer duplex with the sequence 5'-d (CCTCTGGTCTCC) has been reported (Spingler *et al.*, 2001). The platinum atom forms a 1,2-intrastrand cross-link between two adjacent guanosine residues bending the double helix by approximately 30° toward the major groove. Crystallography has provided structural evidence for the importance of chirality in mediating the interaction between oxaliplatin and duplex DNA (Spingler *et al.*, 2001). With oxaliplatin, like cisplatin, adduct lesions are repaired by the nucleotide excision repair system. Oxaliplatin, like cisplatin, is detoxified by glutathione (GSH)-related enzymes. ERCC1 and xeroderma pigmentosum, complementation group A (XPA) expressions were predictive of oxaliplatin sensitivity in six colon cell lines *in vitro* (Arnould *et al.*, 2003). Oxaliplatin combined with 5-FU and folinic acid improved the response rate and progression-free and overall survival of patients with advanced colorectal cancer (De Vita *et al.*, 2005). The dose-limiting adverse reaction of oxaliplatin is neurotoxicity (sodium channel inactivation) and the kinetics are altered after exposure of animals to oxaliplatin. The results from preliminary clinical studies indicate that the sodium channel blockers carbamazepine and gabapentin may be effective in preventing neurotoxicity (Lersch *et al.*, 2002).

Oxaliplatin produces the same type of inter- and 1,2-GG intrastrand cross-links as cisplatin but has a spectrum of activity and mechanisms of action and resistance different from those of cisplatin and carboplatin. The cellular and molecular aspects of the mechanism of action of oxaliplatin have not yet been fully elucidated. However, the intrinsic chemical and steric characteristics of the non-hydrolysable diaminocyclohexane (DACH)-platinum adducts on DNA appear to contribute to the lack of cross-resistance with cisplatin and carboplatin (reviewed by Di Francesco *et al.*, 2002). The anticancer effects of oxaliplatin are optimized when it is administered in combination with other anticancer agents, such as 5-FU, gemcitabine, cisplatin, carboplatin, topoisomerase I inhibitors, and taxanes (reviewed by Ranson and Thatcher, 1999; Raymond *et al.*, 2002). Oxaliplatin has a unique pattern of side effects and besides neurotoxicity they include haematological and gastrointestinal tract toxicity. Grade 3/4 neutropenia occurred in 41.7 % of patients in a phase III clinical trial. Nausea and vomiting is usually mild to moderate and readily controlled with standard antiemetics. Nephrotoxicity is mild allowing administration of oxaliplatin without hydration (Cassidy and Misset, 2002). Sporadically, severe side effects may be observed such as tubular necrosis (Pinotti and Martinelli, 2002). Oxaliplatin, in combination with 5-FU, has been recently approved in Europe, Asia, Latin America and later in the USA (2003) for the treatment of metastatic colorectal cancer.

5.5 Lipoxal

Liposomal encapsulation of oxaliplatin was achieved using Regulon's platform technology into a new formulation, Lipoxal. The drug finished stability test and preclinical studies and was approved for phase I evaluation to determine its dose-limiting toxicity (DLT) and maximum tolerated dose (MTD) (Stathopoulos *et al.*, 2006). Twenty-seven patients with advanced disease of the gastrointestinal system (stage IV gastrointestinal cancers including colorectal, gastric and pancreatic) who had failed previous standard chemotherapy were treated with escalating doses of Lipoxal once weekly for 8 weeks. No serious side effects were observed at $100-250 \, mg/m^2$ whereas at doses of 300 and $350 \, mg/m^2$ of Lipoxal monotherapy mild myelotoxicity, nausea and peripheral neuropathy were observed.

Lipoxal gastrointestinal tract toxicity was negligible. Without antiemetics (ondansetron), nausea or mild vomiting was observed, but with ondansetron administration, no nausea/vomiting or diarrhoea was observed. Mild, grade 1 myelotoxicity (neutropenia) was only seen in two patients (7.4 %) at the highest dose level ($350 \, mg/m^2$). There was no hepatotoxicity, renal toxicity, cardiotoxicity or alopecia. Mild asthenia was observed in three patients.

Neurotoxicity from Lipoxal was observed after at least three infusions of Lipoxal; grade 1 neurotoxicity was seen at 200 and $250 \, mg/m^2$, grade 2 at $300 \, mg/m^2$. Grade 2–3 peripheral neuropathy was observed in all four patients treated at $350 \, mg/m^2$ and this dose level was therefore, considered as DLT and the $300 \, mg/m^2$ level as the MTD. On the basis of these results, grade 2–3 neurotoxicity was considered as the DLT which was observed in 100 % of patients treated with $350 \, mg/m^2$ of Lipoxal; therefore $300 \, mg/m^2$ was defined as the MTD.

Of the 27 patients treated, three achieved partial response and 18 had stable disease for 4 months, (range 2–9 months). Of the three out of 27 patients (11.1 %) that achieved a partial response, two had gastric cancer, one of whom had pleural effusion and the other had bone metastases; the third was a patient with liver metastases from colon carcinoma. Reduction in bone metastases in this patient was observed after Lipoxal monotherapy that coincided with pain reduction (Figure 5.4). The determination of a partial response was based on a computerized tomography (CT) scan for the first patient, a bone scan for the second patient and a CT scan and a bilirubin serum level value for the third patient. The third patient was treated while the serum bilirubin level was 51 mg/dl and after two courses of treatment the level dropped to 8 mg/dl and lasted for 5 weeks.

The duration of response was 4, 7 and 2 months for each of the above patients, respectively. Eighteen (66.7 %) patients achieved stable disease with a median duration of 4–6 months (range 2–9 months). Six patients showed disease progression. In all three responders there was also a reduction of 50 % or more of the marker CA-19-9 and performance status improved from 2 to 1. With respect to effectiveness, the 11 % response rate observed in pretreated patients refractory to previously established tumours could be meaningful in future trials in a combined chemotherapy modality. It is also important to point out that the cancer types selected for this trial are not those which are the most sensitive to chemotherapy.

Lipoxal is a well-tolerated agent. Whereas the main adverse reactions of oxaliplatin are neurotoxicity, haematological and gastrointestinal toxicity; its liposomal encapsulation (Lipoxal) has reduced the haematological and gastrointestinal toxicity and the main side effect was neurotoxicity. The dose of $300 \, mg/m^2$ was established as the MTD but further investigation is needed, particularly with other agents in combination. Gastrointestinal tract and bone marrow toxicities are very much reduced compared to the standard form of oxaliplatin. The only adverse reaction was neurotoxicity which defined DLT. Thus, Lipoxal is a liposomal oxaliplatin, which reduces the cytotoxic agent's adverse reactions without reducing effectiveness. Future studies are aimed at demonstrating the tumour targeting

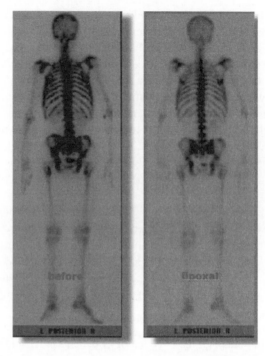

Figure 5.4 Improvement in bone metastases after Lipoxal monotherapy. *Cancer Therapy*, 2007, **5**, 537–83, reproduced with permission from Gene Therapy Press

properties of the drug which are expected based on the fact the liposome shell is similar to that of Lipoplatin.

5.6 New Platinum Compounds

The success of cisplatin has triggered intensive work for discovery of new platinum-based anticancer drugs. However, from over 3000 compounds tested *in vitro* only about 30 have entered into clinical trials (1 %). Considering that less than 10 % of these will make it to registration, the success in new platinum drug design is less than 1 in 1000 compounds.

The medicinal use and application of metals and metal complexes are of increasing clinical and commercial importance. Specifically, metal-based drugs, imaging agents and radionuclides have over the years claimed an increased portion of an annual US$5 billion budget for the whole field.

The clinical success of cisplatin [*cis*-diamminedichloroplatinum(II)] has been the main impetus for the evolution of the family of platinum compounds, which currently holds a vital role in metal-based cancer chemotherapy. The use of cDDP and its effectiveness in cancer chemotherapy has been thoroughly documented (Lippert 1999; Wong and Giandomenico 1999; Farrell 2000). It is used as a principal component for the treatment of testicular, ovarian and bladder cancer. Despite this success, there is still a limited range of tumours sensitive to cisplatin intervention, due to inherent as well as acquired resistance after treatment. Furthermore the side-effects of cisplatin treatment are severe and include the dose-limiting nephrotoxicity, neurotoxicity, ototoxicity and emetogenesis (O'Dwyer *et al.*, 1999; Highley *et al.*, 2000).

DNA is accepted to be the cellular target of cisplatin and the antitumour effects of platinum complexes are believed to result from their ability to form various types of adducts with DNA, which block replication and transcription and induce cell death (Johnson NP *et al.*, 1989; Johnson SW *et al.*, 1998). Also, the nature of DNA adducts affects a number of transduction pathways and triggers apoptosis and necrosis in tumour cells (Fuertes *et al.*, 2003), therefore mediating the cytotoxicity of platinum compounds and resistance. Platinum drug resistance is a multifactorial process and can occur through several mechanisms such as increased drug efflux, drug inactivation, evasion of apoptosis and processing damage and repair (Morin 2003; Siddik, 2003), the latter of which involves well described repair mechanisms such as nucleotide excision repair, mismatch repair, homologous recombination repair and translesion synthesis. The elucidation of the molecular mechanisms that mediate cisplatin mode of action, resistance and sensitivity provided the necessary background for the design and synthesis of new platinum compounds with improved toxicity profile, circumvention of resistance and expansions of tumour panel.

One strategy to overcome cisplatin resistance is to design platinum complexes that specifically deal with some or even all of the above-mentioned resistance mechanisms. However, after more than 30 years of intensive research, no more than 30 compounds have exhibited adequate pharmacological advantages relative to cisplatin, in order to be tested in clinical trials (Fuertes *et al.*, 2002) and only four registered for clinical use (Judson and Kelland, 2000) (Figure 5.5), thus proving that the search for novel platinum compounds remains a difficult task.

Carboplatin entered the clinic in 1998 in an effort to reduce the toxic side effects of the parent drug and exhibits activity in the same set of tumours as cisplatin. From a mechanistic DNA-binding point of view, it is not too surprising that the introduction of these new platinum antitumour drugs did not represent a fundamental breakthrough in the treatment of cancer with platinum agents due to the similarity in the induced DNA conformation changes.

Furthermore it is established that the future of cancer chemotherapy must be directed to look for adjuvant drugs that affect general biochemical mechanisms that can bypass drug resistance rather than to exclusively search for specific drugs which target particular cellular

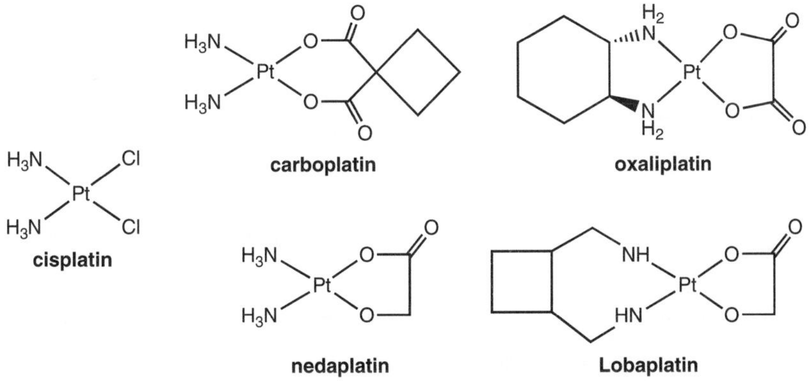

Figure 5.5 Structures of the clinically used platinum anticancer drugs

constituents and one way to achieve that is to increase the spectrum of modes of interaction of the new compound. Over the years the alternative strategies for the design of new class of platinum antitumour compounds included *trans*-geometry, Pt(IV), polynuclear and conjugate compounds with interesting results.

The original empirical structure–activity relationships considered the trans isomer of cisplatin and other transplatin analogues to be inactive (Reedijk, 1996). However, several groups have shown that some *trans* compounds are active *in vitro* and *in vivo*. A distinct difference between cisplatin and its *trans* isomer is that the latter is chemically more reactive than cisplatin and more susceptible to deactivation. Several new analogues of transplatin, which exhibit a different spectrum of cytostatic activity including activity in tumour cells resistant to cisplatin, have been identified (Natile and Collucia, 2001). Examples are analogues containing iminoether groups, heterocyclic amine ligands or aliphatic ligands.

Another alternative strategy is to manipulate chemical and biological properties through oxidation number. A number of Pt (IV) compounds have undergone clinical trials including *cis*-[PtCl$_4$(1,2-dach)] (tetraplatin) and *cis,cis,cis*-[PtCl$_2$(OH)$_2$(iPrNH$_2$)$_2$] (Iproplatin) before being abandoned (Lebwohl and Canetta, 1998). Currently, another Pt(IV) compound is being screened, JM-216 (Hall and Hambley, 2002; Carr *et al*, 2002; Kurata *et al*, 2000; Kelland, 2000) with promising results (Figure 5.6). Another approach that has emerged, due to the understanding of the mechanisms of platinum resistance, is to insert steric bulk at the platinum centre in order to reduce the kinetics of substitution compared to cisplatin. The compound AMD473 has been the most successful result of this effort as it manages to form cytotoxic adducts with DNA while at the same time avoiding the reaction with GSH (Holford *et al.*, 1998) (Figure 5.6).

One unique class of new anticancer platinum agents with distinct chemical and biological properties different from mononuclear platinum drugs are the polynuclear platinum compounds. The first example of this class to advance to clinical trials is BBR3464 (1998) (Figure 5.7), which was also the first platinum drug deviating from cisplatin structure. The structure is notable for the presence of the central platinum moiety, which contributes to DNA affinity through electrostatic and hydrogen bonding interactions (Qu *et al.*, 1993). The 4+ charge, the presence of at least two Pt coordination units capable of DNA binding and the consequences of such DNA binding are remarkable departures from the cisplatin structural paradigm.

It has been demonstrated that polynuclear platinum complexes produce adducts on DNA whose character is different when compared with conventional mononuclear platinum compounds (Brabec *et al.*, 1999; Kasparkova *et al.*, 1999, 2000, 2002; Zehnulova *et al.*,

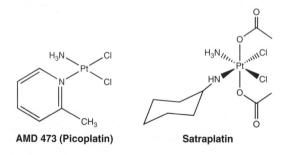

AMD 473 (Picoplatin)   **Satraplatin**

Figure 5.6   Structures of AMD473 (Picoplatin) and JM216 (Satraplatin)

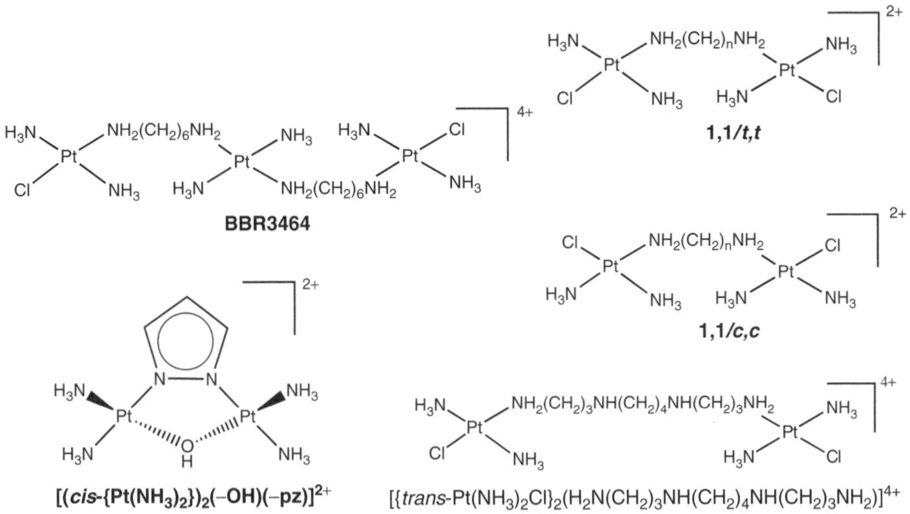

Figure 5.7 Structure of BBR3464 and of binuclear platinum compounds

2001). This suggests that these platinum compounds may escape, at least in part, the conventional mechanism of cisplatin resistance related to DNA damage recognition and repair. Despite the fact that these compounds are relatively new, some data on the structure of their DNA adducts, repair of these adducts, their recognition by specific proteins and cytotoxicity of these new platinum compounds in tumour cell lines are already available. Nonetheless, much less details on the molecular mechanisms underlying their biological effects are available in comparison with mononuclear platinum compounds. For instance, no data on how polynuclear platinum compounds inhibit DNA polymerization by DNA polymerases have been obtained so far in spite of the fact that the relative efficiency of the biological action of platinum compounds (also including DNA replication and antitumour effects) is dependent on the specific structure of the Pt-DNA adduct formed; or even more specifically that the extent and specificity of resistance for platinum compounds have been shown to correlate with the ability of cells to elongate DNA chains that contain platinum adducts. The class as a whole represents a second, distinctly new structural group of platinum-based anticancer agents.

One strategy to improve the antitumour efficacy of new platinum drugs is to conjugate DNA-targeting groups such as intercalators to the metal moiety. Various compounds have been produced during the last decade (Sundquist *et al.*, 1988, 1990; Bowler *et al.*, 1989) that have exhibited enhanced antitumour activity. Renewed interest in platinum complexes with appended intercalators has produced some promising results. A series of cis-ethylenediamineplatinum(II) complexes with tethered 9-aminoacridine-4-carboxamides was able to overcome cross-resistance in human ovarian carcinoma cell lines *in vitro* (Holmes *et al.*, 2001). Altered DNA sequence specificity and increased DNA binding rates compared with those of cisplatin were observed for these intercalator–platinum conjugates (Temple *et al.*, 2000, 2002). Quite recently, a cytotoxic platinum(II) complex {[Pt(en) Cl(ACRAMTU-S)](NO$_3$)$_2$ (en = ethane-1,2-diamine, ACRAMTU = 1-[2-(acridin-9-ylamino)ethyl]-1,3-dimethylthiourea)} (Figure 5.8) has been demonstrated to form

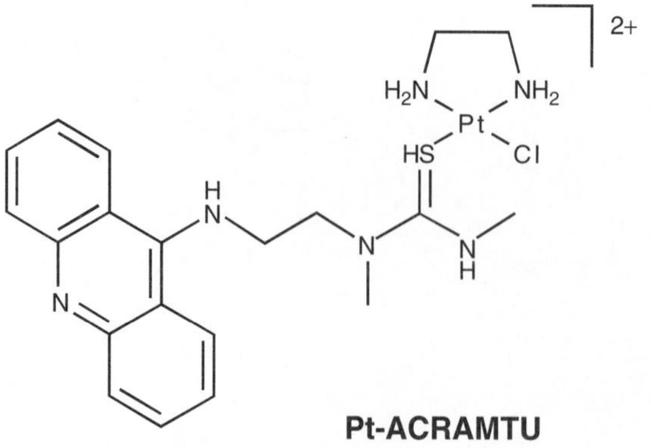

Pt-ACRAMTU

Figure 5.8   Structure of Pt-ACRAMTU

adducts in the minor groove of DNA through platination of the adenine-N3 endocyclic nitrogen (Barry *et al.*, 2005). This finding breaks a longstanding paradigm in platinum-DNA chemistry, namely the requirement for nucleophilic attack of guanine-N7 as the principal step in CL formation. Attachment of intercalating agents to platinum complexes has proven to enhance rates of DNA-platination while minimizing exposure of platinum to inactivating cellular agents such as thiols. These complexes bind covalently and intercalatively to DNA to form novel adducts endowed with the capability of evading cellular DNA repair mechanisms.

The use of water soluble polymers as macromolecular carriers for low-molecular weight conventional drugs has recently arisen as a very promising strategy. The advantages include improved body distribution, prolonged blood circulation, reduced systematic toxicity and localization of more drug molecules to the tumour through enhanced permeability and retention (Matsumara and Maeda, 1986). A significant number of polymer-bound antitumour agents have been developed so far including taxol, doxorubicin, camptothecin and cisplatin that have been conjugated to polymers such as *N*-(2-hydroxypropyl)methacrylamide and PEG. The latter is a fully biocompatible molecule, deprived of toxicity and immunogenicity, which is able to solubilize and vehiculate amphipathic molecules into the bloodstream. It has also been used as a carrier of Pt derivatives in an effort to improve its pharmacological properties (Furin *et al.*, 2003; Aronov *et al.*, 2003) (Figure 5.9).

5.6.1 Picoplatin

Picoplatin (*cis*-PtCl$_2$(NH$_3$)(2-pic), previously AMD473; Figure 5.10) is a new generation sterically hindered platinum cytotoxic compound that provides a differentiated spectrum of activity against a wide range of human tumour cell lines and an improved safety profile. It is designed to overcome acquired resistance to cisplatin *in vitro* and in human tumour xenografts.

In a single-agent dose-escalating phase I study that was performed, picoplatin was initially administered intravenously as a 1 h infusion every 21 days to patients with advanced solid tumours and it demonstrated encouraging activity including those with prior platinum exposure. Neutropenia and thrombocytopenia proved dose limiting while other toxicities included moderate nausea, vomiting, anorexia, and a transient metallic taste. (Beale *et al.*, 2003).

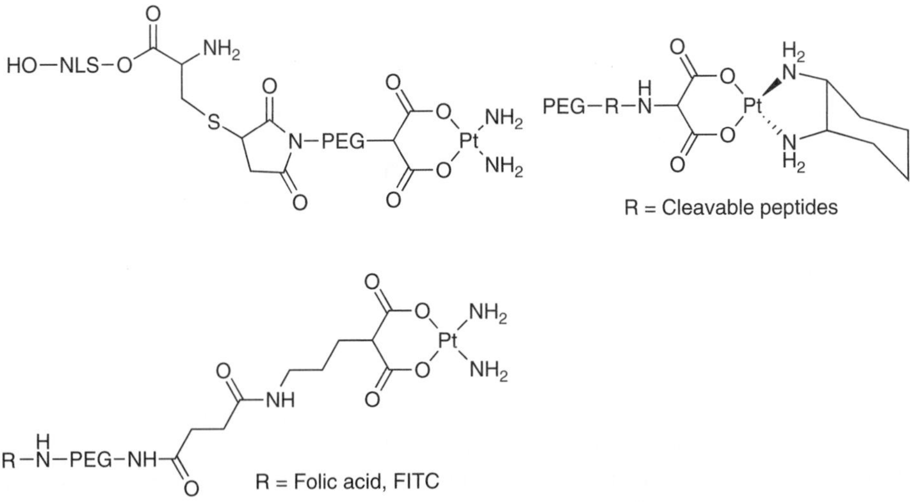

Figure 5.9 Conjugated platinum compounds

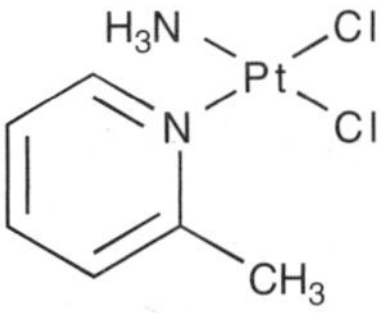

Figure 5.10 The structure of picoplatin

Currently it is being studied in an ongoing multicentre phase II clinical trial in patients with small cell lung cancer and in phase I/II clinical trials in patients with colorectal cancer and prostate cancer. The phase II trial is designed to confirm the clinical activity of picoplatin as second-line therapy in patients with platinum-sensitive or -refractory small cell lung cancer who have failed a prior platinum-based chemotherapy. Picoplatin (Figure 5.10) will also enter phase I trials for colorectal and prostate cancer.

5.6.2 Aroplatin

L-NDDP (Aroplatin; Figure 5.11) is a liposomal formulation of *cis*-bis-neodecanoato-*trans*-R,R-1,2-diaminocyclohexane platinum (II), a structural analogue of oxaliplatin. It is a lipophilic non-cross-resistant platinum compound formulated in large multilamellar liposomes (1–3 μm). The MTD of liposomal-entrapped NDDP (L-NDDP) administered i.v. in humans is 300 mg/m^2 with myelosuppresion as the dose-limiting toxicity.

To determine pathological response rates to liposome-entrapped L-NDDP in the context of phase II studies, the drug was administered intrapleurally in patients with malignant pleural mesothelioma. Intrapleural L-NDDP therapy in this patient population is feasible with significant but manageable toxicity. Although pathological responses are highly encouraging, areas

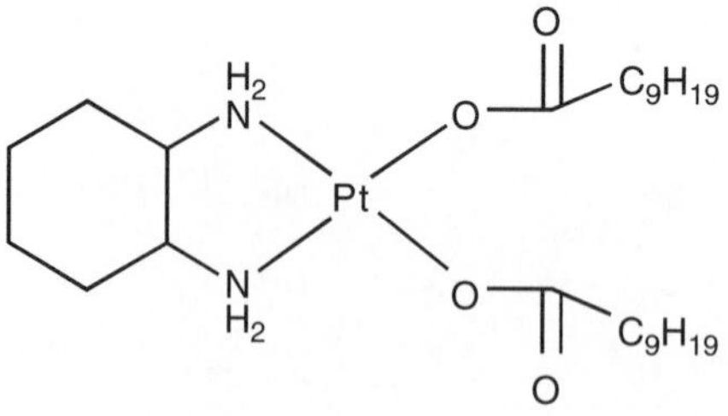

Figure 5.11 The structure of Aroplatin

of mesothelioma that are not in direct communication with the pleural space will evade drug exposure and limit efficacy in some patients. The optimal role of intrapleural L-NDDP therapy currently remains to be determined (Perez-Soler *et al.*, 2005).

Currently various clinical studies are being conducted in an effort to test the safety and efficacy of Aroplatin in a wide range of cancers, including colorectal (Dragovich *et al.*, 2006) and kidney cancers. Initial results suggest that Aroplatin treatment appears to be associated with tumour response and may be a promising approach to therapy for a range of different cancers.

5.7 Cisplatin Resistance and Chemotherapy

The major limitation in the clinical applications of cisplatin has been the development of cisplatin resistance by tumours. Development of platinum drug resistance by tumours is often linked to resistance to other chemotherapy drugs (doxorubicin, taxanes) from cross-talk of mechanisms involving import of these molecules across the cell membrane. Cisplatin resistance arises by clonal expansion of tumour cells in the heterogenous tumour cell population with acquired resistance to cisplatin (with mutations, overexpression or other mechanisms in specific genes that confer resistance) during treatment after killing of the sensitive cells by the drug. The mutagenic effect of cisplatin would favour the development of mutations in various genes and the selection of cells with proliferation advantage in patients that undergo cisplatin chemotherapy. This proliferation advantage could be conferred by limited uptake of the drug by resistant cells and absence of damage-induced apoptosis or by an enhanced repair of the DNA lesions and signalling in favour of cell survival rather than apoptosis.

The probability of response to second-line chemotherapy following platinum-based treatments is usually related to the platinum-free interval. Patients can be classified as either platinum-sensitive or platinum-resistant depending on whether they have relapsed or progressed within 26 weeks of completing first-line platinum-based chemotherapy (Gore *et al.*, 2002). Expression of the mitogen-activated protein kinase phosphatase-1 (MKP-1), involved in inactivation of MAP-kinase pathways, regulation of stress-responses, and suppression of apoptosis, was a prognostic marker for shorter progression-free survival of patients with invasive ovarian carcinomas (Denkert *et al.*, 2002). Apoptotic index can be predictive of treatment outcome in ovarian cancer (Mattern *et al.*, 1998). Salvage monochemotherapy is generally used, but when the platinum-free interval is longer than 24 months, re-treatment with platinum compounds and/or taxanes is indicated.

Several mechanisms can contribute to cisplatin resistance, which are presented below:

5.7.1 Cisplatin resistance at the cell membrane barrier

The reduction in cisplatin accumulation inside cancer cells because of barriers across the cell membrane is considered a major mechanism of the acquired cisplatin resistance (Sharp *et al.*, 1995; Shen *et al.*, 2000). The copper transporter CTR1 appears to control the accumulation of cisplatin in *Saccharomyces cerevisiae* (Figure 5.12). Deletion of CTR1 resulted in a 16-fold reduction in the uptake of copper and an 8-fold reduction in the uptake of cisplatin. CTR1-deficient cells also demonstrated impaired accumulation of the cisplatin analogs carboplatin, oxaliplatin, and ZD0473 (Lin *et al.*, 2002).

Suppression of the multidrug resistance-associated protein (MRP) and multidrug resistance 1 (MDR1) gene expression in human colon cancer cell lines using hammerhead ribozymes, designed to cleave the *MRP* and *MDR1* mRNAs, was sufficient to reverse multidrug resistance

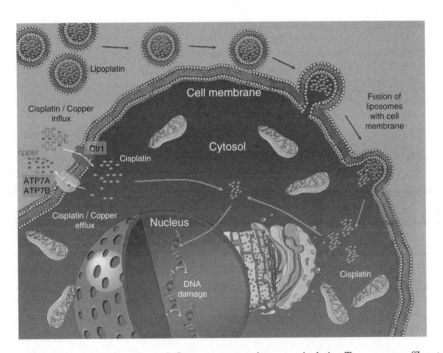

Figure 5.12 Ctr1, the major copper influx transporter, imports cisplatin. Two copper efflux transporters, ATP7A and ATP7B, situated at the periphery of the cell membrane regulate the efflux of cisplatin. Instead, Lipoplatin bypasses Ctr1 thanks to the fusogenic DPPG lipid which commands direct fusion with the cell membrane and cisplatin deliver across the membrane barrier. In addition, because of its 110-nm particle size (compared to cisplatin with a molecular dimension of less than 1 nm) Lipoplatin is taken up by phagocytosis. Tumour cells are known to be more actively engaged in phagocytosis than normal tissue. Thus, Lipoplatin acts as a Trojan Horse for tumour cells. Lipoplatin is proposed to be able to bypass cisplatin resistance. From: Boulikas T. Molecular mechanisms of cisplatin and its liposomally encapsulated form, Lipoplatin™. Lipoplatin™ as a chemotherapy and antiangiogenesis drug. *Cancer Therapy*, 2007, **5**, 349–76, reproduced with permission from Gene Therapy Press

to doxorubicin and etoposide (VP-16) but did not affect resistance to cisplatin, methotrexate and 5-fluorouracil (Nagata *et al.*, 2002).

Mechanisms of resistance to cisplatin include: (i) decreased drug uptake; (ii) more efficient repair of cisplatin-induced DNA lesions; (iii) increased glutathione levels in resistant tumour cells; (iv) increased metallothionein levels; (v) increased BCL2 levels; (vi) loss of the DNA mismatch repair protein MLH1, and others.

5.7.2 Cisplatin resistance from inactivation of cisplatin inside the cytoplasm

Metallothionein is a thiol-containing protein is linked with tumour resistance to cisplatin. Overexpression of metallothionein in a cell line by stable gene transfection resulted in seven-fold protection from cisplatin (Holford *et al.*, 2000).

5.7.3 Cisplatin resistance from faster repair of DNA lesions

Transcript abundance levels of twelve selected DNA repair or multi-drug resistance genes (*LIG1, ERCC2, ERCC3, DDIT3, ABCC1, ABCC4, ABCC5, ABCC10*, GTF2H2, *XPA, XPC* and *XRCC1*) were examined to determine their association with cisplatin chemore-sistance. Two-gene ratio expression profiles (including ERCC2/XPC, ABCC5/GTF2H2, ERCC2/GTF2H2, XPA/XPC and XRCC1/XPC) were compared as a single variable to chemoresistance in eight NSCLC cell lines. The two-variable model with the highest correlation identified the pairs ABCC5/GTF2H2 and ERCC2/GTF2H2 whose expression profilings were proposed as markers suitable to identify cisplatin- resistant tumours in fine needle aspirate biopsies (Weaver *et al.*, 2005).

5.7.4 Gene expression and cisplatin resistance

Several studies have been done to identify the expression of genes related to cisplatin resistance. The presence of a functional wild-type *p53* gene renders cancer cells sensitive to cisplatin. Epithelial ovarian cancer patients undergoing platinum-base chemotherapy showed marked differences in p53 levels and mutations; 83 % of non-responders to chemotherapy had mutations in the *p53* gene compared with 16 % for responders (Kigawa *et al.*, 2002). The DNA mismatch repair genes, and hMSH2 in combination with one of its heterodimer partners binds specifically to cisplatin adducts (Niedner *et al.*, 2001).

NPRL2 is one of the novel candidate tumour suppressor genes identified in the human chromosome 3p21.3 region. Cationic lipid-mediated *NPRL2* gene transfer significantly re-sensitized the response to cisplatin, yielding a 40 % greater inhibition of tumour cell viability and resulting in a two- to three-fold increase in induction of apoptosis by activation of multiple caspases (Ueda *et al.*, 2006).

A cisplatin-resistant cell line, Tca/cisplatin, was established from a cisplatin-sensitive cell line. Development of techniques to disrupt the expression of single genes in engineered cells has identified a number of previously unsuspected genes that control sensitivity to cisplatin. These include those to DNA mismatch repair, the sphingosine-1-phosphate lyase 1, the Golgi vesicular membrane-golvesin, cAMP-specific phosphodiesterases, the regulatory subunit of the cAMP-dependent protein kinase (PKA), the Lyn tyrosine kinase, and a photolyase (reviewed by Niedner *et al.*, 2001). DNA damage leads to simultaneous activation of proapoptotic and survival pathways in a time-dependent, hierarchical manner. Every transition in this response network results from a perturbation of the steady-state levels of intracellular second messengers such as Ca^{2+}, cAMP, cGMP, sphingosine 1-phosphate/ceramide and inositol polyphosphates (Niedner *et al.*, 2001). Upregulation of activator protein-1 (AP-1) transcription factor has also been linked to chemotherapeutic resistance. Thus, inhibition of AP-1 DNA binding may be of clinical value in treating chemotherapeutic resistance (Bonovich *et al.*, 2002). Contrary to AP-1, upregulation of upstream binding factor (UBF), an RNA polymerase I-specific transcription factor, increased sensitivity to cisplatin in cell cultures. Downregulation of UBF with antisense oligodeoxynucleotides caused weak apoptosis without DNA laddering or cleavage of poly (ADP-ribose) polymerase and altered the expression of 30 genes (Huang *et al.*, 2002). Additional genes that might be involved directly or indirectly to cisplatin resistance include *BclXL, bcl-2, Pms2, ERCC1* and *XPA*. Frequently observed genetic changes using comparative genomic hybridization in human tumour cell lines (predominantly ovarian) with acquired resistance to cisplatin included amplification of 4q and 6q, followed by amplification of 5q; amplification of 12q was observed in cell lines made resistant to JM216 or ZD-0473 (platinum analogues, see Boulikas and Vougiouka, 2003 for references) in which increased DNA repair appeared to be the major mechanism of resistance for both agents (Leyland-Jones *et al.*, 1999).

The experimental strategies under investigation aimed at overcoming cisplatin resistance such as introduction of the functional *p53* and *p21* genes (Di Felice *et al.*, 1998) usually mutated during carcinogenesis, or genes that intervene with apoptotic pathways (such as bax, BclX$_L$, bcl-2) are likely to contribute to limiting the disease in combination with regimens using platinum drugs (reviewed by Boulikas and Vougiouka, 2003). For example, *p53* is frequently mutated in late-stage cancer and the introduction of a functional wild-type *p53* gene in gene therapy applications renders cancer cells sensitive to cisplatin (Buller *et al.*, 2002a, b; reviewed by Boulikas, 1997).

References

Ali BH, Al Moundhri MS. Agents ameliorating or augmenting the nephrotoxicity of cisplatin and other platinum compounds: a review of some recent research. *Food Chem Toxicol* 2006, **44**(8), 1173–83.

Al-Sarraf M. Chemotherapeutic management of head and neck cancer. *Cancer Metastasis Rev* 1987, **6**, 181–98.

Arany I, Safirstein RL. Cisplatin nephrotoxicity. *Semin Nephrol* 2003, **23**, 460–4.

Arnould S, Hennebelle I, Canal P, Bugat R, Guichard S. Cellular determinants of oxaliplatin sensitivity in colon cancer cell lines. *Eur J Cancer* 2003, **39**, 112–119.

Aronov A, Horowitz A, Gabizon A, Gibson D. Folate-Targeted PEG as a Potential Carrier for Carboplatin Analogs. Synthesis and *in vitro* Studies. *Bioconugate Chem* 2003, **14**, 563–74.

Baas P, Belderbos JS, Senan S, *et al.* Concurrent chemotherapy (carboplatin, paclitaxel, etoposide) and involved-field radiotherapy in limited stage small cell lung cancer: a Dutch multicenter phase II study. *Br J Cancer* 2006, **94**(5), 625–30.

Barry CG, Day CS, Bierbach U. Duplex-promoted platination of adenine-N3 in the minor groove of DNA: challenging a longstanding bioinorganic paradigm. *J Am Chem Soc* 2005, **127**, 1160–9.

Beale P, Judson I, O'Donnell A, *et al.* A Phase I clinical and pharmacological study of cis-diamminedichloro(2-methylpyridine) platinum II (AMD473). *Br J Cancer* 2003, **88**, 1128–34.

Bonovich M, Olive M, Reed E, O'Connell B, Vinson C. Adenoviral delivery of A-FOS, an AP-1 dominant negative, selectively inhibits drug resistance in two human cancer cell lines. *Cancer Gene Ther* 2002, **9**, 62–70.

Boulikas T. Gene therapy of prostate cancer: p53, suicidal genes, and other targets. *Anticancer Res* 1997, **17**(3A), 1471–5.

Boulikas T, Vougiouka M. Cisplatin and platinum drugs at the molecular level. *Oncol Rep* 2003, **10**, 1663–82.

Boulikas T, Vougiouka M. Recent clinical trials using cisplatin, carboplatin and their combination chemotherapy drugs. *Oncol Rep* 2004, **11**, 559–95.

Boulikas T, Stathopoulos GP, Volakakis N, Vougiouka M. Systemic Lipoplatin infusion results in preferential tumor uptake in human studies. *Anticancer Res* 2005, **25**, 3031–40.

Boulikas T, Mylonakis N, Sarikos G, *et al*. Lipoplatin plus gemcitabine versus cisplatin plus gemcitabine in NSCLC: Preliminary results of a phase III trial. *J Clin Oncol* 2007 ASCO Annual Meeting Proceedings Part I, **25**, 18028.

Bowler BE, Ahmed KJ, Sundquist WI, Hollis LS, Whang EE, Lippard SJ. Synthesis, characterization, and DNA-binding properties of (1,2-diaminoethane)platinum(II) complexes linked to the DNA intercalator acridine orange by trimethylene and hexamethylene chains. *J Am Chem Soc* 1989, **111**, 1299–306.

Brabec V, Kasparkova J, Vrana O, *et al*. DNA modifications by a novel bifunctional trinuclear platinum Phase I anticancer agent. *Biochemistry* 1999, **38**, 6781–90.

Buller RE, Runnebaum IB, Karlan BY, *et al*. A phase I/II trial of rAd/p53 (SCH 58500) gene replacement in recurrent ovarian cancer. *Cancer Gene Ther* 2002a, **9**, 553–66.

Buller RE, Shahin MS, Horowitz JA, *et al*. Long term follow-up of patients with recurrent ovarian cancer after Ad p53 gene replacement with SCH 58500. *Cancer Gene Ther*, 2002b, **9**, 567–72.

Caglar K, Kinalp C, Arpaci F, *et al*. Cumulative prior dose of cisplatin as a cause of the nephrotoxicity of high-dose chemotherapy followed by autologous stem-cell transplantation. *Nephrol Dial Transplant* 2002, **17**, 1931–5.

Carr JL, Tingle MD, McKeage M. Rapid biotransformation of satraplatin by human red blood cells *in vitro*. *Cancer Chemother Pharmacol* 2002, **50**(1), 9–15.

Cassidy J, Misset JL. Oxaliplatin-related side effects: characteristics and management. *Semin Oncol* 2002, **29**(Suppl 15), 11–20.

De Vita F, Orditura M, Matano E, *et al*. A phase II study of biweekly oxaliplatin plus infusional 5-fluorouracil and folinic acid (FOLFOX-4) as first-line treatment of advanced gastric cancer patients. *Br J Cancer* 2005, **92**(9), 1644–9.

Denkert C, Schmitt WD, Berger S, *et al*. Expression of mitogen-activated protein kinase phosphatase-1 (MKP-1) in primary human ovarian carcinoma. *Int J Cancer* 2002, **102**, 507–13.

Devarajan P, Savoca M, Castaneda MP, Park MS, Esteban-Cruciani N, Kalinec G, Kalinec F. Cisplatin-induced apoptosis in auditory cells: role of death receptor and mitochondrial pathways. *Hear Res* 2002, **174**, 45–54.

Devarajan P, Tarabishi R, Mishra J, Ma Q, Kourvetaris A, Vougiouka M, Boulikas T. Low renal toxicity of Lipoplatin compared to cisplatin in animals. *Anticancer Res* 2004, **24**, 2193–2200.

Di Felice V, Lauricella M, Giuliano M, Emanuele S, Vento R, Tesoriere G. The apoptotic effects of cisplatin and carboplatin in retinoblastoma Y79 cells. *Int J Oncol* 1998, **13**, 225–32.

Di Francesco AM, Ruggiero A, Riccardi R. Cellular and molecular aspects of drugs of the future: oxaliplatin. *Cell Mol Life Sci* 2002, **59**, 1914–27.

Dragovich T, Mendelson D, Kurtin S, Richardson K, Von Hoff D, Hoos A. A Phase 2 trial of the liposomal DACH platinum L-NDDP in patients with therapy-refractory advanced colorectal cancer. *Cancer Chemother Pharmacol* 2006, **58**, 759–64.

El Maalouf G, Rodier JM, Faivre S, Raymond E. Could we expect to improve survival in small cell lung cancer? *Lung Cancer* 2007, **57**(Suppl 2), S30–4.

Farrell NP. *Platinum-based Drugs in Cancer Therapy*; Humana Press, Totawa, NJ, 2000.

Fuertes MA, Castilla J, Alonso C, Perez JM. Novel concepts in the development of platinum antitumor drugs. *Curr. Med. Chem. Anticancer Agents* 2002, **2**(4), 539–51.

Fuertes MA, Castilla J, Alonso C, Perez JM. Cisplatin biochemical mechanism of action: From cytotoxicity to induction of cell death through interconnections between apoptotic and necrotic pathways. *Curr Med Chem* 2003, **10**, 257–66.

Furin A, Guiotto A, Baccichetti F, Pasut G, Bertani R., Veronese F. Synthesis, characterization and preliminary cytotoxicity assays of poly(ethylene glycol) – malonato-Pt-DACH conjugates. *Eur J Med Chem* 2003, **38**, 739–49.

Gore ME, Atkinson RJ, Thomas H, *et al*. A phase II trial of ZD0473 in platinum-pretreated ovarian cancer. *Eur J Cancer* 2002, **38**, 2416–20.

Hall MD, Hambley TW. Platinum(IV) antitumor compounds: their bioinorganic chemistry. *Coord Chem Rev* 2002, **232**, 49–67.

Highley MS; Calvert AH. In: *Clinical Experience with Cisplatin and Carboplatin*, edited by LR Kelland and NP Farrell. Humana Press, Totowa NJ, 2000, pp 89–113.

Hill JM, Speer RJ. Organo-platinum complexes as antitumor agents (review). *Anticancer Res* 1982, **2**, 173–86.

Holford J, Beale PJ, Boxall FE, Sharp SY, Kelland LR. Mechanisms of drug resistance to the platinum complex ZD0473 in ovarian cancer cell lines. *Eur J Cancer* 2000, **36**, 1984–90.

Holford J, Raynaud F, Murrer BA, Grimaldi K, Abrams M, Kelland LR. Chemical, biochemical and pharmacological activity of the novel sterically hindered platinum co-ordination complex, *cis*-[amminedichloro(2-methylpyridine)] platinum(II) (AMD473). *Anticancer Drug Des* 1998, **13**, 1–18.

Holmes RJ, McKeage MJ, Murray V, Denny WA, McFadyen WD. *cis*-Dichloroplatinum(II) complexes tethered to 9-aminoacridine-4-carboxamides: synthesis and action in resistant cell lines *in vitro*. *J Inorg Biochem* 2001, **85**, 209–17.

Huang R, Wu T, Xu L, Liu A, Ji Y, Hu G. Upstream binding factor up-regulated in hepatocellular carcinoma is related to the survival and cisplatin-sensitivity of cancer cells. *FASEB J* 2002, **16**(3), 293–301.

Humes HD. Insights into ototoxicity. Analogies to nephrotoxicity. *Ann NY Acad Sci* 1999, **884**, 15–18.

Jerremalm E, Eksborg S, Ehrsson H. Hydrolysis of oxaliplatin-evaluation of the acid dissociation constant for the oxalato monodentate complex. *J Pharm Sci* 2003, **92**, 436–8.

Johnson NP, Butour JL, Villani G, Wimmer FL, Defais M, Pierson V. Metal antitumor compounds: the mechanism of action of platinum complexes. *Prog Clin Biochem Med* 1989, **10**, 1–24.

Johnson SW, Ferry KV, Hamilton TC. Recent insights into platinum drug resistance in cancer. *Drug Resist* 1998, Update 1, 243–54.

Judson I, Kelland LR. New developments and approaches in the platinum arena. *Drugs* 2000, **59**(Suppl 4), 29–36; discussion 37–38.

Kasparkova J, Novakova O, Vrana O, Farrell N, Brabec V. Effect of geometric isomerism in dinuclear platinum antitumor complexes on DNA interstrand cross-linking. *Biochemistry* 1999, **38**, 10997–1005.

Kasparkova J, Farrell N, Brabec V. Sequence specificity, conformation, and recognition by HMG1 protein of major DNA interstrand cross-links of antitumor dinuclear platinum complexes. *J Biol Chem* 2000, **275**(21), 15789–98.

Kasparkova J, Zehnulova J, Farrell N, Brabec V. DNA interstrand cross-links of the novel antitumor trinuclear platinum complex BBR3464. Conformation, recognition by high mobility group domain proteins, and nucleotide excision repair. *J Biol Chem* 2002, **277**, 48076–86.

Kelland LR. An update on satraplatin: the first orally available platinum anticancer drug. *Exp Opin Invest Drugs* 2000, **9**: 1373–82.

Kigawa J, Sato S, Shimada M, Takahashi M, Itamochi H, Kanamori Y, Terakawa N. p53 gene status and chemosensitivity in ovarian cancer. *Hum Cell* 2002, **14**, 165–71.

Kikuno N, Urakami S, Nakamura S, *et al*. Phase-II study of docetaxel, estramustine phosphate, and carboplatin in patients with hormone-refractory prostate cancer. *Eur Urol* 2007, **51**(5), 1252–8.

Kosmidis P, Mylonakis N, Nicolaides C, *et al*. Paclitaxel plus carboplatin versus gemcitabine plus pacli-taxel in advanced non-small-cell lung cancer: a phase III randomized trial. *J Clin Oncol* 2002, **20**, 3565–7.

Kurata T., Tamura T., Sasaki Y., Fujii H., Negoro S., Fukuoka M., Saijo N., (2000) Pharmacokinetic and pharmacodynamic analysis of bis-acetato-ammine-dichloro-cyclohexylamine-platinum(IV) (JM216) administered once a day for five consecutive days: a phase I study. *Jpn J. Clin. Oncol.* **30**: 377–84.

Lebwohl D, Canetta R. Clinical development of platinum complexes in cancer therapy: an historical perspective and an update. *Eur J Cancer* 1998, **34**, 1522–34.

Lersch C, Schmelz R, Eckel F, *et al*. Prevention of oxaliplatin-induced peripheral sensory neuropathy by carbamazepine in patients with advanced colorectal cancer. *Clin Colorectal Cancer* 2002, **2**, 54–8.

Leyland-Jones B, Kelland LR, Harrap KR, Hiorns LR. Genomic imbalances associated with acquired resistance to platinum analogues. *Am J Pathol* 1999, **155**, 77–84.

Lin AM, Rosenberg JE, Weinberg VK, *et al*. Clinical outcome of taxane-resistant (TR) hormone refrac-tory prostate cancer (HRPC) patients (pts) treated with subsequent chemotherapy (ixabepilone (Ix) or mitoxantrone/prednisone (MP). *J Clin Oncol* 2006, **24**(suppl), 231s (abstract 4558).

Lin X, Okuda T, Holzer A, Howell SB. The copper transporter CTR1 regulates cisplatin uptake in *Sac-charomyces cerevisiae*. *Mol Pharmacol* 2002, **62**(5), 1154–9.

Lippert B (ed.) *Cisplatin: Chemistry and Biochemistry of a Leading Anticancer Drug*. Wiley-VCH, New York, 1999.

Martin F, Boulikas T. The challenge of liposomes in gene therapy. *Gene Ther Mol Biol* 1998, **1**, 173–214.

Matsumara Y, Maeda H. A new concept for macromolecular therapeutics in cancer chemotherapy: mechanism of tumoritropic accumulation of proteins and the antitumor agent smancs. *Cancer Res* 1986, **46**, 6387–92.

Mattern J, Stammler G, Koomagi R, Wallwiener D, Kaufmann M, Volm M. Spontaneous apoptosis in ovarian cancer: an unfavorable prognostic factor. *Int J Oncol* 1998, **12**, 351–4.

McKeage MJ. New-generation platinum drugs in the treatment of cisplatin-resistant cancers. *Expert Opin Investig Drugs* 2005, **14**(8), 1033–46.

Morin PJ. Drug resistance and the microenvironment: nature and nurture. *Drug Resist* 2003, Update 6, 169–72.

Nagata J, Kijima H, Hatanaka H, *et al*. Reversal of drug resistance using hammerhead ribozymes against multidrug resistance-associated protein and multidrug resistance 1 gene. *Int J Oncol* 2002, **21**, 1021–6.

Natile G, Coluccia M. Current status of trans-platinum compounds in cancer therapy. *Coord Chem Rev* 2001, 216–217, 383–410.

Niedner H, Christen R, Lin X, Kondo A, Howell SB. Identification of genes that mediate sensitivity to cisplatin. *Mol Pharmacol* 2001, **60**, 1153–60.

O'Dwyer PJ, Stevenson JP, Johnson SW. Clinical status of cisplatin, carboplatin, and other platinum-based antitumor drugs. In: *Cisplatin: Chemistry and Biochemistry of a Leading Anticancer Drug* etited by B Lippert. Wiley-VCH, New York, 1999, pp 31–72.

Pein F, Michon J, Valteau-Couanet D, *et al*. High-dose melphalan, etoposide, and carboplatin followed by autologous stem-cell rescue in pediatric high-risk recurrent Wilms' tumor: a French Society of Pediatric Oncology study. *J Clin Oncol* 1998, **16**(10), 3295–301.

Perez-Soler R, Piperdi B, Walsh GL, *et al*. Phase II study of a liposome-entrapped cisplatin analog (L-NDDP) administered intrapleurally and pathologic response rates in patients with malignant pleural mesothelioma. *J Clin Oncol* 2005, **23**, 3495–34501.

Pinotti G, Martinelli B. A case of acute tubular necrosis due to oxaliplatin. *Ann Oncol* 2002, **13**, 1951–2.

Qu Y, Appleton TG, Hoeschele JD, Farrell N. Cisplatin as a synthon. Synthesis and characterization of triplatinum complexes containing three cis-(Pt(amine)$_2$ units linked in a linear fashion, *Inorg Chem* 1993, **32**, 2591.

Ramalingam S, Belani CP. Taxanes for advanced non-small cell lung cancer. *Exp Opin Pharmacother* 2002, **3**(12), 1693–709.

Ranson M, Thatcher N. Paclitaxel: a hope for advanced non-small cell lung cancer? *Exp Opin Investig Drugs* 1999, **8**(6), 837–48.

Raymond E, Faivre S, Chaney S, Woynarowski J, Cvitkovic E. Cellular and molecular pharmacology of oxaliplatin. *Mol Cancer Ther* 2002, **1**, 227–35.

Reck M, von Pawel J, Macha HN, *et al.* Efficient palliation in patients with small-cell lung cancer by a combination of paclitaxel, etoposide and carboplatin: quality of life and 6-years'-follow-up results from a randomised phase III trial. *Lung Cancer* 2006, **53**(1), 67–75.

Reedijk J. Improved understanding in platinum antitumor chemistry. *Chem Commun* 1996, 801–806.

Rosell R, Gatzemeier U, Betticher DC, *et al.* Phase III randomised trial comparing paclitaxel/carboplatin with paclitaxel/cisplatin in patients with advanced non-small-cell lung cancer: a cooperative multinational trial. *Ann Oncol* 2002, **13**, 1539–49.

Rosenberg B. Noble metal complexes in cancer chemotherapy. *Adv Exp Med Biol* 1977, **91**, 129–50.

Sandercock J, Parmar MK, Torri V, Qian W. First-line treatment for advanced ovarian cancer: paclitaxel, platinum and the evidence. *Br J Cancer* 2002, **87**(8), 815–24.

Sharp SY, Rogers PM, Kelland LR. Transport of cisplatin and bis-acetato-ammine- dichlorocyclohexylamine Platinum(IV) (JM216) in human ovarian carcinoma cell lines: identification of a plasma membrane protein associated with cisplatin resistance. *Clin Cancer Res* 1995, **1**, 981–9.

Shen DW, Goldenberg S, Pastan I, Gottesman MM. Decreased accumulation of [14C]carboplatin in human cisplatin-resistant cells results from reduced energy-dependent uptake. *J Cell Physiol* 2000, **183**, 108–16.

Siddik ZH. Cisplatin: mode of cytotoxic action and molecular basis of resistance. *Oncogene* 2003, **22**, 7265–79.

Spingler B, Whittington DA, Lippard SJ. 2.4 crystal structure of an oxaliplatin 1,2-d (GpG) intrastrand cross-link in a DNA dodecamer duplex. *Inorg Chem* 2001, **40m** 5596–602.

Stathopoulos GP, Boulikas T, Vougiouka M, *et al.* Pharmacokinetics and adverse reactions of a new liposomal cisplatin (Lipoplatin): Phase I study. *Oncol Rep* 2005, **13**, 589–95.

Stathopoulos GP, Boulikas T, Kourvetaris A, Stathopoulos I. Liposomal oxaliplatin in the treatment of advanced cancer: a phase I study. *Anticancer Res* 2006, **26**, 1489–94.

Stathopoulos GP, Michalopoulou P, Antoniou D, *et al.* Liposomal cisplatin and paclitaxel versus cisplatin and paclitaxel in advanced NSCLC. *J Clin Oncol* 2007, ASCO Annual Meeting Proceedings Part I **25**, 7684.

Sternberg CN, Yagoda A, Scher HI, *et al.* Methotrexate, vinblastine, doxorubicin, and cisplatin for advanced transitional cell carcinoma of the urothelium. Efficacy and patterns of response and relapse. *Cancer* 1989, **64**, 2448–58.

Sundquist WI, Bancroft DP, Chassot L, Lippard SJ. DNA promotes the reaction of cis-diamminedichloroplatinum(II) with the exocyclic amino groups of ethidium bromide. *J Am Chem Soc* 1988, **110**, 8559–60.

Sundquist WI, Bancroft DP, Lippard SJ. Synthesis, characterization, and biological activity of cisdiammineplatinum(II) complexes of the DNA intercalators 9-aminoacridine and chloroquine. *J Am Chem Soc* 1990, **112**, 1590–6.

Temple MD, McFadyen WD, Holmes RJ, Denny WA, Murray V. Interaction of cisplatin and DNA-targeted 9-aminoacridine platinum complexes with DNA. *Biochemistry* 2000, **39**, 5593–9.

Temple MD, Recabarren P, McFadyen WD, Holmes RJ, Denny WA, Murray V. The interaction of DNA-targeted 9-aminoacridine-4-carboxamide platinum complexes with DNA in intact human cells. *Biochim Biophys Acta* 2002, **1574**, 223–30.

Ueda K, Kawashima H, Ohtani S, *et al.* The 3p21.3 Tumor suppressor NPRL2 plays an important role in cisplatin-induced resistance in human non-small-cell lung cancer cells. *Cancer Res* 2006, **66**, 9682–90.

Veal GJ, Griffin MJ, Price E, *et al.* A phase I study in paediatric patients to evaluate the safety and pharmacokinetics of SPI-77, a liposome encapsulated formulation of cisplatin. *Br J Cancer* 2001, **84**(8), 1029–35.

Weaver DA, Crawford EL, Warner KA, Elkhairi F, Khuder SA, Willey JC. ABCC5, ERCC2, XPA and XRCC1 transcript abundance levels correlate with cisplatin chemoresistance in non-small cell lung cancer cell lines. *Mol Cancer* 2005, **4**(1), 18.

Wong E, Giandomenico CM. Current status of platinum-based antitumor drugs, *Chem Rev* 1999, **99**(9), 2451–66.

Yamamoto M, Oshiro S, Tsugu H, Hirakawa K, Ikeda K, Soma G, Fukushima T. Treatment of recurrent malignant supratentorial astrocytomas with carboplatin and etoposide combined with recombinant mutant human tumor necrosis factor-alpha. *Anticancer Res* 2002, **22**(4), 2447–53.

Zehnulova J, Kasparkova J, Farrell N, Brabec V. Conformation, recognition by high mobility group domain proteins, and nucleotide excision repair of DNA intrastrand cross-links of novel antitumor trinuclear platinum complex BBR3464. *J Biol Chem* 2001, **276**, 22191–9.

6
Antimicrotubule Agents

Iain Brown, Jay N Sangrithi-Wallace and Andrew C Schofield

6.1 Taxanes

Taxanes are natural compounds derived from trees of the family Taxoidaceae, which were discovered to have antitumour properties. Development of the first taxane introduced in cancer therapy – paclitaxel (Taxol), isolated from the bark of the Pacific Yew, *Taxus brevifolia* – was initially hampered by its limited supply and difficulties in large-scale extraction. The search for other taxanes derived from more abundant sources led to the development of the semisynthetic drug docetaxel (Taxotere), formulated from an inactive taxane precursor found in the needles of more abundant yew species such as the European Yew, *Taxus baccata*. Taxanes belong to a chemically diverse group of antimitotic drugs (Figure 6.1) that target microtubules and their dynamics.

6.1.1 Mode of action

The cellular target of the taxanes is the β-tubulin in the microtubule polymer. Taxanes bind preferentially to microtubules rather than to soluble tubulin. The tubulin-binding site of taxanes, at the N-terminal 31 amino acids of β-tubulin, is distinct from the binding sites of other antimitotic drugs such as the vinca alkaloids and colchicines (Sampath *et al.*, 2003). Taxanes are thought to gain access to their binding sites by diffusing through small openings in the microtubule or fluctuations of the microtubule lattice. Binding of taxanes to their site on the inside microtubule surface stabilizes the microtubule and thereby inhibits depolymerization. This, subsequently, disrupts the normal dynamic reorganization of the microtubule network required for mitosis and cell proliferation and results in the arrest of cells in the G_2M phase of the cell cycle.

In mitotic spindles slowing of the growth and shortening and/or treadmilling dynamics of the microtubules blocks mitotic progression. This suppression of dynamics has at least two downstream effects on the spindle: first, it prevents the mitotic spindle from assembling

Anticancer Therapeutics   Edited by Sotiris Missailidis

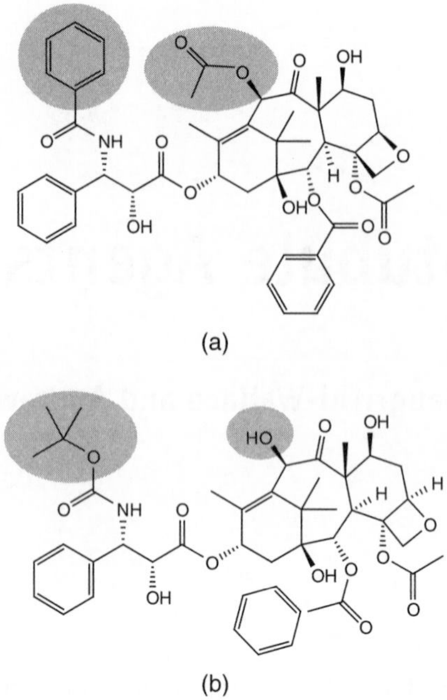

(a)

(b)

Figure 6.1 Comparison of the chemical structure between taxanes. Differences between the structures are highlighted by the shaded circles. (a) Paclitaxel. (b) Docetaxel

normally and second, it reduces the tension at the kinetochores of the chromosomes. Mitotic progress is delayed, with chromosomes often stuck at the spindle poles unable to assemble at the spindle equator. The cell cycle signal to the anaphase-promoting complex, to pass from metaphase into anaphase, is blocked. Drug-blocked cells may eventually exit mitosis, often aberrantly. Crucial for the efficacy of these drugs in cancer chemotherapy, mitotically blocked or mitotically slowed cells eventually die by apoptosis (Drukman and Kavallaris, 2002).

Drugs that increase or decrease microtubule polymerization at high concentrations powerfully suppress microtubule dynamics at 10- to 100-fold lower concentrations and, therefore, kinetically stabilize the microtubules, without changing the microtubule-polymer mass (the effects of the drugs on dynamics are often more powerful than their effects on polymer mass). It was previously thought that the effects of the two classes of drugs on microtubule-polymer mass were the most important actions responsible for their chemotherapeutic properties. The drugs would have to be given and maintained at very high dosage levels to act primarily and continuously on microtubule-polymer mass. Recent evidence, however, suggests that the most important action of these drugs is the suppression of spindle–microtubule dynamics, which results in the slowing, or blocking, of mitosis at the metaphase–anaphase transition and induction of apoptotic cell death (Jordan and Wilson, 2004).

At high concentrations (greater than 200 nM) taxanes increase microtubule polymerization, presumably by the induction of a conformational change in the tubulin structure by an unknown mechanism, increasing its affinity for neighbouring tubulin molecules (microtubule bundling). At lower concentrations (less than 20 nM) taxanes sufficiently stabilize microtubule dynamics without increasing polymerization. For example, just one paclitaxel molecule bound per several hundred tubulin molecules in a microtubule can reduce the rate or extent of microtubule shortening by approximately 50 %. Suppression of microtubule dynamics by taxanes leads to mitotic block even in the absence of significant microtubule bundling (Mollinedo and Gajate, 2003).

6.1.2 Side effects

The most common side effects of taxanes include immunosuppression, bruising or bleeding, anaemia, nausea and vomiting, diarrhoea, hair loss, sore mouth and ulcers, numbness in hands or feet, and pain in joints or muscles. Specific side effects associated with paclitaxel may include low blood pressure, changes in heart rate and abdominal pain whereas other side effects associated with docetaxel may include changes in nail colour and fluid retention.

6.1.3 Differences between the taxanes

Although the taxanes share similar mechanisms of action, differences between paclitaxel and docetaxel have been noted in their molecular pharmacology, pharmacokinetics, and pharmacodynamic profiles (Gligorov and Lotz, 2004). These differences may account for the differences observed between the taxanes in their clinical activity and toxicity.

Even though both paclitaxel and docetaxel bind to the β-subunit of tubulin, they have subtly different binding sites. Paclitaxel binds to the N-terminal 31 amino acids of β-tubulin whereas docetaxel binds to the tau site (Jordan and Wilson, 2004). Considering the molecular pharmacology of the two drugs, docetaxel has a higher affinity for β-tubulin than paclitaxel, and is twice as potent as paclitaxel in inhibiting microtubule depolymerization (Crown, 2001). Docetaxel is primarily active in the S-phase of the cell cycle by targeting centrosome organisation, resulting in incomplete mitosis and eventual apoptosis. Docetaxel is only partially cytotoxic against cells in mitosis and has minimal toxicity against cells in G_1, leading to an accumulation of cells in the G_2M phase of the cell cycle. Paclitaxel, on the other hand, causes cell damage by affecting the mitotic spindle in the G_2 and M phases of the cell cycle (Gligorov and Lotz, 2004). Docetaxel is 10- to 100-fold more potent than paclitaxel in phosphorylating Bcl-2, and this may account for the differential pro-apoptotic activity of docetaxel compared with paclitaxel (Gligorov and Lotz, 2004). The anti-angiogenic effect of docetaxel is four times stronger than that of paclitaxel (Vacca et al., 2002). Other differences between docetaxel and paclitaxel include greater cellular uptake of docetaxel into tumour cells and slower efflux of docetaxel from tumour cells, therefore leading to longer retention times of docetaxel relative to paclitaxel (Herbst and Khuri, 2003).

There are also significant differences between docetaxel and paclitaxel in their pharmacokinetic and pharmacodynamic profiles, which have implications in determining their optimal dosage schedules. Both taxanes are primarily eliminated via hepatic metabolism by the cytochrome P450 enzymes (in particular CYP3A4), and their use is generally precluded in patients with severe hepatic dysfunction. A large fraction of the taxane dose is excreted in faeces as parent drug or hydroxylated metabolites (the known metabolites of both drugs are either inactive or less potent than the parent compounds). Approximately 6 % of either drug is renally excreted. Both agents have wide tissue distribution and avidly bind to plasma proteins (greater than 90 %), however, they are readily eliminated from the plasma compartment, suggesting lower-affinity reversible protein binding. Studies in mice and humans have determined that paclitaxel exhibits non-linear pharmacokinetics, whereas docetaxel exhibits linear pharmacokinetics (McLeod et al., 1998).

6.1.4 Paclitaxel

Paclitaxel (Figure 6.1) is used for the treatment of ovarian cancer, breast cancer and non-small cell lung cancer. In September 2001, the National Institute for Clinical Excellence (NICE) recommended the use of paclitaxel for the treatment of advanced breast cancer after initial treatment (usually anthracycline therapy) has failed. In January 2003, NICE recommended paclitaxel, in combination with platinum-based therapy, as first-line treatment for ovarian cancer. In addition, paclitaxel was also recommended as second-line treatment for women with ovarian cancer who did not receive paclitaxel during their first-line therapy. In September 2006, NICE did not recommend the use of paclitaxel for the treatment of early node-positive breast cancer.

6.1.5 Docetaxel

Docetaxel (Figure 6.1) is used for the treatment of breast cancer, prostate cancer and non-small cell lung cancer. In September 2001, NICE recommended the use of docetaxel for the treatment of advanced breast cancer in women whose initial therapy (often an anthracycline-based therapy) failed. In June 2006, NICE approved the use of docetaxel for the treatment of hormone-refractory prostate cancer. In September 2006, NICE recommended the use of docetaxel, in combination with doxorubicin and cyclophosphamide, for the treatment of women with early node-positive breast cancer. In the United States, however, the use of docetaxel has been approved by the Food and Drug Administration (FDA) for the treatment of advanced gastric cancer (in combination with cisplatin and fluorouracil) and for the treatment of inoperable, advanced head and neck cancer (in combination with cisplatin and fluorouracil).

6.1.6 Other taxane-like antimicrotubule agents

Ixabepilone (BMS-247 550), a synthetic analogue of epothilone B, belongs to a new class of chemotherapeutic agents that disrupt microtubule function and induce cytotoxicity in a similar way to that seen with the taxanes paclitaxel and docetaxel. While its target is similar to that of taxanes, this novel agent has demonstrated significant activity in patients with prostate cancer refractory to taxane-based therapy and has received FDA approval in 2007. The predominant toxicities in the patients treated with ixabepilone alone or in combination with estramustine phosphate included neutropenia, neuropathy, and fatigue.

6.2 Vinca Alkaloids

Vinca alkaloids are a class of antimicrotubule agents, discovered by chance in a species of the herbaceous perennial plant *Catharanthus* (Madagascar Periwinkle). While the sap from the species *C. roseus* has been used historically in folk medicine to treat various conditions, including diabetes and high blood pressure, it was not until the 1950s that it gained interest from the pharmaceutical industry (Noble, 1990). More than 70 alkaloids were identified in the sap of *C. roseus*, two of which – vincristine (Oncovin) and vinblastine (Velbe), were discovered to have antitumour properties (Noble *et al.*, 1958). These two drugs, along with semisynthetic derivatives of vinblastine, namely vindesine (Eldesine), vinorelbine (Navelbine) and vinflunine (Javlor), are used to treat several types of cancer including leukaemia,

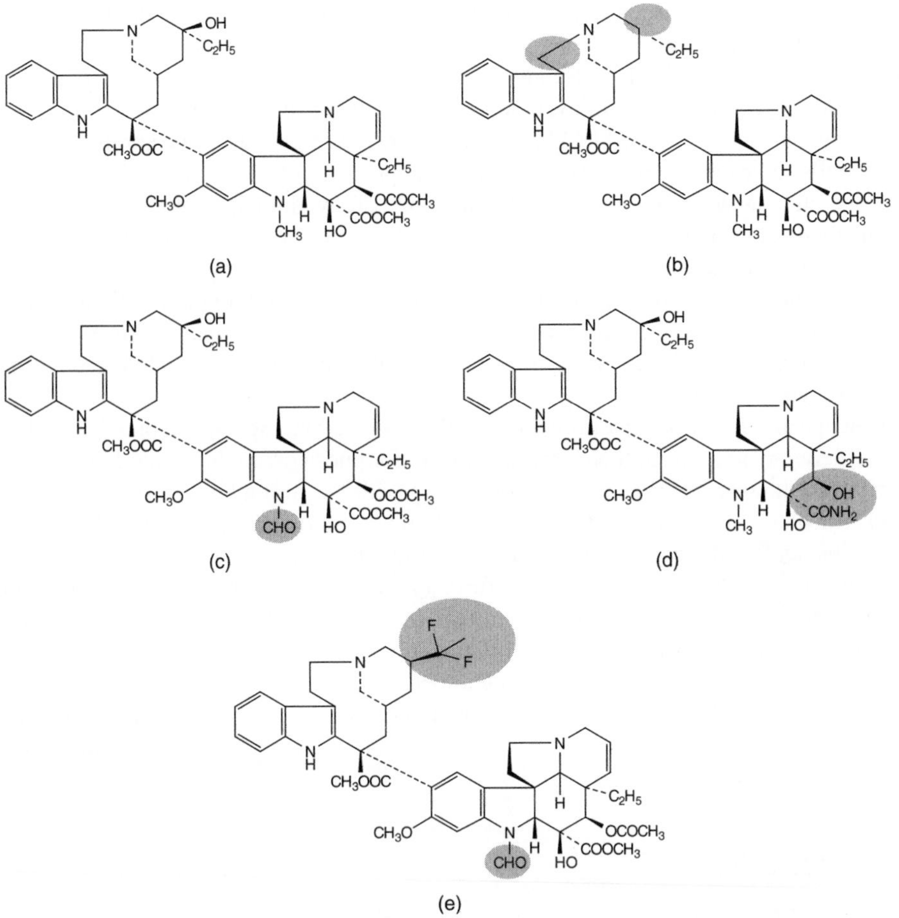

Figure 6.2 Comparison of the chemical structure between vinca alkaloids. Differences between vincristine and the other vinca alkaloids are highlighted by the shaded circles. (a) Vinblastine. (b) Vinorelbine. (c) Vincristine. (d) Vindesine. (e) Vinflunine

lymphoma, melanoma, breast and lung cancers. They share similar chemical structures with minimal changes that result in different activities (Figure 6.2).

6.2.1 Mode of action

Like other antimicrotubule agents, the effects of vinca alkaloids are mediated by their interaction with the tubulin/microtubule system (Jordan *et al.*, 1991). They belong to the tubulin polymerization inhibitors or microtubule-destabilizing class of antimitotic drugs. Vinca alkaloids bind to the β-subunit of tubulin dimers at a distinct region called the vinca-binding domain (Bai *et al.*, 1990).

At high concentrations, vinca alkaloids depolymerize microtubules, destroying mitotic spindles and therefore leaving the dividing cancer cells blocked in mitosis with condensed chromosomes (Jordan *et al.*, 1991). They also bind to soluble tubulin and at high concentrations they favour the formation of spirals, tubules and paracrystals. In every event the normal function of microtubules is impaired (Warfield and Bouck, 1974; Wilson *et al.*, 1982). At low but clinically relevant concentrations, they do not depolymerize spindle microtubules, yet they block mitosis and cells eventually die by apoptosis (Jordan *et al.*, 1991). Studies on the mitotic-blocking action of low vinca alkaloid concentrations in living cancer cells indicate that the block is due to suppression of microtubule dynamics rather than microtubule depolymerization (Jordan, 2002). They diminish microtubule dynamics by suppressing dynamic instability at both ends of microtubules (Toso *et al.*, 1993) and/or treadmilling dynamics. This suppression of dynamics has at least two downstream effects on the spindle: first, it prevents the mitotic spindle from assembling normally and second, it reduces the tension at the kinetochores of the chromosomes. Normal progression through mitosis is delayed. Advancement from metaphase into anaphase is prevented leading to cell death through apoptosis.

Although all the vinca alkaloids interact with the vinca-binding domain on tubulin, there is a clear hierarchy in terms of their overall affinities for purified tubulin. Vinorelbine and, even more so, vinflunine, are weak binders contrasting with the strong binding of vincristine and the intermediate level of vinblastine (Hill, 2001; Kruczynski and Hill, 2001). More recently, vinorelbine and vinflunine have been shown to affect microtubule dynamics very differently from vinblastine (Ngan *et al.*, 2001) and further evidence from studies investigating effects on centromere dynamics also suggest that these three vinca alkaloids display specific mechanisms of action distinct from one another (Okouneva *et al.*, 2003).

Vinca alkaloids, however, are also thought to act on other cellular processes such as RNA, DNA and lipid synthesis and glutathione metabolism and calmodulin-dependent ATPase activity. It is generally accepted, however, that alteration of microtubule dynamics is their primary mechanism of action.

6.2.2 Side effects

Vinca alkaloids are toxic and they produce side effects that are similar to those of many chemotherapeutic agents. Common side effects include alopecia, immunosuppression, constipation and diarrhoea, abdominal cramps, numbness of hands and feet, anaemia, increased bruising and bleeding.

6.2.3 Vinblastine

Vinblastine diminishes microtubule dynamics at both ends of the microtubule. Along with vincristine, vinblastine induces the formation of non-microtubular polymers (Gupta and Bhattacharyya, 2003). It was first approved by the FDA in 1961, and is mainly used for treating Hodgkin's disease, lymphocytic lymphoma, histiocytic lymphoma, advanced testicular cancer, and advanced breast cancer. In Hodgkin's disease, vinblastine is typically used, and found to be effective, in combination with doxorubicin, bleomycin and dacarbazine.

6.2.4 Vinorelbine

Vinorelbine is distinguished from the other vinca alkaloids by the modification of the catharanthine nucleus, instead of the vindoline ring. It is believed that this modification accounts for its difference in antitumour activity and lower toxicity than the other alkaloids (Potier, 1989).

The use of vinorelbine is restricted primarily to breast cancer and non-small cell lung cancer. Vinorelbine can be used as a single agent chemotherapy, although since the discovery of the taxanes, which have a much higher cytotoxic effect, the scope for use of vinorelbine has been somewhat reduced. Nonetheless, vinorelbine is the only single agent recommended for first-line lung cancer in the United States. Vinorelbine was approved by the FDA in 1994, but is approved for use in the UK as a combination agent along with platinum compounds. In December 2002, NICE recommended that vinorelbine could be used for second-line treatment for advanced breast cancer after anthracycline therapy had failed.

In particular, the lesser side effects compared to taxanes, make it an attractive therapeutic option for more vulnerable patients, such as the elderly, and can also be used for patients who choose to avoid alopecia, and in order to maximize quality of life over tumour response (Gridelli, 2001). Unlike the other vinca alkaloids, which are primarily administered intravenously, vinorelbine can be administered either by intravenous bolus injection or by oral administration. The advantages of each method appear to depend on the type of cancer being treated. Oral vinorelbine tends to reduce some of the side effects, without effecting efficacy of the drug. In Scotland, in 2007, vinorelbine was approved in oral capsule form, for use in stage II, III and IV advanced breast cancer after anthracycline treatment had failed (Scottish Medicines Consortium). However, the ELVIS trial demonstrated a limited effect of oral vinorelbine in metastatic non-small cell lung cancer in elderly patients (Gridelli, 2001; Kanard et al., 2004). Intravenous injections of vinorelbine in lung cancer, however, have been shown, by phase III studies, to be more effective (Gridelli, 2001).

6.2.5 Vincristine

Vincristine is used in the treatment of acute leukaemia, lymphoma, breast and lung cancer. Vincristine is rarely used as a single-agent, except in multiple myeloma refractory to melphalan. Combination treatments include CHOP therapy (cyclophosphamide, doxorubicin, vincristine and prednisone) in non-Hodgkin's lymphoma, and MOPP therapy (mustargen, vincristine, procarbazine and prednisone) for Hodgkin's lymphoma.

6.2.6 Vindesine

Vindesine is used primarily to treat acute lymphocytic leukaemia, but is also prescribed for use in breast cancer, colorectal cancer, non-small cell lung cancer, and renal cell cancer. Its toxicity and side effects are similar to those of vinblastine. Although vindesine is used throughout Europe, the FDA does not approve its use in the United States. Vindesine can be used as a single-agent therapy or is particularly useful in combination with platinum compounds, especially in non-small cell lung cancer.

6.2.7 Vinflunine

This drug is made semi-synthetically based on chemical alteration of vinorelbine. Vinflunine has a greater potency than vinorelbine, but has less affinity for tubules involved in axonal transport, suggesting that it may be less neurotoxic than the other vinca alkaloid drugs (Attard *et al.*, 2006). Vinflunine has a weak affinity for P-glycoprotein, which makes it less likely to lead to multi-drug resistance. It does still participate in P-glycoprotein mediated resistance, but not to the same level as other vinca alkaloids.

6.3 Mechanisms of Resistance to Antimicrotubule Agents

There are multiple molecular pathways by which resistance to antimicrotubule agents can occur. One of the main candidates for drug resistance is the multidrug resistance protein, P-glycoprotein, which is a member of the ABC (ATP-binding cassette) superfamily of efflux protein pumps. Vinca alkaloids, like many other anticancer drugs, are subject to resistance through multidrug resistance pathways by cellular efflux *via* P-glycoprotein. Inhibition of P-glycoprotein expression, with a monoclonal antibody, was enough to reverse resistance significantly to vincristine and vinblastine in human tumour xenografts (Rittmann-Grauer *et al.*, 1992). Over-expression of P-glycoprotein has also been associated with resistance to paclitaxel in breast, colon and ovarian cancer cells (Ding *et al.*, 2001; Vredenburg *et al.*, 2001). Furthermore, it has been shown that docetaxel is a substrate for P-glycoprotein in multi-drug-resistant breast cancer cell lines (Ferlini *et al.*, 1999). Resistance to docetaxel in breast cancer cells was associated with increased expression of P-glycoprotein (McDonald *et al.*, 2005). Subsequent inhibition of P-glycoprotein, by verapamil, only partially reversed drug resistance in these docetaxel-resistant breast cancer cells, thus indicating that other mechanisms also contribute to drug resistance (McDonald *et al.*, 2005).

The main cellular target for taxanes and vinca alkaloids are the β-tubulin subunits of the microtubules. Altered expression, or direct gene mutation, may, therefore, affect the efficacy of these antitumour drugs. In particular, altered regulation of microtubule dynamics is thought to lead to changes in β-tubulin expression (Mitchison and Kirschner, 1984). There are six human β-tubulin isotypes, which have a tissue-specific pattern of expression (Drukman and Kavallaris, 2002). The over-expression of different β-tubulin isotypes may be responsible for creating resistance by modifying the microtubule dynamics, which could counteract the microtubule stabilizing and destabilizing effects of the taxanes and vinca alkaloids, respectively. Altered expression of β-tubulin isotypes has been demonstrated in paclitaxel-resistant breast, ovarian and pancreatic cancer cells (Kavallaris *et al.*, 1997; Liu *et al.*, 2001). Alterations of β-tubulin isotypes have also been significantly associated with docetaxel resistance in breast cancer (Hasegawa *et al.*, 2003; Shalli *et al.*, 2005). The evidence concerning the relationship between expression of β-tubulin isotypes and docetaxel resistance is very convincing. However, no one specific isotype can be implicated in resistance (Liu *et al.*, 2001; Shalli *et al.*, 2005). This may make it difficult to use β-tubulin isotypes as a therapeutic target, although recent evidence indicates that over-expression of one particular isotype, class III β-tubulin, is associated with poor response to taxanes and vinca alkaloids in a wide variety of cancer types (Seve and Dumontet, 2008).

Studies have shown that the expression of different β-tubulin isotypes alone is not enough to cause drug resistance (Blade *et al.*, 1999; Kavallaris *et al.*, 1997; Shalli *et al.*, 2005). It is thought that resistance to taxanes could be brought about not by differential β-tubulin isotype expression but by mutations in the β-tubulin genes. Mutations in β-tubulin may have the effect of altering the binding of taxanes to the microtubules. One study demonstrated an association between response to paclitaxel and the presence of somatic β-tubulin mutations in patients with non-small cell lung cancer (Monzo *et al.*, 1999). Recent studies, however, have shown that resistance to taxanes cannot always be correlated with mutation in β-tubulin.

No mutations were found in the class I β-tubulin gene of docetaxel-resistant breast cancer cells (Shalli *et al.*, 2005) or in breast cancer patients treated with paclitaxel (Maeno *et al.*, 2003). Another study, however, only detected a single somatic mutation in the class I β-tubulin gene from a study of 62 breast cancer patients who received taxane therapy (Hasegawa *et al.*, 2002). While mutations in the β-tubulin gene may be correlated with resistance to taxanes and vinca alkaloids (Kavallaris *et al.*, 2001) in some instances, additional or other mechanisms may also be involved. Multiple mechanisms of resistance to pathways have been investigated, usually focusing on specific targets associated with their mechanism of action. For example, bcl-2 and p27 expression are altered upon treatment with docetaxel and paclitaxel and these changes are associated with taxane resistance (Brown *et al.*, 2004; Ferlini *et al.*, 2003).

References

Attard G, Greystoke A, Kaye S, De Bono J. Update on tubulin-binding agents. *Pathol Biol* 2006, **54**, 72–84.

Bai RL, Pettit GR, Hamel E. Binding of dolastatin 10 to tubulin at a distinct site for peptide antimitotic agents near the exchangeable nucleotide and vinca alkaloid sites. *J Biol Chem* 1990, **265**, 17141–9.

Blade K, Menick DR, Cabral F. Overexpression of class I, II or IVb beta-tubulin isotypes in CHO cells is insufficient to confer resistance to paclitaxel. *J Cell Sci* 1999, **112**, 2213–21.

Brown I, Shalli K, McDonald SL, Moir SE, Hutcheon AW, Heys SD, Schofield AC. Reduced expression of p27 is a novel mechanism of docetaxel resistance in breast cancer cells. *Breast Cancer Res* 2004, **6**, R601–7.

Crown J. Docetaxel: overview of an active drug for breast cancer. *Oncologist* 2001, **6**, 1–4.

Ding S, Chamberlainn M, McLaren A, Goh L, Duncan I, Wolf CR. Cross-talk between signalling pathways and the multidrug resistant protein MDR-1. *Br J Cancer* 2001, **85**, 1175–84.

Drukman S, Kavallaris M. Microtubule alterations and resistance to tubulin-binding agents. *Int J Oncol* 2002, **21**, 621–8.

Ferlini C, Raspaglio G, Mozzetti S, *et al.* Bcl-2 down-regulation is a novel mechanism of paclitaxel resistance. *Mol Pharmacol* 2003, **64**, 51–58.

Ferlini C, Distefano M, Pierelli L, *et al.* Cytotoxic effects toward human hematopoietic progenitor cells and tumor cell lines of paclitaxel, docetaxel, and newly developed analogues IDN5109, IDN5111, and IDN5127. *Oncol Res* 1999, **11**, 471–8.

Gligorov J, Lotz JP. Preclinical pharmacology of the taxanes: implications of the differences. *Oncologist* 2004, **9**, 3–8.

Gridelli C. The ELVIS trial: a phase III study of single-agent vinorelbine as first-line treatment in elderly patients with advanced non-small cell lung cancer. Elderly Lung Cancer Vinorelbine Italian Study. *Oncologist* 2001, **6** (Suppl 1), 4–7.

Gupta S, Bhattacharyya B. Antimicrotubular drugs binding to vinca domain of tubulin. *Mol Cell Biochem* 2003, **253**, 41–7.

Hasegawa S, Miyoshi Y, Egawa C, *et al.* Prediction of response to docetaxel by quantitative analysis of class I and III beta-tubulin isotype mRNA expression in human breast cancers. *Clin Cancer Res* 2003, **9**, 2992–7.

Hasegawa S, Miyoshi Y, Egawa C, Ishitobi M, Tamaki Y, Monden M, Noguchi S. Mutational analysis of the class I beta-tubulin gene in human breast cancer. *Int J Cancer* 2002, **101**, 46–51.

Herbst RS, Khuri FR. Mode of action of docetaxel – a basis for combination with novel anticancer agents. *Cancer Treat Rev* 2003, **29**, 407–15.

Hill BT. Vinflunine, a second-generation novel vinca alkaloid with a distinctive pharmacological profile, now in clinical development and prospects for future mitotic blockers. *Curr Pharm Des* 2001, **7**, 1199–212.

Jordan MA. Mechanism of action of antitumor drugs that interact with microtubules and tubulin. *Curr Med Chem Anticancer Agents* 2002, **2**, 1–17.

Jordan MA, Wilson L. Microtubules as a target for anticancer drugs. *Nat Rev Cancer* 2004, **4**, 253–65.

Jordan MA, Thrower D, Wilson L. Mechanism of inhibition of cell proliferation by vinca alkaloids. *Cancer Res* 1991, **51**, 2212–22.

Kanard A, Jatoi A, Castillo R, *et al.* Oral vinorelbine for the treatment of metastatic non-small cell lung cancer in elderly patients: a phase II trial of efficacy and toxicity. *Lung Cancer* 2004, **43**, 345–53.

Kavallaris M, Kuo DY, Burkhart CA, Regl DL, Norris MD, Haber M, Horwitz SB. Taxol-resistant epithelial ovarian tumors are associated with altered expression of specific beta-tubulin isotypes. *J Clin Invest* 1997, **100**, 1282–93.

Kavallaris M, Tait AS, Walsh BJ, He L, Horwitz SB, Norris MD, Haber M. Multiple microtubule alterations are associated with *vinca* alkaloid resistance in human leukemia cells. *Cancer Res* 2001, **61**, 5803–9.

Kruczynski A, Hill BT. Vinflunine, the latest vinca alkaloid in clinical development: a review of its preclinical anticancer properties. *Crit Rev Oncol Hematol* 2001, **40**, 159–73.

Liu B, Staren E, Iwamura T, Appert H, Howard J. Taxotere resistance in SUIT Taxotere resistance in pancreatic carcinoma cell line SUIT 2 and its sublines. *World J Gastroenterol* 2001, **7**, 855–9.

Maeno K, Ito K, Hama Y, *et al.* Mutation of the class I beta-tubulin gene does not predict response to paclitaxel for breast cancer. *Cancer Lett* 2003, **198**, 89–97.

McDonald SL, Stevenson DA, Moir SE, Hutcheon AW, Haites NE, Heys SD, Schofield AC. Genomic changes identified by comparative genomic hybridisation in docetaxel-resistant breast cancer cell lines. *Eur J Cancer* 2005, **41**, 1086–94.

McLeod HL, Kearns CM, Kuhn JG, Bruno R. Evaluation of the linearity of docetaxel pharmacokinetics. *Cancer Chemother Pharmacol* 1998, **42**, 155–9.

Mitchison T, Kirschner M. Dynamic instability of microtubule growth. *Nature* 1984, **312**, 237–42.

Mollinedo F, Gajate C. Microtubules, microtubule-interfering agents and apoptosis. *Apoptosis* 2003, **8**, 413–50.

Monzo M, Rosell R, Sanchez JJ, *et al.* Paclitaxel resistance in non-small-cell lung cancer associated with beta-tubulin gene mutations. *J Clin Oncol* 1999, **17**, 1786–93.

Ngan VK, Bellman K, Hill BT, Wilson L, Jordan MA. Mechanism of mitotic block and inhibition of cell proliferation by the semisynthetic vinca alkaloids vinorelbine and its newer derivative vinflunine. *Mol Pharmacol* 2001, **60**, 225–32.

Noble RL. The discovery of the vinca alkaloids-chemotherapeutic agents against cancer. *Biochem Cell Biol* 1990, **68**, 1344–51.

Noble RL, Beer CT, Cutts JH. Role of chance observation in chemotherapy: Vinca rosea. *Ann N Y Acad Sci* 1958, **76**, 882–894.

Okouneva T, Hill BT, Wilson L, Jordan MA. The effects of vinflunine, vinorelbine, and vinblastine on centromere dynamics. *Mol Cancer Ther* 2003, **2**, 427–36.

Potier P. The synthesis of navelbine prototype of a new series of vinblastine derivatives. *Semin Oncol* 1989, **16** (2 Suppl 4), 2–4.

Rittmann-Grauer LS, Yong MA, Sanders V, Mackensen DG. Reversal of vinca alkaloid resistance by anti-P-glycoprotein monoclonal antibody HYB-241 in a human tumor xenograft. *Cancer Res* 1992, **52**, 1810–16.

Sampath D, Discafani CM, Loganzo F, *et al.* MAC-321, a novel taxane with greater efficacy than paclitaxel and docetaxel *in vitro* and *in vivo*. *Mol Cancer Ther* 2003, **2**, 873–84.

Seve P, Dumontet C. Is class III β-tubulin a predictive factor in patients receiving tubulin-binding agents? *Lancet Oncol* 2008, **9**, 168–75.

Shalli K, Brown I, Heys SD, Schofield AC. Alterations of beta-tubulin isotypes in breast cancer cells resistant to docetaxel. *FASEB J* 2005, **19**, 1299–301.

Toso RJ, Jordan MA, Farrell KW, Matsumoto B, Wilson L. Kinetic stabilization of microtubule dynamic instability *in vitro* by vinblastine. *Biochemistry* 1993, **32**, 1285–93.

Vacca A, Ribatti D, Iurlaro M, Merchionne F, Nico B, Ria R, Dammacco F. Docetaxel versus paclitaxel for antiangiogenesis. *J Hematother Stem Cell Res* 2002, **11**, 103–18.

Vredenburg MR, Ojima I, Veith J, *et al*. Effects of orally active taxanes on P-glycoprotein modulation and colon and breast carcinoma drug resistance. *J Natl Cancer Inst* 2001, **93**, 1234–45.

Warfield RK, Bouck GB. Microtubule-macrotubule transitions: intermediates after exposure to the mitotic inhibitor vinblastine. *Science* 1974, **186**, 1219–20.

Wilson L, Jordan MA, Morse A, Margolis RL. Interaction of vinblastine with steady-state microtubules *in vitro*. *J Mol Biol* 1982, **159**, 125–49.

REFERENCES

7
Antimetabolites in Cancer Therapy

Jessica Scaife and David Kerr

7.1 Introduction

Metabolite is a general term for the organic compounds that are required for normal biochemical reactions in cells. An antimetabolite, therefore, is a chemical with a similar structure to a metabolite, yet different enough to interfere with the normal function of cells including cell division. Antimetabolites are a family of drugs which are a mainstay in cancer chemotherapy. They have chemical structures similar to either folate or to nucleotides, which become the building blocks of DNA.

Folic acid is a growth factor that provides single carbons to the precursors used to form nucleotides in the synthesis of DNA and RNA. Folate antagonists inhibit one or more folate-dependent enzymes, resulting in decreased DNA, RNA and protein synthesis. The pyrimidine antagonists act to block the synthesis of nucleotides containing cytosine and thymine in DNA and cytosine and uracil in RNA. The purine analogues inhibit the production of adenine and guanine found in DNA and RNA. Antimetabolites may be incorporated into a DNA or RNA molecule during synthesis and may directly inhibit crucial enzymes. These mechanisms cause the cell to be unable to synthesize DNA and RNA and therefore it cannot repair or replicate. Actively proliferating tissues such as malignant cells are more sensitive to this effect and the drugs impair tumour growth, with less irreversible damage to normal tissues.

Antimetabolites were first found to be clinically active against cancer 50 years ago with Farber's discovery that aminopterin caused remission in acute leukaemia. In the following 10 years, methotrexate, fluorouracil (5-FU) and mercaptopurine entered clinical practice. Until recently, other significant developments have involved optimising the efficacy of existing antimetabolites, including the use of folinic acid with methotrexate or 5-FU. Development of new antimetabolites has been an active area for research over the past decade

Anticancer Therapeutics   Edited by Sotiris Missailidis
© 2008 John Wiley & Sons, Ltd

and drugs such as permetrexed, capecitabine and fludarabine are being used in clinical practice to treat a variety of tumours.

7.2 Folate Antagonists

Despite the isolation of multiple other agents, methotrexate remains the most commonly used folate antagonist as a chemotherapeutic agent. It is used alone or in combination with other cytotoxics in the treatment of a wide range of cancers including leukaemias, lymphomas, gestational choriocarcinoma, chorioadenoma destruens, hydatidiform mole, carcinomatous meningitis, osteosarcoma and other solid cancers including breast, head and neck, lung, bladder and oesophagus. It is currently under investigation in 172 clinical trials. Phase IV trials include acute lymphoblastic leukaemia (ALL) and acute promyelocytic leukaemia (APL), Burkitt's non-Hodgkin's lymphoma (NHL) and other high-grade lymphoma and primary central nervous system (CNS) lymphoma. Leukaemias, lymphomas, osteosarcoma and other cancers including bladder, breast, head and neck and brain are being evaluated in phase III trials. Phase I and II trials are studying use in a variety of other cancers.

The natural folates are water-soluble members of the B class of vitamins that are essential for cell proliferation and tissue regeneration. In cells, folic acid is reduced to dihydrofolate and then to active tetrahydrofolate by the enzyme dihydrofolate reductase (DHFR). These active folates are co-enzymes necessary for methylation in various metabolic processes, in which they donate methyl groups to specific target molecules. This is a necessary step in the synthesis of purine nucleotides and thymine. Thymidylate synthase (TS) catalyses transfer of the carbon from the tetrahydrofolate to the target molecules by oxidizing the folate ring of the tetrahydrofolate, reverting it back into a dihydrofolate. For this process to repeat, cells must repeatedly use DHFR to reduce the dihydrofolate into the active tetrahydrofolate form, requiring continuous DHFR activity (Figure 7.1).

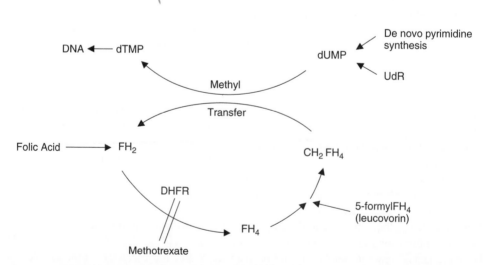

Figure 7.1 Folic acid cycle: inhibition by methotrexate. dTMP, thymidylate; dUMP:uridylate; UdR, uridine; FH_2, dihydrofolate; FH^4, tetrahydrofolate; $CH_2 FH_4$: N^5, N^{10}, methylene tetrahydrofolate; DHFR, dihydrofolate reductase; DNA, deoxy ribonucleic acid

7.2.1 Methotrexate

Methotrexate enters the cell *via* specific folate receptors, the low pH folate transporter, or by reduced folate carriers. Once inside, it binds to the active site of DHFR, preventing synthesis of tetrahydrofolate. The cell cannot create new purine and thymidine nucleotides and therefore cannot synthesize DNA and RNA.

Methotrexate

Methotrexate can be given by many routes: oral, intravenous, subcutaneous, intramuscular, intra-arterial and intrathecal. Oral absorption is dose-dependent in adults with peak serum levels reached in one to two hours. At doses of $30 \, mg/m^2$ or less, methotrexate is well absorbed with a mean bioavailability of approximately 60 percent (FDA Professional Drug Information, 2004). The absorption at doses greater than $80 \, mg/m^2$ is significantly less, possibly due to a saturation effect. Methotrexate is generally completely absorbed from parenteral routes of injection with peak serum concentrations occurring 30–60 minutes after intramuscular administration. With intravenous injection, the initial volume of distribution is approximately 0.18 L/kg (18 % body weight) and steady-state volume of distribution is approximately 0.4–0.8 L/kg (40–80 % body weight) (FDA Professional Drug Information, 2004). It competes with reduced folates for active transport across cell membranes by means of a single carrier-mediated process and at serum concentrations greater than 100 µM, passive diffusion ensures effective intracellular concentrations are achieved.

Methotrexate in serum is approximately 50 % protein-bound and laboratory studies show that it may be displaced from albumin in plasma by various compounds including sulfonamides, salicylates, tetracyclines, chloramphenicol and phenytoin (FDA Professional Drug Information, 2004). It does not penetrate the blood–cerebrospinal fluid (CSF) barrier in therapeutic amounts when given orally or parenterally and high CSF concentrations of the drug are attained by intrathecal administration.

After absorption, methotrexate undergoes hepatic and intracellular metabolism to polyglutamated forms, which can be converted back to methotrexate by hydrolase enzymes. These polyglutamates inhibit DHFR and TS. Small amounts of methotrexate polyglutamates may remain in tissues for extended periods. The retention and prolonged drug action of these active metabolites vary among different cells, tissues and tumours. A small amount of metabolism to 7-hydroxy-methotrexate may occur at doses commonly prescribed. The aqueous solubility of this compound is three- to five-fold lower than methotrexate. Methotrexate is partially metabolized by intestinal flora after oral administration.

The terminal half life for methotrexate is approximately 3 to 10 hours for patients receiving low dose antineoplastic therapy (less than $30 \, mg/m^2$) (FDA Professional Drug Information, 2004). Renal excretion is the primary route of elimination and is dependent on dosage and route of administration. With intravenous administration, 80–90 % of the dose is excreted unchanged in the urine within 24 hours (FDA Professional Drug Information, 2004). Excretion occurs by glomerular filtration and active tubular secretion. Impaired renal function, as well as concurrent use of drugs that also undergo tubular secretion (e.g. salicylate), can markedly increase methotrexate serum levels. Clearance rates vary widely and are decreased at higher doses suggesting non-linear pharmacokinetics. Toxicity for normal tissues is dependent on the duration of exposure to the drug rather than the peak level achieved. When there is delayed elimination due to compromised renal function, a third space effusion, or other causes, methotrexate serum levels may remain elevated for prolonged periods, leading to toxicity. The drug has been detected in human breast milk.

The most frequent side effects of methotrexate are ulcerative stomatitis, leucopenia, nausea, abdominal pain, diarrhoea, malaise, fatigue, chills, fever, dizziness and decreased resistance to infection. It has been reported to cause fetal death and congenital abnormalities, therefore is not recommended for women of childbearing potential and is contraindicated in breastfeeding mothers. It also causes oligospermia. Haematological, renal and liver function must be closely monitored. Methotrexate causes hepatotoxicity, fibrosis and cirrhosis, but only after prolonged use. Acute transient asymptomatic elevations in liver enzymes are frequently seen. Methotrexate-induced lung disease may occur acutely at any time and is not always fully reversible. Symptoms include fever, cough, dyspnoea, hypoxia and an infiltrate on chest X-ray. Malignant lymphomas have been described in patients receiving methotrexate. The drug may induce tumour lysis syndrome, can suppress haematopoiesis and cause pancytopenia. Severe occasionally fatal dermatological reactions including toxic epidermal necrolysis, Stevens–Johnson syndrome, exfoliative dermatitis, skin necrosis and erythema multiforme have been seen.

Methotrexate is the only cytotoxic drug for which there is a role for routine pharmacokinetic monitoring, following administration in high or moderate dose. Bleyer (1989) showed that there was a strong correlation between serum concentrations of methotrexate and the likelihood of serious consequent toxicity and promulgated the concept of "folinic acid rescue" given the potential for leucovorin or folinic acid to bypass inhibited DHFR. This replenishes intracellular reduced folate pools and therefore moderates the degree of subsequent side effects (Figure 7.1).

7.2.2 Permetrexed

Permetrexed is a multitargeted folate antagonist used in malignant pleural mesothelioma (MPM) and non-small cell lung cancer (NSCLC). It is a disodium salt of a synthetic pyrimidine-based antifolate which binds to and inhibits TS. Side effects include suppression of bone marrow function, nausea and vomiting and fatigue and in order to reduce toxicity, patients must receive folate and vitamin B12 supplementation.

Permetrexed

The National Institute for Clinical Excellence (NICE) has recommended permetrexed as a treatment option for MPM in people who have a World Health Organization (WHO) performance status of 0 or 1, who are considered to have advanced disease and for whom surgical intervention is considered inappropriate (National Institute for Clinical Excellence, 2008). There is no standard chemotherapy treatment for MPM and permetrexed in combination with cisplatin is the only regimen licensed for this indication. The EMPHACIS trial suggests that permetrexed plus cisplatin confers a survival benefit of approximately 3 months compared with cisplatin alone (National Institute for Clinical Excellence, 2008). It also demonstrates advantages in terms of one-year survival, median time to progressive disease, tumour response rate and quality of life (National Institute for Clinical Excellence, 2008).

NICE also advises permetrexed as a treatment for relapsed patients with NSCLC (National Institute for Clinical Excellence, 2007a). However, it is not recommended for locally advanced or metastatic disease due to overall survival not being significantly greater than with docetaxel and the results of non-inferiority testing not formally excluding the possibility of a marginal loss of efficacy of permetrexed when compared with docetaxel (National Institute for Clinical Excellence, 2007a).

Pralatrexate

Permetrexed is an agent in 108 clinical trials. There is currently one phase IV and various phase III trials in NSCLC as well as a phase III trial evaluating use in head and neck cancer. There are trials at phase I and II in a variety of tumour types. A recent phase III trial assessed use in pancreatic cancer.

7.2.3 Ralitrexed

Raltitrexed is a quinazoline folate analogue which is transported into cells *via* the reduced folate carrier, where it blocks the folate-binding site of TS. It is licensed in the UK for palliative treatment of advanced colorectal cancer where 5-FU/folinic acid-based regimens are either not tolerated or inappropriate. It causes more nausea and vomiting but less diarrhoea and mucositis. NICE does not recommend its use for patients with advanced colorectal cancer and suggests it be confined to appropriately designed clinical studies (National Institute for Clinical Excellence, 2005a). Raltitrexed did not improve overall survival or progression-free survival when compared with 5-FU/FA and no significant improvement in response rate was seen (National Institute for Clinical Excellence, 2005a). In addition there was a statistically significant decrease in quality of life. Phase III trials have demonstrated that the combination of raltitrexed and cisplatin improves overall survival compared with cisplatin alone in MPM without a harmful effect on quality of life (van Meerbeeck *et al.*, 2005).

Raltitrexed

7.2.4 Nolatrexed

Nolatrexed hydrochloride

Nolatrexed is a water-soluble lipophilic quinazoline folate analogue, which again occupies the folate-binding site of TS. A recent phase III randomized controlled trial compared the survival of patients with unresectable hepatocellular carcinoma treated with nolatrexed or doxorubicin (Gish *et al.*, 2007). However, there was significant toxicity in the nolatrexed arm and the drug showed minimal activity in this trial.

7.2.5 Pralatrexate

Pralatrexate (10-propargyl-10-deazaaminopterin) is a synthetic derivative of pterins. It competes for the folate-binding site of DHFR. It is currently being studied in phase I and II trials in lymphoproliferative malignancies. It has demonstrated limited toxicity and proven efficacy in NSCLC (Krug *et al.*, 2003).

7.2.6 Talotrexin

Talotrexin is an analogue of aminopterin and is a DHFR inhibitor. Membrane transport is a critical determinant of the antineoplastic activity of methotrexate, and impaired uptake of antifolates is a frequent mode of drug resistance. Talotrexin is actively transported into cells by the reduced folate carrier and, therefore, is unlikely to be associated with P-glycoprotein-mediated multidrug resistance. It has been approved for phase I trial in recurrent solid tumours or recurrent or refractory leukaemia.

Talotrexin ammonium

7.3 Pyrimidine Antagonists

Robert Duschinsky synthesized 5-FU in the late 1950s. Uracil is a normal component of RNA, and the rationale behind development of the drug was that cancer cells, with their increased genetic instability, might be more sensitive to molecules similar in structure to the natural compound. 5-FU demonstrated specific uracil antagonism, proved to have antitumour capabilities and remains the most commonly used pyrimidine antagonist.

7.3.1 5-Fluorouracil

5-Fluorouracil

5-FU is usually given in combination with folinic acid to enhance TS inhibition by increasing the pool of intracellular folate. Dosages vary and depend upon many factors including the type and location of the cancer and the health and age of the patient. It is used in a variety of intravenous combination chemotherapies for treating bowel, breast, gastric, oesophageal, pancreatic, head and neck, anal and ovarian cancers. Oral analogues are also used and it may be applied topically to treat multiple actinic or solar keratoses and superficial basal cell carcinoma.

There are different intravenous 5-FU regimens requiring either infusions or bolus injections. There is considerable variability in current UK practice because of a lack of consensus over the optimum regimen. Those requiring infusion are more complex and often require admission to hospital, but are reported to be superior to bolus regimens in terms of progression-free survival periods, safety, toxicity and quality of life, but equally effective in terms of overall survival (National Institute for Clinical Excellence, 2003a). The modified de Gramont infusion regimen is often used. It involves a 2-hour infusion of folinic acid followed by a bolus injection of 5-FU and a continuous infusion of 5-FU over 46 hours, and allows patients to be treated on an outpatient basis.

5-FU inhibits DNA synthesis both by blocking the formation of normal pyrimidine nucleotides *via* enzyme inhibition and by interfering with DNA synthesis after incorporation into a growing DNA molecule. Thymine differs from uracil by the presence of a methyl group on the fifth carbon in the pyrimidine ring. This methyl group is added by TS. If a 5-FU molecule is in this position instead of uracil, the enzyme cannot add a methyl group to the fifth carbon due to the fluoride atom at that location. Without this addition, the thymine nucleotides cannot be made and are not available for DNA synthesis.

In vivo, 5-FU is converted to the active metabolite 5-fluoroxyuridine monophosphate (F-UMP). Competing with uracil, F-UMP is incorporated into RNA, thereby inhibiting RNA synthesis and cell growth. Another active metabolite, 5–5-fluoro-2'-deoxyuridine-5'-O-monophosphate (F-dUMP), inhibits TS, resulting in the depletion of thymidine triphosphate (TTP), a necessary constituent of DNA. Other 5-FU metabolites become incorporated into RNA and DNA, further impairing cellular growth (Figure 7.2).

5-FU is metabolized primarily in the liver, resulting in inactive degradation products. The parent drug is excreted unchanged (7–20 %) in the urine in 6 hours (FDA Professional Drug Information, Fluorouracil). The mean half life is about 16 minutes. Common side effects include tiredness, nausea, diarrhoea, bone marrow suppression, mouth sores and pigmentation changes in the skin. Less common side effects include eye 'grittiness' or blurred vision, hair thinning, loss of appetite, rashes, and soreness of the palms and soles of the feet (palmar–plantar erythrodysesthesia syndrome).

Hospitalization is recommended for the first course of injection therapy because of the possibility of severe toxic reactions. 5-FU should be used with extreme caution in poor-risk patients who have had high-dose pelvic irradiation or previous use of alkylating agents, who have widespread involvement of bone marrow by metastatic tumours, or impaired hepatic or renal function. It is contraindicated in patients with poor nutritional status, pregnancy and

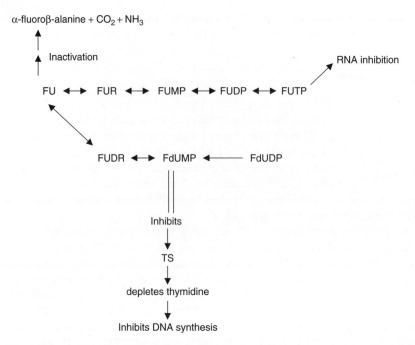

Figure 7.2 Metabolism of 5-fluorouracil (FU). FUR: fluorouridine; FUMP: fluorouridine monophosphate; FDP: fluorouridine diphosphate; FUTP: fluorouridine triphosphate; FUDR: fluorodeoxyuridine; FdUMP: fluorodeoxyuridine monophosphate; FdUP: fluoro deoxyuridine diphosphate; TS: thymidylate synthase. Folinic acid (Leucovorin) stabilizes ternary complex of TS and FdUMP, increasing inhibition of the enzyme

dihydropyrimidine dehydrogenase enzyme (DPD) deficiency. Cimetidine may increase serum concentrations of 5-FU and potentially increase toxicity. Folinic acid may enhance gastrointestinal toxicity of 5-FU and fatalities have occurred because of severe toxic enterocolitis. Exposure to ultraviolet rays should be avoided because the intensity of the reaction may be increased. Rarely, severe toxicity including stomatitis, diarrhoea, neutropenia and neurotoxicity has been attributed to DPD deficiencies, and is dose dependent.

5-FU is extensively used for the treatment of colorectal cancer. NICE guidelines advise adjuvant chemotherapy to all patients with Dukes' C carcinoma in reasonable health whose disease is sufficiently advanced that treatment is likely to be beneficial, based on performance status and comorbidities (National Institute for Clinical Excellence, 2004). The standard course is 5-FU and folinic acid intravenously for 6 months with an absolute increase in 5-year survival rates of 4–13 % (National Institute for Clinical Excellence, 2004). Dukes' B should receive adjuvant 5-FU as part of trials such as NCRN QUASAR1. When Dukes' B have adverse features such as vascular invasion, peritoneal involvement or perforation or if surgical margins are inflamed or contain tumour, patients have a higher disease-related mortality rate and are therefore thought to be more likely to benefit from chemotherapy (Peterson *et al.*, 2002; Lennon *et al.*, 2003).

NICE advises that patients with advanced colorectal cancer who are sufficiently fit should be treated with systemic chemotherapy as first or second-line therapy, typically with an established regimen containing 5-FU (National Institute for Clinical Excellence, 2004). If they relapse or have metastatic disease then they should receive palliative chemotherapy if their performance status is reasonable. Chemotherapy should be either 5-FU/folinic acid or an oral fluropyrimidine (National Institute for Clinical Excellence, 2003a). Meta-analysis of 13 randomized controlled trials (RCTs) showed that palliative chemotherapy increased median survival time by 3.7 months and decreased risk of death by 35 % with an absolute improvement in survival rate of 16 % at 1 year (National Institute for Clinical Excellence, 2004). Meta-analysis of seven trials showed significantly decreased mortality rates with palliative chemotherapy for metastatic colorectal cancer (National Institute for Clinical Excellence, 2004). In 14 of these, quality of life was similar or better with chemotherapy than without. Chemotherapy early in the course of metastatic disease leads to better outcomes than when given after symptoms have become severe. There is an increase in survival by 3–6 months without increase in adverse effects or quality of life (National Institute for Clinical Excellence, 2004).

In anal cancer, NICE advises primary chemoradiotherapy with mitomycin C, 5-FU and radiation in most patients (National Institute for Clinical Excellence, 2004). There are areas of uncertainty and patients should be encouraged to participate in trials such as the CRUK ACT2 trial.

5-FU has been the standard chemotherapy used in the UK for pancreatic cancer with evidence suggesting a small survival advantage and improvements in quality of life in a proportion of patients (National Institute for Clinical Excellence, 2001). It is administered using a variety of doses and schedules but the response rate rarely exceeds 20 % and no consistent effect on disease-related symptoms or survival has been demonstrated (National Institute for Clinical Excellence, 2001a).

There is very strong evidence from systematic reviews of RCTs that multiple-agent chemotherapy decreases annual recurrence rates and overall death from breast cancer (National Institute for Clinical Excellence, 2002). Most trials involved adjuvent CMF (cyclophosphamide, methotrexate and 5-FU). However there is no evidence of a difference in survival rates between CMF and other multiple-agent regimens (National Institute for Clinical Excellence, 2002). CAF (cyclophosphamide, doxorubicin and 5-FU) is better tolerated than CMF and fewer cycles are necessary to produce an equivalent level of benefit. Despite many trials the optimum regimen remains unclear and there are wide variations between UK oncologists in prescribing habits.

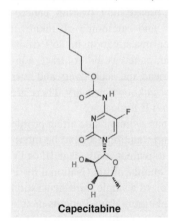

Capecitabine

5-FU is currently an agent in 234 clinical trials. Phase IV trials are investigating colorectal and oesophageal cancers. It is currently in phase III trials in pancreatic, biliary tract, colorectal, breast, anal, bladder, head and neck, oesophageal, gastric and hepatocellular carcinomas. There are a variety of cancers in phase I and II trials.

5-Fluoro-2-deoxycytidine is a prodrug and is converted by intracellular deaminases to the cytotoxic agent 5-FU. It is currently in its initial phase I trial in advanced solid tumours.

7.3.2 Capecitabine

Capecitabine is an oral prodrug of 5-FU and is selectively activated by tumour cells. The enzyme involved in the final conversion, thymidine phosphorylase, is found at higher levels in tumour tissues, thereby reducing systemic exposure to 5-FU. It has an accepted therapeutic use for breast and colorectal cancer.

NICE recommends capecitabine monotherapy as an option for adjuvant treatment of patients with Dukes' C colon cancer and as a first-line treatment (in combination with folinic acid) in metastatic colorectal cancer (National Institute for Clinical Excellence, 2003a, b; National Institute for Clinical Excellence, 2006). Capecitabine monotherapy is recommended as an option for people with locally advanced or metastatic breast cancer who have not previously received capecitabine in combination therapy and for whom anthracycline and taxane-containing regimens have failed or further anthracycline therapy is contraindicated (National Institute for Clinical Excellence, 2003c). In the treatment of locally advanced or metastatic breast cancer, capecitabine in combination with docetaxel is recommended in preference to single agent docetaxel in people for whom anthracycline-containing regimens are unsuitable or have failed. Capecitabine combination therapy is likely to be more effective than docetaxel monotherapy in terms of several outcomes including overall survival (National Institute for Clinical Excellence, 2003a, b).

Capecitabine is currently an agent in 216 trials. Phase IV trials are investigating breast and colorectal cancer. Phase III trials include breast, colorectal, pancreatic, gastric, oesophageal, hepatocellular, biliary tract and head and neck cancers. There are various phase I and II trials.

7.3.3 Gemcitabine

Gemcitabine

Gemcitabine is converted intracellularly to the active metabolites difluorodeoxycytidine di- and triphosphate (dFdCDP and dFdCTP). dFdCDP inhibits ribonucleotide reductase, thereby decreasing the deoxynucleotide pool available for DNA synthesis. dFdCTP is incorporated into DNA, resulting in DNA strand termination and apoptosis. Gemcitabine has a wide spectrum of activity against solid tumours. It is used in bladder, breast, lung, ovarian, pancreatic and renal cell carcinoma and carcinoma of unknown primary. It is currently under investigation in 269 trials. Phase III trials include bladder, pancreas, biliary tract, non-small cell lung, breast, colon, head and neck, ovary and liver cancer, NHL and carcinoma of unknown primary. There are various others at the phase I and II trial stage.

Gemcitabine is administered as an intravenous infusion and has a similar side effect profile to 5-FU. It can induce hepatic, renal and pulmonary toxicity and function needs to be monitored. NICE recommends that chemotherapy should be offered to patients with stage III or IV NSCLC and PS 0 or 1 to improve survival, disease control and quality of life (National Institute for Clinical Excellence, 2005). This should be a combination of a single third-generation drug (docetaxel, gemcitabine, paclitaxel or vinorelbine) plus a platinum drug. Patients unable to tolerate a platinum combination may be offered single-agent chemotherapy with a third-generation drug (National Institute for Clinical Excellence, 2005b). NICE also states that gemcitabine may be considered as a treatment option for patients with advanced or metastatic adenocarcinoma of the pancreas and a Karnofsky performance score of 50 or more, where first line chemotherapy is to be used (National Institute for Clinical Excellence, 2001a). It is also licensed as a second line treatment for patients with 5-FU refractory pancreatic cancer.

Gemcitabine in combination with paclitaxel is recommended as an option for the treatment of metastatic breast cancer when docetaxel monotherapy or docetaxel plus capecitabine are also considered appropriate (National Institute for Clinical Excellence, 2007b). It has UK marketing authorization for treatment of patients with metastatic breast cancer who have relapsed following adjuvant or neo-adjuvant chemotherapy (National Institute for Clinical Excellence, 2007b). It is considered for people who are younger and fitter than the general population of patients with metastatic breast cancer and for those who require a higher level of efficacy than would be achieved with a monotherapy regimen, without the toxicity usually associated with a combination regimen, for example because of visceral metastasis. Capecitabine is an important option in later lines of therapy for metastatic breast cancer, therefore the use of docetaxel plus capecitabine as a first-line choice would reduce the possibility of using capecitabine in later therapy. Gemcitabine plus paclitaxel is a useful option because of its efficacy and low level of toxicity.

7.3.4 Cytarabine

Cytarabine

Cytarabine is an analogue of cytidine with arabinose instead of ribose. It is converted to the triphosphate form within the cell and then competes with cytidine for incorporation into DNA. The arabinose sugar sterically hinders the rotation of the molecule within DNA and replication ceases, specifically during the S phase of the cell cycle. Cytarabine also inhibits DNA polymerase, resulting in a decrease in DNA replication and repair. It is administered as an infusion or as a subcutaneous injection. Cytarabine has been formulated inside small particles of a synthetic lipid material called DepoFoam. This dosage form slowly releases the drug and provides a sustained action when administered intra-thecally. There is also a liposomal formulation containing a combination of cytarabine and daunorubicin.

Cytarabine is used in many leukaemias including ALL, acute non-lymphocytic and chronic myelogenous (CML) forms. It is also used in carcinomatous meningitis, meningitis lymphomatous, myelodysplasia, bone marrow transplantation, Hodgkin's disease (HD), NHL, ovarian cancer and retinoblastoma. Cytarabine is given as part of the DHAP: dexamethasone, cytarabine, cisplatin and R-DHAP (rituximab) regimens and ESHAP and R-ESHAP: etoposide, methylprednisolone and cytarabine for lymphoma inpatients. It is important to monitor blood cell and platelet counts throughout the duration of treatment with frequent blood tests.

Cytarabine is currently in 181 trials. Phase IV trials are investigating bone marrow transplant patients and those with ALL, NHL, acute myeloid leukaemia (AML) and APL. Phase III trials involve AML, myelodysplasia, NHL, ALL, HD and CML. Phase I and II are looking at a variety of cancers.

7.3.5 CP-4055

CP-4055 is the lipophilic 5'-elaidic acid ester of cytarabine and is converted intracellularly into cytarabine triphosphate by deoxycytidine kinase. It shows increased cellular uptake and

CP-4055

Azacitidine

retention, resulting in increased activation by deoxycytidine kinase to cytarabine triphosphate, decreased deamination and deactivation by deoxycytidine deaminase, and increased inhibition of DNA synthesis. CP-4055 also inhibits RNA synthesis, an effect not seen with cytarabine. It is currently in phase II trials in colorectal and metastatic malignant melanoma and phase I and II trials in haematological malignancy.

7.3.6 Azacitidine

Azacitidine is used for myelodysplasia. It is an analogue of cytidine, which is incorporated into DNA, where it reversibly inhibits DNA methyltransferase. Hypomethylation of DNA by azacitidine may activate tumour suppressor genes silenced by hypermethylation, resulting in an antitumour effect. This agent is also incorporated into RNA, thereby disrupting normal RNA function and impairing tRNA cytosine-5-methyltransferase activity. It is currently in 32 trials. Phase III trials are investigating its role in AML and myelodysplasia, and there are a variety of phase I and II trials.

7.3.7 Tegafur

Tegafur

Tegafur is a prodrug that is gradually converted to 5-FU in the liver by the cytochrome P-450 enzyme system. It is given in combination with uracil, which inhibits the degradation of 5-FU, resulting in sustained higher levels of 5-FU in tumour cells. Folinic acid is usually added to the tegafur and uracil (UFT) combination to act as a modulator. UFT is an option for first-line treatment of metastatic colorectal cancer (National Institute for Clinical Excellence, 2003a). Oral treatment with UFT is more convenient than standard intravenous treatment with 5-FU. Both regimens are well tolerated, do not differ in their impact on health- related quality of life and achieve similar disease-free and overall survival (Kopec *et al.*, 2007; Lembersky *et al.*, 2006).

S-1 is an oral preparation of tegafur, combined with two modulators, 5-chloro-2,4-dihydroxypyridine (CDHP) and potassium oxonate, with antineoplastic activity. CDHP is a reversible inhibitor of dihydropyrimidine dehydrogenase (DPD), which is responsible for the rapid catabolism of 5-FU into its inactive metabolites. Potassium oxonate preferentially localizes in the gastrointestinal tract and inhibits orotate phosphoribosyltransferase, the major enzyme responsible for 5-FU activation, thereby reducing local concentrations of 5-FU and decreasing the incidence of 5-FU-related gastrointestinal toxicity. S-1 is currently under investigation in 29 trials including head and neck cancer, gastric, breast, colorectal and pancreatic cancers.

S-1 preparation

7.3.8 Floxuridine

Floxuridine

Floxuridine is a fluorinated pyrimidine monophosphate analogue of 5-fluoro-2′-deoxyuridine-5′-phosphate (FUDR-MP) with antineoplastic activity. It inhibits TS and is also metabolized to 5-FU and other metabolites that can be incorporated into RNA and inhibit the utilization of preformed uracil in RNA synthesis. It is an accepted treatment for gastrointestinal malignancy metastatic to the liver. It is currently an agent in 11 phase II trials for gastric, colorectal, hepatocellular and oesophageal cancers, with some also in phase I. It can be given by hepatic artery infusion, which has been shown to increase overall survival, response rate, THP, and is associated with better physical functioning compared with systemic therapy (Kemeny *et al.*, 2006).

7.3.9 Doxifluridine

Doxifluridine

Doxifluridine is an oral prodrug of 5-FU. It has been designed to circumvent the rapid degradation of 5-FU by dihydropyrimidine dehydrogenase in the gut wall and is converted into 5-FU in the presence of pyrimidine nucleoside phosphorylase. This drug is currently in two Phase III trials in gastric cancer.

7.3.10 Decitabine

Decitabine is a cytidine analogue with potential antineoplastic activity. It incorporates into DNA and inhibits DNA methyltransferase, resulting in hypomethylation of DNA and arrest of DNA replication in the S phase. It is currently

Decitabine

Sapacitabine

Mercaptopurine

in a total of 21 trials. Phase III trials involve myelodysplasia and AML and there are a variety of Phase I and II trials.

7.3.11 Sapacitabine

Sapacitabine is an orally bioavailable pyrimidine analogue with potential antineoplastic activity. It is hydrolysed by amidases to the deoxycytosine analogue CNDAC (2′-cyano-2′-deoxyarabinofuranosylcytosine), which is then phosphorylated into its active triphosphate form. As an analogue of deoxycytidine triphosphate, CNDAC triphosphate molecules are incorporated into DNA strands during replication, thereby leading to single-stranded DNA breaks caused by beta-elimination during the fidelity checkpoint process. As a result, this agent induces cell cycle arrest in the G_2 phase and subsequently triggers apoptosis. This agent has only recently entered phase I and II trial in leukaemia, lymphoma and myelodysplasia patients.

7.4 Purine Antagonists

7.4.1 Mercaptopurine

Mercaptopurine was synthesized and developed by Hitchings, Elion, and associates at the Wellcome Research Laboratories and has been in use since 1953. It is 1, 7-dihydro-6H-purine-6-thione monohydrate, an analogue of the purine bases adenine and hypoxanthine. It is used in the treatment of ALL, AML, Hodgkin's lymphoma in children and 'ymphoblastic lymphoma. It is currently in 40 trials worldwide. ALL and APL are in phase III and IV trial and various lymphomas are in Phase III. There are several phase I and II trials.

Mercaptopurine is metabolized by hypoxanthine-guanine phosphoribosyltransferase (HGPRT), to 6-thioguanosine-5'-phosphate (6-thioGMP) and 6-thioinosine monophosphate (TIMP). TIMP inhibits conversion of IMP to xanthylic acid (XMP) and to adenylic acid (AMP). Methylation of TIMP forms 6-methylthioinosinate (MTIMP) and both TIMP and MTIMP inhibit glutamine-5-phosphoribosylpyrophosphate amidotransferase, the first enzyme unique to the de novo pathway for purine ribonucleotide synthesis. 6-Thiopurine methyltransferase (TPMT) converts mercaptopurine to 6-methylmercaptopurine ribonucleoside (MMPR), which is also a potent inhibitor of *de novo* purine synthesis. Mercaptopurine is also incorporated into DNA in the form of deoxythioguanosine, which results in the disruption of DNA replication. It is not known which of these effects are predominantly responsible for cell death.

Mercaptopurine is usually given orally but when higher doses are required, the drug may be given intravenously. Absorption of an oral dose is incomplete and variable, averaging approximately 50 % of the administered dose and the factors influencing absorption are unknown (FDA Professional Drug Information, 2004b). There is negligible entry into CSF. Monitoring of plasma levels of mercaptopurine during therapy is of questionable value and there is technical difficulty in determining plasma concentrations which are seldom greater than 1 to 2 µg/mL after a therapeutic oral dose. The active intracellular metabolites have longer half-lives than the parent drug, haemodialysis does not reduce toxicity of the drug and there is no known pharmacological antagonist to the biochemical actions of mercaptopurine *in vivo*.

The catabolism of mercaptopurine and its metabolites is complex. After oral administration, urine contains intact mercaptopurine, thiouric acid (formed by direct oxidation by xanthine oxidase, probably *via* 6-mercapto-8-hydroxypurine), and a number of 6-methylated thiopurines. It is advisable to start with smaller dosages in patients with impaired renal function, as this can result in slower elimination of the drug and metabolites and a greater cumulative effect. It is not known whether the drug is excreted in human milk.

Mercaptopurine tablets are indicated for remission induction and maintenance therapy of acute leukaemia. The response depends upon the particular subclassification of leukaemia and whether the patient is paediatric or adult. In ALL, it induces complete remission in approximately 25 % of paediatric patients and 10 % of adults when given as a single agent (FDA Professional Drug Information, 2004b). Combination chemotherapy with vincristine, prednisolone, and L-asparaginase results in more frequent complete remission. Mercaptopurine single agent or combination maintenance therapy is used to prolong remission. Mercaptopurine induces complete remission in approximately 10 % of patients with AML, whether paediatric or adult, and these figures are improved by using combination therapy (FDA Professional Drug Information, 2004b).

The most consistent, dose-related toxicity is bone marrow suppression, presenting with anaemia, leukopenia, thrombocytopenia, or any combination of these. There are individuals with an inherited deficiency of TPMT who may be unusually sensitive to the myelosuppressive effects of mercaptopurine following initiation of treatment. Substantial dosage reductions may be required to avoid the development of life-threatening bone marrow suppression in these patients. This is exacerbated by coadministration with other drugs that inhibit TPMT, such as mesalazine and sulphasalazine. Myelosuppression is often unavoidable during the induction phase of adult acute leukaemia if remission induction is to be successful. Life-threatening infections and haemorrhage have been observed as a consequence. It is recommended that a full blood count is obtained weekly while the patient is on mercaptopurine. In cases where the cause of fluctuations is unclear, bone marrow examination may be useful. Enhanced marrow suppression has been noted in some patients also receiving trimethoprim-sulfamethoxazole (FDA Professional Drug Information, 2004b).

Hyperuricaemia and/or hyperuricosuria may occur in patients receiving mercaptopurine as a consequence of rapid cell lysis accompanying the antineoplastic effect. Adverse effects can be minimized by increased hydration, urine alkalinization, and the prophylactic administration of a xanthine oxidase inhibitor such as allopurinol. Mercaptopurine is hepatotoxic, causing both intrahepatic cholestasis and parenchymal cell necrosis. It is unclear how much of the hepatic damage is due to direct toxicity from the drug and how much may be due to a hypersensitivity reaction. In studies in paediatric patients, the incidence of hepatotoxicity ranged from 0 % to 6 % (FDA Professional Drug Information, 2004b). It is advised to monitor liver function tests at weekly intervals when initiating therapy and at monthly intervals thereafter. Patients who are receiving mercaptopurine with other hepatotoxic drugs or with known pre-existing liver disease may need more frequent testing and lower dosage. Mercaptopurine can cause fetal harm and women should be advised to avoid becoming pregnant.

Mercaptopurine has been associated with intestinal ulceration, skin rashes, hyperpigmentation, alopecia, oligospermia and decreased immunity. Inhibition of the anticoagulant effect of warfarin has been reported. Carcinogenic potential exists in humans, but the extent of the risk is unknown.

The genetic changes that lead to the uncontrolled growth of cancer cells also lead to genetic instability and increased mutation rates and these can reduce effectiveness of mercaptopurine. Animal tumours displaying resistance have often lost the ability to convert mercaptopurine to TIMP. Mutations that alter enzymes in the metabolic pathway leading to the formation of adenine and guanine and to DNA synthesis may limit the effectiveness of the drug and additional drugs may be used in an attempt to overcome resistance.

7.4.2 Fludarabine

Fludarabine monophosphate

Fludarabine is a fluorinated nucleotide analogue of the antiviral agent vidarabine (ara-A). It is administered parenterally as a phosphate salt and is preferentially transported into malignant cells where it is rapidly dephosphorylated to 2-fluoro-ara-A and then phosphorylated intracellularly by deoxycytidine kinase to the active triphosphate, 2-fluoro-ara-ATP. This metabolite inhibits DNA polymerase alpha, ribonucleotide reductase and DNA primase, thereby interrupting DNA synthesis and inhibiting tumour cell growth.

Fludarabine is used to treat acute non-lymphocytic leukaemia (ANLL), CLL and indolent NHL and can be administered orally or intravenously. Common side effects are similar to mercaptopurine. NICE suggests that it can be effective for patients for whom first-line chemotherapy has failed or who cannot tolerate it, and appears to produce less nausea, vomiting and hair loss than combination chemotherapy (National Institute for Clinical Excellence, 2001b). It is currently in 196 trials. Phase IV trials are investigating ALL and AML and phase III trials are evaluating use in CLL, AML, myelodysplasia and NHL. It is in phase I and II trials in a variety of malignancies including CML, myeloma, melanoma and breast, renal and prostate cancers.

7.4.3 Pentostatin

Pentostatin

Pentostatin (2′-deoxycoformycin) is a purine analogue antibiotic isolated from the bacterium *Streptomyces antibioticus*. It binds to and inhibits adenine deaminase (ADA), an enzyme essential to purine metabolism. ADA activity is greatest in cells of the lymphoid system with T-cells having higher activity than B-cells and T-cell malignancies higher

ADA activity than B-cell malignancies. Inhibition of ADA by pentostatin results in elevated intracellular levels of dATP which may block DNA synthesis through the inhibition of ribonucleotide reductase. It may also inhibit RNA synthesis and selectively deplete CD26+ lymphocytes. It is used to treat CLL, Cutaneous T-cell lymphoma and Hairy cell leukaemia. It is under investigation in Phase I-IV trials in CLL, as well as several Phase I and II trials in other haematological malignancies.

7.4.4 Cladribine

Cladribine

Cladribine is a purine analogue. Its metabolite, cladribine triphosphate, incorporates into DNA, resulting in single-strand breaks in DNA, depletion of nicotinamide adenine dinucleotide (NAD) and adenosine triphosphate (ATP), and apoptosis. This agent is resistant to adenosine deaminase, an enzyme that inactivates some antineoplastic agents and is selectively toxic to lymphocytes and monocytes, which exhibit little deoxynucleotide deaminase activity. It is an accepted therapeutic use for AML, CLL, hairy cell leukaemia, cutaneous T-cell lymphoma, NHL and Waldenstrom's macroglobulinaemia. It is currently under investigation in seven clinical trials. Use in ALL is being studied in phase II and IV, AML in phase I, II and III, hairy cell leukaemia in phase II and CML in phase I trial.

7.4.5 Thioguanine

Thioguanine

Thioguanine is a synthetic guanosine analogue. It is phosphorylated by hypoxanthine-guanine phosphoribosyl transferase and incorporates into DNA and RNA, resulting in inhibition of DNA and RNA synthesis and cell death. It also inhibits glutamine-5-phosphoribosylpyrophosphate amido transferase, thereby inhibiting purine synthesis. It is used to treat ANLL. It is in phase II, III and IV trial in ALL, phase III in NHL, phase II and III in AML and phase II trial in glioma.

7.4.6 Nelarabine

Nelarabine

Nelarabine is an arabinonucleoside antimetabolite. It is demethoxylated by adenosine deaminase to become biologically active 9-beta-D-arabinosylguanine (ara-G), which incorporates into DNA, thereby inhibiting DNA synthesis and inducing apoptosis of tumour cells during S phase. It is used to treat T-cell lymphoblastic lymphoma and phase I, II and III trials are assessing use in this disease further.

7.4.7 Cordycepin

Cordycepin is an antibiotic isolated from the fungus *Cordyceps militaris*. It is an adenosine analogue, which is

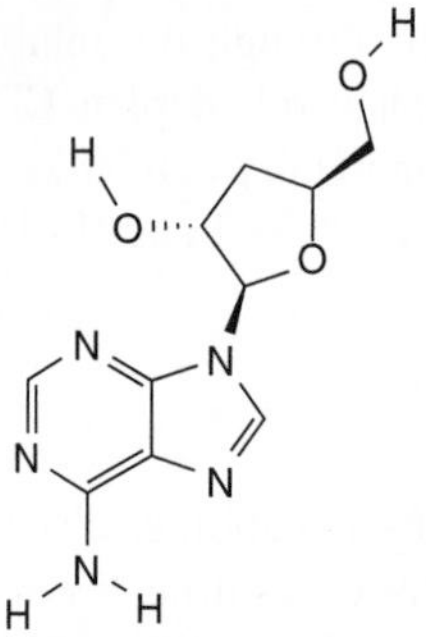

Cordycepin

phosphorylated to its mono-, di-, and triphosphate intracellularly. Triphosphate cordycepin incorporates into RNA, and inhibits transcription elongation and RNA synthesis due to the absence of a hydroxyl moiety at the 3′ position. It is converted to an inactive metabolite by adenosine deaminase and therefore must be administered with an adenosine deaminase inhibitor in order to be effective. It has displayed cytotoxicity against some leukaemic cell lines *in vitro*. It is an accepted therapeutic use for TdT-positive ALL.

7.4.8 Clofarabine

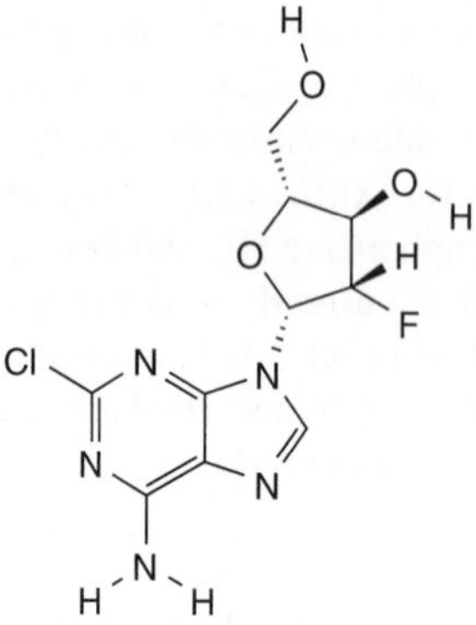

Clofarabine is phosphorylated intracellularly to the active 5′-triphosphate metabolite, which inhibits ribonucleotide reductase and DNA polymerase, resulting in inhibition of DNA repair and synthesis of DNA and RNA. This purine analogue also disrupts mitochondrial function and membrane integrity, resulting in the release of various factors, including cytochrome C and apoptosis-inducing factors. It is in trials at phase I to III in AML and ALL, and in phase I and II trials in NHL, myelodysplasia and CML.

Clofarabine

7.4.9 Triciribine phosphate

Triciribine phosphate is a tricyclic nucleoside, which inhibits amidophosphoribosyltransferase and inosine monophosphate dehydrogenase, resulting in decreased purine nucleotide biosynthesis, decreased DNA and protein synthesis, and cell cycle arrest. It has completed phase II trials in a variety of malignancies and is currently in one phase I study in patients with metastatic cancers.

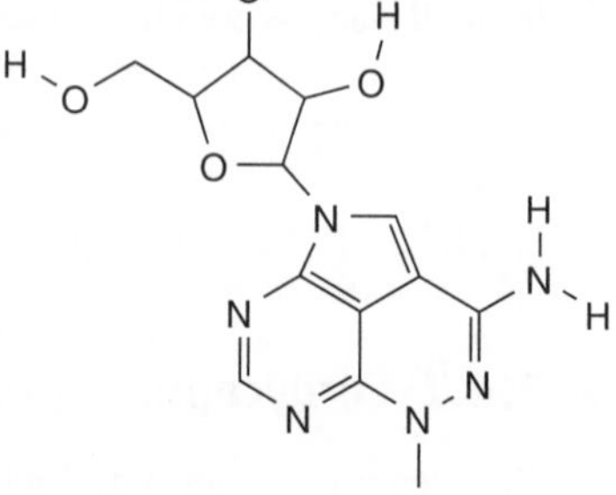

Triciribine

AG2037 (pelitrexol)

7.4.10 Pelitrexol

Pelitrexol inhibits glycinamide ribonucleotide formyltransferase (GARFT), the first folate-dependent enzyme of the *de novo* purine synthesis pathway essential for cell proliferation. Inhibition reduces the purine nucleotide pool required for DNA replication and RNA transcription. As a result, pelitrexol causes cell cycle arrest in S-phase, and ultimately inhibits tumour cell proliferation. Phase I and II trials have recently assessed its potential.

7.4.11 Dezaguanine

Dezaguanine

Dezaguanine replaces guanine to incorporate into DNA and inhibits de novo purine synthesis, thereby inducing cell death. It has completed one phase I trial.

7.5 Summary

It is interesting to note that although some antimetabolites are more than 50 years old, they remain the therapeutic mainstay for a range of solid and haematological malignancies. They are an active focus of clinical research, and new analogues have widened the spectrum of disease activity, whilst reducing toxicity.

References

Food and Drug Administration Professional Drug Information. Fluorouracil. http://www.drugs.com/ppa/fluorouracil.html

Food and Drug Administration Professional Drug Information. Methotrexate. Mylan Pharmaceuticals Inc. 2004a. http://www.drugs.com/pro/methotrexate.html

Food and Drug Administration Professional Drug Information. Mercaptopurine. Mylan Pharmaceuticals Inc. 2004b. http://www.drugs.com/pro/mercaptopurine.html

Gish RG, Porta C, Lazar L, *et al*. Phase III randomized controlled trial comparing the survival of patients with unresectable hepatocellular carcinoma treated with nolatrexed or doxorubicin. *J Clin Oncol* 2007, **25** (21), 3069–75.

Kemeny NE, Niedzwiecki D, Hollis DR, *et al*. Hepatic arterial infusion versus systemic therapy for hepatic metastases from colorectal cancer: a randomized trial of efficacy, quality of life, and molecular markers (CALGB 9481). *J Clin Oncol* 2006, **24**(9), 1395–403.

Kopec JA, Yothers G, Ganz PA, *et al*. Quality of life in operable colon cancer patients receiving oral compared with intravenous chemotherapy: results from National Surgical Adjuvant Breast and Bowel Project Trial C-06. *J Clin Oncol* 2007, **25**(4), 424–30.

Krug LM, Azzoli CG, Kris MG, *et al*. 10-propargyl-10-deazaaminopterin: an antifolate with activity in patients with previously treated non-small cell lung cancer. *Clin Cancer Res* 2003, **9**(6), 2072–8.

Lembersky BC, Wieand HS, Petrelli NJ, *et al*. Oral uracil and tegafur plus leucovorin compared with intravenous fluorouracil and leucovorin in stage II and III carcinoma of the colon: results from National Surgical Adjuvant Breast and Bowel Project Protocol C-06. *J Clin Oncol* 2006, **24**(13), 2059–64.

Lennon AM, Mulcahy MD, Hyland JMP, *et al*. Peritoneal involvement in stage II colon cancer. *Am J Clin Pathol* 2003, **119**, 108–13.

National Institute for Clinical Excellence. The use of gemcitabine for the treatment of pancreatic cancer. Technology Appraisal No. 25. NICE, London, 2001a. www.nice.org.uk

National Institute for Clinical Excellence. Fludarabine for the treatment of B-cell chronic lymphocytic leukaemia. Technology Appraisal No. 29. NICE, London, 2001b. www.nice.org.uk

National Institute for Clinical Excellence. Improving outcomes in breast cancer. Cancer service guidance. NICE, London, 2002. www.nice.org.uk

National Institute for Clinical Excellence. Capecitabine and tegafur uracil for metastatic colorectal cancer. Technology appraisal No. 61. NICE, London, 2003a. www.nice.org.uk

National Institute for Clinical Excellence. Guidance on the use of capecitabine and tegafur with uracil for metastatic colorectal cancer. Technology Appraisal No. 61. NICE, London, 2003b. www.nice.org.uk

National Institute for Clinical Excellence. Guidance on the use of capecitabine for the treatment of locally advanced or metastatic breast cancer. Technology Appraisal No. 62. NICE, London, 2003c. www.nice.org.uk

National Institute for Clinical Excellence. Improving outcomes in colorectal cancer. Cancer service guidance. NICE, London, 2004. www.nice.org.uk

National Institute for Clinical Excellence. Colorectal cancer (advanced) – irinotecan, oxaliplatin and raltitrexed (review). Technology Appraisal No. 93. NICE, London, 2005a. www.nice.org.uk

National Institute for Clinical Excellence. Lung cancer: diagnosis and treatment. Clinical guideline No. 24. NICE, London, 2005b. www.nice.org.uk

National Institute for Clinical Excellence. Capecitabine and oxaliplatin in the adjuvant treatment of stage III (Dukes' C) colon cancer. Technology Appraisal No. 100. NICE, London, 2006. www.nice.org.uk

National Institute for Clinical Excellence. Permetrexed for the treatment of non-small-cell lung cancer. Technology Appraisal No. 124. NICE, London, 2007a. www.nice.org.uk

National Institute for Clinical Excellence. Gemcitabine for the treatment of metastatic breast cancer. Technology Appraisal No. 116. NICE, London, 2007b. www.nice.org.uk

National Institute for Clinical Excellence. Permetrexed disodium for the treatment of malignant pleural mesothelioma. Technology Appraisal No. 135. NICE, London, 2008. www.nice.org.uk

Peterson VC, Baxter KJ, Love SB, Shepherd NA. Identification of objective pathological prognostic determinants and models of prognosis in Dukes' B colon cancer. *Gut* 2002, **51**, 65–9.

van Meerbeeck JP, Gaafar R, Manegold C, *et al*.: Randomized phase III study of cisplatin with or without raltitrexed in patients with malignant pleural mesothelioma: an intergroup study of the European Organisation for Research and Treatment of Cancer Lung Cancer Group and the National Cancer Institute of Canada. *J Clin Oncol* 2005, **23**(28), 6881–9.

8
Antitumour Antibiotics

Manuel M. Paz

8.1 Introduction

The term 'antitumour antibiotic' is employed to classify substances produced by micro-organisms that exert their anticancer activity by interacting with DNA. In this chapter, the definition has been expanded to also include clinically used compounds isolated from plants, as well as one antibiotic, trabectedin, produced by a marine invertebrate. Aside from their origin, all share in common to have DNA as the most plausible primary biological target. The manner in which each of the antibiotics presented here interacts with DNA differs considerably, and we will encounter alkylating agents, non-covalent binding agents and DNA-cleaving agents. The efforts dedicated during the last decades by the scientific community towards understanding the mechanisms underlying the cytotoxicity of antitumour antibiotics resulted in enormous advances, but in many cases the precise mode of action of the drugs remains unclear. We provide here a concise summary of its clinical applications and side effects, and a brief review of the present knowledge on their biomolecular mode of action.

8.2 Actinomycin

The actinomycins are a family of antibiotics structurally characterized by a phenoxazinone chromophore linked to two pentapeptide lactones (Figure 8.1). The first members of the actinomycin family were isolated from cultures of *Actinomyces antibioticus* in the early 1940s (Waksman, 1943). The clinically used actinomycin D, also called dactinomycin, was isolated in the mid 1950s and is produced currently from *Streptomyces parvulus* (Williams and Katz, 1954).

8.2.1 Clinical applications

Actinomycin D is indicated for the treatment of Wilm's tumour, rhabdomyosarcoma, Ewing's sarcoma, trophoblastic neoplasms and testicular carcinomas. The most common side effects

Anticancer Therapeutics   Edited by Sotiris Missailidis

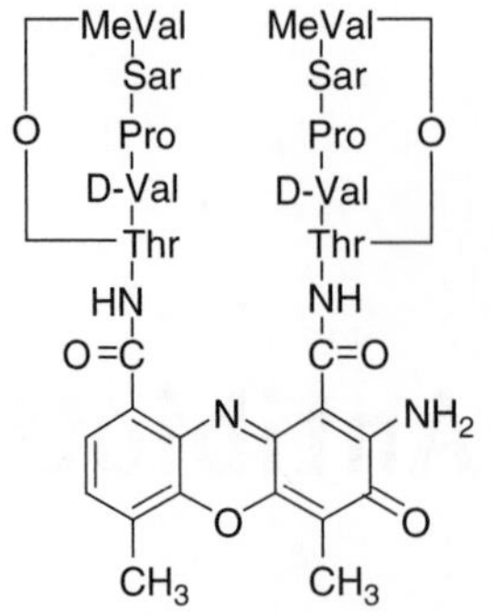

Figure 8.1   Structure of actinomycin D

are bone marrow toxicity, gastrointestinal ailments, skin reactions, alopecia and immuno-suppression (Mauger and Lackner, 2005). Recent reports showed its efficacy in gestational trophoblastic neoplasia (Covens *et al.*, 2006). A combination of dactinomycin with melphalan is efficient for the treatment of melanoma and soft tissue sarcoma (Brady *et al.*, 2006).

8.2.2 Mechanism of action

The cytotoxic effects of actinomycins have been attributed to inhibition of DNA-dependent RNA synthesis (Sobell, 1985). The commonly accepted theory proposes that this inhibition originates from the strong binding of actinomycin to specific structural DNA motifs near the transcriptional complex, blocking the elongation of the RNA chain (Philips and Crothers, 1986). Actinomycin binds DNA by intercalation of the chromophore, with the two peptide lactones residing in the minor groove. This binding is sequence selective, showing a strong preference for GpC base pairs. The structural origin of this selectivity is the formation of hydrogen bonds connecting the N2 amino group of guanine with carbonyl groups in the peptide side chains of the drug (Kamitori and Takusagawa, 1992). Other binding modes have been reported: non GpC, high affinity binding sites for dactinomycin have been detected in duplex DNA (Snyder *et al.*, 1989); binding to parallel double helical DNA (Li *et al.*, 2006a), binding to single strand DNA hairpins that do not contain GpC sites (Chou *et al.*, 2002). This last binding mode involves the disruption of two base pairs by the actinomycin chromophore, with two DNA bases looped out, interacting perpendicularly with the chromophore. Alternative modes of action involving inhibition of topoisomerases or blocking DNA binding sites of transcription factors have also been proposed (Mauger and Lackner, 2005).

8.3 Mitomycin C

The first antibiotics of the mitomycin family were discovered in Japan the mid 1950s. Mitomycin C was isolated in 1958 from *Streptomyces caespitosus* (Hata *et al.*, 1956) and has been used for the treatment of certain types of cancer since the early 1960s. Its structure is shown in Figure 8.2.

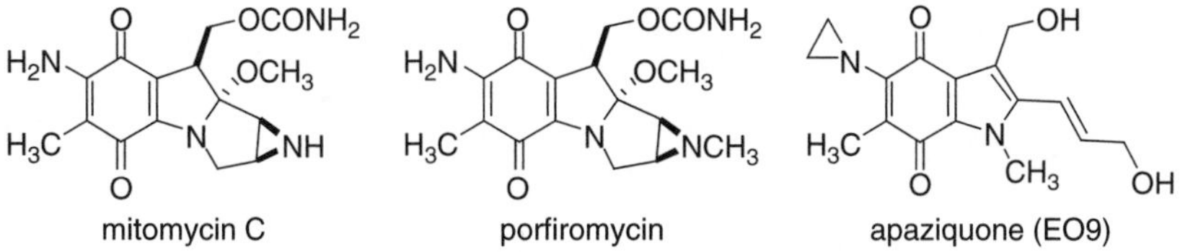

Figure 8.2 Structure of some mitomycin antibiotics

8.3.1 Clinical applications

Mitomycin C is not used as a single-agent therapy. In combination with other antitumour drugs it is effective for the treatment of a number of solid tumours, such as adenocarcinoma of the stomach or pancreas, superficial bladder cancer, epidermoid anal carcinomas and esophagus carcinomas. It is also employed in the palliative treatment of advanced cancers (Begleiter, 2000). Mitomycin C has seen an increased ophthalmological use as adjunctive therapy in various ocular surgeries (Abraham *et al.*, 2006), and also in otolaryngologic procedures (Tabaee *et al.*, 2007).

One of the causes for the limited clinical use of mitomycin C is the severity of its secondary effects. The most common and serious side effect is myelosuppression, a dose-related, cumulative effect that appears several weeks after administration. Also common and severe toxicities are thrombocytopenia, leucopenia and anaemia. Other frequent side effects of mitomycin C include anorexia, nausea, vomiting and diarrhoea (Begleiter, 2000).

Despite its significant toxicity, mitomycin C continues to be the subject of many clinical trials. For example, a phase II trial has shown the efficacy of a therapy of mitomycin C, irinotecan and cisplatin for relapsed small lung cancer (Fennell *et al.*, 2007). Another trial showed that capecitabine and mitomycin are efficient for the treatment of metastatic breast cancer resistant to paclitaxel and anthracyclines (Maisano *et al.*, 2007). Promising results were observed in a phase II trial of mitomycin C, doxorubicin and 5-fluorouracil for the treatment of advanced gastric cancer (Gnad-Vogt *et al.* 2005). A mitomycin C prodrug encapsulated in liposomes showed a better activity and reduced toxicity compared to MC in tumour models (Gabizon *et al.*, 2006).

8.3.2 Mechanism of action

Early experiments in the mid 1960s indicated that the cytotoxic effect of mitomycin C was associated to inhibition of DNA synthesis. This effect was attributed to the formation of a covalent crosslink joining together two complementary strands of DNA (Iyer and Szybalski, 1964). The same authors noted that mitomycin C itself was not reactive towards DNA, but was converted to alkylating species by activation with reducing enzymes. Extensive investigations performed in the next decades provided profuse data on the complex mode of action of mitomycin C. A number of recent reviews covered this research (Tomasz and Palom, 1997; Wolkenberg and Boger, 2002; Galm *et al.*, 2005).

Mitomycin C can be activated both by one-electron (e.g. NADPH:cytochrome P450 reductase) and two-electron (e.g. DT diaphorase) donating enzymes (Sartorelli *et al.*, 1994). The reduction of the quinone ring initiates a cascade of reactions that convert mitomycin C into a highly reactive intermediate (Figure 8.3).

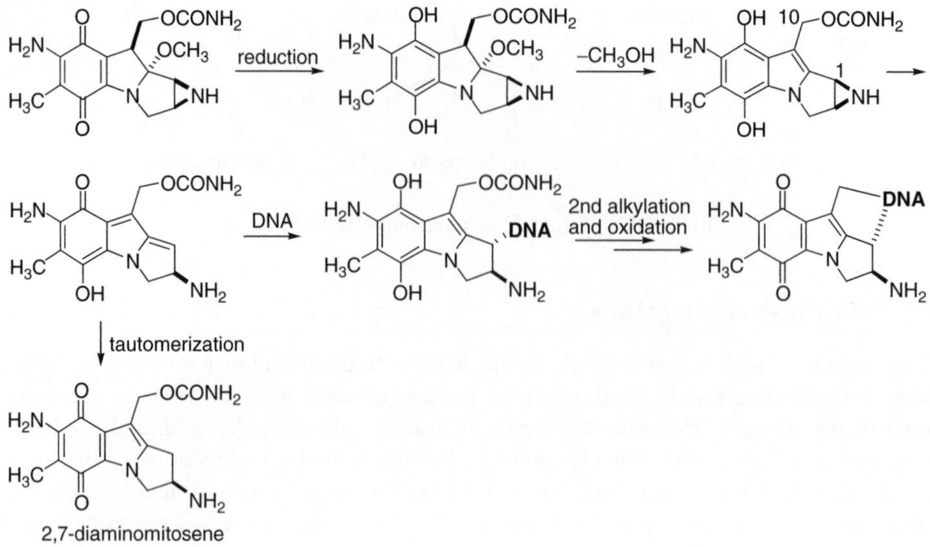

Figure 8.3 Reductive activation cascade of mitomycin C

This activation starts with the elimination of methanol to form an indole hydroquinone, where the positions 1 and 10 are highly reactive towards nucleophiles. This intermediate reacts with the 2-amino group of guanines in complementary strands of duplex DNA to form the key interstrand crosslink. The reaction proceeds in two steps: the initial monoadduct formation occurs at the 1 position of activated mitomycin, and the second arm reaction links the 10 position to a guanine in the opposite DNA strand.

The reaction is sequence-specific, linking the guanines in opposite strands at 5′-CpG sequences, while 5′-GpC sequences are not crosslinked. Two factors account for this selectivity: the increased reactivity of guanine at CpG steps, and the orientation of the initial monoadduct towards the 3′ end of the opposite strand. As opposed to other DNA crosslinks, the mitomycin C interstrand crosslink (Figure 8.4) does not induce significant structural perturbations of the original DNA double helix (Tomasz and Palom, 1997).

In addition to the interstrand crosslink adduct, the reaction of activated mitomycin C with DNA also forms intrastrand crosslinks (at GpG sites) and monoadducts. The major metabolite

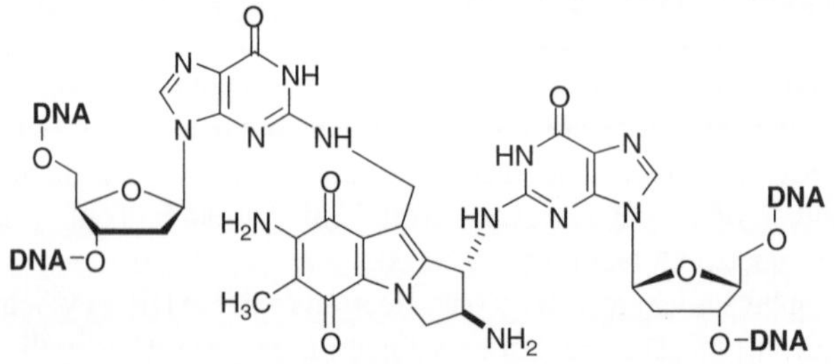

Figure 8.4 Structure of the mitomycin C-interstrand crosslink

formed *in vivo* by reductive activation of mitomycin C is 2,7-diaminomitosene (Chirrey *et al.*, 1995) and this compound, after reductive activation, can generate two additional monoadducts by reaction with DNA (Palom *et al.*, 2000). The crosslinks and several monoadducts were identified in tumour cells treated with mitomycin C (Bizanek *et al.*, 1993). How each of these covalent adducts contribute to the biological responses to mitomycin C treatment is still unclear, however indirect evidence indicate that the interstrand crosslink is the lesion responsible for the cytotoxic effects (Palom *et al.*, 2002). The guanine-N7 monoadduct of 2,7-diaminomitosene was shown to be an innocuous DNA lesion (Utzat *et al.*, 2005). The biological effects of mitomycin C could also be triggered in part by reactive oxygen species, generated by the redox cycling of the quinone in the presence of oxygen, as with other quinone-containing antitumour drugs (Doroshow, 1986).

The requirement of a reductive activation to form DNA adducts helps to explain the observed selective toxicity of mitomycin C towards solid tumours, characterized by low oxygen levels in some cells. These hypoxic cells have an increased capacity to reduce mitomycin C to form the active intermediates capable of generating lethal DNA adducts. For this reason, mitomycin C is frequently labelled as the 'prototype bioreductive drug' (Sartorelli *et al.*, 1994). The high potency of mitomycin C, combined with the limitations of its clinical use, prompted the development of hundreds of analogues, but only a few of them reached clinical trials (Begleiter, 2000).

8.3.3 Other mitomycins

Porfiromycin, a natural *N*-methyl derivative of mitomycin C, has been studied in a number of clinical trials as an adjuvant to radiotherapy (Begleiter, 2000). The most advanced, a phase III trial of porfiromycin for the treatment of squamous cell cancer of the head and neck, was halted in 2002.

Apaziquone (EO9), a synthetic mitomycin analogue that produces DNA crosslinks and strand breaks after activation by one or two-electron reducing enzymes, showed promising results in preclinical trials (Hargreaves *et al.*, 2000). Phase II trials have been conducted for advanced colorectal, breast, gastric, pancreatic and non small cell lung cancer. Phase III trials are in progress for the treatment of non-invasive bladder cancer (McKeown *et al.*, 2007).

8.4 Bleomycin

Bleomycins A2 and B2 are linear glycosylated peptide antibiotics that were isolated from *Streptomyces verticellus* in the mid-1960s (Umezawa *et al.*, 1966). Today, more than 200 members of the bleomycin family have been isolated (Hecht, 1995). The clinically used Blenoxane is a mixture that contains predominantly bleomycins A2 and B2. The structure of bleomycin A2 is shown in Figure 8.5.

8.4.1 Clinical applications

Blenoxane is used to treat a variety of tumours, mostly in combination with other anti-cancer agents. The most frequent indications are Hodgkin's and non-Hodgkin's lymphomas, squamous cell carcinomas, testicular carcinomas and malignant pleural effusions. Other therapeutic uses are renal carcinomas, soft tissue sarcomas, and the treatment of Kaposi's sarcoma related to acquired immune deficiency syndrome (AIDS) (Sikic *et al.*, 1985).

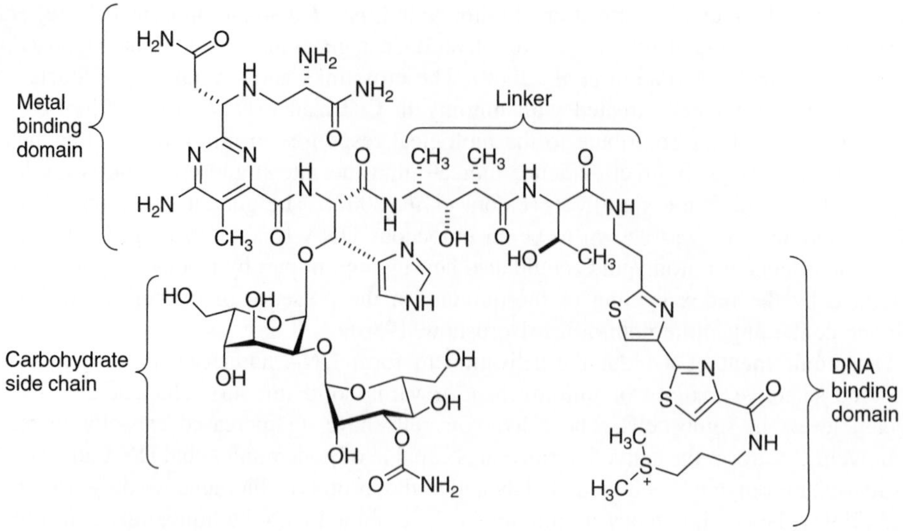

Figure 8.5 Structure of bleomycin A2, showing the four different structural regions

The most serious secondary effect of bleomycin administration is a cumulative pulmonary toxicity that may lead to lung fibrosis, and causes mortality in 1–2 % of bleomycin-treated patients. Other frequent adverse effects are nausea, vomiting, loss of appetite, alopecia, allergic reactions, and several forms of skin damage (Azambuja *et al.*, 2005).

The clinical applications of bleomycin are limited by its lack of permeability through cell membranes. To enter the target cell, bleomycin must first associate with a carrier plasma membrane protein, and this process is limited by the low number of cell surface proteins (Pron *et al.*, 1999). The use of electroporation techniques to increase the permeability of cell membranes, has shown to enhance the efficacy of bleomycin by more than 1000-fold. Clinical studies of electrochemotherapy with bleomycin showed a high response rate in the treatment of a number of tumours: cutaneous and subcutaneous nodes of melanoma, breast cancer, colon cancer, squamous cell carcinoma of skin and cervix and other, with minor side effects, and a good acceptability by the patients (Sersa *et al.*, 2007; Gothelf *et al.*, 2003).

8.4.2 Mechanism of action

There is ample evidence supporting that bleomycin-induced DNA damage is at the origin of its cytotoxic effects, in particular through the direct generation of double strand DNA breaks (Burger, 1998; Boger and Cai, 1999; Claussen and Long, 1999). However, DNA degradation might not be the exclusive source of the cytotoxic effects of bleomycin, as all major classes of RNAs can also be cleaved by bleomycin-induced oxidative damage (Abraham *et al.*, 2003).

The cleavage of nucleic acids by bleomycin (see Figure 8.6) is dependent on the presence of three cofactors: a metal ion (usually Fe^{2+}/Fe^{3+}), oxygen, and a one-electron reducing agent. The structure of bleomycin can be divided into four regions: a N-terminus domain constitutes the metal-chelating and oxygen activation domain; a C-terminal domain provides the DNA binding affinity, both by intercalation and minor groove-binding; a flexible

$$Fe(III) \cdot Bm + H_2O_2 \text{ or } \cdot O_2^- / e^-$$

$$\downarrow$$

$$Fe(II) + Bm \longrightarrow Fe(II) \cdot Bm \xrightarrow{O_2} \cdot O_2^- \text{-} Fe(III) \cdot Bm \xrightarrow{e^-} HOO \cdot Fe(III) \cdot Bm \xrightarrow{DNA} DNA \text{ cleavage}$$

"activated bleomycin"

Figure 8.6   Activation of bleomycin to DNA-cleaving species

central linker connecting the DNA-binding region and the metal coordinating domain, and a carbohydrate domain. The function of the flexible linker is presumably to give the metal-chelating domain of bleomycin the necessary flexibility to cleave both strands without dissociation from the DNA (Chen and Stubbe, 2004). The role of the glycoside region is still unclear, but it has been shown to be crucial for the DNA and RNA cleaving efficiency of the drug (Thomas *et al.*, 2002).

Although the administered form of bleomycin is metal-free, the absolute requirement of metal ions to induce nucleic acid damage *in vitro* suggests a metallobleomycin as the active form. The most studied metal *in vitro* is Fe(II), but complexes with Cu, Mn, Co, Ru and Ni also induce DNA cleavage, and could be relevant for the physiological activity of bleomycin (Petering *et al.*, 1990). While metal-free bleomycin is a flexible molecule with low DNA binding affinity, the metallated forms are structurally more constrained and present a higher affinity for DNA (Povirk *et al.*, 1981). The complex Fe(II)·bleomycin reacts with molecular oxygen to give a diamagnetic species, that is subsequently reduced by a one-electron reductant or by disproportionation to 'activated bleomycin', the last detectable intermediate before DNA degradation (Claussen and Long, 1999). This 'activated bleomycin' can also be generated *in vitro* by reaction of Fe(III)·bleomycin with peroxide or superoxide, opening the possibility of several routes of nucleic acid damage by bleomycin (Burger, 1998).

Several mechanisms have been proposed for the generation of the species that actually reacts with DNA, the most favoured hypothesis proposing a concerted reaction of activated bleomycin with DNA (Chen and Stubbe, 2004). Other postulated mechanisms involve a heterolytic or homolytic cleavage of the peroxide preceding DNA oxidation. The DNA oxidation occurs specifically at C4 of a deoxyribose residue. Activated bleomycin abstracts the C4' hydrogen to give two types of lesions, a C4' hydroxyl or hydroperoxyl substituted deoxyribose. The C4' hydroperoxyl lesion undergoes a series of reactions (Figure 8.7) that result in the cleavage of the DNA strand, with formation of the propenal derivative of the excised base, a 3'-phosphoglycolate terminus and a 5'-phosphate terminus. The C4'-hydroxy lesion does not induce DNA breaks directly, it releases a free base to form an abasic site, which can afterward be cleaved in alkaline conditions.

The DNA cleavage occurs more efficiently in duplex form, and presents some sequence-selectivity: the pyrimidines at GpC and GpT sites are the preferred positions, while nucleotides that are substituents at the 3' of cytosine are the less favoured sites. Single strand breaks are more frequent (80–90 %) than double strand breaks. The experimental evidence indicates that the latter are not the consequence of two independent single strand breaks, but arise from two consecutive single strand breaks that occur without dissociation of the DNA-bound drug. Several mechanisms have been postulated to explain the double-strand DNA cleavage by a single bleomycin molecule (Chen and Stubbe, 2004). The mechanisms of RNA cleavage by bleomycin are analogous to those of DNA, with the addition of RNA fragments that arise from abstraction of C-1 hydrogen (Hecht, 2005). The primary sites of RNA cleavage are the junctions of single and double-stranded regions (Hecht, 2000).

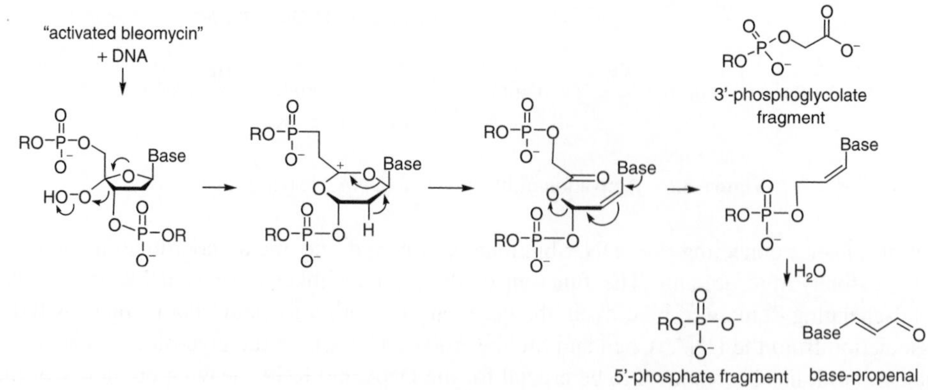

Figure 8.7 Mechanism of DNA cleavage by activated bleomycin

Recently, a mammalian tRNA was found to be an efficient target of bleomycin cleavage (Tao *et al.*, 2006). These observations suggest that the biological effects of bleomycin could alternatively be exerted by inhibition of protein synthesis due to the degradation of key transfer RNAs.

8.5 Anthracyclines

The anthracyclines are a class of antibiotics, structurally characterized by a tetracyclic chromophore that contains an anthraquinone motif (Figure 8.8). The members of this family present an ample range of biological activity: antibacterial, immunosuppressants, antiparasitics and antitumour. Two anticancer anthracyclines, doxorubicin and daunorubicin, have been used clinically in the last four decades, and are among the most efficient antitumour drugs.

8.5.1 Doxorubicin

Doxorubicin was isolated in Italy in the 1960s from cultures of *Streptomyces peucetius*, showing a wide spectrum of antitumour activity, in particular against solid tumours. Its large-scale

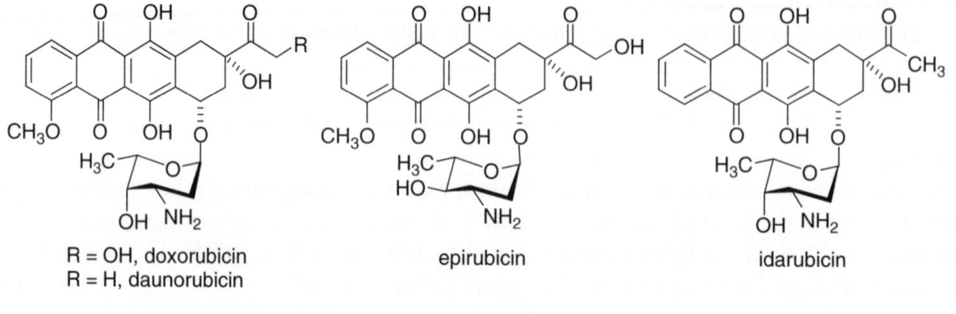

R = OH, doxorubicin
R = H, daunorubicin

epirubicin

idarubicin

Figure 8.8 Structure of the anthracycline drugs in clinical use

preparation involves a semisynthetic process starting from daunorubicin that was developed in the early 1970s (Arcamone, 2005).

Clinical applications

Doxorubicin is extensively used for the treatment of a wide spectrum of tumours, especially for solid tumours. Frequent indications include breast cancer, ovarian cancer, transitional cell bladder cancer, bronchogenic lung cancer, thyroid cancer, gastric cancer, soft tissue and osteogenic sarcomas, neuroblastoma, Wilms' tumour, malignant lymphoma (Hodgkin's and non-Hodgkin's), acute myeloblastic leukaemia, acute lymphoblastic leukaemia, and Kaposi's sarcoma.

The major limitations for the clinical use of doxorubicin are the development of resistance of tumour cells and cardiotoxicity. Dilative cardiomyopathy and congestive heart failure are cumulative side effects that appear usually after 1 year. To prevent these severe effects, a maximum cumulative dose of $600 \, mg/m^2$ has been established. Less critical secondary effects are nausea, vomiting, loss of appetite, diarrhoea, and hair and skin damage (Minotti et al., 2004).

Mechanism of action

Doxorubicin has classically been considered to mediate its biological effects as a topoisomerase IIα inhibitor (Gewirtz, 1999). Topoisomerases are a family of enzymes involved in the regulation of DNA supercoiling. DNA is usually present in the cells as a supercoiled double helix that must be unwound during replication or transcription, so that a single DNA strand serves as template for polymerases. Topoisomerase activity is crucial for this process, relieving the torsional stress, due to unwinding of the duplex, ahead of the replication fork. The mechanism of this change in DNA topology involves a transient single strand cleavage (in the case of topoisomerase I) or a double strand cleavage (for topoisomerase II). These breaks allow the passing of the intact single strand (topoisomerase I) or double strand (topoisomerase I) through the gap generated by the break. Posterior religation of the cleaved strand(s) restores the DNA chain integrity. The intermediate formed in this process, usually termed as the 'cleavable complex' or 'covalent binary complex', can be stabilized by DNA intercalating agents to form a stable ternary DNA–topoisomerase–drug complex, that results in inhibition of topoisomerase-catalysed DNA relaxation, diminishing DNA synthesis, and eventually causing cell death (Wang, 2002). The primary biological target of the anthracycline, camptothecin and podophyllotoxin families of antitumour antibiotics is considered to be the topoisomerase–DNA covalent complex. Doxorubicin binds strongly to duplex DNA by intercalation of the anthracycline chromophore selectively at CG sequences, with the aminosugar binding the DNA minor groove of the double helix (Braña et al., 2001). This binding affinity alone is not sufficient to explain the cytotoxic effects of the drug, and topoisomerase poisoning, with concomitant inhibition of the religation of the cleaved duplex appears to be the mechanism more closely associated with the biological activity. The formation of the cleavable complex induces the generation of double strand breaks (Tewey et al., 1984). The apoptotic response observed in doxorubicin-treated cells has been attributed to inefficient repair of these double strand breaks. The biological mode of action of anthracyclines remains still unclear, since additional modes of action have been postulated (Gewirtz, 1999). As with all antitumour quinones, radical species, generated by redox cycling of the anthraquinone, could induce DNA damage. Other mechanisms include membrane-related effects, inhibition of DNA or RNA synthesis and the formation of DNA adducts and crosslinks.

A completely different possibility emerged after the serendipitous discovery of the formation of a formaldehyde-mediated adduct between doxorubicin and DNA. The adduct is formed selectively at GCN sequences, with formaldehyde forming an aminal link between the 2-amino group of guanine and the amino group of daunosamine, with the sugar moiety bound to the minor groove and the chromophore intercalated between the CN sequence (Wang *et al.*, 1991). This monoadduct shows physical properties similar to covalent interstrand crosslinks, a consequence of the strong non-covalent interactions with the non modified DNA strand, and is often termed 'virtual crosslink' (Post *et al.*, 2005). The treatment of cancer cells with doxorubicin results in an increase of intracellular formaldehyde levels, supporting a potential biological role for these virtual crosslinks (Kato *et al.*, 2000). Doxoform, a doxorubicin dimer that incorporates three formaldehyde molecules in its structure showed an increased cytotoxicity (1 to 4-fold) towards tumour cells, and is efficient towards doxorubicin-resistant cell lines (Fenick *et al.*, 1997). A series of investigations showed that the co-administration of doxorubicin and formaldehyde-releasing prodrugs results in a strong synergistic effect (Cutts *et al.*, 2005). Interestingly, the use of the same prodrugs that activate doxorubicin towards cancer cells protects against cardiotoxicity (Rephaeli *et al.*, 2007). Two recent reports, using either doxorubicin in combination with a prodrug of formaldehyde (Swift *et al.*, 2006), or a formaldehyde-containing doxorubicin derivative (Kalet *et al.*, 2007) indicate that the formation of doxorubicin-formaldehyde-DNA adducts induces apoptosis, but by a mechanism different than topoisomerase II inhibition. These results open new strategies to develop more efficient doxorubicin-based anticancer therapies.

8.5.2 Liposomal doxorubicin

The liposomal encapsulation is a novel method of targeting antitumour drugs to tumours, improving their efficacy and limiting their secondary effects. Liposomal anthracyclines present a series of advantages over the free drugs: extended half life, a slower conversion to inactive metabolites and a preferential accumulation in tissues with increased microvascular permeability, as is often the case in tumours (Haag and Kratz, 2006; Abraham *et al.*, 2005).

Pegylated liposomal doxorubicin is one of the most active agents for the treatment of AIDS-related Kaposi's sarcoma and is efficient in the management of recurrent ovarian cancer and advanced breast cancer (Gabizon, 2001). The liposomal formulation of doxorubicin significantly reduces toxicity, specifically the cardiotoxic effects of doxorubicin. Other common adverse effects as nausea, vomiting, alopecia, and stomatitis, are less severe than with doxorubicin. A frequent side effect of liposomal doxorubicin is a hand and foot syndrome, but this can be managed by reducing the dose and increasing the cycle duration (Vail *et al.*, 2004).

8.5.3 Daunorubicin

Daunorubicin was isolated together with doxorubicin from *Streptomyces* cultures (Arcamone *et al.*, 1969). Structurally, the only difference from doxorubicin is the absence of the hydroxyl group in the chromophore side chain. Daunorubicin was the first anthracycline antibiotic used in chemotherapy. The spectrum of antitumour activity is narrower than doxorubicin. The main indications are acute myeloid leukaemia and acute lymphocytic leukaemia, and it is also used to treat neuroblastoma and chronic myelogenous leukaemia. Its side effects are similar to those of doxorubicin, with the addition of a significant haematological toxicity (Arcamone, 2005). The proposed mechanisms to explain the biological activity of daunorubicin are analogous to those of doxorubicin (Minotti *et al.*, 2004). A liposomal formulation of daunorubicin has also been developed, showing more efficacy than free daunorubicin against multidrug resistance cell lines, lower toxicity profile, and the capability to overcome the blood brain barrier (Krown *et al.*, 2004). Liposomal daunorubicin is used mainly to treat AIDS-related Kaposi's sarcoma, and is also employed to treat specific types of leukaemia and non-Hodgkin's lymphoma (Fassas and Anagnostopoulos, 2004).

8.5.4 Epirubicin

Epirubicin is a semisynthetic anthracycline that differs from doxorubicin in the stereochemistry at C4′ of the sugar moiety. Like doxorubicin, it presents a broad range of antitumour activity, and has been used since 1982 for the treatment of many types of cancer, including breast cancer, lung cancer, ovary cancer, stomach cancer, and lymphoma. Side effects are similar to doxorubicin, with a slightly lower cardiotoxicity and myelosupression, allowing the use of dose intensive regimens with high response rates (Arcamone, 2005).

8.5.5 Idarubicin

Idarubicin is a semisynthetic analogue of doxorubicin lacking the aromatic methoxy group, which presents a similar mechanism of action. Its main indication is to treat acute myelocytic leukaemia, and is also active towards lymphomas and breast cancer. The side effects are similar to those of doxorubicin, but cardiotoxicity is less frequently reported with idarubicin than with other anthracyclines. Unlike the other currently available anthracyclines, it has a significant oral bioavailability, with side effects similar to intravenous administration. Oral idarubicin is marketed in some countries for the treatment of acute myeloblastic leukaemia in elder patients, but its impact on patients' quality of life is unclear (Crivellari *et al.*, 2004).

8.6 Trabectedin (Ecteinascidin, ET-743)

Trabectedin (Figure 8.9) is a tetrahydroisoquinoline alkaloid isolated in the early 1990s from the marine tunicate *Ecteinascidia turbinata*.

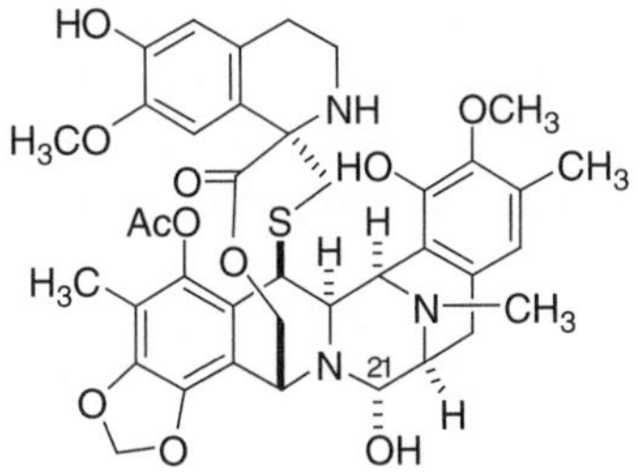

Figure 8.9 Structure of trabectedin

8.6.1 Clinical applications

Initial studies on trabectedin performed in the late 1990s showed a remarkable activity both *in vivo* and *in vitro*, and clinical trials soon started. Trabectedin is especially efficient towards soft tissue sarcomas (Grosso *et al.*, 2007), and was approved in Europe in 2007 for this indication. It has also been shown to be efficient as a single agent in advanced breast cancer (Zelek *et al.*, 2006) and advanced ovarian carcinoma (Sessa *et al.*, 2005), but not for colorectal cancers (Paz-Ares *et al.*, 2007). Clinical trials are ongoing for advanced prostate cancer, relapsed solid tumours, and sarcomas.

8.6.2 Mechanism of action

Trabectedin binds in the minor groove of DNA and alkylates at the N2 position of guanine residues through the 21-carbinolamine group. The formation of trabectedin adducts shows sequence selectivity, with the central guanine in 5′-PuGC and 5′-PyGG sequences being the preferred site for alkylation. This selectivity has been attributed to the reversibility of alkylation reaction, allowing the formation of the most stable end product by migration of adducts formed at less favoured sequences. Adducts at the preferred sequences are stabilized by a more favourable array of intermolecular bonds (Seaman and Hurley., 1998). The formation of the adduct imparts significant perturbations on the original double helix, widening the minor groove, and resulting in a DNA bend towards the major groove with an extrahelical protruding portion of the bound drug. The unique features of this unusual DNA distortion are likely to contribute to the biological effects of trabectedin (Zewail-Foote *et al.*, 2001). The biological consequences of the formation of these adducts are not fully understood, and many studies suggest that trabectedin is operating through a unique mode of action, different from the other DNA-interacting drugs (Erba *et al.*, 2001). The first proposed mode of action was topoisomerase I poisoning, but this effect required concentrations of the drug 100 times larger than those required for cytotoxicity (Takebayashi *et al.*, 1999). A number of subsequent studies pointed to a novel, transcription-targeted mechanism of action (Henríquez *et al.*, 2005). Trabectedin appears to target some genes (*p21*, *MDR1*) in a promoter-specific manner (Minuzzo *et al.*, 2000). Other reports indicated a mode of action involving interactions with the transcription-coupled nucleotide excision repair (NER) to generate lethal DNA strand breaks (Takebayashi *et al.*, 2001). Trabectedin showed decreased activity in NER-deficient cell lines (Damia *et al.*, 2001), an effect opposed to other DNA alkylators. This repair-dependent toxicity has been explained proposing the formation of a ternary complex with a protein during the NER processing of trabectedin-DNA adducts. This trapped intermediate has been postulated to be analogous to the cleavable complex generated by topoisomerase poisons (Zewail-Foote *et al.*, 2001). The effect of trabectedin adducts in the generation of double strand breaks is unclear. Initial studies in cells exposed to IC_{50} concentrations of trabectedin failed to detect single or double-strand breaks (Erba *et al.*, 2001). However, recent reports did show that the treatment of cancer cells with pharmacologically relevant concentrations of trabectedin induced double strand DNA breaks as key intermediates in the biological processing of trabectedin-DNA adducts (Soares *et al.*, 2007).

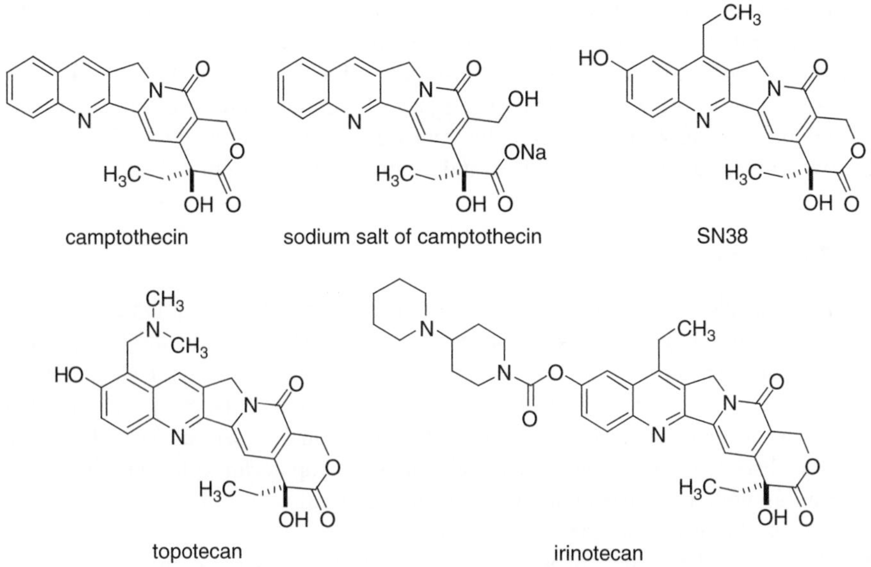

Figure 8.10 Structure of camptothecin and its analogues

8.7 Camptothecins

Camptothecin is a pentacyclic antitumour antibiotic (Figure 8.10) isolated in 1966 from alkaloid extracts of the tree *Camptotheca acuminata* during a screening programme at the US National Cancer Institute, showing remarkable activity in some murine leukaemia models (Wall, 1966). A number of recent reviews analyse in detail the mode of action (Rahier, 2005; Thomas *et al.*, 2004; Rothenberg, 1997) and clinical applications (Adams and Burke, 2005; Sriram *et al.*, 2005; Pizzolato *et al.*, 2003) of camptothecin and its analogues. A brief summary is presented here.

Camptothecin presents very low water solubility. The sodium salt of camptothecin was employed in clinical trials to overcome this problem, but opening the lactone ring compromised its activity, being 10 times less active. Clinical trials for melanomas and gastrointestinal cancer started in the late 1960s, but poor antitumour activity was observed, together with serious side effects (myelosuppression and haemorrhagic cystitis), consequently the trials were discontinued (Moertel *et al.*, 1972).

8.7.1 Mechanism of action

Topoisomerase I was identified as the probable biological target of camptothecin in the late 1980's, providing a model to predict the activity of camptothecin analogues *in vitro* (Jaxel *et al.*, 1989). The mechanism of inhibition of topoisomerases by antitumour drugs has been described previously in the anthracyclines section of this chapter. Camptothecin binds and stabilizes the covalent binary complex DNA–topoisomerase I, inhibiting the religation of the cleaved strand, and eventually DNA synthesis and cell viability (Rahier *et al.*, 2005). In addition to the inhibition of DNA relaxation, other biochemical effects of camptothecin have been observed, most of them related to topoisomerase I activity (Thomas *et al.*, 2004). In the early 1990's, a series of water-soluble analogues of camptothecin were prepared, showing *in vitro* inhibition of topoisomerase I, cytotoxicity and antitumour activity (Kingsbury *et al.*, 1991). Two of these analogues, topotecan and irinotecan, reached clinical use.

8.7.2 Topotecan

Topotecan is a semisynthetic camptothecin analogue, with a dimethylaminomethyl substituent in the A ring, that confers the desired water solubility. Topotecan was the first analogue of camptothecin to get approval for clinical use. It is mainly used to treat advanced ovarian cancer, small cell lung cancer and acute myeloid leukaemia. Topotecan may cause severe blood disorders (myelosuppression, neutropenia). Common, less serious, side effects are gastrointestinal disorders and alopecia. One advantage of topotecan is the possibility of oral administration, with a number of studies showing no significant differences when compared to intravenous administration (Pizzolato and Saltz, 2003).

The X-ray crystal structure of topotecan bound to the covalent complex topoisomerase I–DNA duplex has been resolved, showing that topotecan mimics a DNA base pair, binding the complex by DNA intercalation at the cleaved site, with only one direct hydrogen bond to the enzyme (Staker *et al.*, 2002).

The most accepted theory in the biological consequences of topoisomerase poisoning proposes that topoisomerase poisoning of the covalent complex generates cytotoxic DNA lesions after colliding with an advancing replication fork (Pommier, 2006). However, recent studies indicate an alternative mechanism: the binding of topotecan to the covalent complex results in the selective inhibition of removal of positive supercoils by topoisomerase I. The accumulation of positive supercoils was also observed in yeast cultures, and presented a direct correlation with topotecan toxicity (Koster *et al.*, 2007).

8.7.3 Irinotecan

Irinotecan is a water soluble camptothecin analogue that incorporates a bispiperidine substituent in the A ring. Irinotecan does not stabilize the covalent complex, but acts as a prodrug of the efficient topoisomerase poison SN38, after carboxylesterase mediated cleavage of the bispiperidinyl side chain (Thomas *et al.*, 2004).

Irinotecan was approved in Japan in 1994 for lung, cervix and ovary cancers, and later in Europe and the U.S. for the treatment of colorectal cancers (Sriram *et al.*, 2005). The most common side effects of irinotecan are diarrhoea (early and late) and neutropenia (Rahier, 2005).

A number of other camptothecin analogues are under clinical evaluation, including liposome-encapsulated versions (Rahier, 2005).

8.8 Podophyllotoxins

Podophyllotoxin (Figure 8.11) is a cytotoxic aryltetralinlactone lignan first isolated from the plant *Podophyllum peltatum* in 1880, and other members of the family were isolated later.

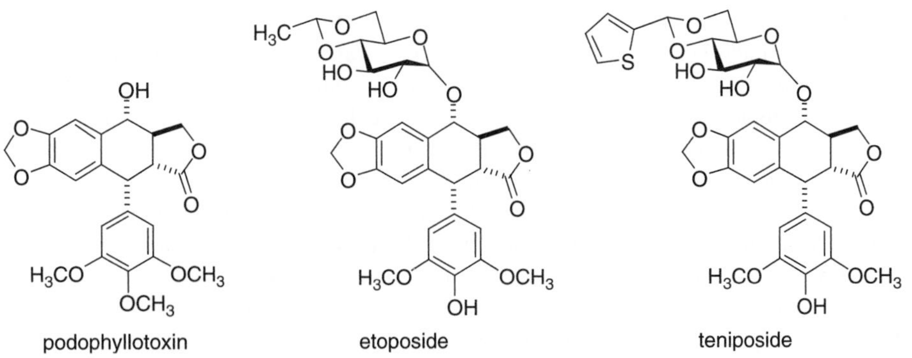

Figure 8.11 Structure of podophyllotoxin and its analogues

Despite their strong antitumour activity, none of them found clinical use due to the severity of their side effects. Subsequent efforts in the search of useful podophyllotoxin analogues led to the discovery of etoposide and teniposide, two clinically relevant antitumour agents. Interestingly, these analogues present a biological mode of action divergent from the original drug. Podophyllotoxin exerts its biological action by binding tubulin, the protein subunit that forms microtubules. This binding inhibits the assembly of the mitotic spindle and induces cell cycle arrest at mitosis (Lee and Xiao, 2005). On the other hand, the molecular target of etoposide and teniposide is the topoisomerase II–DNA covalent complex (Baldwin and Osheroff, 2005).

8.8.1 Etoposide

Etoposide is a podophyllotoxin analogue that contains a glycoside substituent in the C ring. Etoposide was first synthesized in the early 1970's (Keller-Juslén *et al.*, 1971), and approved by the FDA in 1983. It can be administrated intravenously or orally.

Clinical applications

The main indications for etoposide are testicular and lung cancers, and is also used to treat non-Hodgkin's lymphomas, mycosis fungoides, Hodgkin's disease, acute myelogenous leukaemia, acute lymphocytic leukaemia, chronic myelogenous leukaemia, Wilms' tumour, neuroblastoma, AIDS-related Kaposi's sarcoma, gestational trophoblastic tumours, ovarian germ-cell tumours, hepatoma, Ewing's sarcoma, rhabdomyosarcoma, brain tumours, and refractory advanced breast cancer. The most common side effects are gastrointestinal disorders, changes in taste and hair damage. Myelosuppression is dose related and dose limiting. Bone marrow recovery is usually complete after three weeks, and no cumulative toxicity has been reported.

Mechanism of action

The biological responses induced by etoposide treatment have been reviewed (Montecuco and Biamonti, 2007). Its mode of action is similar to the one described previously in this chapter for doxorubicin. The primary cellular target of etoposide is the covalent complex of topoisomerase II and DNA. Poisoning of the binary covalent complex by etoposide results in the formation of true double strand breaks, and the accumulation of DNA lesions activates a series of cellular events that eventually result in cell death (Montecucco and Biamonti, 2007). A number of studies indicate that, as opposed to other topoisomerase inhibitors, the binding of etoposide to the covalent complex is governed by the interactions of etoposide with the enzyme, not by interactions with DNA (Burden *et al.*, 1996). Etoposide binds very weakly to DNA in the absence of topoisomerase II, but binds topoisomerases in the absence of DNA (Wilstermann *et al.*, 2007). A decreased topoisomerase-etoposide binding affinity has been suggested as a possible mechanism of drug resistance (Kingma *et al.*, 1999).

Etopophos, a phenolic phosphate ester of etoposide, is a water-soluble prodrug developed to increase the low water solubility of etoposide. It can be converted efficiently *in vivo* by phosphatases to the active drug. Etopophos is more active and less toxic than etoposide in tumour models and shows better bioavailability. From the results of clinical trials it was concluded that it is preferable to etoposide for routine clinical use (Greco and Hainsworth, 1996). Etopophos was approved by the FDA in 1996 and is indicated for testicular tumours and small cell lung cancer.

8.8.2 Teniposide

Teniposide is a podophyllotoxin derivative containing a thioethyl-substituted glycoside side chain. It is used in the treatment of childhood acute lymphocytic leukaemia, non-Hodgkin's leukaemia, refractory neuroblastoma, and other types of cancer. The most frequent side effects are myelosuppression, gastrointestinal disorders, hypotension, and allergic reactions.

The mode of action of teniposide is similar to that of etoposide. It induces cell apoptosis, mediated by topoisomerase II inhibition (Li *et al.*, 2006b; Roy *et al.*, 1992).

The success of etoposide and teniposide stimulated the search for new podophyllotoxin analogues with improved therapeutic properties. Several potential drug candidates such as GL-331, TOP 53, NK611, and azatoxin are in various stages of development (You, 2005).

References

Abraham AT; Lin JJ; Newton DL; Rybak S; Hecht SM. RNA cleavage and inhibition of protein synthesis by bleomycin. *Chem Biol* 2003, **10**, 45–52.

Abraham LM, Selva D, Casson R, Leibovitch I. Mitomycin: clinical applications in ophthalmic practice. *Drugs* 2006, **66**, 321–40.

Abraham SA, Waterhouse DN, Mayer LD, Cullis PR, Madden TD, Bally MB. The liposomal formulation of doxorubicin. *Methods Enzymol* 2005, **391**, 71–97.

Adams VR, Burke TG (eds). *Camptothecins in Cancer Therapy*. Humana Press, Totowa, NJ, 2005.

Arcamone F, Franceschi G, Penco S, Selva A. Adriamycin (14-hydroxydaunomycin), a novel antitumor antibiotic. *Tetrahedron Lett* 1969, **10**, 1007–10.

Arcamone FA. Anthracyclines. In: *Anticancer Agents from Natural Products*, edited by GM Cragg, DGI Kingston and DJ Newman. Taylor & Francis, Boca Raton, FL, 2005, p 299–320.

Azambuja E, Fleck JF, Batista RG, Menna Barreto SS. Bleomycin lung toxicity: who are the patients with increased risk? *Pulm Pharmacol Ther* 2005, **18**, 363–6.

Baldwin EL, Osheroff N. Etoposide, topoisomerase II and cancer *Curr Med Chem Anticancer Agents* 2005, **5**, 363–72.

Begleiter A. Clinical applications of quinone containing alkylating agents. *Front Biosci* 2000, **5**, 153–71.

Bizanek R, Chowdary D, Arai H, Kasai M, Hughes CS, Sartorelli AC, Rockwell S, Tomasz M. Adducts of mitomycin C and DNA in EMT6 mouse mammary tumour cells: effects of hypoxia and dicumarol on adduct patterns. *Cancer Res* 1993, **53**, 5127–34.

Boger DL; Cai H. Bleomycin: synthetic and mechanistic studies. *Angew Chem Int Ed* 1999, **38**, 448–76.

Brady MS, Brown K, Patel A, Fisher C, Marx W. A phase II trial of isolated limb infusion with melphalan and dactinomycin for regional melanoma and soft tissue sarcoma of the extremity. *Ann Surg Oncol* 2006, **8**, 1123–9.

Braña MF, Cacho M, Gradillas A, de Pascual-Teresa B, Ramos A. Intercalators as anticancer drugs. *Curr Pharm Des* 2001, **7**, 1745–80.

Burden DA, Kingma PS, Froelich-Ammon SJ, Bjornsti MA, Patchan MW, Thompson RB, Osheroff N. Topoisomerase II· etoposide interactions direct the formation of drug-induced enzyme-DNA cleavage complexes *J Biol Chem* 1996, **271**, 29238–44.

Burger RM. Cleavage of nucleic acids by bleomycin. *Chem Rev* 1998, **98**, 1153–70.

Chen J, Stubbe J. Bleomycins: new methods will allow reinvestigation of old issues. *Curr Opin Chem Biol* 2004, **8** 175–81.

Chen J, Stubbe J. Bleomycins: towards better therapeutics. *Nat Rev Cancer* 2005, **5**, 102–12.

Chirrey L, Cummings J, Halbert GW, Smyth JF. Conversion of mitomycin C to 2,7-diaminomitosene and 10-decarbamoyl 2,7-diaminomitosene in tumour tissue *in vivo*. *Cancer Chemother Pharmacol* 1995, **35**, 318–22.

Chou SH, Chin KH, Wang AH. Unusual DNA duplex and hairpin motifs *Nucleic Acids Res* 2003, **31**, 2461–74.

Claussen CA, Long EC. Nucleic acid recognition by metal complexes of bleomycin. *Chem Rev* 1999, **99**, 2797–816.

Covens A, Filiaci VL, Burger RA, Osborne R, Chen MD. Phase II trial of pulse dactinomycin as salvage therapy for failed low-risk gestational trophoblastic neoplasia: a Gynecologic Oncology Group study. *Cancer* 2006, **107**, 1280–6.

Crivellari D, Lombardi D, Spazzapan S, Veronesi A, Toffoli G. New oral drugs in older patients: a review of idarubicin in elderly patients. *Crit Rev Oncol Hematol* 2004, **49**, 153–63.

Cutts SM, Nudelman A, Rephaeli A, Phillips DR. The power and potential of doxorubicin-DNA adducts. *IUBMB Life* 2005, **57**, 73–81.

Damia G, Silvestri S, Carrassa L, Filiberti L, Faircloth GT, Liberi G, Foiani M, D'Incalci M. Unique pattern of ET-743 activity in different cellular systems with defined deficiencies in DNA-repair pathways, *Int. J. Cancer* 2001, **92**, 583–8.

Doroshow JH. Role of hydrogen peroxide and hydroxyl radical formation in the killing of Ehrlich tumour cells by anticancer quinones. *Proc Natl Acad Sci U S A* 1986, **83**, 4514–18.

Erba E, Bergamaschi D, Bassano L, Damia G, Ronzoni S, Faircloth GT, D'Incalci M. Ecteinascidin-743 (ET-743), a natural marine compound, with a unique mechanism of action, *Eur J Cancer* 2001, **37**, 97–105.

Fassas A, Anagnostopoulos A. The use of liposomal daunorubicin (DaunoXome) in acute myeloid leukemia. *Leuk Lymphoma* 2005, **46**, 795–802.

Fenick DJ, Taatjes DJ, Koch TH. Doxoform and daunoform: Anthracycline-formaldehyde conjugates toxic to resistant tumour cells. *J Med Chem* 1997, **40**, 2452–61.

Fennell DA, Steele JP, Shamash J, Slater SE, Sheaff MT, Wells P, Rudd RM, Stebbing J. Phase II trial of irinotecan, cisplatin and mitomycin for relapsed small cell lung cancer. *Int J Cancer* 2007, **121**, 2575–7.

Gabizon AA, Tzemach D, Horowitz AT, Shmeeda H, Yeh J, Zalipsky S. Reduced toxicity and superior therapeutic activity of a mitomycin C lipid-based prodrug incorporated in pegylated liposomes. *Clin. Cancer Res* 2006, **12**, 1913–20.

Gabizon AA. Pegylated liposomal doxorubicin: metamorphosis of an old drug into a new form of chemotherapy. *Cancer Invest* 2001, **19**, 424–36.

Galm U, Hager MH, Van Lanen SG, Ju J, Thorson JS, Shen B. Antitumour antibiotics: bleomycin, enediynes, and mitomycin, *Chem Rev* 2005, **105**, 739–58.

Gewirtz DA. A critical evaluation of the mechanisms of action proposed for the antitumour effects of the anthracycline antibiotics Adriamycin and daunorubicin. *Biochem Pharmacol* 1999, **57**, 727–41.

Gnad-Vogt SU, Hofheinz RD, Saussele S, *et al.* Pegylated liposomal doxorubicin and mitomycin C in combination with infusional 5-fluorouracil and sodium folinic acid in the treatment of advanced gastric cancer: results of a phase II trial. *Anticancer Drugs* 2005, **16**, 435–40.

Gothelf A, Mir LM, Gehl J. Electrochemotherapy: results of cancer treatment using enhanced delivery of bleomycin by electroporation. *Cancer Treat Rev* 2003, **5**, 371–87.

Greco FA, Hainsworth JD. Clinical studies with etoposide phosphate. *Semin Oncol* 1996, **23**, 45–50.

Grosso F, Jones RL, Demetri GD. Efficacy of trabectedin (ecteinascidin-743) in advanced pretreated myxoid liposarcomas: a retrospective study. *Lancet Oncol* 2007, **8**, 595–602.

Haag R, Kratz F. Polymer therapeutics: concepts and applications *Angew Chem Int Ed Engl* 2006, **458**, 1198–215.

Hargreaves RH, Hartley JA, Butler J. Mechanisms of action of quinone-containing alkylating agents: DNA alkylation by aziridinylquinones. *Front Biosci* 2000, **5**, 172–80.

Hata T, Sano Y, Sugawara R, Matsumae A, Kanomori K, Shima T, Hoshi T. Mitomycin, a New Antibiotic from *Streptomyces*. I. *J Antibiot, Ser A* 1956, **9**, 141–6.

Hecht SM. Bleomycin group antitumor agents. In: *Cancer Chemotherapeutic Agents*, edited by WO Foye. American Chemical Society: Washington, DC, 1995, p 369.

Hecht SM. Bleomycin: new perspectives on the mechanism of action. *J Nat Prod* 2000, **63**, 158–68.

Hecht SM. Bleomycin group antitumor agents. In: *Anticancer Agents from Natural Products*. edited by GM Cragg, DGI Kingston and DJ Newman. Taylor & Francis Group, Boca Raton, FL, 2005, p 357–81.

Henríquez R, Faircloth G, Cuevas C. Ecteinascidin 743 (ET-743; Yondelis^{TM}), Aplidin and Kahalide F. In: *Anticancer Agents from Natural Products*, edited by GM Cragg, DGI Kingston and DJ Newman. Taylor & Francis, Boca Raton, FL, 2005, pp 215–40.

Iyer VN, Szybalski W. Mitomycins and porfiromycin: chemical mechanism of activation and cross-linking of DNA. *Science* 1964, **145**, 55–8.

Jaxel C, Kohn KW, Wani MC, Wall ME, Pommier Y. Structure-activity study of the actions of camptothecin derivatives on mammalian topoisomerase I: evidence for a specific receptor site and a relation to antitumour activity *Cancer Res* 1989, **49**, 1465–9.

Kalet BT, McBryde MB, Espinosa JM, Koch TH. Doxazolidine induction of apoptosis by a topoisomerase ii independent mechanism *J Med Chem* 2007, **50**, 4493–500.

Kamitori S, Takusagawa F. Crystal structure of the 2:1 complex between d(GAAGCTTC) and the anticancer drug actinomycin D. *J Mol Biol* 1992, **225**, 445–56.

Kato S, Burke PJ, Fenick DJ, Taatjes DJ, Bierbaum VM. Mass spectrometric measurement of formaldehyde generated in breast cancer cells upon treatment with anthracycline antitumour drugs. *Chem Res Toxicol* 2000, **13**, 509–16.

Keller-Juslén C, Kuhn M, Stähelin H, von Wartburg A. Synthesis and antimitotic activity of glycosidic lignan derivatives related to podophyllotoxin. *J Med Chem* 1971, **14**, 936–40.

Kingma PS, Burden DA, Osheroff N. Binding of etoposide to topoisomerase II in the absence of DNA: decreased affinity as a mechanism of drug resistance. *Biochemistry* 1999, **38**, 3457–61.

Kingsbury WD, Boehm JC, Jakas DR, *et al.* Synthesis of water-soluble (aminoalkyl)camptothecin analogues: inhibition of topoisomerase I and antitumour activity. *J Med Chem* 1991, **34**, 98–107.

Koster DA, Palle K, Bot ES, Bjornsti MA, Dekker NH. Antitumour drugs impede DNA uncoiling by topoisomerase I. *Nature* 2007, **448**, 213–17.

Krown SE, Northfelt DW, Osoba D, Stewart JS. Use of liposomal anthracyclines in Kaposi's sarcoma. *Semin Oncol* 2004, **31**, 36–52.

Lee K-H, Xiao Z. Podophyllotoxin and analogs. In: *Anticancer Agents from Natural* Products edited by GM Cragg, DGI Kingston and DJ Newman. Taylor & Francis, Boca Raton, FL, 2005, pp 5–21.

Leroy D, Kajava AV, Frei C, Gasser SM. Analysis of etoposide binding to subdomains of human DNA topoisomerase II alpha in the absence of DNA. *Biochemistry* 2001, **40**, 1624–34.

Li H, Peng X, Leonard P, Seela F. Binding of actinomycin C_1 (D) and actinomin to base-modified oligonucleotide duplexes with parallel and antiparallel chain orientation *Bioorg Med Chem* 2006a, **14**, 4089–100.

Li J, Chen W, Zhang P, Li N. Topoisomerase II trapping agent teniposide induces apoptosis and G2/M or S phase arrest of oral squamous cell carcinoma. *World J Surg Oncol* 2006b, **4**, 41.

Maisano R, Caristi N, Mare M, Raffaele M, Iorfida M, Mafodda A, Zavettieri M, Nardi M. Mitomycin C plus capecitabine (mixe) in anthracycline- and taxane-pretreated metastatic breast cancer. A multicenter phase II study. *Anticancer Res* 2007, **27**, 2871–5.

Mauger AB, Lackner H. The actinomycins. In: *Anticancer Agents from Natural Products*, edited by GM Cragg, DGI Kingston and DJ Newman. Taylor & Francis, Boca Raton, FL, 2005, pp 281–97.

McKeown SR, Cowen RL, Williams KJ. Bioreductive drugs: from concept to clinic. *Clin Oncol (R Coll Radiol)* 2007, **6**, 427–42.

Minotti G, Menna P, Salvatorelli E, Cairo G, Gianni L. Anthracyclines: molecular advances and pharmacologic developments in antitumour activity and cardiotoxicity. *Pharmacol Rev* 2004, **56**, 185–229.

Minuzzo M, Marchini S, Broggini M, Faircloth G, D'Incalci M, Mantovani R. Interference of transcriptional activation by the antineoplastic drug ET-743. *Proc Natl Acad Sci U SA* 2000, **97**, 6780–84.

Moertel CG, Schutt AJ, Reitemeier RJ, Hahn RG. Phase II study of camptothecin (NSC-100880) in the treatment of advanced gastrointestinal cancer. *Cancer Chemother Rep* 1972, **56**, 95–101.

Montecucco A, Biamonti G. Cellular response to etoposide treatment. *Cancer Lett* 2007, **252**, 9–18.

Palom Y, Belcourt MF, Musser SM, Sartorelli AC, Rockwell S, Tomasz M. Structure of adduct X, the last unknown of the six major DNA adducts of mitomycin C formed in EMT6 mouse mammary tumour cells. *Chem Res Toxicol* 2000, **13**, 479–88.

Palom Y, Suresh Kumar G, Tang LQ, Paz MM, Musser SM, Rockwell S, Tomasz M. Relative toxicities of DNA cross-links and monoadducts: new insights from studies of decarbamoyl mitomycin C and mitomycin C. *Chem Res Toxicol* 2002, **15**, 1398–406.

Paz-Ares L, Rivera-Herreros F, Diaz-Rubio E, *et al*. Phase II study of trabectedin in pretreated patients with advanced colorectal cancer. *Clin Colorectal Cancer* 2007, **7**, 522–8.

Petering DH, Byrnes RW, Antholine WE. The role of redox-active metals in the mechanism of action of bleomycin. *Chem Biol Interact* 1990, **73**, 133–82.

Phillips DR, Crothers DM. Kinetics and sequence specificity of drug–DNA interactions: an *in vitro* transcription assay *Biochemistry* 1986, **25**, 7355–62.

Pizzolato JF, Saltz LB. The camptothecins. *Lancet* 2003, **361**, 2235–42.

Pommier Y. Topoisomerase I inhibitors: camptothecins and beyond. *Nat Rev Cancer* 2006, **6**, 789–802.

Post GC, Barthel BL, Burkhart DJ, Hagadorn JR, Koch TH. Doxazolidine, a proposed active metabolite of doxorubicin that cross-links DNA. *J Med Chem* 2005, **48**, 7648–57.

Povirk LF, Hogan M, Dattagupta N, Buechner M. Copper (II)·bleomycin, iron (III)·bleomycin, and copper (II)·phleomycin: comparative study of deoxyribonucleic acid binding. *Biochemistry* 1981, **20**, 665–71.

Pron G, Mahrour N, Orlowski S; *et al*. Internalisation of the bleomycin molecules responsible for bleomycin toxicity: a receptor-mediated endocytosis mechanism. *Biochem Pharmacol* 1999, **57**, 45–56.

Rahier NJ, Thomas CJ, Hecht SM. Camptothecin and its analogs. In: *Anticancer Agents from Natural Products*, edited by GM Cragg, DGI Kingston and DJ Newman. Taylor & Francis, Boca Raton, FL, 2005, pp 5–21.

Rephaeli A, Waks-Yona S, Nudelman A, Tarasenko I, Tarasenko N, Phillips DR, Cutts SM, Kessler-Icekson G. Anticancer prodrugs of butyric acid and formaldehyde protect against doxorubicin-induced cardiotoxicity. *Br J Cancer* 2007, **96**, 1667–74.

Rothenberg ML. Topoisomerase I inhibitors: review and update. *Ann Oncol* 1997, **8**, 837–55.

Roy C, Brown DL, Little JE, Valentine BK, Walker PR, Sikorska M, Leblanc J, Chaly N. The topoisomerase II inhibitor teniposide (VM-26) induces apoptosis in unstimulated mature murine lymphocytes. *Exp Cell Res* 1992, **200**, 416–24.

Sartorelli AC, Hodnick WF, Belcourt MF, Tomasz M, Haffty B, Fischer JJ, Rockwell S. Mitomycin C: a prototype bioreductive agent. *Oncol Res* 1994, **6**, 501–8.

Seaman FC, Hurley LH. Molecular basis for the DNA sequence selectivity of ecteinascidin 736 and 743: evidence for the dominant role of direct readout via hydrogen bonding. *J Am Chem Soc* 1998, **120**, 13028–41.

Sersa G, Miklavcic D, Cemazar M, Rudolf Z, Pucihar G, Snoj M. Electrochemotherapy in treatment of tumours. *Eur J Surg Oncol* 2008, **34**, 232–40.

Sessa C, De Braud F, Perotti A, *et al.* Trabectedin for women with ovarian carcinoma after treatment with platinum and taxanes fails. *J Clin Oncol* 2005, **23**, 1867–74.

Sikic BI, Rosencweig M, Carter SK. *Bleomycin Chemotherapy.* Academic Press, Orlando, FL, 1985.

Snyder JG, Hartman NG, D'Estantott BL, Kennard O, Remeta DP, Breslauer KJ. Binding of actinomycin D to DNA: Evidence for a nonclassical high-affinity binding mode that does not require GpC sites. *Proc Natl Acad Sci USA* 1989, **86**, 3968–72.

Soares DG, Escargueil AE, Poindessous V, Sarasin A, de Gramont A, Bonatto D, Henriques JA, Larsen AK. Replication and homologous recombination repair regulate DNA double-strand break formation by the antitumour alkylator ecteinascidin 743. *Proc Natl Acad Sci U S A* 2007, **104**, 13062–7.

Sobell H. Actinomycin and DNA transcription. *Proc Natl Acad Sci U S A* 1985, **82**, 5328–31.

Sriram D, Yogeeswari P, Thirumurugan R, Bal TR. Camptothecin and its analogues: a review on their chemotherapeutic potential. *Nat Prod Res* 2005, **4**, 393–412.

Staker BL, Hjerrild K, Feese MD, Behnke CA, Burgin AB, Stewart L. The mechanism of topoisomerase I poisoning by a camptothecin analog. *Proc Natl Acad Sci U S A* 2002, **99**, 15387–92.

Sugiyama M, Kumagai T, Hayashida M, Maruyama M, Matoba Y. The 1.6-A° crystal structure of the copper(II)-bound bleomycin complexed with the bleomycin-binding protein from bleomycin-producing *Streptomyces verticillus*. *J Biol Chem* 2002, **277**, 2311–20.

Swift LP, Rephaeli A, Nudelman A, Phillips DR, Cutts SM. Doxorubicin-DNA adducts induce a non-topoisomerase II-mediated form of cell death. *Cancer Res* 2006, **66**, 4863–71.

Tabaee A, Brown SM, Anand VK. Mitomycin C and endoscopic sinus surgery: where are we? *Curr. Opin. Otolaryngol. Head Neck Surg* 2007, **1**, 40–3.

Takebayashi Y, Pourquier P, Yoshida A, Kohlhagen G, Pommier Y. Poisoning of human DNA topoisomerase I by ecteinascidin 743, an anticancer drug that selectively alkylates DNA in the minor groove. *Proc Natl Acad Sci U S A* 1999, **96**, 7196–201.

Takebayashi Y, Pourquier P, Zimonjic DB, *et al.* Antiproliferative activity of ecteinascidin 743 is dependent upon transcription-coupled nucleotide-excision repair. *Nat Med* 2001, **7**, 961–6.

Tao Z-F, Konishi K, Keith G; Hecht SM. An efficient mammalian transfer RNA target for bleomycin. *J Am Chem Soc* 2006, **128**, 14806–7.

Tewey KM, Rowe TC, Yang L, Halligan BD, Liu LF. Adriamycin-induced DNA damage mediated by mammalian DNA topoisomerase II. *Science* 1984, **22**, 466–8.

Thomas CJ, Chizhov AO, Leitheiser CJ, Rishel MJ, Konishi K, Tao ZF, Hecht SM. Solid-phase synthesis of bleomycin A5 and three monosaccharide analogues: exploring the role of the carbohydrate moiety in RNA cleavage. *J Am Chem Soc* 2002, **124**, 12927–8.

Thomas CJ, Rahier NJ, Hecht SM. Camptothecin: current perspectives. *Bioorg Med Chem* 2004, **12**, 1585–1604.

Tomasz M, Palom Y. The mitomycin bioreductive antitumour agents: Cross-linking and alkylation of DNA as the molecular basis of their activity. *Pharmacol Ther* 1997, **76**, 73–87.

Umezawa H, Maeda K, Takeuchi T, Okami Y. New antibiotics, bleomycin A and B.Y. *J Antibiot (Tokyo) Ser A* 1966, **19**, 200–9.

Utzat CD, Clement CC, Ramos LA, Das A, Tomasz M, Basu AK. DNA adduct of the mitomycin C metabolite 2,7-diaminomitosene is a nontoxic and nonmutagenic DNA lesion *in vitro* and *in vivo*. *Chem Res Toxicol* 2005, **18**, 213–23.

Vail DM, Amantea MA, Colbern GT, Martin FJ, Hilger RA, Working PK. Pegylated liposomal doxorubicin: proof of principle using preclinical animal models and pharmacokinetic studies. *Semin Oncol* 2004, **31**(Suppl 13), 16–35.

Waksman SA. Production and activity of streptothricin. *J Bacteriol* 1943, **46**, 299–310.

Wall ME, Wani MC, Cook CE, Palmer KH. Plant antitumour agents, I: the isolation and structure of camptothecin, a novel alkaloidal leukemia and tumour inhibitor from *Camptotheca acuminata*. *J Am Chem Soc* 1966, **88**, 3888–990.

Wang J, Gao YG, Liaw YC, Li YK. Formaldehyde crosslinks daunorubicin and DNA efficiently: HPLC and X-ray diffraction studies. *Biochemistry* 1991, **30**, 3812–15.

Wang JC. Cellular roles of DNA topoisomerases: a molecular perspective. *Nat Rev Mol Cell Biol* 2002, **6**, 430–40.

Williams WK, Katz E. Development of a chemically defined medium for the synthesis of actinomycin D by *Streptomyces parvulus*. *Antimicrob Agents Chemother* 1977, **2**, 281–90.

Wilstermann AM, Bender RP, Godfrey M, Choi S, Anklin C, Berkowitz DB, Osheroff N, Graves DE. Topoisomerase II - drug interaction domains: identification of substituents on etoposide that interact with the enzyme. *Biochemistry* 2007, **46**, 8217–25.

Wolkenberg SE, Boger DL. Mechanisms of *in situ* activation for DNA-targeting antitumour agents. *Chem Rev* 2002, **102**, 2477–95.

You Y. Podophyllotoxin derivatives: current synthetic approaches for new anticancer agents. *Curr Pharm Des* 2005, **11**, 1695–717.

Zelek L, Yovine A, Brain E, *et al*. A phase II study of Yondelis (trabectedin, ET-743) as a 24-h continuous intravenous infusion in pretreated advanced breast cancer. *Br J Cancer* 2006, **94**, 1610–4.

Zewail-Foote M, Li VS, Kohn H, Bearss D, Guzman M, Hurley LH. The inefficiency of incisions of ecteinascidin 743-DNA adducts by the UvrABC nuclease and the unique structural feature of the DNA adducts can be used to explain the repair-dependent toxicities of this antitumour agent, *Chem Biol* 2001, **8**, 1033–49.

9
Alkylating Agents

Ana Paula Francisco, Maria de Jesus Perry, Rui Moreira
and Eduarda Mendes

9.1 Introduction

Alkylating agents are the oldest class of anticancer drugs and are a major cornerstone in the treatment of leukaemia, lymphomas and solid tumours. The introduction of nitrogen mustard in the 1940s can be considered the origin of antineoplastic chemotherapy targeting all tumour cells. The discovery that the reagent formed a covalent bond with DNA was made through later studies that demonstrated specific sites of alkylation on purine bases, leading to crosslinking of strands and induction of apoptosis. At present, five major types of alkylating agents are used in the chemotherapy of neoplastic diseases:

9.2 Nitrogen Mustards

The nitrogen mustards get their name because they are related to the sulfur-containing mustard gases used during the First World War. The N-mustard compound chlormethine was the first alkylating agent to be used medicinally, in 1942 (Papac, 2001). The nitrogen atom is able to displace a chloride ion intramolecularly to form the highly electrophilic aziridinium ion (Figure 9.1). Alkylation of DNA can take place and the lesion formed by mustards in cells results from cross-linking between different DNA bases. N-mustards are bifunctional alkylating agents (i.e., they have two electrophilic sites) and the DNA undergoes intrastrand and interstrand cross-linking (Balcome *et al.*, 2004; Silverman *et al.*, 2004). The most frequent site of attachment on DNA is N^7 site of guanine but adducts are formed at the O^6 and N^3 sites of guanine. Other sites include the N^7, N^3, and N^1 of adenine; and N^1, N^3 of cytosine (Figure 9.2) (Silverman *et al.*, 2004). There is also evidence that the inhibition of DNA replication caused by DNA crosslinks results in termination of transcription (Bauer and Povirk, 1997; Masta *et al.*, 1994).

Anticancer Therapeutics   Edited by Sotiris Missailidis
© 2008 John Wiley & Sons, Ltd

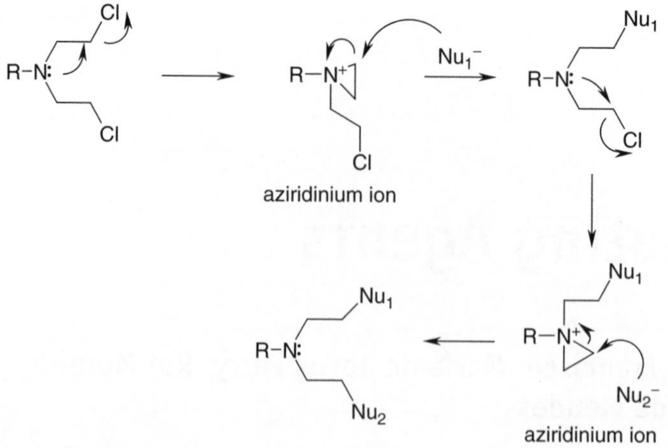

Figure 9.1 Alkylations by nitrogen mustards (adapted from Silverman, 2004)

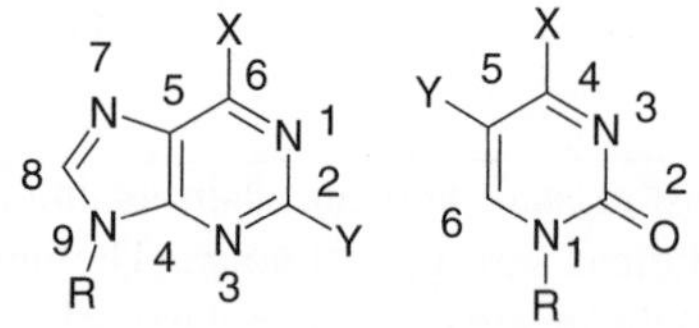

Figure 9.2 Numbering system of purines (adenine and guanine) and pyrimidines (cytosine and thymine)

9.2.1 Chlormethine

Chlormethine (Figure 9.3), which chemically is the tertiary amine methylbis (2-chloroethyl) amine, is an alkylating agent that as been in use as part of combination regimens in treatment of Hodgkin's disease, non-Hodgkin's lymphoma, as palliative chemotherapy in lung and breast cancers and as a lotion to skin lesions of mycosis fungoides (cutaneous T-cell lymphoma) (Chabner *et al.*, 2006; Kim *et al.*, 2003). The major use is in Hodgkin's disease and in the MOPP regimen (mechlorethamine, vincristine, procarbazine, and prednisone). The observation that approximately 20 % of the treated patients failed to achieve complete remission of their lymphoma, suggested that the primary cause of treatment failure was the presence and overgrowth of cells resistant to the drugs in the MOPP regimen (Bonadonna *et al.*, 2005; De Vita, 2003). Subsequently, a large number of alternative chemotherapy regimens have been developed for this disease. In many of these regimens, chormethine has been replaced by another more chemically stable alkylating agent, while in others, alternating cycles of MOPP and other combinations have been used. In addition, several combination regimens such as ABVD (doxorubicin, bleomycin, vinblastine, dacarbazine) or BEACOPP (bleomycin, etoposide, adriamycine, cyclophosphamide, vincristine, procarbazine, and prednisone) have been shown to be at least as effective and most probably more effective than mustine containing regimens (Bonadonna *et al.*, 2005; Diehl *et al.*, 1997). In the case of other lymphomas, mustine-containing regimens are rarely used today in order to avoid toxicological

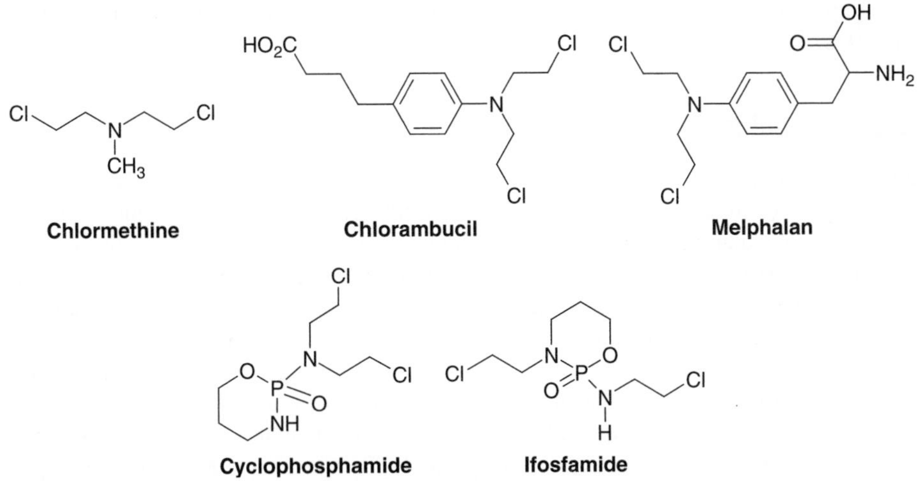

Figure 9.3 Nitrogen mustards currently in use

problems such as bone marrow suppression (leukopenia and thrombocytopenia), resistance due to increased drug inactivation and decrease drug uptake, and irritation at the injection site (chlormethine hydrochloride is usually given by slow infusion into the veins) (Chabner *et al.*, 2006). Chlormethine is so reactive with water that is marketed as a dry solid (HCl salt) and aqueous solutions are prepared immediately prior to injection.

9.2.2 Chlorambucil

Chlorambucil (Leukeran) was first synthesized by Everett and colleagues in 1953 (Everett *et al.*, 1953) and chemically is the 4-[bis(2-chlorethyl) amino]benzenebutanoic acid (Efthimiou *et al.*, 2007). As can be seen from the structures (Figure 9.3), chlorambucil and melphalan, have electron-withdrawing groups substituted on the nitrogen atom. This alteration reduces the nucleophilicity of the nitrogen and renders the molecules less reactive. Chlorambucil is indicated in the treatment of chronic lymphocytic leukaemia (CLL) and primary (Waldenstrom's) macroglobulinaemia, and may be used for follicular lymphoma (Chabner *et al.*, 2006; Sweetman, 2007). In treating CLL the standard initial daily dosage of chlorambucil is 0.1–0.2 mg/kg, given once daily and continued for 3 to 6 weeks. Maintenance therapy (usually 2 mg daily) often is required to maintain clinical remission (Chabner *et al.*, 2006). Chlorambucil, either alone or in combination, has been widely accepted as treatment of choice for patients with advanced stage disease for many years. For B-cell chronic lymphocytic leukaemia (B-CLL) the mainstay of therapy has been the alkylating agents, especially chlorambucil or cyclophosphamide, used alone or with prednisone. These agents, when given in an adequate dose for a sufficient duration, frequently yield response rates over 70 %, although complete response remains the exception (Lister *et al.*, 1978; Nicolle *et al.*, 2004; Oken *et al.*, 2004). The optimum dose of chlorambucil has not been defined and there are numerous different dosing schedules available. Pharmacokinetic studies suggest decreased bioavailability with successive cycles, probably due to accelerated metabolism (Silvennoinen *et al.*, 2000). The most common adverse effect of chlorambucil is myelosuppression and seizures are rarely reported as another form of acute toxicity (Chabner *et al.*, 2006).

9.2.3 Melphalan

Melphalan chemically is 4-[bis(2-chloroethyl)amino]-L-phenylalanine and it is a phenylalanine derivative of nitrogen mustard that acts as a bifunctional alkylating agent (Figure 9.3). Because L-phenylalanine is a precursor to melanin, it was thought that L-phenylalanine nitrogen mustard might accumulate in melanomas. Although melphalan is active against selective human neoplastic diseases it is not active against melanomas (Silverman *et al.*, 2004). Melphalan is used mainly in the treatment of multiple myeloma and in doses of 6–8 mg daily for a period of 4 days, in combination with other agents (a period of up to 4 weeks should occur). The usual intravenous dose is 15 mg/m^2 infused over 15–20 min. Doses are repeated at 2-week intervals for four doses and then at 4-week intervals based on response and tolerance (Chabner *et al.*, 2006). In 1968, Alexanian and colleagues reported that an association of oral melphalan and prednisone could improve the prognosis of multiple myeloma. This combination has been a standard treatment for myeloma for the past 40 years and results in a response rate of 50 % (Alexanian *et al.*, 1968; Smith and Newland, 1999). More recent studies show that low-dose thalidomide and dexamethasone have a true anti-tumour effect and that this is superior to that achieved by oral melphalan and prednisone (Palumbo *et al.*, 2002). Melphalan has also been given to patients with carcinoma of the breast and ovary, neuroblastoma, Hodgkin's disease, and in polycythaemia vera, and has been given by intra-arterial regional perfusion for malignant melanoma and soft-tissue sarcomas (Pinguet *et al.*, 2000; Chabner *et al.*, 2006; Grootenboers *et al.*, 2006). Melphalan is also used in the treatment of amyloidosis (Iggo *et al.*, 2000; Merlini, 2006; Sanchorawala *et al.*, 2001). The adverse effect of melphalan is mostly haematological and is similar to that of others alkylating agents.

9.2.4 Bendamustine

Bendamustine hydrochloride was developed in the 1960s in the former East Germany, but was never systematically studied in patients until the 1990s. The drug (4-{5-[bis-(2-chloroethyl)amino]-1-methyl-1H-benzoimidazol-2-yl}butanoic acid belongs to the group of bifunctional alkylating agents. By adding a nitrogen mustard group to position 5 of the benzimidazol nucleus and a butanoic acid residue to position 2, bendamustine hydrochloride combines the features of alkylating agents with those of purine- and amino acid analogues (Schwanen *et al.*, 2002) (Figure 9.4). This drug is given intravenously as the hydrochloride in lymphomas, including Hodgkin's disease, chronic lymphocytic leukaemia, multiple myeloma, and breast cancer. It had partial cross-resistance to other alkylating agents such as ifosfamide or cyclophosphamide (Balfour and Goa, 2001; Weidmann *et al.*, 2002). A phase III study was performed to compare the efficacy of bendamustine/prednisolone with melphalan/prednisolone, a standard regimen, in 136 patients with multiple myeloma. This randomized trial suggested that a bendamustine-containing regimen might replace melphalan/prednisolone for multiple myeloma (Ponisch *et al.*, 2000).

Currently, a new phase II study of bendamustine and rituximab in treating patients with relapsed chronic lymphocytic leukaemia is ongoing and a phase III trial of safety and efficacy of bendamustine in patients with indolent non-Hodgkin's lymphoma (NHL), who are

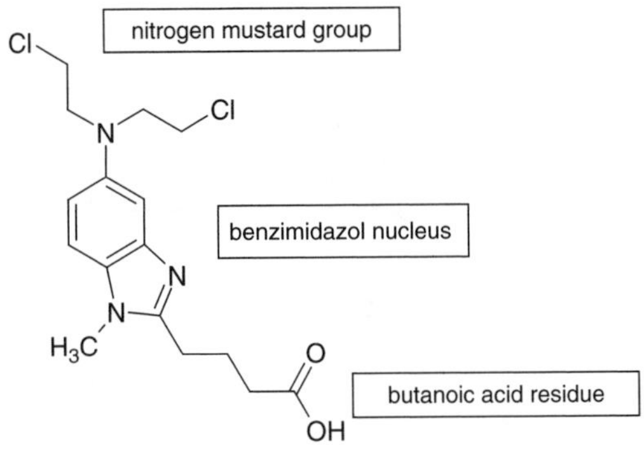

Figure 9.4   Chemical structure of bendamustine

refractory to rituximab, was performed (http://www.cancer.gov/). Bendamustine has been approved and used in Germany for about 30 years but has not been approved outside of Germany. In reality it is being studied as an investigational drug in the United States.

9.2.5 Cyclophosphamide

Cyclophosphamide, (2-[bis(2-chloroethyl)amino]perhydro-1,3,2-oxazaphosphorinan2-oxide-monohydrate), is an alkylating agent chemically related to nitrogen-mustard (Figure 9.3) and is also an immunosuppressive agent. Cyclophosphamide is one of the most widely used cytotoxic agents, often in combination or sequentially with other antineoplastic agents. It is used in the treatment of malignant diseases of the brain, breast, endometrium, lung, bladder, cervical, testis and ovary; it is given for Burkitt's and other non-Hodgkin's lymphomas, multiple myeloma, and mycosis fungoides. It is also used in gestational trophoblastic tumours in childhood malignancies such as neuroblastoma, retinoblastoma, Wilms' tumour; Ewing's sarcoma and several leukaemias (acute and chronic myelogenous leukaemias, acute and chronic lymphoblastic leukaemia). Other uses will be on osteosarcoma, soft-tissue sarcoma, thymoma, and cutaneous T-cell lymphoma (Sweetman, 2007; Cyclophosphamide AHFS, 2007). As an immunosuppressant, cyclophosphamide has been used to control rejection following hepatic, renal, cardiac and bone-marrow transplantation (Cyclophosphamide AHFS, 2007). It has also been used in the management of disorders thought to have an auto-immune component including amyloidosis, Beh͈cet's syndrome, glomerular kidney disease, idiopathic thrombocytopenic purpura, aplastic anaemia, cryptogenic fibrosing alveolitis, polymyositis, scleroderma, systemic lupus eryhtematosus, and vasculitic syndromes including Churg–Strauss syndrome, polyarteritis nodosa, and Wegener's granulomatosis (Sweetman, 2007).

Cyclophosphamide is a prodrug that requires biotransformation by a group of P450 cytochrome enzymes to exert its cytotoxicity. In the initial activation reaction, the carbon-4 of the oxazaphosphorine ring is hydroxylated. The product – 4-hydroxycyclo-phosphamide (4-OH-CPA) – exists in equilibrium with its acyclic tautomer aldophosphamide, which breaks down by spontaneous β elimination, to release phosphoramide mustard and acrolein (Figure 9.5). While phosphoramide mustard is thought to be the active alky-lating species, acrolein is an unwanted byproduct, responsible for haemorrhagic cystitis

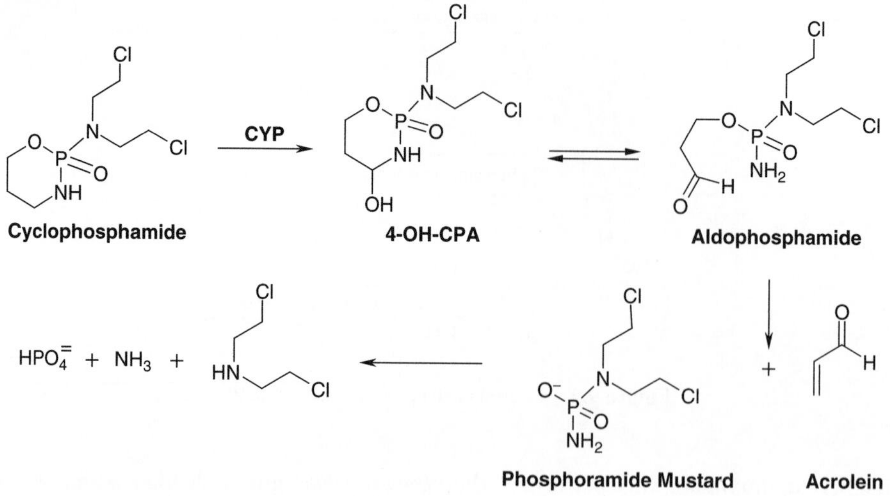

Figure 9.5   Metabolism of cyclophosphamide by its bioactivation pathway

(Figure 9.5). Alternatively, aldophosphamide may be oxidized to the inactive metabolite carboxyphosphamide by aldehyde dehydrogenase. The other principal inactive metabolite, dechloroethylcyclophosphamide, is produced by a separate oxidative N-dealkylation reaction, also catalysed by CYP3A4 (Boddy and Yule, 2000). Cyclophosphamide's cytotoxic effect is mainly due to cross-linking of strands of DNA and RNA of tumour cell. The major dose-limiting effect is myelosuppression. In comparison with many other anticancer drugs cyclophosphamide exhibits relatively little non-haematopoietic toxicity. Clinically important side effects are cardiac dysfunctions at high-doses; haemorrhagic cystitis may develop after high or prolonged dosage, and can be life-threatening; alopecia occurs in about 40 % of patients. Other adverse effects are nausea and vomiting (can be reduced by prophylactic antiemetics), mucositis; disturbances of carbohydrate metabolism and gonadal suppression. Cyclophosphamide, as other alkylating agents, has carcinogenic, mutagenic and teratogenic potential. Secondary malignancies have developed in some patients, often several years after administration. Because the activation and elimination pathways of cyclophosphamide are dependent on drug metabolism, there is wide scope for drug interaction and a large number have been reported in humans (Sweetman, 2007; Cyclophosphamide BC Cancer, 2006). Cyclophosphamide is usually administered intravenously in either 5 % dextrose or 0.9 per cent saline (Sweetman, 2007; Cyclophosphamide BC Cancer, 2006). Numerous dosing schedules exist and depend on disease, condition of the patient, response and concomitant therapy (radiotherapy or other chemotherapy). The white cell count is usually used to guide the dose. Oral administration of cyclophosphamide is usually restricted to its use as an immunosuppressive and in some adjuvant breast cancer regimens (Sweetman, 2007; Boddy and Yule, 2000).

9.2.6 Ifosfamide

Ifosfamide   (Figure 9.3)   (3-(2-chloroethyl)-2-(2-chloroethylamino)perhydro-1,3,2-oxaza-phosphorinane 2-oxide) is a structural analogue of cyclophosphamide. The clinical use of

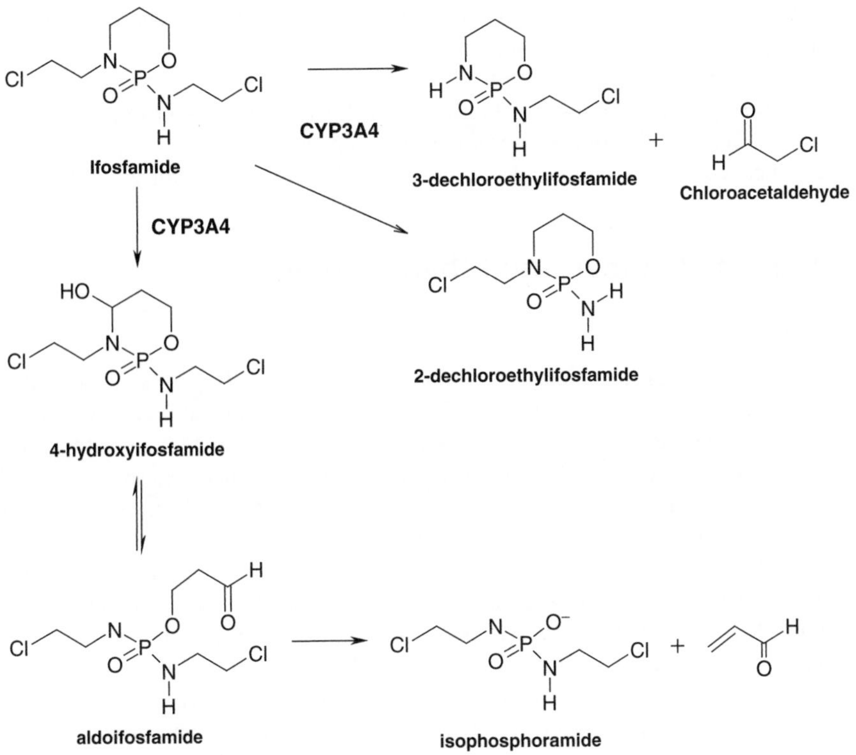

Figure 9.6 Metabolism of ifosfamide

ifosfamide includes adult and paediatric solid tumours (Groninger *et al.*, 2004). In adults it is less common as first-line treatment (except for soft-tissue sarcoma). It is used in the treatment of cervical cancer, testicular cancer and soft tissue sarcoma. Other uses include breast and ovarian cancer, lung cancer, neuroblastoma, Ewing's sarcoma, Hodgkin's disease and paediatric brain tumours. In the treatment of paediatric tumours ifosfamide is usually combined in multi-drug regimens (Boddy and Yule, 2000).

Ifosfamide is cell cycle phase non-specific and its mechanism of action (Figure 9.6) is presumed to be identical to the cyclophosphamide, that is activation by hepatic microsomal enzymes to form the DNA reactive mustard species (Boddy and Yule, 2000).

CYP3A4-mediated hydroxylation and subsequent spontaneous degradation result in the activation of the alkylating agent isophosphoramide mustard and another CYP3A4-mediated route yields the inactive metabolites 2- and 3-dechloroethylifosfamide and an equimolecular quantity of the metabolite chloroacetaldehyde which is supposed to be neuro- and nephro-toxic (Groninger *et al.*, 2004; Zhang *et al.*, 2006).

Toxic effects on the urinary tract may be more severe with ifosfamide than with cyclophosphamide, like haemorrhagic cystitis, dysuria and haematuria. Ifosfamide should not be administered without the use of an uroprotective agent such as mesna (ethanesulfonic acid), a mercaptan, which scavenges and inactivates reactive molecules such acrolein produced by ifosfamide activation (Remington, 2005; Ifosfamide BC Cancer, 2007). Other side effects are nausea and vomiting (dose-related), alopecia and myelosuppression. CNS adverse effects have been reported, especially confusion, drowsiness, depressive psychosis, hallucinations and rarely, seizures (Sweetman, 2007).

Drug interactions resulting in modification of metabolism of ifosfamide are potentially important (Groninger *et al.*, 2004). Ifosfamide is usually administered intravenously, by intermittent or continuous infusion. Numerous dosing schedules exist and depend on disease, response and concomitant therapy (Sweetman, 2007; Ifosfamide BC Cancer, 2007).

9.3 Methylmelamines and Ethylenimines

9.3.1 Altretamine

Altretamine is known as hexamethylmelamine and is structurally similar to the alkylating agent triethylenemelamine (tretamine) (Figure 9.7). Although altretamine structurally resembles an alkylating agent, it has not been found to have alkylating activity *in vitro*. *N*-Demethylation of altretamine may produce reactive intermediates, which covalently bind to DNA, resulting in DNA damage (Keldsen *et al.*, 2003). There is some evidence that it may inhibit DNA and RNA synthesis but the exact mechanism of action is poorly understood (Morimoto *et al.*, 1980). It is used as a palliative treatment for persistent or recurrent ovarian cancer following treatment failure with a cisplatin- or alkylating agent-based combination (Malik, 2001; Chan *et al.*, 2004). Altreatmine is orally administrated and well tolerated (Chabner *et al.*, 2006; Alberts *et al.*, 2004). The usual dose of altretamine as a single agent in ovarian cancer is $260 \, mg/m^2$ daily in four divided doses, for 14 or 21 consecutive days out of a 28-day cycle, for up to 12 cycles. Side effects are few and primarily related to the gastrointestinal tract and there is less need for laboratory monitoring and use of growth factors (Keldsen *et al.*, 2003; Malik, 2001).

9.3.2 Thiotepa

Thiotepa or Tris(1-aziridinyl)phosphine sulfide, belongs to the chemical family of ethyleneimines (Figure 9.7). It is an alkylating agent used in the treatment of breast, ovarian, and bladder carcinomas and was approved by the Food and Drug Administration (FDA) in 1959 (Gill *et al.*, 1996). Because of its dose-limiting toxicity (myelosuppression) thiotepa has largely been replaced by the nitrogen mustards (Soloway and Ford, 1983), although there is a renewed interest in this drug as it appears to be one of the most effective anticancer agents in high dose regimens. One particularly well-established regimen in high-dose chemotherapy for solid tumours is the CTCb regimen which consist of a combination of cyclophosphamide

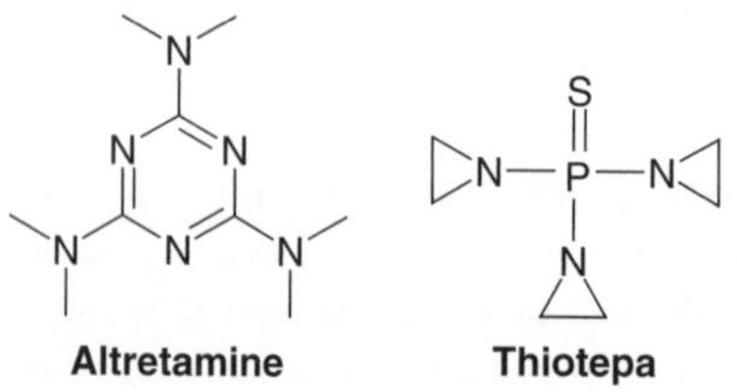

Altretamine   **Thiotepa**

Figure 9.7 Chemical structures of altretamine and thiotepa

(6000 mg/m^2), thiotepa (500 mg/m^2) and carboplatin (800 mg/m^2) all given as continuous 96 h infusions (Huitema *et al.*, 2002; Bergh *et al.*, 2000).

9.4 Methylhydrazine Derivatives

Procarbazine (N-isopropyl-α-(2-methylhydrazino)-p-toluamide hydrochloride) was initially synthesized as a potential monoamine oxidase (MAO) inhibitor, but later found to be an antineoplastic drug (Figure 9.8).

Procarbazine is used for the treatment of Hodgkin's disease, in combination with other anticancer drugs such as MOPP regimen, where it has made a particularly important contribution to the remarkable progress achieved in terms of long-term survival of patients (Pletsa *et al.*, 1997). Other uses are non-Hodgkin's lymphoma, small-cell lung cancer and some other malignant neoplasms, including tumours of the brain (Sweetman, 2007).

The drug is well absorbed orally and partially excreted in the urine. Procarbazine is a unique antineoplastic agent with multiple sites of action, despite the fact that the exact mode of cytotoxic action has not been elucidated. Figure 9.9 presents a simplified scheme for the bioactivation pathways of procarbazine, which is known to be complex and not fully understood. The first step involves oxidation by erythrocyte and hepatic microsomal enzymes that rapidly convert procarbazine (**1** Figure 9.9) in azoprocarbazine (**2** Figure 9.9) with the release of hydrogen peroxide. Several pathways of chemical and enzymatic nature have been proposed for the conversion of azoprocarbazine to the major urinary metabolite – N-isopropylterephthalamic acid (**6** Figure 9.9). In pathway B (Figure 9.9) the azo-procarbazine isomerizes to the hydrazone (**3** Figure 9.9) and following hydrolysis splits into a benzaldehyde derivative (**4** Figure 9.9) and methylhydrazine (**5** Figure 9.9). This is further degraded to CO_2 and CH_4 (both detected as expired products derived from the N-methyl group of procarbazine) and possibly hydrazine, whereas the aldehyde is oxidized to the acid (**6** Figure 9.9). In pathway A azo-procarbazine is N-oxidized in two possible isomeric azoxyderivatives (**7** methylazoxy and **8** benzylazoxy, Figure 9.9). Methylazoxy metabolite, a relatively stable intermediate, can spontaneously give rise to a methylating species, probably the methyldiazonium ion ($CH_3N_2^+$) that can directly damage DNA through an alkylation reaction (Dunn *et al.*, 1979). Free radical intermediates may also be involved in cytotoxicity (Moloney *et al.*, 1985; Chabner *et al.*, 2006). During oxidative breakdown in the body, hydrogen peroxide is formed, which may account for some of the drug's action. There is evidence that procarbazine may inhibit transmethylation of methyl groups of methionine into t-RNA. The absence of functional t-RNA could cause a cessation of protein synthesis and consequently DNA and RNA synthesis (FDA Matulane). Procarbazine is cell-phase specific for the S-phase of cell division.

The most common toxic effects include leucopenia (white blood count under 4000/ mm^3) and thrombocytopenia (platelets under 100 000/ mm^3). Gastrointestinal symptoms such as

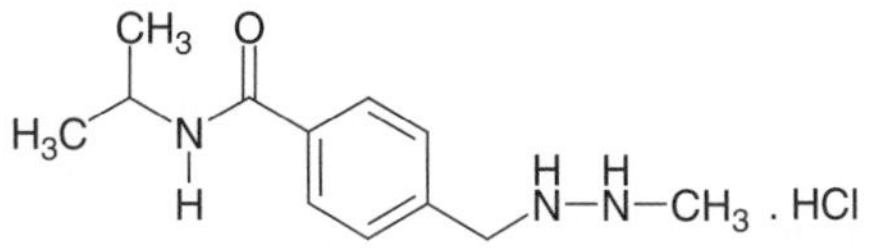

Figure 9.8 Chemical structure of procarbazine hydrochloride

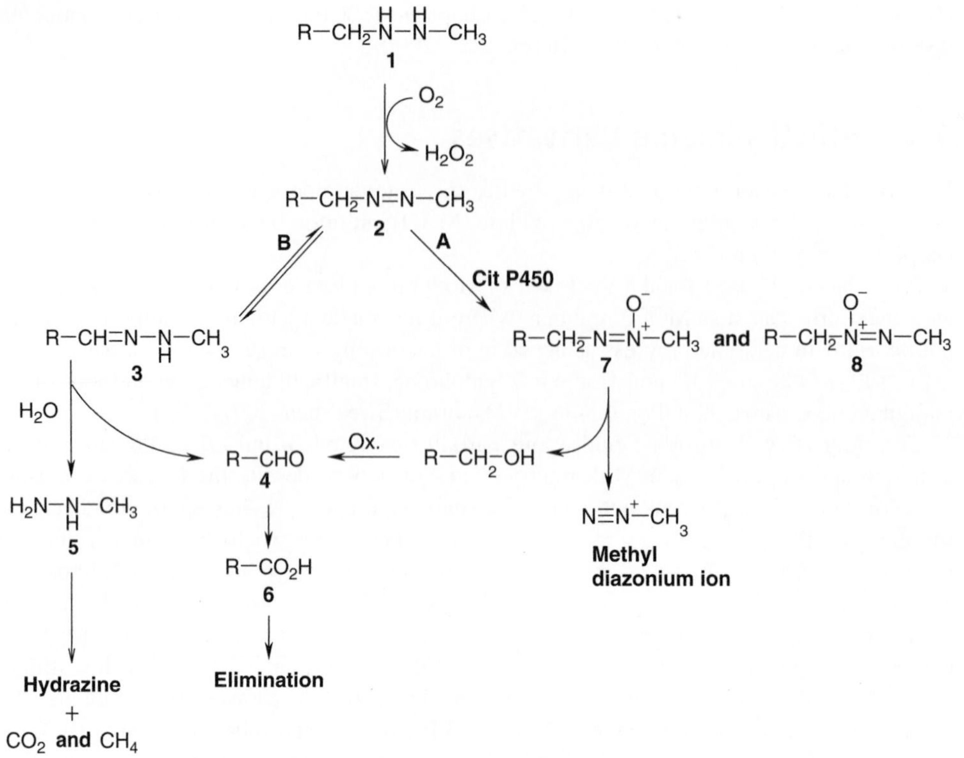

Figure 9.9   Bioactivation mechanism of procarbazine

mild nausea and vomiting occur in most patients (>50 %) (Chabner *et al.*, 2006). Stomatitis, diarrhoea, abdominal pain and dry mouth are other possible effects. (Procarbazine BC Cancer, 2006; Sweetman, 2007). Because procarbazine is a weak monoamine oxidase inhibitor that crosses the blood–brain barrier, it presents some neurotoxic effects. These effects may take the form of altered levels of consciousness, peripheral neuropathy, ataxia or effects of MAO inhibition. Particularly distressing for the patients are frequent nightmares, depression, insomnia, headache, nervousness and hallucinations, which occur in 10–30 per cent of patients (Procarbazine BC Cancer, 2006).

Procarbazine can cause azoospermia, which is often irreversible, and amenorrhoea in females. Procarbazine has serious long-term toxicities, namely infertility and secondary acute non-lymphocytic leukaemia (Massoud *et al.*, 2004). Procarbazine is carcinogenic, mutagenic and teratogenic (Sweetman, 2007). Procarbazine being an inhibitor of MAO, it is necessary to pay attention to the possibility of reactions with foods containing high tyramine contents and with drugs such as sympathomimetic agents and tricyclic antidepressant drugs.

For the administration of the drug as a single agent (Matulane®), it is recommended single or divided doses of 2 to 4 mg/kg/day for the first week. This will minimize nausea and vomiting. Daily dosage should then be maintained at 4 to 6 mg/kg/day until maximum response is obtained. When this is achieved, the dose may be maintained at 1–2 mg/kg/day. If there is evidence of haematological or other toxicity, the drug should be discontinued until there has been satisfactory recovery. When used in combination with other anticancer drugs, it should be appropriately reduced, e.g., in the MOPP regimen; the dose is 100 mg/m² daily for 14 days (Matulane Sigma-tau, 2004).

9.5 Alkylsulfonates

Busulfan (tetramethylene di(methanesulphonate) (Figure 9.10) is a bifunctional alkylating agent in which two labile methanesulfonate groups are attached to an alkyl chain. In aqueous media, busulfan hydrolyzes to release the methanesulfonate groups. This rapidly produces reactive carbonium ions capable of alkylating DNA. This leads to breaks in the DNA molecule as well as intrastrand cross-linkages, resulting in interference of DNA replication and transcription of RNA. The antitumour activity of busulfan is cell cycle phase-non-specific. Because busulfan has a selective immunosuppressive effect on bone marrow, it has been used in the palliative treatment of chronic myeloid leukaemia (CML). It provides symptomatic relief with a reduction in spleen size and improves the clinical state of the patient. The fall in leukocyte count is usually accompanied by a rise in the haemoglobin concentration. Permanent remission is not induced and resistance to its beneficial effects gradually develops (Sweetman, 2007). Currently, busulfan is one of the most frequently used chemotherapy agents in high dose preparative chemotherapy regimens for patients undergoing haemopoietic stem cell transplantation, in combination with other chemotherapeutic agents, usually cyclophosphamide (Slattery et al., 2000).

There are several studies comparing busulfan, interferon-alpha, hydroxyurea and bone marrow transplantation in treating the chronic phase of CML (Ohnishi et al., 1995; Silver et al., 1999; Lin et al., 2003). Myelosuppression is the dose-limiting side effect, which manifests as leucopenia, thrombocytopenia, and sometimes, anaemia. Busulfan may cause hyperpigmentation (darkening of the skin). With bone marrow transplant dosing, the following adverse effects are common: mucositis/stomatitis (85 %), fever (83 %), nausea and vomiting (72 %), rash (67 %), diarrhoea (58 %) and infection (31 %) (Busulfan BC Cancer). Rarely, progressive interstitial pulmonary fibrosis, known as 'busulfan lung', can occur on prolonged treatment. Other rare adverse effects include liver damage (veno-occlusive disease), gynaecomastia, cataract formation, and, at high doses, CNS effects including convulsions (Sweetman, 2007). Pubertal development and gonadal function may be adversely influenced by high-dose busulfan therapy in children and adolescents. As with others alkylating agents, it is potentially carcinogenic, mutagenic, and teratogenic. Busulfan has been associated with the development of acute leukaemia in humans (Bishop and Wassom, 1986).

In treating CML, the initial oral dose of busulfan varies with the total leukocyte count and the severity of the disease. Dosing on a weight basis is the same for paediatric patients and adults, approximately 60 µg/kg of body weight or 1.8 mg/m² of body surface, daily. The initial oral dose is then adjusted appropriately to subsequent haematological and clinical responses,

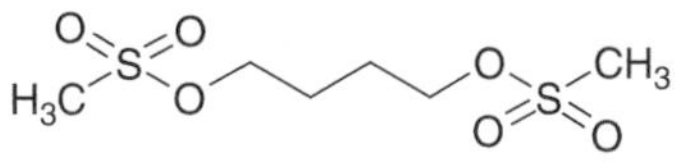

Figure 9.10 Chemical structure of busulfan

with the aim of reduction of the total leukocyte count to approximately 15 000 cells/mm^3 (15 × 10^9/L) and platelets <400 × 10^9/L (Chabner *et al.*, 2006; NHS Cancer Network, 2006).

A decrease in the leukocyte count is not usually seen during the first 10 to 15 days of treatment. As a matter of fact, it may actually increase during this period and that should not be interpreted as drug resistance, nor should the dose be increased. Because leukocyte count may fall for more than a month after discontinuing the drug, it is recommended that busulfan be withdrawn when the total leukocyte count has declined to ~15000 cells/mm^3. A normal blood count is usually achieved within 12 to 20 weeks (Chabner *et al.*, 2006; Busulfan BC Cancer). Because total leukocyte count declines exponentially when using a constant dose of busulfan, it is possible to predict the time when therapy should be discontinued (Myleran® Glaxo Smith Kline, 2003).

In conditioning regimens for bone marrow transplantation, busulfan has been given in usual doses of 3.5 to 4 mg/kg daily in divided doses for 4 days by mouth (total dose 14 to 16 mg/kg), with cyclophosphamide, for ablation of the recipient's bone marrow (Sweetman, 2007).

9.6 Nitrosoureas

The anticancer activity of this class of compounds was first discovered in a large-scale random screening program done by the US National Cancer Institute in 1959. The nitrosoureas are cytotoxic alkylating agents, with a wide range of activities against solid and non-solid tumours. The lipophilicity of most nitrosoureas allow for efficient crossing of the blood–brain barrier providing higher CSF to plasma ratios of these drugs in comparison to other alkylating agents (Schabel, 1976; McCormick and McElhinney, 1990; Lemoine *et al.*, 1991).

Nitrosoureas possess alkylating and carbamoylating activities. Upon introduction to the body, the nitrosoureas rapidly undergo a spontaneous hydrolytic decomposition, resulting in the release of electrophiles that can react with cellular components. The generally accepted mechanism by which these agents decompose in aqueous solution in order to generate the chloroethylating and the carbamoylating species is depicted in Figure 9.11. Gnewuch and Sosnovsky have done an extensively review of the mechanistic aspects and fragmentation pathways of the *N*-nitrosoureas (Gnewuch and Sosnovsky, 1997).

Despite their frequent use, the therapeutic efficacy of these agents is limited by the development of resistance. The mechanism of resistance is complex and involves multiple DNA repair pathways (Rabik *et al.*, 2006; Mishina *et al.*, 2006; Chen *et al.*, 2007). The first defence

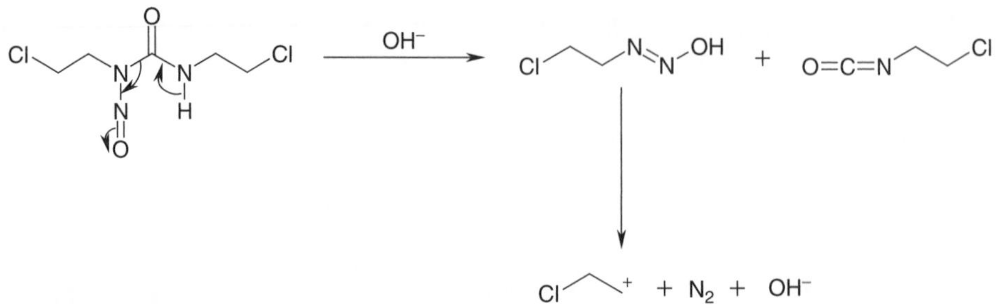

Figure 9.11 Decomposition of carmustine in buffered aqueous solution (adapted from Gnewuch and Sosnovsky, 1997)

mechanism against DNA lesions is the repair of adducts by the enzyme O^6-methylguanine DNA-methyltransferase (MGMT). MGMT produces the transfer of O^6-alkyl from DNA to the cysteine 145 moiety of the enzyme in a stoichiometric manner, leading to protein degradation. The activity of this enzyme virtually eliminates apoptosis. In the absence of MGMT O^6-methylguanine cytotoxicity is mediated mainly by the MMR system (Passagne *et al.*, 2003; Gerson, 2002).

9.6.1 Carmustine

Carmustine, 1,3-bis(2-chloroethyl)-1-nitrosourea (Figure 9.12) (BCNU), was the first nitrosourea to be subjected to extensive development. It has been used alone or in adjuvant therapy to treat brain tumours, colon cancer, lung cancer, Hodgkin and Non-Hodgkin's lymphomas, melanoma, multiple myeloma, and mycosis fungoides (McEvoy, ASHSP, 2007).

Carmustine is highly lipophilic and crosses the blood−brain barrier readily. The antineoplastic and toxic effects of carmustine are thought to be caused by active metabolites. One metabolite, the chloroethyl carbonium ion, leads to the formation of DNA cross-links during all phases of the cell cycle, resulting in cell cycle arrest and apoptosis (Lown *et al.*, 1981).

Carmustine is administered by intravenous (i.v.) infusion over a period of 1−2 hours and intracranially as wafer biodegradable implants following surgical resection of brain tumours. The usual intravenous dosage of carmustine, used as a single agent is 150 to 200 mg/m². Lower doses are given in combination therapy. For combination regimens numerous dosing schedules exist and depend on disease, response and concomitant therapy. Dosage may be reduced, delayed or discontinued in patients with bone marrow depression (McEvoy, ASHSP, 2007). Some of the most common combination chemotherapy regimens used are the Dartmouth regimen (cisplatin, BCNU, DTIC, and tamoxifen), to treat malignant melanoma (Boehnke-Michaud, 1996); the BEAM regimen (BCNU, etoposide, Ara-C, and

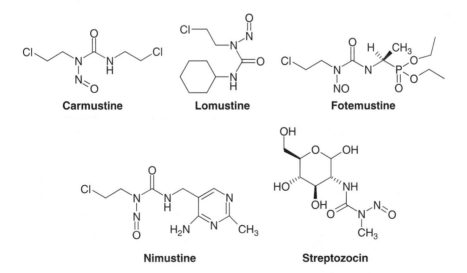

Carmustine   **Lomustine**   **Fotemustine**

Nimustine   **Streptozocin**

Figure 9.12   Chemical structure of some nitrosourea derivatives in clinical use

melphalan) to treat aggressive non-Hodgkin lymphoma (Stewart *et al.*, 2006); the VBAD (vincristine, BCNU, adriamycin, and dexamethasone) and the VBMCP regimen (vincristine, BCNU, cyclophosphamide, melphalan and prednisone) to treat multiple myeloma (Rubia *et al.*, 2006); the BAVEC-MiMA regimen (carmustine, doxorubicin, etoposide, vincristine and cyclophosphamide plus mitoxantrone or cytarabine and methotrexate with citrovorum factor), used for the treatment of aggressive non-Hodgkin's lymphoma (Rigacci *et al.*, 2005); and the Dexa-BEAM regimen (dexamethasone, carmustine, melphalan, etoposide, and cytarabine) to treat refractory aggressive non-Hodgkin's lymphoma (Atta *et al.*, 2007).

The most frequent and serious adverse effect of intravenous carmustine is delayed and cumulative bone-marrow depression. Other adverse effects include pulmonary fibrosis, renal and hepatic damage and neuroretinitis. Nausea and vomiting are common but can be reduced by prophylactic antiemetic therapy. Hypotension, tachycardia, chest pain, headache, and hypersensitivity reactions have been reported (Sweetman, 2007).

In recent years, novel approaches to improve the effectiveness of the treatment of malignant gliomas have been explored. Local therapy with gradual release carmustine wafers circumvents the obstacle of blood–brain barrier and it is able to deliver a high dose directly into the affected region. Convulsions, cerebral oedema, brain swelling, amnesia, aphasia, ataxia, abnormalities of wound healing at the site of implantation, and an increased incidence of intracranial infection have all been reported with implantable carmustine-impregnated wafers (Lawson *et al.*, 2007; Dyke *et al.*, 2007).

Tumour resistance to carmustine has been found to be associated with intracellular expression of O^6-methylguanine-DNA methyltransferase (MGMT). Administration of O^6-benzilguanine, a substrate that inactivates MGMT, may overcome chemotherapy resistance. Combination chemotherapeutic regimens with carmustine and O^6-benzylguanine have been tested in clinical trials in patients with advanced soft tissue sarcoma and malignant melanoma. However, the results were disappointing because of significant myelosuppression without improvement in the clinical outcome (Gajewski *et al.*, 2005; Ryan *et al.*, 2006).

Carmustine is generally not cross-resistant with other alkylating agents, but cross-resistance between carmustine and lomustine has occurred (Carmustine AHFS, 2007).

As are other alkylating agents carmustine is potentially carcinogenic, mutagenic and teratogenic (Sanderson and Shield, 1996).

9.6.2 Lomustine

Lomustine, 1-(2-chloroethyl)-3-cyclohexyl-1-nitrosourea (Figure 9.12) (CCNU), is used as a component of combination chemotherapy in addition to surgical and radiotherapeutic procedures for the treatment of primary and metastatic brain tumours, breast cancer, lung cancer, Hodgkin's lymphoma and melanoma. It alkylates DNA and RNA, can cross-link DNA and inhibits several enzymes by carbamoylation (Lomustine AHFS, 2007).

Unlike carmustine, it is administered orally. When given as a single agent, lomustine is licensed for use in adults and children as a single dose of 120 to $130 \, \text{mg/m}^2$. Doses are generally reduced when lomustine is given as part of a combination regimen. The schedule in those cases depends on the disease, response and concomitant therapy. The most common combination chemotherapeutic regimens are the PCV regimen (lomustine, procarbazine and vincristine) used to treat astrocytic tumours after surgery and radiation therapy, medulloblastoma and oligodendroglioma (Cairncross *et al.*, 2006); the BOLD regimen (bleomycin, vincristine, lomustine and dacarbazine) to treat metastatic melanoma (Larkin *et al.*, 2007), and the '8 in 1' regimen (vincristine, CCNU, procarbazine, hydroxyurea, cisplatin, Ara-C, cyclophosphamide or dacarbazine and methylprednisolone) (Gardner and Martingano, 1996).

The most frequent and serious toxicity of lomustine is delayed myelossuppresion. Other adverse effects include pulmonary toxicity characterized by pulmonary infiltrates and fibrosis, nausea and vomiting. Hepatic toxicity has been reported in a small percentage of patients and usually is reversible. Adverse nervous system effects include disorientation, lethargy, ataxia and dysarthria (Lomustine BC Cancer).

At the moment, clinical trials are running to test lomustine's therapeutic potential in acquired immune deficiency syndrome-related non-Hodgkin's lymphoma, in brain tumours, central nervous system, lymphoma, malignant glioma and medulloblastoma (NCI/clinical trials).

9.6.3 Fotemustine

Fotemustine, diethyl {1-[3-(2-chloroethyl)-3-nitrosoureido]ethyl}phosphonate (Figure 9.12), is a third-generation chloroethylnitrosourea that has demonstrated significant antitumour effects in malignant melanoma, particularly where cerebral metastases are present (Avril *et al.*, 2004).

When used as a single agent it is licensed for intravenous or intra-arterial infusion in usual doses of $100\,mg/m^2$ weekly for 3 weeks (Sweetman, 2007).

Haematological toxicity is characterized by delayed and reversible neutropenia and/or thrombocytopenia.

9.6.4 Nimustine

Nimustine, 3-[(4-amino-2-methylpyrimidin-5-yl)methyl]-1-(2-chloroethyl)-1-nitrosourea (Figure 9.12) (ACNU), was discovered in 1974 as the first water-soluble nitrosourea compound (Sugiyama *et al.*, 2007). The actions and uses of ACNU are similar to those of carmustine. It is licensed for use in the treatment of brain tumours and brain metastases from lung cancers and colorectal cancer.

Nimustine is given as a single dose of 90 to $100\,mg/m^2$ by i.v. injection. It may also be given intra-arterially (Imbesi *et al.*, 2006). In patients with recurrent glioblastomas it has been administered in a combination regimen with cisplatin and the results indicate that this regimen could be an effective salvage therapy for these patients (Choi *et al.*, 2002; Gwak *et al.*, 2005). Another chemotherapeutic regimen tested in patients with high-grade gliomas consisted of procarbazine, nimustine and vincristine (Ogawa *et al.*, 2006) or nimustine, carboplatin, vincristine and interferon-beta with radiotherapy. The results indicated that these regimens are safe, well tolerated, and may prolong survival (Aoki *et al.*, 2006).

9.6.5 Streptozocin

Streptozocin, 2-deoxy-2-[[(methylnitrosoamino)carbonyl]amino]-D-glucose (Figure 9.12), is a glucosamine–nitrosourea compound, isolated from *Streptomyces achromogenes* in 1956. This substance displayed significant antimicrobial action and antitumour activity (Schein *et al.*, 1974). During preclinical toxicological studies to characterize the antitumour effect, it became apparent that streptozocin produced hyperglycaemia in rats and dogs within a few hours of treatment. Histological subsequent studies showed that streptozocin induced rapid degranulation of islet beta cells and could produce permanent diabetes mellitus (Evans *et al.*, 1965).

Streptozotocin has a potent diabetogenic effect in several animal species, including mice, dogs, and monkeys. However, the diabetogenic action exhibits species variability and the drug has not been shown to produce a clinically important diabetogenic effect in humans. Streptozocin is used alone or with other antineoplastics mainly in the treatment of pancreatic endocrine (islet-cell) tumours. It has been tried in other tumours including exocrine pancreatic cancer and prostate cancer (Streptozocin AHFS, 2007).

The mechanism of action of streptozocin is not completely clear, but it is known that it inhibit DNA synthesis, interfere with biochemical reactions of nicotinamide adenine dinucleotide (NAD) and reduced NAD (NADH), and inhibit some enzymes involved in gluconeogenesis. It is cell cycle non-specific. *In vivo*, streptozocin undergoes spontaneous decomposition to produce reactive methylcarbonium ions that alkylate DNA and cause interstrand cross-linking (Bolzán and Bianchi, 2002). Unlike other currently available nitrosoureas that possess bifunctional alkylating activity, streptozocin lacks a chloroethyl group and is therefore considered a monofunctional alkylating agent. Like other nitrosoureas, streptozocin has the ability to carbamoylate various proteins (mainly the lysine aminoacid) and nucleic acids following *in vivo* decomposition and subsequent formation of isocyanate. The presence of D-glucopyranose moiety is responsible for the enhanced uptake of streptozocin by pancreatic islet cells. Streptozocin itself does not cross the blood–brain barrier but its metabolites are found in the cerebrospinal fluid.

Streptozocin is licensed for intravenous injection or infusion in doses of 1 g/m^2, weekly, increased, if necessary, after 2 weeks to up 1.5 g/m^2. It has also been given by intra-arterial infusion, but this route of administration is associated with increased risk of nephrotoxicity (Streptozocin, AHFS, 2007). In patients with progressive malignant neuroendocrine tumours of the pancreas a survival benefit has been shown for streptozocin combination regimens. Treatment regimens are effective in functioning and non-functioning tumours. However, chemotherapy is almost ineffective in patients with well-differentiated neuroendocrine tumours originated in the gastrointestinal tract (carcinoids) (Arnold *et al.*, 2005). Streptozocin plus doxorubicin (500 mg/m^2 per day as intravenous infusion and 50 mg/m^2 of doxorubicin as i.v. injection) protocol has been tested to treat well-differentiated metastatic endocrine pancreatic carcinoma. Streptozocin plus 5-fluorouracil (500 mg/m^2 per day and 400 mg/m^2 of 5-fluorouracil as i.v. infusion) regimen appears to be superior to streptozocin alone. Streptozocin should not be given with other potentially nephrotoxic drugs.

Cumulative nephrotoxicity is common and may be severe and irreversible. Other effects include severe nausea and vomiting and alterations of the liver function or occasionally severe hepatotoxicity. Streptozocin may affect glucose metabolism. A diabetogenic effect has been reported. Unlike other nitrosoureas, streptozocin causes little myelosuppression.

9.6.6 Novel nitrosourea derivatives of clinical interest

Attempts to develop more effective and less toxic antineoplastic nitrosoureas lead to the emergence of several novels analogs with improved toxicity profiles, and higher anticancer activity.

Cloretazine, 1,2-bis(methylsulfonyl)-1-(2-chloroethyl)-2-(methylaminocarbonyl)hydrazine (Figure 9.13), is a sulfonyl hydrazine prodrug recently reported to display significant

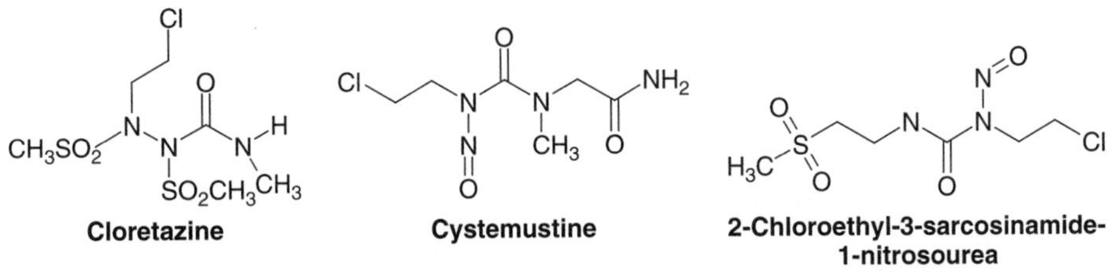

Figure 9.13 Novels nitrosourea derivatives underwent clinical trials

activity against both solid and haematological malignancy-derived cell lines, and broad-spectrum anti-tumour activity in animal models (Finch *et al.*, 2001).

Cloretazine is able to alkylate the O^6 position of guanine, resulting in DNA cross-linking, strand breaks, chromosomal aberrations and disruption of DNA synthesis. Intracellular metabolism of this compound, also release methyl isocyanate, which inhibits O^6-alkylguanine transferase. In contrast to BCNU, cloretazine does not generate hydroxyethylating, vinylating or aminoethylating species, and alkylation is relatively specific. Because of the different species generated, *in vitro*, cloretazine may have a distinct pattern of activity or toxicity in comparison to standard alkylating agents (Ishiguro *et al.*, 2006; Rabik *et al.*, 2006).

Cystemustine, 1-(2-chloroethyl)-3-(2-methylsulfonylethyl)-1-nitrosourea (Figure 9.13) (CNCC), is a third-generation nitrosourea, selected for its high solubility and good coefficient diffusion through the blood–brain barrier and into tumour cells (Madelmont *et al.*, 1985; Godeneche *et al.*, 1990; Roche *et al.*, 2000). Cystemustine was a more potent inducer of apoptosis than BCNU.

2-Chloroethyl-3-sarcosinamide-1-nitrosourea (SarCNU) (Figure 9.13) is a novel analogue of the chloroethylnitrosoureas. It has advantages over other nitrosoureas for its greater anti-tumour activity, less toxicity and its ease of administration (Webster *et al.*, 2005). All clinically available CNUs enter tumour cells *via* passive diffusions. SarCNU is unique in that the presence of the sarcosinamide group allows the drug to enter cells *via* extraneuronal monoamine transporter (EMT). SarCNu represents a potentially new anticancer agent that displays increased antitumour activity by exploiting a physiological aspect of tumour cells, the presence of high levels of EMT in certain types of tumours.

The National Cancer Institute supported several clinical trials to assess response rate, survival and effects of treatment, in malignant gliomas and metastatic colorectal cancer (Wong *et al.*, 2006).

9.7 Triazenes

The drugs dacarbazine and temozolomide (Figure 9.14) are structurally unique because they contain three adjacent nitrogen atoms that confer versatile physicochemical properties and antitumour activity based on the methylation of DNA.

9.7.1 Dacarbazine

Dacarbazine, (5-(3, 3-dimethyltriazeno)imidazol-4-carboxamide, DTIC) (Figure 9.14), is an alkylating agent used mainly in the treatment of metastatic malignant melanoma. It has

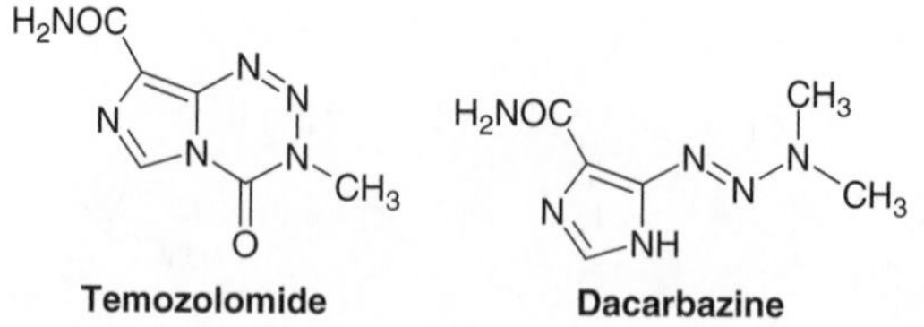

Figure 9.14   Chemical structure of temozolomide and dacarbazine

been on the market for three decades and remains the most effective single agent for the therapy of metastatic melanoma, with a response rate ranging from 15 to 25 %. Complete responses are rare and short in duration (Middleton *et al.*, 2000). Numerous studies have evaluated whether different schedules and dacarbazine-based combinations (combination chemotherapy and combination with immunotherapy) improve clinical outcomes in the survival (Eggermont and Kirkwood, 2004).

It is also given to patients with Hodgkin's disease, notably with doxorubicin, bleomycin, and vinblastine (ABVD). Dacarbazine is used with other drugs in the treatment of soft-tissue sarcoma, and may be given in neuroblastoma, Kaposi's sarcoma, and other tumours.

Dacarbazine is a cell-cycle non-specific antineoplastic that is thought to function as an alkylating agent after it has been activated in the liver. Dacarbazine is extensively metabolized in the liver by the cytochrome P450 isoenzyme system to its active metabolite 5-(3-methyl-1-triazeno)imidazole-4-carboxamide (MTIC). This spontaneously decomposes to the major metabolite 5-aminoimidazole-4-carboxamide (AIC), which is known to be an intermediate in purine and nucleic acid biosynthesis, and to methyldiazonium ion, which is believe to be the active alkylating species (Figure 9.15), producing methyl adducts at the accessible nucleophilic atoms in DNA by an S_N2 reaction. The most common lesion produced in DNA is methylation at the N^7 position of guanine, followed by methylation at the N^3 position of adenine and the O^6 position of guanine (Newlands *et al.*, 1997).

About half of a dose of dacarbazine is excreted unchanged in the urine by tubular secretion. Anorexia, nausea, and vomiting occur in more than 90 %of patients initially, but tolerance may develop after the first few days of successive treatment when the drug is given on a 5-day schedule (Dacarbazine-Mayne Pharma, 2003). Less frequent adverse effects include diarrhoea, skin reactions, phototoxicity, alopecia, a flu-like syndrome, facial flushing and paraesthesia, headache, blurred vision, seizures, and rare but potentially fatal hepatotoxicity. In fact, dacarbazine has been associated with fatal hepatic vascular toxicity, caused by thrombosis of the hepatic veins, necrosis, and extensive haemorrhage (Ceci *et al.*, 1988). There may be local pain at the injection site and extravasation produces tissue damage. Anaphylaxis has occurred occasionally. Leucopenia and thrombocytopenia, although usually moderate, may be severe, so haematological monitoring is required during therapy. Dacarbazine is potentially carcinogenic, mutagenic and teratogenic (Sweetman, 2007).

Numerous dosing schedules exist and depend on disease, response and concomitant therapy. Dacarbazine is given by the i.v. route and is licensed for use as a single agent for metastatic melanoma in doses of 2 to 4.5 mg/kg daily for 10 days, repeated at intervals of 4 weeks, or 200 to 250 mg/m^2 daily for 5 days, repeated at intervals of 3 weeks. It can also be given in a dose of 850 mg/m^2 by intravenous infusion at 3-week intervals. In the treatment of Hodgkin's disease doses of 150 mg/m^2 daily for 5 days repeated every 4 weeks, or 375 mg/m^2 every 15 days have been given with other agents. In the treatment of soft-tissue sarcoma,

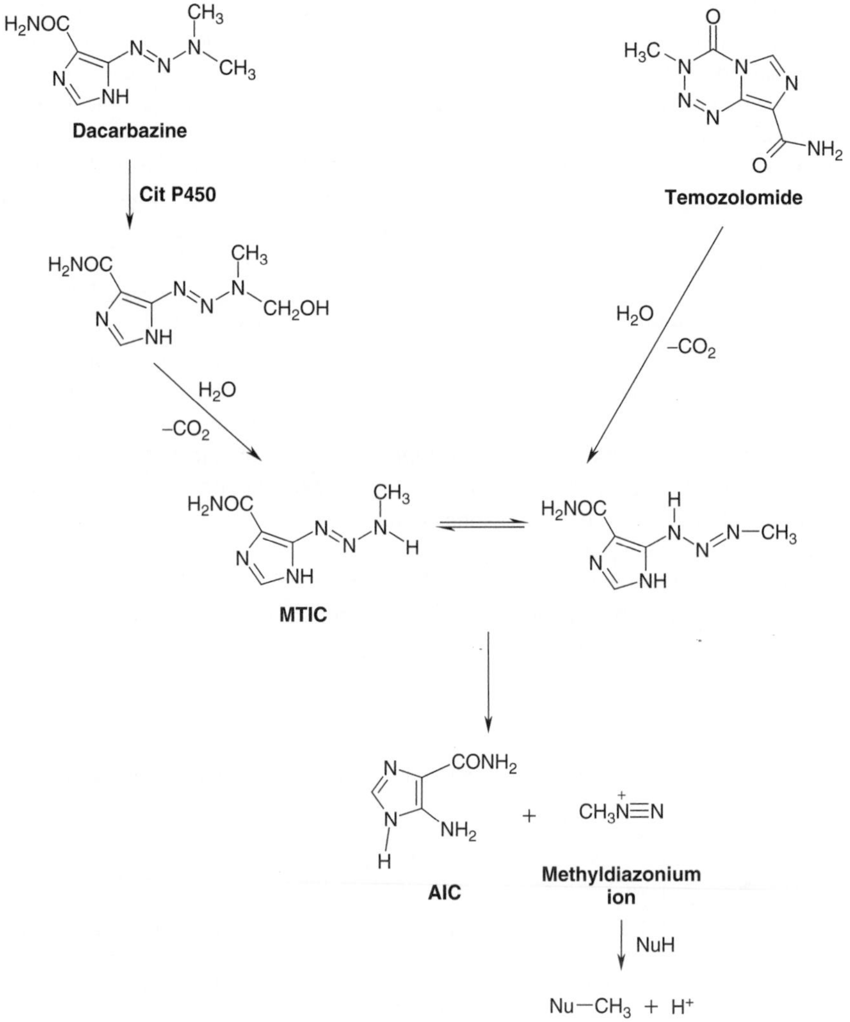

Figure 9.15 Activation of the triazene compounds dacarbazine (metabolic oxidation) and temozolomide (decomposition in aqueous solution)

dacarbazine $250\,mg/m^2$ is given daily for 5 days repeated every 3 weeks; it is usually given with doxorubicin (Sweetman, 2007).

9.7.2 Temozolomide

Temozolomide (3-methyl-4-oxoimidazo[5,1-d][1,2,3,5]tetrazine-8-carboxamide) (Figure 9.14) is a monofunctional alkylating agent, currently licensed for the treatment of any refractory high-grade glioma, namely recurrent anaplastic astrocytoma and glioblastoma multiforme (Friedman *et al.*, 2000). Studies have also been done to demonstrate the efficacy of temozolomide as first-line chemotherapy in patients with newly diagnosed high-grade gliomas (Friedman *et al.*, 1998) and activity in the treatment of malignant melanoma and brain metastases from solid tumours (Agarwala and Kirkwood, 2000). Despite demonstrating efficacy equal to that of dacarbazine, patients with malignant melanoma treated with temozolomide had better physical functioning, less fatigue and sleep disturbances (Middleton *et al.*, 2000, 2003; Bleehen *et al.*, 1995).

Temozolomide is structurally and functionally related to dacarbazine, and both are prodrugs. Dacarbazine requires enzymatic N-demethylation in the liver to generate the highly methylating species (MTIC); whereas temozolomide undergoes spontaneous chemical degradation at physiological pH to MTIC (Figure 9.15) (Newlands *et al.*, 1997).

Temozolomide is rapidly absorbed after oral administration, and because it is lipophilic, has good blood–brain barrier permeability (Newlands *et al.*, 1997). Temozolomide has relatively little toxicity. The most common non-haematological toxic effects during treatment are nausea and vomiting (generally well controlled by antiemetics), asthenia, headache, fatigue, fever, diarrhoea, abdominal pain, pruritis, pulmonary infection, rash and constipation. With regards to haematological toxicities, myelosuppression (neutropenia and thrombocytopenia) is the most serious adverse effect and is dose limiting. However, it does not appear to be cumulative and is relatively easily treated (Dinnes *et al.*, 2000). Temozolomide has carcinogenic, mutagenic, and teratogenic potential.

Temozolomide is administered orally in the form of hard capsules. The usual dose for recurrent or progressive malignant gliomas in adults and children over 3 years of age (and previously untreated with chemotherapy) is $200\,\text{mg/m}^2$ daily by mouth for 5 days, repeated every 28 days. In patients who have received previous courses of chemotherapy the dose should be reduced to $150\,\text{mg/m}^2$ for the first cycle of therapy, but may be increased to $200\,\text{mg/m}^2$ for subsequent courses if there is no haematological toxicity. A dose of $200\,\text{mg/m}^2$ daily for 5 days every 28 days is also used for metastatic malignant melanoma (Sweetman, 2007).

References

Agarwala SS, Kirkwood JM. Temozolomide, a novel alkylating agent with activity in the central nervous system, may improve the treatment of advanced metastic melanoma. *Oncologist* 2000, **5**, 144–51.

Alberts DS, Jiang C, Liu PY, Wilczynski S, Markman M, Rothenberg ML. Long-term follow-up of a phase II trial of oral altretamine for consolidation of clinical complete remission in women with stage III epithelial ovarian cancer in the Southwest Oncology Group. *Int J Gynecol Cancer* 2004, **14**, 224–8.

Alexanian R, Bergsagel DE, Migliore PJ, Vaughn WK, Howe CD. Melphalan therapy for plasma cell myeloma. *Blood* 1968, **31**, 1–10.

Aoki T, Takahashi JA, Ueba T, *et al.* Phase II study of nimustine, carboplatin, vincristine, and interferon-beta with radiotherapy for glioblastoma multiforme: experience of the Kyoto Neuro-Oncology Group. *J Neurosurg* 2006, **105**, 385–91.

Arnold R, Rinke A, Schmidt Ch, Hofbauer L. Endocrine tumours of the gastrointestinal tract: Chemotherapy. *Best Pract Res Clin Gastroenterol* 2005, **19**, 649–56.

Atta J, Chow KU, Weidmann E, Mitrou PS, Hoelzer D, Martin H. Dexa-BEAM as salvage therapy in patients with primary refractory aggressive non-Hodgkin lymphoma. *Leuk Lymphoma* 2007, **48**, 349–56.

Avril MF, Aamdal S, Grob JJ, *et al.* Fotemustine compared with dacarbazine in patients with disseminated malignant melanoma: a phase III study. *J Clin Oncol* 2004, **22**, 1118–25.

Balcome S, Park S, Dorr DR, Hafner L, Philips L, Tretyakova N. Adenine-containing DNA–DNA cross-links of antitumor nitrogen mustards. *Chem Res Toxicol* 2004, **17**, 950–62.

Balfour JA, Goa KL. Bendamustine. *Drugs* 2001, **61**, 631–8.

Bauer GB, Povirk LF. Specificity and kinetics of interstrand and intrastrand bifunctional alkylation by nitrogen mustards at a G-G-C sequence. *Nucl Acids Research* 1997, **25**, 1211–18.

Bergh J, Wiklund T, Erikstein B, *et al.* Tailored fluorouacil, epirubicin, and cyclophosphamide compared with marrow-supported high-dose chemotherapy as adjuvant treatment for high-risk breast cancer: a randomised trial. *Lancet* 2000, **356**, 1384–91.

Bishop JB, Wassom JS. Toxicological review of busulfan (Myleran). *Mutat Res* 1986, **168**, 15–45.

Bleehen NM, Newlands ES, Lee SM, *et al.* Cancer Research Campaign Phase II trial of temozolomide in metastatic melanoma. *J Clin Oncol* 1995, **13**, 910–13.

Boddy AV, Yule SM. Metabolism and pharmacokinetics of oxazaphosphorines. *Clin Pharmacokinet* 2000, **38**, 291–304.

Boehnke-Michaud L. Skin cancers and melanomas. *In Textbook of Therapeutics: Drug and Disease Management*, 6th edn, edited by Herfindal ET, Gourley DR, Williams & Wilkins, Baltimore, 1996, p 1678.

Bolzan AD, Bianchi MS. Genotoxicity of streptozotocin. *Mutat Res* 2002, **512**, 121–34.

Bonadonna G, Viviani S, Bonfante V, Gianni A, Valagussa P. Survival in Hodgkin's disease patients – report of 25 years of experience at the Milan Cancer Institute. *Eur J Cancer* 2005, **41**, 998–1006.

Busulfan BC Cancer Agency Drug. The Cancer Drug Manual©. Vancouver. Available from http://www.bccancer.bc.ca/HPI/DrugDatabase/default.htm

Cairncross G, Berkey B, Shaw E, *et al.* Phase III trial of chemotherapy plus radiotherapy compared with radiotherapy alone for pure and mixed anaplastic oligodendroglioma: Intergroup Radiation Therapy Oncology Group Trial 9402. *J Clin Oncol* 2006, **24**, 2707–14.

Carmustine AHFS Drug information, Bethesda, MD: American Society of Health-System Pharmacists. Electronic version 2007. Available from http://www.medecinescomplete.com/. [accessed 19 July 2007].

Ceci G, Bella M, Melissari M, Gabrielli M, Bocchi P, Cocconi G. Fatal hepatic vascular toxicity of DTIC: is it really a rare event? *Cancer* 1988, **61**, 1988–91.

Chabner BA, Amrein PC, Druker BJ, *et al.* Antineoplastic agents. In *Goodman & Gilman's The Pharmacological Basis of Therapeutics*, 11th edn, edited by. Laurence L. Brunton, John S. Lazo and Keith L. Parker. McGraw Hill, New York, NY, 2006, pp 1315–35.

Chan J, Loizzi V, Manetta A, Berman M. Oral altretamine used as salvage therapy in recurrent ovarian cancer. *Gynecol Oncol* 2004, **92**, 368–71.

Chen CC, Taniguchi T, D'Andrea A. The Fanconi anemia (FA) pathway confers glioma resistance to DNA alkylating agents. *J Mol Med* 2007, **85**, 497–509.

Choi IS, Lee SH, Kim TY, *et al.* Phase II study of chemotherapy with ACNU plus cisplatin followed by cranial irradiation in patients with newly diagnosed glioblastoma multiforme. *J Neurooncol* 2002, **60**, 171–6.

Cyclophosphamide AHFS Drug information, Bethesda, MD: American Society of Health-System Pharmacists. Electronic version 2007. Available from http://www.ashp.org/ahfs/first_rel/PDF/cyclophosphamide_essential.pdf [accessed 2 October 2007].

Cyclophosphamide BC Cancer Agency. The Cancer Drug Manual©, Vancouver. Available from http://www.bccancer.bc.ca/NR/rdonlyres/37BE9D9E-8E5A-48C2-ABA1-953637C4A5C0/19503/CyclophosphamideMonograph2Nov06.pdf [accessed 25 April 2008].

Dacarbazine Mayne Pharma (Canada) Inc. DACARBAZINE FOR INJECTION product monograph. Montreal, Quebec; 25 July 2003.

De Vita VT. A selective history of the therapy of Hodgkin's disease. *Br J Haematol* 2003, **122**, 718–27.

Diehl V, Sieber M, Rüffer U, *et al.* BEACOPP: An intensified chemotherapy regimen in advanced Hodgkin's disease. *Ann Oncol* 1997, **8**, 143–48.

Dinnes J, Cave C, Huang S, Major K, Milne R. The effectiveness and cost-efectiveness of temozolomide for the treatment of recurrent malignant glioma. Draft report to NICE, Wessex, Institute of Health Research and Development, 2000.

Dunn DL, Lubet RA, Prough RA. Oxidative metabolism of *N*-isopropyl-α-(2-methylhydrazino)-*p*-toluamide hydrochloride (Procarbazine) by rat liver microsomes. *Cancer Res* 1979, **39**, 4555–63.

Dyke JP, Sanelli PC, Voss HU, *et al.* Monitoring the effects of BCNU chemotherapy Wafers (Gliadel) in glioblastoma multiforme with proton magnetic resonance spectroscopic imaging at 3.0 Tesla. *J Neurooncol* 2007, **82**, 103–10.

Efthimiou M, Andriaanopoulos C, *et al.* Aneugenic potential of the nitrogen mustards analogues melphalan, chlorambucil and p-N,N-bis(2-chloroethyl)aminophenylacetic acid in cell cultures *in vitro*. *Mutat Res* 2007, **617**, 125–37.

Eggermont AMM, Kirkwood JM. Re-evaluating the role of dacarbazine in metastatic melanoma: what have we learned in 30 Years? *Eur J Cancer*, 2004, **40**, 1825–36.

Evans JS, Gerritsen GC, Mann KM, Owen SP. Antitumor and hyperglycemic activity of streptozotocin (NSC-37917) and its cofactor, U-15,774. *Cancer Chemother Rep* 1965, **48**, 1–6.

Everett, JL, Roberts, JJ, Ross, WCJ. Aryl-2-halogenoalkylamines. XII. Some carboxylic derivatives of N,N-Di-2-chloroethylaniline. *J. Chem. Soc.*, 1953, **3**, 2386–92.

FDA Matulane – U.S. Food and Drug Administration. Matulane label. http://www.accessdata.fda.gov/scripts/cder/onctools/labels.cfm?GN=procarbazine [accessed in 6 June 2008]

Finch RA, Shyam K, Penketh PG, Sartorelli AC. 1,2-Bis(methylsulfonyl)-1-(2-chloroethyl)-2-(methylamino)carbonylhydrazine (101M): a novel sulfonylhydrazine prodrug with broad-spectrum antineoplastic activity. *Cancer Res* 2001, **61**, 3033–8.

Friedman HS, Kerby T, Calvert H. Temozolomide and treatment of malignant glioma. *Clin Cancer Res* 2000, **6**, 2585–97.

Friedman HS, McLendon RE, Kerby T, *et al.* DNA mismatch repair and O⁶-alkylguanine-DNA alkyltransferase analysis and response to Temodal in newly diagnosed malignant glioma. *J Clin Oncol* 1998, **16**, 3851–7.

Gajewski TF, Sosman J, Gerson SL, Liu L, Dolan E, Lin S, Vokes EE. Phase II trial of the O⁶-alkylguanine DNA alkyltransferase inhibitor O⁶-benzylguanine and 1,3-bis(2-chloroethyl)-1-nitrosourea in advanced melanoma. *Clin Cancer Res* 2005, **11**, 7861–5.

Gardner ML and Martingano EC. Pediatric solid tumors of childhood. In: *In Textbook of Therapeutics: Drug and Disease Management*, 6th edn, edited by Herfindal ET, Gourley DR, Williams & Wilkins, Baltimore, 1996, p 1625.

Gerson SL. Clinical relevance of MGMT in the treatment of cancer. *J Clin Oncol* 2002, **20**, 2388–99.

Gill RD, Cussac C, Souhami RL, Laval F. Increased Resistence to *N*, *N'*, *N''*-triethylenethiophosphoramide (thiotepa) in cells expressing the *Escherichia coli* formamidopyrimidine-DNA glycosylase. *Cancer Res* 1996, **56**, 3721–4.

Gnewuch CT, Sosnovsky G. A critical appraisal of the evolution of *N*-nitrosoureas as anticancer drugs. *Chem Rev* 1997, **97**, 829–1014.

Godeneche D, Rapp M, Thierry A, Laval F, Madelmont JC, Chollet P, Veyre A. DNA damage induced by a new 2-chloroethyl nitrosourea on malignant melanoma cells. *Cancer Res* 1990, **50**, 5898–903.

Groninger E, Proost JH, de Graaf SSN. Pharmacokinetic studies in children with cancer. *Crit Rev Oncol/Hematol* 2004, **52**, 173–97.

Grootenboers M, Heeren J, Putte B, *et al.* Isolated lung perfusion for pulmonary metastases, a review and work in progress. *Perfusion* 2006, **21**, 267–76.

Gwak HS, Youn SM, Kwon AH, Lee SH, Kim JH, Rhee CH. ACNU-cisplatin continuous infusion chemotherapy as salvage therapy for recurrent glioblastomas: phase II study. *J Neurooncol* 2005, **75**, 173–80.

Huitema AD, Spaander M, Mathjt RA, Tibben MM, Holtkamp MJ, Beijnen JH, Rodenhuis S. Relationship between exposure and toxicity in high-dose chemotherapy with cyclophosphamide, thiotepa and carboplatin. *Ann Oncol* 2002, **13**, 374–84.

Ifosfamide BC Cancer Agency. The Cancer Drug Manual©, Vancouver. Available from http://www.bccancer.bc.ca/NR/rdonlyres/3EA37D74–6C87–4AD3-AFFC-8E93E0995DFB/21354/Ifosfamidehandout20Feb08.pdf

Iggo N, Littlewood T, Winearls CG. Prospects for effective treatment of AL amyloidosis? *Q J Med* 2000, **93**, 257–60.

Imbesi F, Marchioni E, Benericetti E, Zappoli F, Galli A, Corato M, Ceroni M. A randomized phase III study: comparison between intravenous and intraarterial ACNU administration in newly diagnosed primary glioblastomas. *Anticancer Res* 2006, **26**, 553–8.

Ishiguro K, Seow HA, Penketh PG, Shyam K, Sartorelli AC. Mode of action of the chloroethylating and carbamoylating moieties of the prodrug cloretazine. *Mol Cancer Ther* 2006, **5**, 969–76.

Keldsen N, Havsteen H, Vergote I, Bertelsen K, Jakobsen A. Altretamine (hexamethylmelamine) in the treatment of platinum-resistant ovarian cancer: a phase II study. *Gynecol Oncol* 2003, **88**, 118–22.

Kim YH, Martinez G, Varghese A, Hoppe RT. Topical nitrogen mustard in the management of mycosis fungoides. *Arch Dermatol* 2003, **139**, 165–73.

Larkin JM, Hughes SA, Beirne DA, *et al*. A phase I/II study of lomustine and temozolomide in patients with cerebral metastases from malignant melanoma. *Br J Cancer* 2007, **96**, 44–8.

Lawson HC, Sampath P, Bohan E, *et al*. Interstitial chemotherapy for malignant gliomas: the Johns Hopkin experience. *J Neurooncol* 2007, **83**, 61–70.

Lemoine A, Lucas C, Ings RM. Metabolism of the chloroethylnitrosoureas. *Xenobiotica* 1991, **21**, 775–91.

Lin K-H, Lin D-T, Jou S-T, Lu M-Y, Chang Y-H, Tien H-F. Forty-seven children suffering from chronic myeloid leukemia at a center over a 25-year period. *Pediatr Hematol Oncol* 2003, **20**, 505–15.

Lister TA, Cullen MH, Beard ME, *et al*. Comparison of combined and single-agent chemotherapy in non-Hodgkin's lymphoma of favourable histological type. *BMJ* 1978, **1**, 533–7.

Lomustine AHFS Drug information, Bethesda, MD:American Society of Health-System Pharmacists. Electronic version 2007. Available from http://www.medecinescomplete.com/. [accessed 19 July 2007].

Lomustine BC Cancer Available from htpp://bccancer.bc.ca/HPI/DrugDatabase/DrugIndexPro/lomustine.htm. [accessed 27 July 2007].

Lown JW, Chauhan SM. Mechanism of action of (2-haloethyl)nitrosoureas on DNA. Isolation and reactions of postulated 2-(alkylimino)-3-nitrosooxazolidine intermediates in the decomposition of 1,3-bis(2-chloroethyl)-, 1-(2-chloroethyl)-3-cyclohexyl-, and 1-(2-chloroethyl)-3-(4′-trans-methylcyclohexyl)-1-nitrosourea. *J Med Chem* 1981, **24**, 270–9.

Madelmont JC, Godeneche D, Parry D, Duprat J, Chabard JL, Plagne R, Mathe G, Meyniel G. New cysteamine (2-chloroethyl)nitrosoureas. Synthesis and preliminary antitumor results. *J Med Chem* 1985, **28**, 1346–50.

Malik I. Altretamine is an effective palliative therapy of patients with recurrent epithelial ovarian cancer. *Jpn J Clin Oncol* 2001, **31**, 69–73.

Massoud M, Armand JP, Ribrag V. Procarbazine in haematology: an old drug with a new life? *Eur J Cancer* 2004, **40**, 1924–7.

Masta A, Gray PJ, Philips R. Molecular basis of nitrogen mustards effects on transcription processes: role of depurination. *Nucl Acids Res* 1994, **22**, 3880–6.

Matulane – Sigma-tau Pharmaceuticals, Inc., Gaithersburg, MD 20877 (Revised: February 2004). Available from http://www.matulane.com/pdf/prescribe-info.pdf. [accessed 17 October 2007].

McCormick JE, McElhinney RS. Nitrosoureas from chemist to physician: classification and recent approaches to drug design. *Eur J Cancer* 1990, **26**, 207–21.

McEvoy GK. *American Hospital Formulary Service: drug information 2007/ed*. American Society of Health-System Pharmacists, Bethesda: ASHSP, 2007, pp 953–8.

Merlini G. Refining therapy for AL amyloidosis. *Blood* 2006, **108**, 3632–3.

Middleton M, Grob J, Aaronson N, *et al*. Randomized phase III study of temozolomide versus dacarbazine in the treatment of patients with advanced metastatic malignant melanoma. *J Clin Oncol* 2000, **18**, 158–66.

Middleton M, Jonas DL, Kiebert GM. Health-related quality of life in patients with advanced metastatic melanoma: results of a randomized phase III study comparing temozolomide with dacarbazine. *Cancer Investig*, 2003, **6**, 821–9.

Mishina Y, Duguid EM, He C. Direct reversal of DNA alkylation damage. *Chem Rev* 2006, **106**, 215–32.

Moloney SJ, Wiebkin P, Cummings SW, Prough RA. Metabolic activation of the terminal N-methyl group of N-isopropyl- α-(2-methylhydrazino)-p-toluamide hydrochloride (procarbazine). *Carcinogenesis* 1985, **6**, 397–401.

Morimoto M, Green D, Rahman A, Goldin A, Schein S. Comparative pharmacology of pentamethylmelamine and Hexamethylmelamine in Mice. *Cancer Res* 1980, **40**, 2762–7.

Myleran®, GlaxoSmithKline I. Myleran product monograph. Research Triangle Park, North Carolina, 2003.

NCI/clinical trials. Available from htpp://nciterms.nci.nih.gov/. [accessed 26 July 2007].

Newlands ES, Stevens MFG, Wedge SR, Wheelhouse RT, Brock C. Temozolomide: a review of its discovery, chemical properties, pre-clinical development and clinical trials. *Cancer Treat Rev* 1997, **23**, 35–61.

NHS Cancer Network. Available from http://www.swshcn.nhs.uk/healthcare-professionals/clinical-policies-and-protocols/haematology_protocols/myeloproliferative-dysplastic-disorders/Busulfan%20V1%204.06.pdf [accessed in 6 June 2008]

Nicolle A, Proctor SJ, Summerfield GP. High dose chlorambucil in the treatment of lymphoid malignancies. *Leuk Lymphoma* 2004, **45**, 271–5.

Ogawa K, Yoshii Y, Toita T, *et al.* Hyperfractionated radiotherapy and multi-agent chemotherapy (procarbazine, ACNU and vincristine) for high-grade gliomas: a prospective study. *Anticancer Res* 2006, **26**, 2457–62.

Ohnishi K, Ohno R, Tomonaga M, *et al.* A randomized trial comparing Interferon-α with busulfan for newly diagnosed chronic myelogenous leukemia in chronic phase. *Blood* 1995, **86**, 906–16.

Oken MM, Lee S, Kay NE, Knospe W, Cassileth PA. Pentostatin. chlorambucil and prednisone therapy for b-chronic lymphocytic leukemia: a phase I/II study by the Eastern Cooperative Oncology Group Study E1488. *Leuk Lymphoma* 2004, **45**, 79–84.

Palumbo A, Giaccone L, Bertola A. Low-dose thalidomide plus dexamethasone is an effective salvage therapy for advanced myeloma. *Haematologica* 2002, **86**, 399–403.

Papac RJ. Origins of cancer therapy. *Yale J Biol Med* 2001, **74**, 391–8.

Passagne I, Evrard A, Winum JY, Depeille P, Cuq P, Montero JL, Cupissol D, Vian L. Cytotoxicity, DNA damage, and apoptosis induced by new fotemustine analogs on human melanoma cells in relation to O^6-methylguanine DNA-methyltransferase expression. *J Pharmacol Exp Ther*, 2003, **307**, 816–23.

Pinguet F, Culine S, Bressolle F, Astre C, Serre MP, Chevillard C, Fabbro M. A Phase I and pharmacokinetic study of melphalan using a 24-hour contínuos infusion in patients with advanced malignancies. *Clin Cancer Res* 2000, **6**, 57–63.

Pletsa V, Valavanis C, van Delft HM, Steenwinkel M-J ST, Kyrtopoulos SA. DNA damage and mutagenesis induced by procarbazine in λlacZ transgenic mice: Evidence that bone marrow mutations do not arise primarily trough miscoding by O^6-methylguanine. *Carcinogenesis* 1997, **11**, 2191–6.

Ponisch W, Mitrou P, Merkle K, *et al.* Multicentric prospective randomized trial of bendamustine/prednisone versus melphalan/prednisone in 136 patients with multiple myeloma. *Proc Am Soc Clin Oncol* 2000; **19**: 9a (Abstr 25).

Procarbazine BC Cancer Agency Drug. The Cancer Drug Manual©]. Vancouver. Available from http://www.bccancer.bc.ca/NR/rdonlyres/7B97C0C8-FFF9–4317–8397–81E6268746A9/19521/Procarbazinemonograph3Nov06.pdf. [accessed 10 October 2007].

Rabik CA, Njoku MC, Dolan ME. Inactivation of O^6-alkylguanine DNA alkyltransferase as a means to enhance chemotherapy. *Cancer Treat Rev* 2006, **32**, 261–76.

Remington. *The Science and Practice of Pharmacy*, 21st edn. Lippincott Williams & Wilkins, Baltimore, 2005, Chapter 86, p 1580. Scholtz JM (Chapter author); Hendrickson R (Editor)

Rigacci L, Carrai V, Nassi L, Alterini R, Longo G, Bernardi F, Bosi A. Combined chemotherapy with carmustine, doxorubicin, etoposide, vincristine, and cyclophosphamide plus mitoxantrone, cytarabine and methotrexate with citrovorum factor for the treatment of aggressive non-Hodgkin lymphoma: a long-term follow-up study. *Cancer* 2005, **103**, 970–7.

Roche H, Cure H, Adenis A, Fargeot P, Terret C, Lentz MA, Madelmont JC, Fumoleau P, Hanausk A, Chollet P. Phase II trial of cystemustine, a new nitrosourea, as treatment of high-grade brain tumors in adults. *J Neurooncol* 2000, **49**, 141–5.

Rubia J de la, Blade J, Lahuerta JJ, *et al*. Effect of chemotherapy with alkylating agents on the yield of CD34+ cells in patients with multiple myeloma. Results of the Spanish Myeloma Group (GEM) Study. *Haematologica*, 2006, **91**, 621–7.

Ryan CW, Dolan ME, Brockstein BB, McLendon R, Delaney SM, Samuels BL, Agamah ES, Vokes EE. A phase II trial of O6-benzylguanine and carmustine in patients with advanced soft tissue sarcoma. *Cancer Chemother Pharmacol* 2006, **58**, 634–9.

Sanchorawala V, Wright D, Seldin D, *et al*. An overview of the use of high-dose melphalan with autologous stem cell transplantation for the treatment of AL amyloidosis. *Bone Marrow Transpl* 2001, **28**, 637–42.

Sanderson BJ, Shield AJ. Mutagenic damage to mammalian cells by therapeutic alkylating agents. *Mutat Res* 1996, **355**, 41–57.

Schabel FM Jr. Nitrosoureas: a review of experimental antitumor activity. *Cancer Treat Rep* 1976, **60**, 665–98.

Schein PS, O'Connell MJ, Blom J, *et al*. Clinical antitumor activity and toxicity of streptozotocin (NSC-85998). *Cancer* 1974, **34**, 993–1000.

Schwanen C, Hecker T, Hubinger G, Rittgen W, Bergmann L, Karakas T. *In vitro*evaluation of bendamustine induced apoptosis in B-chronic lymphocytic leukemia. *Leukemia* 2002, **16**, 2096–105.

Silvennoinen R, Malminiemi K, Maliminiemi O, Seppala E, Vilpo J. Pharmacokinetics of chlorambucil in patients with chronic lymphocytic leukaemia: comparison of different days, cycles, doses. *Pharmacol Toxicol* 2000, **87**, 223–8.

Silver RT, Woolf SH, Hehlmann R, *et al*. An evidence based analysis of the effect of busulfan, hydroxyurea, interferon and allogeneic bone marrow transplantation in treating the chronic phase of chronic myeloid leukemia: developed for the American Society of Hematology. *Blood* 1999, **94**, 1517–36.

Silverman, R. *The Organic Chemistry of Drug Design and Drug Action*. Academic Press, San Diego, 2004, pp 353–68.

Slattery JT, Gibbs JP, McCune JS. Plasma concentration monitoring of busulfan. *Clin Pharmacokin* 2000, **39**, 155–65.

Smith ML, Newland AC. Treatment of myeloma. *Q J Med* 1999, **92**, 11–14.

Soloway MS, Ford KS. Thiotepa-induced myelosupression: review of 670 bladder instillations. *J Urol* 1983, **130**, 889–91.

Stewart DA, Bahlis N, Valentine K, *et al*. Upfront double high-dose chemotherapy with DICEP followed by BEAM and autologous stem cell transplantation for poor-prognosis aggressive non-Hodgkin lymphoma. *Blood* 2006, **107**, 4623–7.

Streptozocin AHFS Drug information, Bethesda, MD: American Society of Health-System Pharmacists. Electronic version 2007. Available from http://www.medecinescomplete.com/. [accessed 19 July 2007].

Sugiyama S, Yamashita Y, Kikuchi T, Saito R, Kumabe T, Tominaga T. Safety and efficacy of convection-enhanced delivery of ACNU, a hydrophilic nitrosourea, in intracranial brain tumor models. *J Neurooncol*, 2007, **82**, 41–7.

Sweetman SC (ed.) *Martindale: The Complete Drug Reference*. Pharmaceutical Press, London. Electronic version, September 2007.

Webster M, Cairncross G, Gertler S, Perry J, Wainman N, Eisenhauer E. Phase II trial of SarCNU in malignant glioma: unexpected pulmonary toxicity with a novel nitrosourea: a phase II trial of the national cancer institute of canada clinical trials group. *Invest New Drugs* 2005, **23**, 591–6.

Weidmann E, Kim S-Z, Rost A, *et al.* Bendamustine is effective in relapsed or refractory aggressive non-Hodgkin's lymphoma. *Ann Oncol* 2002, **13**, 1285–9.

Wong RP, Baetz T, Krahn MJ, Biagi J, Wainman N, Eisenhauer E. National Cancer Institute of Canada Clinical Trials Group. SarCNU in recurrent or metastatic colorectal cancer: a phase II study of the National Cancer Institute of Canada Clinical Trials Group. *Invest New Drugs* 2006, **24**, 347–51.

Zhang J, Tian Q, Zhou S-F. Clinical pharmacoplogy of cyclophosphamide and ifosfamide. *Curr Drug Ther* 2006, **1**, 55–84.

10
Hormone Therapies

George C. Zografos, Nikolaos V. Michalopoulos and Flora Zagouri

10.1 Introduction

Hormone therapy or hormonal therapy is the use of hormones in medical treatment. Treatment with hormone antagonists may also referred to as hormonal therapy. More specifically, hormone therapy is the treatment that adds, blocks, or removes hormones. For certain conditions (such as diabetes or menopause), hormones are given to adjust low hormone levels. To slow or stop the growth of certain cancers (such as prostate and breast cancer), synthetic hormones or other drugs may be given to block the body's natural hormones. Moreover, hormone therapies are used in Klinefelter's or Turner's syndromes.

Hormonal therapy is one of the major modalities of medical treatment for cancer, others being cytotoxic chemotherapy and targeted therapy (biotherapeutics). It involves the manipulation of the endocrine system through exogenous administration of specific hormones, particularly steroid hormones, or drugs that inhibit the production or activity of such hormones (hormone antagonists). Because steroid hormones are powerful drivers of gene expression in certain cancer cells, changing the levels or activity of certain hormones can cause certain cancers to cease growing, or even undergo cell death. Surgical removal of endocrine organs, such as orchectomy and oophorectomy can also be employed as a form of hormonal therapy.

Hormonal therapy is used for several types of cancers derived from hormonally responsive tissues, including the breast, prostate, endometrium, and adrenal cortex. Hormonal therapy may also be used in the treatment of paraneoplastic syndromes or to ameliorate certain cancer- and chemotherapy-associated symptoms, such as anorexia. Perhaps the most familiar example of hormonal therapy in oncology is the use of the selective oestrogen-response modulator tamoxifen for the treatment of breast cancer, although another class of hormonal agents, aromatase inhibitors, now have an expanding role in that disease.

Anticancer Therapeutics   Edited by Sotiris Missailidis
© 2008 John Wiley & Sons, Ltd

10.2 Oestrogen Receptor Targeted Therapeutics

Oestrogens influence many physiological processes in mammals, including but not limited to reproduction, cardiovascular health, bone integrity, cognition, and behaviour. Given this widespread role for oestrogens in human physiology, it is not surprising that oestrogens are also implicated in the development or progression of numerous diseases, which include but are not limited to various types of cancer (breast, ovarian, colorectal, prostate, endometrial), osteoporosis, neurodegenerative diseases, cardiovascular disease, insulin resistance, lupus erythematosus, endometriosis, and obesity. In many of these diseases, oestrogens mediate their effects through the oestrogen receptor (ER), which serves as the basis for many therapeutic interventions.

The ER exists as two isoforms, ERα and ERβ, which are encoded by two different genes. They have distinct tissue expression patterns (Mueller and Korach, 2001) in both humans and rodents, and gene-targeted animal models lacking these receptors exhibit distinct phenotypes and provide some of the most definitive experimental models for evaluating the role of the ER in disease and normal physiology (Couse and Korach, 1999). ERα and ERβ are encoded by separate genes, *ESR1* and *ESR2*, respectively, found at different chromosomal locations, and numerous mRNA splice variants exist for both receptors in both diseased and normal tissue (Deroo and Korach, 2006). Like other steroid receptors, the ER has a structure consisting of a DNA binding domain (DBD) flanked by two transcriptional activation domains (AF-1 and AF-2). The receptor binds its ligand estradiol in the ligand-binding domain (Cui *et al.*, 2005).

Oestrogens induce cellular changes through several different mechanisms. Central to these mechanisms is the protein to which oestrogens bind, the oestrogen receptor. In the 'classical' mechanism of oestrogen action, oestrogens diffuse into the cell and bind to the ER, which is located in the nucleus. This nuclear oestrogen–ER complex binds to oestrogen response element sequences directly or indirectly through protein–protein interactions with activator protein 1 (AP1) or SP1 sites in the promoter region of oestrogen-responsive genes, resulting in recruitment of coregulatory proteins (coactivators or corepressors) to the promoter, increased or decreased mRNA levels and associated protein production, and a physiological response. This classical, or 'genomic,' mechanism typically occurs over the course of hours. In contrast, oestrogen can act more quickly (within seconds or minutes) *via* 'non-genomic' mechanisms, either through the ER located in or adjacent to the plasma membrane, or through other non-ER plasma membrane-associated oestrogen-binding proteins (Figure 10.1), resulting in cellular responses such as increased levels of Ca^{2+} or NO, and activation of kinases. The ER may be targeted to the plasma membrane by adaptor proteins such as caveolin-1 or Shc.

Hormonal therapy is a rapidly progressing molecular-targeted therapy for oestrogen-dependent disease, using drugs such as selective oestrogen receptor modulators (SERMs) or oestrogen receptor downregulators (ERDs). In parallel with the development of therapeutic strategies based on the modulation of ER activity, compounds that inhibit the production of oestrogen have also been generated and are currently in use clinically. Inhibitors of aromatase, the enzyme that converts androgens to oestrogens, have been successfully used in treatment of breast cancer and other diseases. Thus, targeting both the ER and the production of its activating ligand provides complementary strategies for the management of oestrogen-regulated diseases.

Three major classes of therapeutic intervention are classically considered: agents that directly target oestrogen receptor through molecules that bind ER and alter ER function; oestrogen deprivation through aromatase inhibition or ovarian ablation or suppression; and

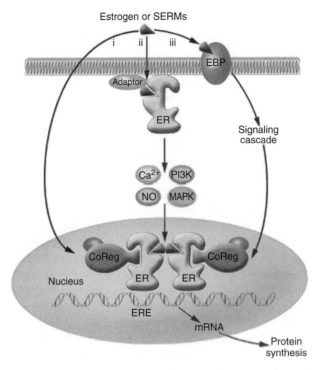

Figure 10.1 Models of oestrogen action. In the 'classical' pathway of oestrogen action (i), oestrogen or other selective oestrogen receptor modulators (SERMs) bind to the oestrogen receptor (ER), a ligand-activated transcription factor that regulates transcription of target genes in the nucleus by binding to oestrogen response element (ERE) regulatory sequences in target genes and recruiting coregulatory proteins (CoRegs) such as coactivators. Rapid or 'nongenomic' effects of oestrogen may also occur through the ER located in or adjacent to the plasma membrane (ii), which may require the presence of 'adaptor' proteins, which target the ER to the membrane. Activation of the membrane ER leads to a rapid change in cellular signaling molecules and stimulation of kinase activity, which in turn may affect transcription. Lastly, other non-ER membrane-associated oestrogen-binding proteins (EBPs) may also trigger an intracellular response (iii). Reproduced from Deroo BJ, Korach KS. Oestrogen receptors and human diesease. *J Clin Invest* 2006, **116**, 561–70, with permission

sex steroid therapies, including oestrogen, progestins and androgens. Since these approaches have quite different mechanisms of action, they can be used in sequence, recapturing tumour control with each class switch. In addition the side effects of these agents are quite distinct, with very important consequences for women's health particularly when they are used for long-term adjuvant therapy.

Since oestrogen receptors are implicated in several diseases, the feasibility of anti-hormone therapy is determined by the level of ER and its functional state. ER and progesterone receptor (PR) levels are routinely measured in breast cancer biopsy specimens. ER and PR constitute significant molecular markers, since chemotherapy activity is different in patient cohorts defined to hormone receptors (Biganzoli *et al.*, 2007).

In breast cancer, the presence of oestrogen receptor α (ERα) has been widely used as a predictive marker for endocrine therapy together with clinicopathological factors. However, it is known that approximately 30–40 % of ERα-positive patients do not respond to endocrine therapy, while some ERα-negative patients are responsive to this therapy (Hayashi and Yamaguchi, 2006). The remaining ER-positive and -negative breast cancers constitute a substantial fraction that does not respond to antihormone therapy. The signal for enhanced proliferation of ER-negative breast cancer cells is proposed to be initiated by the interaction of epidermal growth factor (EGF) with epidermal growth factor receptor (EGFR) and transmitted *via* a pathway in which active nuclear factor-κB (NF-κB) plays an important role by regulating the level of the cell cycle regulatory proteins. Thus, NF-κB is a potential target of therapy for ER-negative breast cancer patients (Biswas *et al.*, 2001).

Compared to breast cancer, relatively little is known about the responsiveness of ovarian carcinomas to oestrogens. Both steroid hormones and gonadotropins contribute to the aetiology of ovarian cancer in humans. The ER is present in approximately two-thirds of human ovarian tumours, with ERα being expressed in tumours of both epithelial and stromal origin, whereas ERβ is abundantly expressed predominantly in granulosa cell–derived tumours and to a lesser extent in mucinous tumours of epithelial origin (Chu *et al.*, 2000). Oestrogens stimulate proliferation of ovarian cancer cell lines and normal ovarian surface epithelial cells in culture. In contrast, growth is inhibited by ER antagonists (Syed *et al.*, 2001). Anti-oestrogens have been used clinically to treat ovarian cancer, with an overall response rate in the range of 15 % (Kali *et al.*, 2004). Despite these findings, clinical trials indicate that tamoxifen does not effectively inhibit recurrence of ovarian cancers (Jager *et al.*, 1995).

ERβ is the predominant ER form in both normal and malignant human colon tissue. ERβ protein levels are reportedly lower in colon tumours compared with normal colon tissue, and loss of ERβ is associated with advanced stages of colon cancer and tumour cell dedifferentiation, suggesting a protective role for ERβ in colon tumourigenesis (Jassam *et al.*, 2005; Konstantinopoulos *et al.*, 2003). Development of an ERβ-specific agonist for treatment of inflammatory bowel disease (i.e. Crohn's disease) and rheumatoid arthritis has been reported (Harris *et al.*, 2003). *In vitro*, both tamoxifen and raloxifene inhibit proliferation and viability of colon cancer cell lines, which may indicate the value of SERMs for the prevention and treatment of colon cancer (Picariello *et al.*, 2003).

Regarding prostate cancer, the clinical evidence linking increased serum oestrogen levels or an increased oestrogen/androgen ratio with elevated risk of prostate cancer remains inconclusive (Bosland, 2000). Studies have suggested a protective role for ERβ-specific agonists, as treatment of rodents with these compounds results in prostate atrophy due to apoptosis (Krishnan *et al.*, 2004).

Several reports of sex differences in lung cancer risk and disease presentation suggest that oestrogen may be involved in the aetiology of this disease. In this respect, we have shown recently that both ERα and β are expressed in non-small cell lung cancer (NSCLC) cell lines, tumour tissues, and cells derived from normal lung. Additionally, 17h-oestradiol acts as a mitogen for NSCLC cells *in vitro* and *in vivo* and can modulate the expression of genes in NSCLC cell lines that are important for control of cell proliferation. Anti-oestrogens have therapeutic value to treat or prevent lung cancer. Especially, ER antagonists with no agonist effects, inhibit lung tumour xenograft growth in severe combined immunodeficient mice by approximately 40 % (Stabile *et al.*, 2005).

There are clinical and experimental data to support a protective effect of oestrogens against neurodegenerative disease in humans. Several recent reviews described the various mechanisms by which oestrogens may provide neuroprotection (Garcia-Segura *et al.*, 2001; Wise *et al.*, 2001; Kajta and Beyer, 2003; Amantea *et al.*, 2005). In stroke, experimental evidence from animal and cell culture models show that oestrogen treatment protects against neuronal cell death due to insult (Amantea *et al.*, 2005). Experimental data also suggest that oestrogens protect against stroke. Both ERα and ERβ are found in various regions of the human and rodent brain, including the hypothalamus, hippocampus, cerebral cortex, midbrain, brainstem, and forebrain, and ER-mediated effects are thought to provide neuroprotection. Even though there is evidence that supports the participation of ER in Parkinson's disease and Alzheimer's disease, further clinical and experimental research is required to determine whether SERMs or oestrogen therapy will be a useful therapeutic approach for ameliorating the severity or incidence of Parkinson's disease and Alzheimer's disease.

10.2.1 Selective oestrogen receptor modulators (SERMs)

SERMs are a new category of therapeutic agents that bind with high affinity to oestrogen receptors and mimic the effect of oestrogens in some tissues but act as oestrogen antagonists in others. SERM–ER complexes appear to modulate the signal transduction pathway to oestrogen-responsive genes (through the oestrogen response elements, EREs) by binding fewer or different coactivators or by binding a corepressor protein (Jordan *et al.*, 2001).

There are currently several lines of investigation to elucidate the molecular mechanism of SERM through ERs. The complex is interpreted as an inhibitory signal at some sites but as a stimulatory complex at others. The SERM–ER complex has several options to produce a multiplicity of effects through gene activation. The ERs may be modulated by different levels or types of coactivator or corepressor protein in target cells. Since SERMs are known to have different actions at target genes through either ERα–SERM or ERβ–SERM complexes (Barkhem *et al.*, 1998), it is possible that one complex modulates the other (Hall and McDonnell, 1999). Clearly, the relative concentrations of ERα and ERβ (Fabian and Kimler, 2005; Enmark *et al.*, 1997) at different sites could ultimately control the actions of SERMs. Alternatively, the SERM–ER complexes could activate genes by a novel protein–protein interaction with fos/jun at AP-1 sites (Webb *et al.*, 1999) that is not available to oestrogen–ER complexes. Finally, it is equally possible that SERM action may be modified through non-genomic effects in specific tissues (Jordan *et al.*, 2001).

Tamoxifen

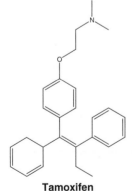

Tamoxifen

Tamoxifen is a non-steroidal triphenylethylene first synthesized in 1966. It is metabolized in a biphasic manner with a plasma half-life of approximately 7 days by the cytochrome P450/CYP2D6 pathway to *N*-desmethyl tamoxifen, an active metabolite with a half-life of approximately 14 days, and 4-hydroxyl tamoxifen and excreted in feces (Dehal and Kupfer, 1997, 1999). Tamoxifen was the first clinically available SERM. It is a potent anti-oestrogen in the breast, and its use in breast cancer patients has made it the most prescribed antineoplastic drug worldwide. It has oestrogen-like activity on bone metabolism, and it also reduces cholesterol. However, its ability to produce proliferation, polyp formation, and even carcinomas in the endometrium is well known.

Tamoxifen was originally developed as an oral contraceptive but, in an early clue to its complicated endocrine properties, tamoxifen was found to induce ovulation. Activity in metastatic breast cancer was first described in the 1970s (Westerberg *et al.*, 1976; Morgan *et al.*, 1976). Response rates ranging from 16 to 56%, and a better toxicity profile than alternative therapies, such as high dose oestrogen therapy or adrenalectomy, resulted in rapid adoption of tamoxifen as the treatment of choice for advanced disease (Muss *et al.*, 1994). Prospective randomized clinical trials of tamoxifen as adjuvant therapy also demonstrated considerable efficacy. Five years of adjuvant tamoxifen therapy reduces the risk of death by 26% and nearly halves the 10-year recurrence risk for patients with oestrogen receptor 'rich' tumours (Early Breast Cancer Trialists' Collaborative Group, 1998). For patients with metastatic disease treated with tamoxifen the overall mean time to disease progression is about 6 months, the duration of response is between 12 and 18 months, and responses in some patients may persist for a number of years (Sawka *et al.*, 1997). In the National Surgical Adjuvant Breast and Bowel Project P-1 Study (NSABP) placebo controlled prevention trial P01 the excess incidence of pulmonary embolus, deep venous thrombosis, cerebrovascular accident, cataract and endometrial cancer for patients receiving tamoxifen therapy was around 0.5% (Fisher *et al.*, 1998). For patients with advanced disease, thrombosis risk is ten times that experienced by healthy women receiving tamoxifen. At least 40% of tamoxifen treated women will develop postmenopausal-type symptoms, including hot flushes, vaginal dryness/dyspareunia, and vaginal discharge (Heldermon and Ellis, 2006).

Models to assess the risks and benefits of tamoxifen in women at varying ages and levels of breast cancer risk have been developed (Gail *et al.*, 2000). In general, older women require a higher level of breast cancer risk to clearly benefit from tamoxifen, particularly if they have a uterus. For white women under the age of 50 years with a uterus, a net benefit for tamoxifen was seen with a 5-year risk of breast cancer development of 1.5%. For those aged 50–59 years, this increases to a 4.0%–5.9% risk for a moderate probability of benefit (0.60 to 0.89) or a 6.0% or greater risk for a high probability of benefit (0.90 to 1.00) (Heldermon and Ellis, 2006).

Tamoxifen reduced the incidence of ER-positive tumours by 69% but had no effect on ER-negative tumours. It is likely that some of the reduction in cancer incidence in the tamoxifen treated group was due to the treatment of clinically occult disease. However, benefit was observed for each year of follow-up in the study, with the 33% risk reduction observed in year 1 increasing to 69% in year 5. Mathematical modelling suggests that these results are best explained by a combination of both treatment and prevention. The observation from the overview analysis that the reduction in contralateral breast cancer incidence persists 5 or more years after tamoxifen is stopped further supports the idea that tamoxifen not only treats but also prevents breast cancer (Jordan *et al.*, 2001).

Based on the successive analysis of accumulative randomized worldwide clinical trials, it is possible to summarize the main conclusions for tamoxifen therapy (Jordan VC and Brodie A. 2007).

- Five years of adjuvant tamoxifen enhances disease free survival. There is a 50% decrease in recurrences observed in ER positive patients 15 years after diagnosis.

- Five years of adjuvant tamoxifen enhances survival with a decrease in mortality 15 years after diagnosis.

- Adjuvant tamoxifen does not provide an increase in disease free or overall survival in ER negative breast cancer.

- Five years of adjuvant tamoxifen alone is effective in premenopausal women with ER.

- The benefits of tamoxifen in lives saved from breast cancer, far outweighs concerns about an increased incidence of endometrial cancer in postmenopausal women.

- Tamoxifen does not increase the incidence of second cancers other than endometrial cancer.

- No non-cancer related overall survival advantage is noted with tamoxifen when given as adjuvant therapy.

Toremifene

An additional agent in the SERM class and structurally related to tamoxifen is toremifene. Toremifene has similar efficacy and side effects to tamoxifen in the treatment of advanced disease. Toremifene is not approved for adjuvant therapy in the US (Robertson, 2004; Heldermon and Ellis, 2006).

Toremifene (Fareston®)

Droloxifene

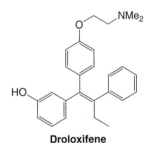

Droloxifene is also structurally similar to tamoxifen, although its relative binding affinity for the ER is between 10- and 60-fold higher than the binding affinity of tamoxifen. Phase III trial showed that in both pre- and postmenopausal women with ER-positive and/or progesterone receptor (PgR)-positive breast cancer, both objective response (OR) and median time to progression (TTP) were significantly better for tamoxifen compared with droloxifene (Robertson, 2004).

Droloxifene

Idoxifene

Like droloxifene, idoxifene has an increased binding affinity for the ER compared with tamoxifen. As first-line treatment, a phase III study in postmenopausal women with metastatic breast cancer showed there was no difference in OR or median duration of response (DoR) between idoxifene and tamoxifen (Robertson, 2004).

Idoxifene

The development of the 'fixed-ring' SERMs in the early 1980s stimulated optimism that these new anti-oestrogens, with their well-tolerated side effect profiles, would be of greater clinical usefulness compared with the triphenylethylenes. Structurally, the second-generation SERMs resemble the benzothiophene raloxifene.

Raloxifene

Raloxifene was initially developed as LY156 758 (Black *et al.*, 1982). Although raloxifene (originally named keoxifene) was developed initially for breast cancer treatment, its use was abandoned in the late 1980s because clinical trials showed no activity in tamoxifen-resistant patients (Buzdar *et al.*, 1988).

The current clinical use of raloxifene is the direct result of the concept that SERMs could be developed for the prevention of osteoporosis or atherosclerosis but reduces the risk of breast cancer as a beneficial side effect. Oestrogens regulate skeletal homeostasis in both men and women. Osteoporosis is due to increased bone resorption in both females and males and is associated with oestrogen deficiency. Oestrogens inhibit bone turnover by reducing osteoclast-mediated bone resorption and enhancing osteoblast-mediated bone formation. The role of raloxifene was tested successfully in clinical trials of raloxifene for the treatment and prevention of osteoporosis (Delmas *et al.*, 1997; Ettinger *et al.*, 1999; Cummings *et al.*, 1999). In a prospective, randomized trial of 7705 postmenopausal women with osteoporosis, raloxifene at a dose of 60 or 120 mg given daily reduced the risk of vertebral fractures by 30–50 %, at a mean follow-up of 36 months (Ettinger *et al.*, 1999). This reduction occurred despite the fact that raloxifene does not reduce bone turnover and does not increase bone density as much as a conjugated equine oestrogen (Prestwood *et al.*, 2000). Raloxifene also increased bone density in the femoral neck, but no difference in the rate of non-vertebral fractures was noted. In this study, raloxifene reduced the incidence of invasive breast cancer by 76 % (Cummings *et al.*, 1999).

Raloxifene increased the incidence of hot flashes from 6.4 % in the placebo group to 9.7 % in the group receiving 60 mg of raloxifene ($P< 0.001$). A small increase in the occurrence of leg cramps was also noted (Ettinger *et al.*, 1999). Limited information is available on the effect of raloxifene on mood and cognition (Jordan *et al.*, 2001).

Raloxifene has only 2 % bioavailability; unlike tamoxifen, which accumulates, raloxifene is rapidly excreted. Thus, the actions of raloxifene as a chemopreventive could become undermined by poor compliance which would explain the inability of raloxifene to control invasive and non-invasive breast cancer optimally. Clearly, new long-acting SERMs need to be developed and tested in the clinic.

Arzoxifene

Arzoxifene is a benzothiophene analogue of raloxifene. Even though arzoxifene might have advantages over both tamoxifen and raloxifene in the prevention of bone loss in postmenopausal women, a phase III trial, which compared arzoxifene with tamoxifen in postmenopausal women with advanced disease, was terminated and development of arzoxifene discontinued for this indication (Robertson, 2004). Arzoxifene in doses ranging from 10 to 200 mg/d produced a 19 % clinical benefit rate in subjects previously treated with tamoxifen and chemotherapy in a phase I trial. Significant reductions in bone turnover markers, low-density lipoproteins and total cholesterol, follicle-stimulating hormone, and luteinizing hormone (LH) were seen. Response rates of 26–43 % in women with tamoxifen-sensitive metastatic breast cancer have been reported for arzoxifene with a 10 % objective response rate in tamoxifen-resistant patients.

**Arzoxifene
(LY353381•HCl; LY353355)**

Raloxifene

ERA-923

ERA-923 is a second-generation SERM currently under development for the treatment of tamoxifen-refractory metastatic breast cancer. A randomized, placebo-controlled study evaluated the safety and tolerability of ERA-923 in 50 healthy postmenopausal women. ERA-923 was demonstrated to be safe and well tolerated with no incidence of vaginal discharge or bleeding, suggesting that this agent had no uterotrophic effect on the endometrium (Cotreau *et al.*, 2002).

ERA-923

Acolbifene

Acolbifene is a fourth-generation SERM of the benzopyrans class. In preclinical models, both drugs are more potent than tamoxifen or raloxifene for inhibiting growth of tamoxifen sensitive tumours. Both agents are able to suppress growth of tumour cells in some tamoxifen-resistant xenograft models. Both arzoxifene and acolbifene have demonstrated activity in metastatic breast cancer. Regarding acolbifene, phase II trials showed that the overall objective response rate was 18 % in tamoxifen-sensitive and 5 % in tamoxifen-resistant patients. Both drugs were well tolerated in phase I and II trials with no evidence of uterine hypertrophy. Hot flushes, muscle and bone pain, fatigue, asthenia, and nausea were the most frequent complaints (Fabian and Kimler, 2005).

Acolbifene

Other SERMs

Other structural analogues of tamoxifen have been synthesized, including TAT-59, which has a tenfold higher affinity for ER than tamoxifen and was more effective at inhibiting human breast cancer xenograft growth *in vivo*. However, it was equivalent to tamoxifen in a late phase II trial and its further development has been abandoned (Howell *et al.*, 2004).

The prodrug EM-800 is a potent anti-oestrogen. This is an orally active, so-called pure non-steroidal anti-oestrogen, which is a pro-drug of the active benzopyrene derivative EM-652 (SCH 57 068). A phase II study of EM-800, undertaken in 43 postmenopausal women who had failed tamoxifen either in the metastatic or adjuvant setting, showed that there was one complete response (CR) and five partial responses (PRs) (response rate 14 %), with most of the responses occurring in those who had received at least 3 years adjuvant tamoxifen.

An additional 23 % (10 patients) patients had stable disease for more than 6 months (Howell *et al.*, 2004).

10.2.2 Oestrogen receptor downregulators (ERDs)

Anti-oestrogens are divided into non-steroidal SERMs and 'pure' steroidal anti-oestrogens. 'Pure' anti-oestrogens are, by definition, anti-oestrogenic in all target tissues because there is destruction of the oestrogen receptor in target sites. The pure anti-oestrogens are effective in laboratory models as second-line treatment for tamoxifen-stimulated breast and endometrial cancer (Bentrem and Jordan, 2002). The selective oestrogen receptor down-regulators like fulvestrant have a higher affinity for the ER compared with tamoxifen, but none of its agonistic activities due to the fact that they can bind, block, and then degrade the ER. Fulvestrant has only just entered the clinic and this new agent is showing promising activity in the treatment of advanced breast cancer.

Fulvestrant inhibits dimerization of ER and promotes accelerated receptor turnover. This property suppresses receptor protein levels, reduces shuttling of the receptor from the cytoplasm to the nucleus and inhibits oestrogen-dependent transcription (Morris and Wakeling, 2002). Unfortunately the compound is insoluble in water which severely limits oral bioavailability and is therefore administered as a 250 mg intramuscular monthly depot injection. In this form the half-life is approximately 40 days and the drug is metabolized by a combination of various pathways with ultimate hepatobiliary excretion in feces. Two phase III randomized controlled trials in which fulvestrant was compared to the aromatase inhibitor anastrozole in the setting of tamoxifen-resistant disease were prospectively designed for combined analysis (Robertson *et al.*, 2003). The overall conclusion was that the two agents were equivalent in terms of efficacy and safety. Median times to progression and overall survival in the combined analysis were not statistically different (5.5 and 4.1 months for fulvestrant and anastrozole, respectively). Side effects were similar except joint disorders were significantly lower with fulvestrant (5.4 % versus 10.6 %). Fulvestrant was compared to tamoxifen as first line endocrine therapy for advanced breast cancer with no statistically significant difference in median times to progression for fulvestrant and tamoxifen: 6.8 and 8.3 months, respectively. In hormone receptor positive tumours the median time to progression was 8.2 months for fulvestrant and 8.3 months for tamoxifen and overall responses were 33.2 % versus 31.1 % (Morris and Wakeling, 2002). It is unclear where to place fulvestrant in the endocrine therapy algorithm since alternative oral therapies have equivalent efficacy and are more convenient. Since fulvestrant has been reported to have efficacy after the sequence of tamoxifen followed by an aromatase inhibitor (Perey *et al.*, 2002), third line therapy is the most likely application of this agent for the time being. Research questions currently being addressed include whether a higher dose would be of benefit (500 mg monthly) and also if a loading dose schedule would prevent early progression due to inadequate drug levels.

Fulvestrant

To date, at least five ERDs have undergone or are currently in pre-clinical testing. Pre-clinical data for SR 16 234 and ZK 191 703 show a three- to four-fold increase in duration of MCF-7, ZR-75 and T47D xenograft tumour growth control compared to SERMs, and improved antitumour characteristics were seen when compared to fulvestrant. RU 58 668 remains in preclinical testing and, as with fulvestrant and ZD 164 834, does not have significant oral bioavailability. However, there is some evidence that novel nanoparticle delivery systems and pegylation of this molecule can improve its *in vivo* potency when administered intravenously (Howell *et al.*, 2004).

GW5638 is a tamoxifen analogue, although, like fulvestrant, this agent downregulates the ER. In pre-clinical studies in athymic mice, GW5638 has been compared with fulvestrant and raloxifene. Both GW5638 and fulvestrant were effective in blocking tamoxifen-stimulated breast tumour growth (Dardes *et al.*, 2002). TAS-108 and ZK 191 703 are pharmacologically related to fulvestrant and TAS-108 may have similar effects to fulvestrant on the ER. The development of new drugs that work by downregulation of the ER, rather than selective ER modulation, offers the advantages of a reduced side-effect profile and also provides the option of further treatment choice after failure on tamoxifen. As resistance to tamoxifen may be due, in part, to agonist effects, anti-oestrogens that use ER downregulation may also delay the emergence of resistance to hormonal therapy (Robertson, 2004).

The partial oestrogen agonist profile of tamoxifen and other SERMs may promote the health of other several oestrogen-responsive organs, but it may also enable the development of resistance *via* upregulation of the MAP kinase and AKT pathways. Upregulation of these pathways in turn results in an increase in transcriptional activity of coactivators and ER (Clarke *et al.*, 2003; Schiff *et al.*, 2003; Lee *et al.*, 2001). Lack of ERα expression, as well as reduction in the level of corepressors, are other potential mechanisms of tamoxifen resistance. Combination therapy employing a SERM plus another agent which might reduce MAP kinase and phosphoinositol-3 kinase signalling would be predicted to reduce the development of tamoxifen-resistant breast cancer (Sporn, 2002; Sun *et al.*, 2004). On the other hand, evidence has started to emerge that the various signalling pathways 'cross-talk' at several levels with the ER pathway, and that this interaction becomes the dominant pathway when tumours become resistant to endocrine therapy. Thus, emphasis in drug development has focused on both growth factor tyrosine kinase and ras-mediated signalling, although kinases in cell cycle regulation also present potential targets (Lo and Johnston, 2003).

10.2.3 Oestrogen deprivation therapy

The initial endocrine therapy of breast cancer was removal of the ovaries. This can now also be achieved though pharmacological intervention. In the premenopausal woman with hormone receptor positive breast cancer, one of these forms of ovarian ablation is often recommended as primary treatment to allow the patient to be treated according to postmenopausal guidelines. Randomized trials support the combination of ovarian suppression or ablation and tamoxifen as the first line therapy for hormone receptor positive advanced breast cancer in premenopausal women (Heldermon and Ellis, 2006).

LH releasing hormone analogues and aromatase inhibitors constitute the chemical method for oestrogen deprivation therapy.

Oral aromatase inhibitors have been shown to reduce aromatase activity and estradiol concentrations within tumoural tissue of patients with primary breast cancer.

Formestane

Exemestane (Aromasin)

Anastrozole (Arimidex)

Letrozole (Femara)

Representative of the second generation of aromatase inhibitors, 4-hydroxyandrostenedione, or formestane, was developed in 1981 as a steroidal, suicide substrate inhibitor of aromatase. Formestane has shown good results in metastatic disease and as neoadjuvant therapy (Dowsett and Coombes, 1994; Gazet *et al.*, 1996). However, a limit to its use is the need for parenteral application (Gazet *et al.*, 1996). Third-generation aromatase inhibitors are potent inhibitors of the aromatase enzyme.

There are two types of aromatase inhibitors, irreversible steroidal activators (e.g. exemestane) and reversible non-steroidal imidazole-based inhibitors (e.g. anastrozole, letrozole) (Campos 2004). A third non-steroidal triazole, vorozole, has recently been withdrawn from clinical development (Buzdar and Howell, 2001). In postmenopausal women with advanced disease, both steroidal and non-steroidal aromatase inhibitors have shown good clinical efficacy (Brueggemeier 2004), without cross-resistance between the two groups, and similarly acceptable short-term toxicity profiles.

However, emerging data suggest that there may be differences in their effects on end organs, which may become evident with longer term use and longer observation. Steroidal agents appear to have beneficial effects on lipid and bone metabolism, due to their androgenic effects, whereas non-steroidal agents may have neutral or unfavorable effects (Campos, 2004).

In advanced breast cancer, letrozole proved to be significantly better than tamoxifen regarding response rate, clinical benefit, time to progression and time to treatment failure. Exemestane was also significantly more effective than tamoxifen as first-line therapy in postmenopausal women with advanced breast cancer.

Once it was evident that aromatase inhibitors were more effective than tamoxifen, the focus of clinical trials soon moved to the use of the agents in the adjuvant setting for the treatment of early breast cancer. The trials studied the effectiveness of aromatase inhibitors following tamoxifen, of aromatase inhibitors alone, and/or of the combination of aromatase inhibitors and tamoxifen in adjuvant therapy. They indicated that letrozole alone was better than tamoxifen or combined treatments. This result was analogous to results later reported for the ATAC trial. In addition, when tamoxifen treatment was no longer effective, tumour growth was significantly reduced in mice switched to letrozole treatment. Similar conclusions were reached in the MA-17 trial with letrozole and the IES trial with exemestane following tamoxifen in early breast cancer. Based on data from these and other trials, it was recommended by the American Society of Clinical Oncology (ASCO) technology assessment panel that optimal adjuvant hormonal therapy for a postmenopausal woman with receptor-positive breast cancer include an aromatase inhibitor as initial therapy or after treatment with tamoxifen (Jordan and Brodie, 2007).

The aromatase inhibitors are all well tolerated. Patients experienced less gynaecological symptoms such as endometrial cancer, vaginal bleeding, and vaginal discharges. There were fewer cerebrovascular and venous thromboembolic events in patients receiving aromatase inhibitors than in those on tamoxifen. However, a low incidence of bone toxicity and musculoskeletal effects are associated with aromatase inhibitors. The latter includes small but significant increases in arthritis, arthralgia and/or myalgia with aromatase inhibitors compared to tamoxifen (Jordan and Brodie, 2007).

Overall, the aromatase inhibitors are proving to be superior agents to tamoxifen in the treatment of postmenopausal women with all stages of breast cancer. However, the pharmacology of tamoxifen and other non-steroidal anti-oestrogens was to provide new therapeutic opportunities for improving women's health.

10.2.4 Sex steroid therapies

Triphenylchlorethylene, triphenylmethylethylene and stilboestrol were the first synthetic oestrogens to be used to treat carcinoma of the breast, with favourable results. It is only relatively recently, however, that high-dose oestrogens have been included in the category of SERMs, as it is now apparent that these agents have anti-oestrogenic effects on breast tumours. It may seem rather paradoxical that both the reduction of oestrogen levels and the administration of high-dose oestrogen can cause tumour regression. Although the mechanism by which high-dose oestrogen induces tumour regression is not entirely understood, this therapy has proven to be effective in breast cancer treatment. Diethylstilboestrol (DES) and Ethinyloestradiol (EE2) have been tested in several studies presenting efficacy for breast cancer (Robertson et al., 2004).

In a randomized trial of diethyl stilboestrol (DES) and tamoxifen, tamoxifen was initially favoured because high dose oestrogen was associated with more adverse cardiac events, oedema, nausea and vaginal bleeding. However, long-term follow up demonstrated a significant survival advantage for patients who received DES as their initial treatment (Peethambaram et al., 1999). Obtaining DES is difficult in the USA so oestradiol is a viable alternative at doses up to 30 mg/day. Generic oestradiol is available as 2 mg tablets and the treatment of advanced breast cancer is included in the package label. Progestins have activity in advanced breast cancer, by unclear mechanisms that may involve aromatase inhibition, increased oestrogen turnover, or direct actions on steroid receptors. Available agents for breast cancer include the orally available megesterol acetate (Megace) and the intramuscular injectable medroxyprogesterone, which are equally efficacious. Testosterone and its analogues have demonstrated response rates of 20% in metastatic breast cancer but are of historical interest only (Muss, 1992). Androgen side effects include virilization, oedema, hot flashes, and jaundice. They have reduced efficacy after the use of prior endocrine therapy and are inferior to oestrogens and tamoxifen (Schifeling et al., 1992). Figures 10.2 and 10.3 present a treatment algorithm for metastatic endocrine receptor

Diethylstilboestrol (DES)

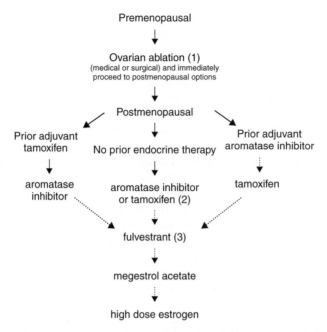

Figure 10.2   Treatment algorithm for metastatic endocrine receptor positive breast cancer. Reproduced from Heldermon C, Ellis M. Endocrine therapy from breast cancer. *Update Cancer Therapuetics* 2006, **I**, 285–97, with permission

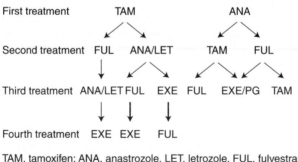

TAM, tamoxifen; ANA, anastrozole, LET, letrozole, FUL, fulvestrant;
EXE, examestane; PG, progestine

Figure 10.3   Suggested treatment schema following first treatment with tamoxifen or anastrozole. Reproduced from Robertson JFR. Selective oestrogen receptor modulators/new antioestrogens: a clinical perspective. *Cancer Treat Rev* 2004, **30**, 695–706, with permission

positive breast cancer and a suggested schema in the endocrine treatment sequence for post-menopausal patients following treatment with either adjuvant tamoxifen or adjuvant anastro-zole, respectively.

Loss of oestrogen, or its receptor(s), contributes to the development or progression of various diseases. Both activation (*via* oestrogen agonists) and inhibition (*via* oestrogen antagonists) of ER action are therapeutic strategies currently used in the clinical setting.

ER antagonists are effective for the treatment of breast cancer, and their efficacy in the treatment of other hormone-dependent cancers awaits further study. However, despite supporting evidence from animal models, the utility of oestrogen treatment to reduce the risk or severity of neurodegenerative or cardiovascular disease remains unsettled, and studies usually disagree on the optimum dose and frequency of treatment, as well as the optimum patient age at treatment initiation. Correct timing of treatment initiation during the 'window of opportunity' may be required to determine the true efficacy of oestrogen treatment. In addition, tailoring the type and duration of hormone therapy based on genetic profile may provide a means to optimize treatment for each patient. Development of tissue-specific SERMs might alleviate some of the risks associated with traditional oestrogen therapy complicated by effects in other tissues. A role for oestrogen and ERα in male fertility has also been demonstrated, and further investigation of these mechanisms may provide novel therapeutic strategies for male infertility. Finally, diseases exist for which oestrogen has been implicated in their pathogenesis but a definitive role for the ER has yet to be established. These include endometriosis and polycystic ovary syndrome, and diseases with sex-specific biases, such as lupus erythematosus, which is more common in females than in males. What role the ER may play in the risk or severity of these and other diseases will no doubt increase our ever-expanding knowledge of the relationship among oestrogen, ERs, and disease.

10.3 Progesterone-Targeted Therapy

Progesterone and its cognate receptor play essential roles in systems as diverse as the reproductive tract, the mammary gland and the nervous system. The best understood role of ovarian steroid hormones is probably their participation in reproductive tract physiology. Oestrogens drive the proliferation of the endometrium after menses, and induce progesterone receptor (PR) expression, whereas progesterone plays a role in proliferation, differentiation and maintenance of the endometrial epithelium and stroma in preparation for implantation.

PR is expressed as two isoforms (PR-A and PR-B) from a single gene. Like ER, PR contains a DBD, LBD, and multiple AFs (Li *et al.*, 2004). Studies in the rodent mammary gland have shown that PR is critical for lobuloalveolar development of the gland (Lydon *et al.*, 1996) and that the ratio of PR-A to PR-B is critical for proper mammary gland development (Shyamala *et al.*, 2000). Interestingly, an overabundance of PR-A in human breast cancers has recently been reported to be associated with resistance to tamoxifen (Hopp *et al.*, 2004), whereas a functional promoter polymorphism that results in increased production of PR-B is associated with an increased risk of breast cancer (De Vivo *et al.*, 2003). The dramatic increase in breast cancer incidence in women taking both oestrogen and progesterone for hormone replacement therapy, compared with oestrogen alone, emphasizes the importance of progesterone and the PR in breast cancer (Rossouw *et al.*, 2002). Approximately 75 % of primary breast cancers express ER, and more than half of these cancers also express PR (McGuire., 1978). PR status is a good predictor of tumour responsiveness to therapy. Nearly 50 % of all ER-positive tumours also are reported to be PR-positive and approximately 75 % of these ER/PR-positive tumours respond positively to endocrine therapy. ER-positive, PR-negative tumours are reported to be less responsive to therapy, perhaps suggesting that PR may be necessary for positive therapeutic outcomes with hormone therapy. Alternatively, because ER is a key transcription factor for the activation of PR, lack of PR expression in these ER-positive/PR-negative cells also could suggest that the oestrogen response pathway may not be functional in these tumours. To our knowledge only a small fraction of tumours are ER-negative/PR-positive (approx. 5 %) and they demonstrate an intermediate response to endocrine therapy (Keen and Davidson, 2003).

Both ER and PR are prognostic factors, although both are weak and lose their prognostic value after long-term follow up (Bardou *et al.*, 2003). PR is an oestrogen-regulated gene, and its synthesis in normal and cancer cells requires oestrogen and ER. Therefore, it is not surprising that ER-positive/PR-positive tumours are more common than ER-positive/PR-negative tumours. The aetiology of ER-positive/PR-negative tumours is currently unclear. Some studies have shown that ER and PR status can change over the natural history of the disease or during treatment. For instance, sequential breast cancer biopsies have shown that ER levels are reduced slightly with intervening endocrine therapy, although complete loss is uncommon. In contrast, PR levels decrease more dramatically during tamoxifen therapy, with up to half of tumours completely losing PR expression when resistance develops. These ER-positive/PR-negative metastatic tumours then display a much more aggressive course after loss of PR compared with tumours retaining PR, and patients then have a worse overall survival, indicating a change in tumour cell-regulatory mechanisms. Whether and how the loss of PR affects the poor clinical course of these tumours is at present unclear.

Although ER-positive/PR-negative primary untreated tumours may simply evolve by loss of PR from subclinical ER-positive/PR-negative tumours, the differences in the biology and outcome of ER-positive/PR-negative tumours suggest that some of these tumours may initially evolve separately as ER-positive/PR-negative tumours, representing their own individual stable phenotype from the outset. Indeed, recent prospective studies have shown that ER-positive/PR-negative tumours have their own unique epidemiologic risk factors. For instance, the incidence of each of the four receptor tumour subtypes (ER-positive/PR-positive, ER-positive/PR-negative, ER-negative/PR-positive, and ER-negative/PR-negative) differs with age, pregnancy history, postmenopausal hormone use, and body mass index after menopause (Cui *et al.*, 2005).

ER-positive/PR-negative breast cancers respond less well to selective ER modulator (SERM) therapy than ER-positive/PR-positive tumours. The predictive value of PR has long been attributed to the dependence of PR expression on ER activity, with the absence of PR reflecting a non-functional ER and resistance to hormonal therapy. However, recent clinical and laboratory evidence suggests that ER-positive/PR-negative breast cancers may be specifically resistant to SERMs, whereas they may be less resistant to oestrogen withdrawal therapy with aromatase inhibitors, which is a result inconsistent with the non-functional ER theory. Novel alternative molecular mechanisms potentially explaining SERM resistance in ER-positive/PR-negative tumours have been suggested by recent experimental indications that growth factors may downregulate PR levels. Thus, the absence of PR may not simply indicate a lack of ER activity, but rather may reflect hyperactive cross talk between ER and growth factor signaling pathways that downregulate PR even as they activate other ER functions. Therefore, ER-positive/PR-negative breast tumours might best be treated by completely blocking ER action *via* oestrogen withdrawal with aromatase inhibitors, by targeted ER degradation, or by combined therapy targeting both ER and growth factor signalling pathways.

Progesterone and its related signalling pathways play important role in endometrial carcinoma, which is one of the most common female pelvic malignancies. The physiological roles of progesterone in the regulation of the glandular epithelium of the endometrium are, in general, considered to antagonize oestrogen-mediated cell proliferation and to induce cellular differentiation. Progesterone has been clinically demonstrated to provide some protection against the stimulatory effects of oestrogenic agents. For example, hormone replacement therapy using combinations of oestrogen and progestin yields a lower risk of endometrial carcinoma, despite an increment in the incidence of breast carcinoma.

Reduced expression of either one or both of the PR isoforms has been observed in the great majority of endometrial carcinomas, compared with hyperplastic or normal endometrium. Case negative for either one or both of the PR isoforms were associated significantly with shorter disease-free and overall survival of the patients. Loss of expression of PR isoforms, especially that of PR-A, may result in more aggressive biological characteristics in human endometrioid endometrial carcinoma (Ito *et al.*, 2007).

New developments in endocrine therapy of endometrial carcinoma consist of progestin therapy, the organic cation transporter SLC22A16, aromatase inhibitors, retinoid and the peroxisome proliferator-activated receptor ligand (Ito *et al.*, 2007).

Atypical hyperplasia and endometrial carcinoma especially of the well-differentiated endometrioid type, often express PR, and their growth is suppressed by progestin. In general, the effect of progestin is considered to be mediated through PR, because the response rate to MPA in PR-positive carcinoma was higher (70 %) compared with PR-negative tumours. *In situ* abundance of 17β-hydroxysteroid dehydrogenases (17-HSD) type 2, which catalyses the conversion of the potent oestrogen, E2, to an inactive form, E1, and PR, especially PR-B, can predict the responsiveness of patients with endometrioid endometrium. In addition, progestin stimulates the expression of 17-HSD type 2 and thereby promoting the regression of endometrium proliferative disease (Ito *et al.*, 2007).

Overall, the rapid development of new approaches to endocrine therapy will enhance treatment options for oestrogen/progesterone mediated disease and advance the emerging clinical strategy of chemoprevention.

10.4 Neuroendocrine Tumours

Neuroendocrine tumours comprise a diverse group of malignancies and, when defined broadly, include not only gastrointestinal neuroendocrine tumours but also pheochromocytoma, thyroid cancer, and even small cell lung cancer (Kulke, 2007). Neuroendocrine tumours are classified according to their differentiation, localization and functionality (Table 10.1). The classification and the stage of the tumour disease have important prognostic implications, and influence therapeutic decisions.

Gastrointestinal neuroendocrine tumours of the foregut include neuroendocrine tumours of the stomach, duodenum and pancreas. Neuroendocrine tumours of the duodenum are rare, the majority producing and secreting gastrin, less often serotonin or somatostatin. While the autonomous secretion of gastrin or serotonin gives rise to well-known hypersecretion syndromes such as Zollinger–Ellison or carcinoid syndrome, most somatostatinomas are clinically indolent and thus referred to as non-functioning tumours. Most of these tumours are either benign or well-differentiated carcinomas. Most neuroendocrine tumours of the pancreas are functioning tumours, while 20–50 % is non-functioning tumours (Venkatesh *et al.*, 1990; Lo *et al.*, 1996). Functioning tumours are diagnosed early in the course of the disease due to the respective hormonal hypersecretion syndromes, and identifying the small primary may be a diagnostic challenge. Most functioning tumours are well-differentiated neuroendocrine carcinomas, with the exception of insulinomas, which are benign in 90 % of cases.

Table 10.1   Criteria for the classification of neuroendocrine tumours

Localization	Differentiation	Proliferation	Functionality
Stomach	Well-differentiated tumour	Ki67 $\leqslant 2\%$	Functioniong
Duodenum	Well-differentiated carcinoma	Ki67 $>3\%$	Non-functioning
Pancreas	Poorly differentiated carcinoma	Ki67 $>20\%$	
Jejunum/ileum			
Colon			
Rectum			

From Plockinger U and Wiedemann B. Biotherapy. *Best Practice and Research Clinical Endocrinology and Metabolism* 2007a, **21**, 146–62. Reproduced with permission from Elsevier Ltd.

Non-functioning tumours, on the other hand are diagnosed due to signs and symptoms of the tumour disease, i.e., a large abdominal mass or liver metastases. In functioning tumours, death is due to malignant disease, since symptomatic therapies offer excellent control for most hormone hypersecretion syndromes. Biotherapy is the primary treatment for patients with hormonal hypersecretion syndromes such as the watery diarrhea syndrome of a rare vasoactive intestinal peptide-oma, or the skin manifestations of a glucagonoma. Neuroendocrine tumours of the midgut include tumours of the lower jejunum and ileum, the appendix (23–28 %, and 19 % of all gastrointestinal endocrine tumours) and the right colon. Most of these tumours are well-differentiated and grow slowly. On average 7.5 years elapse after the first symptoms until a diagnosis is finally made. As a consequence of this indolent clinical course, most patients have widespread disease at diagnosis. Thus, these tumours are rarely cured by surgery, as most tumours have already metastasized at diagnosis.

Considering therapeutic decisions, the prognostic implications of tumour localization, size and differentiation have to be accounted for. With neuroendocrine tumours of the small bowel which are restricted to the bowel wall and have undergone apparently curative resection, recurrences with metastases occurred after a median time of 16 years. After 25 years, only 23 % of these patients were free of disease (Moertel, 1987). Tumour size is an important prognostic parameter in appendiceal tumours. These tumours almost never metastasize if their size is <1 cm. In contrast, those >2 cm do metastasize and should be regarded as aggressive carcinomas (Moertel, 1987). On the other hand, neuroendocrine tumours of the small intestine (appendiceal tumours excluded) of comparable size can behave very aggressively (Oberg, 1993): local lymph-node metastases are present at the time of diagnosis in up to 39 % and non-localized disease is evident in 70.7 % of these patients (Modlin and Sandor, 1997).

Surgery remains a cornerstone in the treatment of neuroendocrine tumours, even when they are not completely resectable. In the recent years, a surgical approach has been more utilized including resection of the primary tumour and, also, lymph nodes metastases together with enucleation of liver metastases.

Debulking procedures are important in managing both severe clinical syndromes and in facilitating medical treatment. The medical treatment of functioning endocrine tumours must be based on the growth properties of the tumour and includes chemotherapy, somatostatin analogues, interferon-alpha (IFN-α) alone and associated with somatostatin analogues, chemoembolization and radiolabelled somatostatin analogues (Tomassetti *et al.*, 2004).

Chemotherapy is a palliative option in both slow-growing, well-differentiated, or rapidly proliferating, poorly differentiated, neuroendocrine tumours. In slow-growing, well-differentiated tumours, tumour growth is unpredictable, and slow tumour progression may alternate with long intervals of stable disease. As the quality of life is good in most patients with metastasized well-differentiated neuroendocrine tumours, antiproliferative therapy should only be initiated whenever progressive disease has been demonstrated according to standard criteria. While different treatment regimens have been shown to be effective in well-differentiated pancreatic neuroendocrine carcinomas, chemotherapeutic options for tumours of the small bowel are poor. Localization of the primary is not as important for poorly differentiated carcinomas, as these rapidly growing tumours respond to different chemotherapeutic agents, irrespective of the localization of the primary. These tumours may grow rapidly, and medical treatment should not be withheld to demonstrate progressive disease.

Biotherapy is defined as the therapy of hormonal hypersecretion syndromes and/or tumour growth with substances, or pharmacological derivatives thereof, occurring naturally in the body. Despite the widespread therapeutic use of biotherapy in neuroendocrine tumours for more than 20 years, data fulfilling the criteria of evidence-based medicine are rare. However, there are indications of the benefit of biotherapy on symptoms of hormone hypersecretion, while on the other hand, definite data on the antiproliferative effectiveness and its positive effect on survival are still lacking (Plockinger and Wiedenmann, 2007a, b).

10.4.1 Somatostatin analogues (SSA)

Endogenous somatostatin (SS) circulates in two biological active forms, SS-14 and SS-28, due to proteolytic processing from a large precursor molecule. SS binds with high affinity to five G protein-coupled membrane receptors (sst1–5). Ligand binding inhibits adenylate cyclase activity, reduces calcium influx, and negatively influences hormone synthesis and secretion. An inhibitory influence on proliferation may be due to the activation of phosphotyrosine phosphatases, the mitogen-activated protein kinase activity (Lamberts *et al.*, 1991), and to the inhibition of the transcription factor complex activator protein 1. *In vitro*, high doses of SSA induce apoptosis in tumour cells, and this could translate into inhibition of tumour growth. Additional antiproliferative effects may be related to the antiangiogenic activity of SS either directly or *via* inhibition of growth factors (Dasgupta, 2004). Most neuroendocrine tumours express a higher density of sst in tumour tissue compared to the normal tissue. This allows for specific, tumour tissue-targeted, therapeutic effects and should reduce the number of side effects, i.e., suppression of physiologically secreted hormones. However, sst-subtype expression varies considerably between different tumour types and among tumours of the same type. Even within a given tumour, sst expression is not homogenously distributed (Hofland and Lamberts, 2003).

The clinically used SSA, octreotide and lanreotide, preferentially bind to sst2 and sst5. For these analogues, serum half-life is increased considerably compared to native somatostatin. The subcutaneous injectable octreotide has to be given three times daily, while long acting preparations, (lanreotide long-lasting and octreotide long-acting repeatable [LAR]), allow for one intramuscular injection every two (lanreotide) to four (octreotide LAR) weeks. In the case of lanreotide autogel, an interval up to 6 weeks between injections and the possibility to be injected by the patients themselves may increase patients' comfort and compliance. Recently, pasireotide has been introduced, a SSA with high affinity for sst1–3 and sst5. It has been shown to be effective in patients who do not respond to the currently available SSA octreotide and lanreotide (Plockinger and Wiedenmann B., 2007a, b).

The latest achievement in this area is the development of a new compound, SOM-230, which is a somatostatin analogue with a broader receptor binding profile. This analogue binds with high affinity to sst subtypes 1, 2, 3 and 5 and it has been used in clinical trials including both patients with acromegaly and carcinoid tumours (Janson, 2006).

Octreotide and lanreotide both effectively inhibit autonomous hormone or neurotransmitter secretion by neuroendocrine gastrointestinal tumours. Thus, control of hypersecretion symptoms – i.e. carcinoid syndrome with flush and diarrhoea, watery diarrhoea syndrome and the necrotic skin lesions of the glucagonoma syndrome – can be achieved in most patients. Unfortunately, tachyphylaxis develops after months or even years of treatment in virtually all patients. Tachyphylaxis may be due to desensitization, homologous agonist-induced down-regulation in sst numbers on the cell surface, heterologous regulation of SS receptor expression, or even SS receptor gene mutations (Hofland and Lamberts, 2003).

SSAs are indicated in patients with symptoms due to excessive hormone release by a neuroendocrine tumour or its metastases. In patients with the carcinoid syndrome, octreotide LAR is equally potent in the control of flushing and diarrhoea if compared to subcutaneous administration. SSAs are indicated in the therapy of the watery diarrhoea syndrome and the therapy of the glucagonoma syndrome. While SSAs are effective in lowering the gastrin concentration in patients with Zollinger–Ellison syndrome, effective symptomatic relief in these patients can be achieved by proton pump inhibitors, and this is currently the preferred medication for the control of autonomous gastrin secretion. In patients with insulin hypersecretion, SSA may reduce insulin concentration. However, as glucagon secretion is also inhibited, patients have to be observed closely at the beginning of therapy to prevent severe hypoglycemia due to the reduced glucagon-dependent counter-regulation. Thus, diazoxid is the preferred primary therapy for inhibition of insulin secretion, effectively reducing the risk of hypoglycemia. SSAs effectively inhibit glucagon secretion in patients with a glucagonoma syndrome (O'Toole et al., 2000; Ricci et al., 2000; Ruszniewski et al., 2004).

There is an excellent effect of SSAs on tumour markers. Octreotide results in remission and/or stabilization of tumour markers in approximately 70 % of the patients (Oberg, 1994). The decline of tumour markers such as chromogranin A is due to the antisecretory effect of SSAs and should not be interpreted as indication for tumour volume reduction. Unfortunately, the duration of remission is short (median 8–12 months), with tachyphylaxis occurring early in the course of therapy (Arnold et al., 1996).

Clinical studies have, so far, given disappointing results with regard to tumour regression. Tumour shrinkage is demonstrated in less than 10 % of the patients. However, stabilization of tumour growth, after computerized tomography scan (CT)-documented progression prior to treatment, occurs in up to 50 % of the patients with neuroendocrine tumours of various locations (Saltz et al., 1993). Stable disease is observed in 37–45 % of the patients with documented tumour progression before SSA therapy. The median duration of stabilization was 18–26.5 months (Panzuto et al., 2006). In a highly select group of patients with progressive disease, 47 % of the patients demonstrated at least stable disease when treated with a high dose of lanreotide (Faiss et al., 1996). This has been confirmed recently in 75 % of the patients with advanced midgut carcinoids, with stabilization for 6–24 months (Welin et al., 2004). In neuroendocrine tumours of the small intestine, the therapeutic effect of high-dose treatment may be slightly better (stable disease in eight of ten patients) than with conventional dosage (Faiss et al., 1999).

Predictors of the clinical outcome of SSA therapy have been analysed. A pancreatic primary, with no previous surgical therapy and distant, extrahepatic metastases, indicates a poor response to treatment in multivariate analysis, while age, size of the primary, and Ki67 did not influence the response rate to SSA therapy (Panzuto *et al.*, 2006). Patients achieving a positive response (stabilization) after 6 months of treatment maintain it throughout long-term follow-up and live longer than patients unresponsive to therapy (Arnold *et al.*, 2005; Panzuto *et al.*, 2006).

Frequently occurring side-effects such as abdominal discomfort, bloating, and steatorrhea, due to the inhibition of pancreatic enzymes, are mostly mild and subside spontaneously within the first weeks of therapy. Persistent steatorrhoea can be treated with supplementation of pancreatic enzymes. Cholestasis with subsequent cholecystolithiasis does occur in up to 60 % of the patients, due to inhibition of cholecystokinin and production of lithogenic bile (Trendle *et al.*, 1997). Prophylactic therapy with chenodeoxycholic acid and ursodeoxycholic acid may be able to prevent the occurrence of gallstone disease in patients on long-term SSA therapy (Plockinger U. *et al.*, 1990). Due to steatorrhoea, malabsorption with reduction of serum vitamin D concentration and subsequently reduced calcium absorption may occur. In patients on long-term therapy, the vitamin B12 concentration should be monitored. Serum vitamin B12 concentration may decline, possibly due to a direct inhibition of the intrinsic factor secretion at the parietal cell (Plockinger *et al.*, 1990, 1998). In addition a moderate effluvium may occur and is usually reversible when the drug is stopped.

In summary, SSAs effectively control symptoms of hypersecretion syndromes in patients with neuroendocrine tumours of the gastrointestinal tract. Despite the minor effects on tumour volume reduction observed so far, an antiproliferative effect does occur, with stabilization of the disease for up to 25 months. A possible positive effect on tumour volume regression with high-dose treatment has still to be demonstrated. Survival may be prolonged in those patients responding positively to somatostatin analogue therapy. In addition, SSAs significantly increase the quality of life in patients with symptoms related to hormone secretion, while side-effects of SSA therapy are limited.

SSAs are usually used as combined therapy with INF-α. IFN-α has a direct effect on the tumour cells but also trigger the immune system. It inhibits the production of the growth-factor-receptor and other agents secreted by the tumour cells. It is also known to induce class 1 antigens on the cell surface; it is also assumed that IFN-α has an anti-angiogenic effect. One of the reasons for the discrepancy in biochemical and tumour responses might be that interferon induces fibrosis within the tumour which will not decrease the tumour size at CT scan or ultrasound. The treatment should continue for the long-term and some patients should be treated continuously for more than 10 years (Tomassetti *et al.*, 2004). The combination of SSAs and IFN-α has been used in an effort to enhance the antiproliferative effect of interferon therapy, to add the positive effect of SSAs on hypersecretory syndromes, and to reduce the dose of IFN-α and thus the number of IFN-related side effects.

In an early study by Creutzfeldt *et al.* (1991), no additional antiproliferative effect was seen when combination therapy was used, while more recent studies demonstrated an increased number of patients with stable disease for tumour volume (57–75 %) and/or remission or stable biochemical parameters (77–92 %). To summarize, if biotherapy with SSA or IFN fails, due to progression of the disease or to tachyphylaxis (SSA), or side effects are intolerable with IFN-α, the combination of both drugs may be useful in individual patients.

In recent years, some researchers have tried to develop a somatostatin analogue which has a high affinity for somatostatin receptors and which could be linked to a therapeutic beta-emitting radioisotope. Beta particles emitted from a radiolabelled peptide bound to a tumour cell also kill neighbouring cells, because the path length of beta-particles can extend over several cell diameters. The crossfire of beta-particles can, in theory, destroy both somatostatin receptor-positive and-negative tumour cells. A new generation of somatostatin analogues which ensure better stability of the radiometal–peptide complex incorporating the macrocyclic chelator 1,4,7,10-tetraazacyclododecane-N,N',N'',N'''-tetraacetic acid (DOTA) instead of diethylenetriaminepentaacetic acid (DTPA), have been developed. After labelling with ^{90}Y (beta-emitter) or ^{111}In (gamma-emitter), they showed improved biodistribution and tumour uptake in animal models and their clinical utility as diagnostic and therapeutic tools was confirmed in patients. However, the beta-emitters (^{90}Y) are considered preferable to ^{111}In for radiation therapy. In the last 3 years, many reports have appeared in the literature about the therapeutic usefulness of peptide receptor radionuclide therapy (PRRT) in neuroendocrine functioning tumours. Studying several patients, the authors demonstrated that it was possible to obtain a partial remission on a small number of patients and the stabilization of a previously progressive tumour in the majority of patients. The Rotterdam Group reported their results using ^{177}Lu DOTA,Tyr3 octreotate, which shows the highest tumour uptake of all the octreotide analogues tested so far, with excellent tumour–kidney ratios. The antitumoural effects of these two compounds are a promising new treatment modality for patients who have SST 2 receptor-positive tumours. In general, ^{111}In DOTATOC, ^{90}Y, and ^{177}Lu are well-tolerated; renal toxicity is mild and can usually be avoided with the concomitant infusion of protective amino acids. Hence, the majority of studies, show that the antitumoural effects and symptomatic improvement produced by internal radiotherapy are most encouraging (Tomassetti *et al.*, 2004).

Biotherapy is preferentially indicated for treatment of hormone hypersecretion syndromes in patients with neuroendocrine tumours. Excellent symptomatic relief can be achieved by SSAs, IFN-α and other medical treatment strategies. Antiproliferative effects are not convincing, but stabilization of the disease does occur in up to 50 % of the patients with either SSA or interferon therapy. Combined treatment does not add substantial effect and is not recommended. Side effects of SSA therapy are minor, and quality of life is good with SSA therapy, while with interferon the drug has to be individually titrated to balance treatment efficacy and side effects.

Novel agents have been developed for treatment of neuroendocrine tumours. The targets are growth factors, growth factor receptors, angiogenetic molecules, and kinases involved in proliferation pathways. Targeted therapy aims at different levels of these proliferation pathways like antibody-induced neutralization of vascular endothelial growth factor (VEGF) ligands or the epithelial growth factor receptor to prevent ligand–receptor interaction or direct inhibition of the receptor tyrosine kinase by small molecules. The best known examples of multikinase inhibitors are imatinib or sunitinib, with activity against kinases of platelet-derived growth factor receptor (PDGF-R), bcr-abl, c-kit and PDGF-R, VEGF-R, RET, and c-kit, respectively. RAD001 (everolimus) a rapamycin analogue, is an inhibitor of the mammalian target of rapamycin (mTOR). Inhibition of the protein kinase mTOR may reduce cell growth, increase apoptosis, and reduce metastatic spread. It may have synergistic effects with SSA, as the mTOR pathway is stimulated by IGF-1, which is inhibited by SSA (Plockinger and Wiedenmann, 2007b).

References

Amantea D, Russo R, Bagetta G, Corasaniti MT. From clinical evidence to molecular mechanisms underlying neuroprotection afforded by oestrogens. *Pharmacol Res* 2005, **52**, 119–32.

Arnold R, Rinke A, Klose KJ, *et al*. Octreotide versus octreotide plus interferon-alpha in endocrine gastroenteropancreatic tumours: a randomized trial. *Clin Gastroenterol Hepatol* 2005, **3**, 761–71.

Arnold R, Trautmann ME, Creutzfeldt W, *et al*. Somatostatin analogue octreotide and inhibition of tumour growth in metastatic endocrine gastroenteropancreatic tumours. *Gut* 1996, **38**, 430–8.

Bardou VJ, Arpino G, Elledge RM, *et al*. Progesterone receptor status significantly improves outcome prediction over oestrogen receptor status alone for adjuvant endocrine therapy in two large breast cancer databases. *J Clin Oncol* 2003, **21**, 1973–9.

Barkhem T, Carlsson B, Nilsson Y, Enmark E, Gustafsson J, Nilsson S. Differential response of oestrogen receptor alpha and oestrogen receptor beta to partial oestrogen agonists/antagonists. *Mol Pharmacol* 1998, **54**, 105–12.

Bentrem DJ, Jordan VC. Role of antiestrogens and aromatase inhibitors in breast cancer treatment. *Curr Opin Obstet Gynecol* 2002, **14**, 5–12.

Biganzoli L, Claudino WM, Pestrin M, Pozzessere D, Di Leo A. Selection of chemotherapeutic drugs in adjuvant programs based on molecular profiles: Where do we stand? *Crit Rev Oncol/Hematol* 2007, **62**, 1–8.

Biswas DK, Dai SC, Cruz A, Weiser B, Graner E, Pardee AB. The nuclear factor kappa B (NF-kappa B): a potential therapeutic target for estrogen receptor negative breast cancers. *Proc Natl Acad Sci U S A* 2001, **98**, 10386–91.

Black LJ, Jones CD, Clark JH, Clemens JA. LY156758: a unique anti-oestrogen displaying high affinity for oestrogen receptors, negligible oestrogenic activity and near-total oestrogen antagonism *in vivo*. *Breast Cancer Res Treat* 1982, **2**, 279.

Bosland MC. The role of steroid hormones in prostate carcinogenesis. *J Natl Cancer Inst Monogr* 2000, **27**, 39–66.

Brueggemeier RW. Aromatase inhibitors: new endocrine treatment of breast cancer. *Semin Reprod Med* 2004, **22**, 31–43.

Buzdar A, Howell A. Advances in aromatase inhibition: clinical efficacy and tolerability in the treatment of breast cancer. *Clin Cancer Res* 2001, **7**, 2620–35.

Buzdar AU, Marcus C, Holmes F, Hug V, Hortobagyi G. Phase II evaluation of Ly156758 in metastatic breast cancer. *Oncology* 1988, **45**, 344–5.

Campos SM. Aromatase inhibitors for breast cancer in postmenopausal women. *Oncologist* 2004, **9**, 126–36.

Chu S, Mamers P, Burger HG, Fuller PJ. Oestrogen receptor isoform gene expression in ovarian stromal and epithelial tumours. *J Clin Endocrinol Metab* 2000, **85**, 1200–5.

Clarke R, Liu MC, Bouker KB, *et al*. Anti-oestrogen resistance in breast cancer and the role of oestrogen receptor signaling. *Oncogene* 2003, **22**, 7316–39.

Cotreau MM, Stonis L, Dykstra KH, *et al*. Multiple-dose, safety, pharmacokinetics, and pharmacodynamics of a new selective oestrogen receptor modulator, ERA-923, in healthy postmenopausal women. *J Clin Pharmacol* 2002, **42**, 157–65.

Couse JF, Korach KS. Oestrogen receptor null mice: what have we learned and where will they lead us? *Endocr Rev* 1999, **20**, 358–417.

Creutzfeldt W, Bartsch HH, Jacubaschke U *et al*. Treatment of gastrointestinal endocrine tumours with interferon-alpha and octreotide. *Acta Oncol* 1991, **30**, 529–35.

Cui X, Schiff R, Arpino G, Osborne CK, Lee AV. Biology of progesterone receptor loss in breast cancer and its implication for endocrine therapy. *J Clin Oncol* 2005, **23**, 7721–35.

Cummings SR, Eckert S, Krueger KA, *et al*. The effect of raloxifene on risk of breast cancer in postmenopausal women: results from the MORE randomized trial. Multiple Outcomes of Raloxifene Evaluation. *JAMA* 1999, **281**, 2189–97.

Dardes RC, O'Regan RM, Gajdos C, *et al*. Effects of a new clinically relevant anti-oestrogen (GW5638) related to tamoxifen on breast and endometrial cancer growth *in vivo*. *Clin Cancer Res* 2002, **8**, 1995–2001.

Dasgupta P. Somatostatin analogues: multiple roles in cellular proliferation, neoplasia, and angiogenesis. *Pharmacol Ther* 2004, **102**, 61–85.

De Vivo I, Hankinson SE, Colditz GA, *et al.* A functional polymorphism in the progesterone receptor gene is associated with an increase in breast cancer risk. *Cancer Res* 2003, **63**, 5236–38.

Dehal SS, Kupfer D. CYP2D6 catalyzes tamoxifen 4-hydroxylation in human liver. *Cancer Res* 1997, **57**(16), 3402–6.

Dehal SS, Kupfer D. Cytochrome P-450 3A and 2D6 catalyze ortho hydroxylation of 4-hydroxytamoxifen and 3-hydroxytamoxifen (droloxifene) yielding tamoxifen catechol: involvement of catechols in covalent binding to hepatic proteins. *Drug Metab Dispos* 1999, **27**(6), 681–8.

Delmas PD, Bjarnason NH, Mitlak BH, *et al.* Effects of raloxifene on bone mineral density, serum cholesterol concentrations, and uterine endometrium in postmenopausal women. *N Engl J Med* 1997, **337**, 1641–7.

Deroo BJ, Korach KS. Oestrogen receptors and human diesease. *J Clin Invest* 2006, **116**, 561–70.

Dowsett M, Coombes RC. Second generation aromatase inhibitor – 4-hydroxyandrostenedione. *Breast Cancer Res Treat* 1994, **30**, 81–7.

Early Breast Cancer Trialists' Collaborative Group. Tamoxifen for early breast cancer: an overview of the randomised trials. *Lancet* 1998, **351**, 1451–67.

Enmark E, Pelto-Huikko M, Grandien K, *et al.* Human oestrogen receptor beta-gene structure, chromosomal localization, and expression pattern. *J Clin Endocrinol Metab* 1997, **82**, 4258–65.

Ettinger B, Black DM, Mitlak BH, *et al.* Reduction of vertebral fracture risk in postmenopausal women with osteoporosis treated with raloxifene: results from a 3-year randomized clinical trial. Multiple Outcomes of Raloxifene Evaluation (MORE) Investigators. *JAMA* 1999, **282**, 637–45.

Fabian CJ, Kimler BF. Selective oestrogen-receptor modulators for primary prevention of breast cancer. *J Clin Oncol* 2005, **23**, 1644–55.

Faiss S, Rath U, Mansmann U, *et al.* Ultra-high-dose lanreotide treatment in patients with metastatic neuroendocrine gastroenteropancreatic tumours. *Digestion* 1999, **60**, 469–76.

Faiss S, Scherubl H, Riecken EO, *et al.* Drug therapy in metastatic neuroendocrine tumours of the gastroenteropancreatic system. *Recent Results Cancer Res* 1996, **142**, 193–207.

Fisher B, Costantino JP, Wickerham DL, *et al.* Tamoxifen for prevention of breast cancer: report of the National Surgical Adjuvant Breast and Bowel Project P-1 Study. *J Natl Cancer Inst* 1998, **90**, 1371–88.

Gail MH, Costantino JP, Bryant J, *et al.* Weighing the risks and benefits of tamoxifen treatment for preventing breast cancer [published erratum appears in *J Natl Cancer Inst* 2000, **92**, 275]. *J Natl Cancer Inst* 1999, **91**, 1829–46.

Garcia-Segura LM, Azcoitia I, DonCarlos LL. Neuroprotection by estradiol. *Prog Neurobiol* 2001, **63**, 29–60.

Gazet JC, Coombes RC, Ford HT, *et al.* Assessment of the effect of pretreatment with neoadjuvant therapy on primary breast cancer. *Br J Cancer* 1996, **73**, 758–62.

Hall JM, McDonnell DP. The oestrogen receptor beta-isoform (ERbeta) of the human oestrogen receptor modulates ERalpha transcriptional activity and is a key regulator of the cellular response to oestrogens and anti-oestrogens. *Endocrinology* 1999, **140**, 5566–78.

Harris HA, Albert LM, Leathurby Y, *et al.* Evaluation of an oestrogen receptor-beta agonist in animal models of human disease. *Endocrinology* 2003, **144**, 4241–9.

Hayashi S, Yamaguchi Y. Basic research for hormone-sensitivity of breast cancer. *Breast Cancer* 2006, **13**, 123–8.

Heldermon C, Ellis M. Endocrine therapy from breast cancer. *Update Cancer Therapuetics* 2006, **I**, 285–97.

Hofland LJ, Lamberts SW. The pathophysiological consequences of somatostatin receptor internalization and resistance. *Endocr Rev* 2003, **24**, 28–47.

Hopp TA, Weiss HL, Hilsenbeck SG, *et al.* Breast cancer patients with progesterone receptor PR-A-rich tumours have poorer disease-free survival rates. *Clin Cancer Res* 2004, **10**, 2751–60.

Howell SJ, Jonston S, Howell A. The use of selective oestrogen receptor modulators and selective oestrogen receptor down-regulators in breast cancer. *Best Pract Res Clin Endocrinol Metab* 2004, **18**, 47–66.

Ito k, Utsunomiya H., Yaegashi N, Sasano H. Biological roles of oestrogen and progesterone in human endometrial carcinoma – new developments in potential endocrine therapy for endometrial cancer. *Endocr J* 2007, **54**, 667–79.

Jager W, Sauerbrei W, Beck E, *et al.* A randomized comparison of triptorelin and tamoxifen as treatment of progressive ovarian cancer. *Anticancer Res* 1995, **15**, 2639–42.

Janson ET. Treatment of neuroendocrine tumours with somatostatin analogs. *Pituitary* 2006, **9**, 249–56.

Jassam N, Bell SM, Speirs V, Quirke P. Loss of expression of ooestrogen receptor beta in colon cancer and its association with Dukes' staging. *Oncol Rep* 2005, **14**, 17–21.

Jordan VC. Optimising endocrine approaches for the chemoprevention of breast cancer Beyond the Study of Tamoxifen and Raloxifene (STAR) Trial. *Eur J Cancer* 2006, **42**, 2909–13.

Jordan VC, Brodie A. Development and evolution of therapies targeted to the oestrogen receptor for the treatment and prevention of breast cancer. *Steroids* 2007, **72**, 7–25.

Jordan VC, Gapstur S, Morrow M. Selective oestrogen receptor modulation and reduction in risk of breast cancer, osteoporosis, and coronary heart disease. *J Natl Cancer Inst* 2001, **93**, 1449–57.

Kajta M, Beyer C. Cellular strategies of oestrogen-mediated neuroprotection during brain development [review]. *Endocrine* 2003, **21**, 3–9.

Kali KR, Bradley SV, Fuchshuber S, Conover CA. Oestrogen receptor-positive human epithelial ovarian carcinoma cells respond to the antitumour drug suramin with increased proliferation: possible insight into ER and epidermal growth factor signaling interactions in ovarian cancer. *Gynecol Oncol* 2004, **94**, 705–12.

Keen JC, Davidson NE. The biology of breast carcinoma. *Cancer Supplement* 2003, **97**, 825–33.

Konstantinopoulos PA, Kominea A, Vandoros G *et al.* Ooestrogen receptor beta (ERbeta) is abundantly expressed in normal colonic mucosa, but declines in colon adenocarcinoma paralleling the tumour's dedifferentiation. *Eur J Cancer* 2003, **39**, 1251–58.

Krishnan G. Novel ERbeta selective agonists induce prostate atrophy in rodents without affecting the hypothalamo-pituitary-gonadal axis. In: *The Endocrine Society Annual Meeting 2004*. The Endocrine Society Press, Chevy Chase, MD, USA, 2004, pp 180–1.

Kulke MH. Gastrointestinal neuroendocrine tumours: a role for targeted therapies? *Endocr Relat Cancer* 2007, **14**, 207–19.

Lamberts SW, Krenning EP, Reubi JC. The role of somatostatin and its analogs in the diagnosis and treatment of tumours. *Endocr Rev* 1991, **12**, 450–82.

Lee AV, Cui X, Oesterreich S. Cross-talk among oestrogen receptor, epidermal growth factor, and insulin-like growth factor signaling in breast cancer. *Clin Cancer Res* 2001, **7**, 4429s–35s.

Li X, Lonard DM, O'Malley BW. A contemporary understanding of progesterone receptor function. *Mech Ageing Dev* 2004, **125**, 669–78.

Lo CY, van Heerden JA, Thompson GB, *et al.* Islet cell carcinoma of the pancreas. *World J Surg* 1996, **20**, 878–883.

Lo S, Johnston S. Novel systemic therapies for breast cancer. *Surg Oncol* 2003, **12**, 277–87.

Lydon JP, DeMayo FJ, Conneely OM, *et al.* Reproductive phenotpes of the progesterone receptor null mutant mouse. *J Steroid Biochem Mol Biol* 1996, **56**, 67–77.

McGuire WL. Hormone receptors: Their role in predicting prognosis and response to endocrine therapy. *Semin Oncol* 1978, **5**, 428–33.

Modlin IM, Sandor A. An analysis of 8305 cases of carcinoid tumours. *Cancer* 1997, **79**, 813–29.

Moertel CG. Karnofsky memorial lecture. An odyssey in the land of small tumours. *J Clin Oncol* 1987, **5**, 1502–22.

Morgan Jr, LR, Schein PS, Woolley PV, *et al.* Therapeutic use of tamoxifen in advanced breast cancer: correlation with biochemical parameters. *Cancer Treat Rep* 1976, **60**, 1437–43.

Morris C, Wakeling A. Fulvestrant ('Faslodex') – a new treatment option for patients progressing on prior endocrine therapy. *Endocr Relat Cancer* 2002, **9**, 267–76.

Mueller SO, Korach KS. Oestrogen receptors and endocrine diseases: lessons from oestrogen receptor knockout mice. *Curr Opin Pharmacol* 2001, **1**, 613–19.

Muss HB. Endocrine therapy for advanced breast cancer: a review. *Breast Cancer Res Treat* 1992, **21**(1), 15–26.

Muss HB, Case LD, Atkins JN, *et al*. Tamoxifen versus high-dose oral medroxyprogesterone acetate as initial endocrine therapy for patients with metastatic breast cancer: a piedmont oncology association study. *J Clin Oncol* 1994, **12**, 1630–8.

O'Toole D, Ducreux M, Bommelaer G, *et al*. Treatment of carcinoid syndrome: a prospective crossover evaluation of lanreotide versus octreotide in terms of efficacy, patient acceptability, and tolerance. *Cancer* 2000, **88**, 770–6.

Oberg K. Chemotherapy and biotherapy in neuroendocrine tumours. *Curr Opin Oncol* 1993, **5**, 110–20.

Oberg K. Endocrine tumours of the gastrointestinal tract: systemic treatment. *Anticancer Drugs* 1994, **5**, 503–19.

Panzuto F, Di Fonzo M, Iannicelli E, *et al*. Long-term clinical outcome of somatostatin analogues for treatment of progressive, metastatic, well-differentiated entero-pancreatic endocrine carcinoma. *Ann Oncol* 2006, **17**, 461–66.

Peethambaram PP, Ingle JN, Suman VJ, Hartmann LC, Loprinzi CL. Randomized trial of diethylstilbestrol vs. tamoxifen in postmenopausal women with metastatic breast cancer. An updated analysis. *Breast Cancer Res Treat* 1999, **54**, 117–22.

Perey L, Thurlimann B, Hawle H, *et al*. Fulvestrant ('faslodex') as hormonal treatment in postmenopausal patients with advanced breast cancer progressing after treatment with tamoxifen and aromatase inhibitors. *Breast Cancer Res Treat* 2002, **76**, S72.

Picariello L, Fiorelli G, Martineti V, *et al*. Growth response of colon cancer cell lines to selective oestrogen receptor modulators. *Anticancer Res* 2003, **23**, 2419–24.

Plockinger U, Wiedenmann B. Biotherapy. *Best Pract Res Clin Endocrinol Metab* 2007a, **21**, 145–62.

Plockinger U, Wiedenmann B. Treatment of gastroenteropancreatic neuroendocrine tumours. *Virchows Arch* 2007b, **451**, S71–S80.

Plockinger U, Dienemann D, Quabbe HJ. Gastrointestinal side-effects of octreotide during long-term treatment of acromegaly. *J Clin Endocrinol Metab* 1990, **71**, 1658–62.

Plockinger U, Perez-Canto A, Emde C, *et al*. Effect of the somatostatin analog octreotide on gastric mucosal function and histology during 3 months of preoperative treatment in patients with acromegaly. *Eur J Endocrinol* 1998, **139**, 387–94.

Prestwood KM, Gunness M, Muchmore DB, Lu Y, Wong M, Raisz LG. A comparison of the effects of raloxifene and oestrogen on bone in postmenopausal women. *J Clin Endocrinol Metab* 2000, **85**, 2197–202.

Ricci S, Antonuzzo A, Galli L, *et al*. Long-acting depot lanreotide in the treatment of patients with advanced neuroendocrine tumours. *Am J Clin Oncol* 2000, **23**, 412–15.

Robertson JF, Osborne CK, Howell A, *et al*. Fulvestrant versus anastrozole for the treatment of advanced breast carcinoma in postmenopausal women: a prospective combined analysis of two multicenter trials. *Cancer* 2003, **98**, 229–38.

Robertson JFR. Selective ooestrogen receptor modulators/new antiooestrogens: a clinical perspective *Cancer Treat Rev* 2004, **30**, 695–706.

Rossouw JE, Anderson GL, Prentice RL, *et al*. Risks and benefits of oestrogen plus progestin in healthy postmenopausal women: Principal results from the Women's Health Initiative randomized controlled trial. *JAMA* 2002, **288**, 321–33.

Ruszniewski P, Ish-Shalom S, Wymenga M, *et al*. Rapid and sustained relief from the symptoms of carcinoid syndrome: results from an open 6-month study of the 28-day prolonged-release formulation of lanreotide. *Neuroendocrinology* 2004, **80**, 244–251.

Saltz L, Trochanowski B, Buckley M, *et al*. Octreotide as an antineoplastic agent in the treatment of functional and nonfunctional neuroendocrine tumours. *Cancer* 1993, **72**, 244–48.

Sawka CA, Pritchard KI, Shelley W, *et al*. A randomized crossover trial of tamoxifen versus ovarian ablation for metastatic breast cancer in premenopausal women: a report of the National Cancer Institute of Canada Clinical Trials Group (NCIC CTG) trial MA. 1. *Breast Cancer Res Treat* 1997, **44**, 211–5.

Schifeling DJ, Jackson DV, Zekan PJ, Muss HB. Fluoxymesterone as third line endocrine therapy for advanced breast cancer. A phase II trial of the Piedmont Oncology Association. *Am J Clin Oncol* 1992, **15**(3), 233–5.

Schiff R, Massarweh S, Shou J, *et al*. Breast cancer endocrine resistance: How growth factor signaling and oestrogen receptor coregulators modulate response. *Clin Cancer Res* 2003, **9**, 447S–54S.

Shyamala G, Yang X, Cardiff RD, *et al*. Impact of progesterone receptor on cell-fate decisions during mammary gland development. *Proc Natl Acad Sci U S A* 2000, **97**, 3044–9.

Sporn MB. Hobson's choice and the need for combinations of new agents for the prevention and treatment of breast cancer. *J Natl Cancer Inst* 2002, **94**, 242–3.

Stabile LP, Lyker JS, Gubish CT, Zhang W, Grandis JR, Siegfried JM. Combined targeting of the oestrogen receptor and the epidermal growth factor receptor in non-small cell lung cancer shows enhanced antiproliferative effects. *Cancer Res* 2005, **65**, 1459–70.

Suh N, Lamph WW, Glasebrook AL, *et al*. Prevention and treatment of experimental breast cancer with the combination of a new selective oestrogen receptor modulator, arzoxifene, and a new rexinoid, LG 100268. *Clin Cancer Res* 2002, **8**, 3270–5.

Sun SY, Hail N, Jr., Lotan R. Apoptosis as a novel target for cancer chemoprevention. *J Natl Cancer Inst* 2004, **96**, 662–72.

Syed V, Ulinski G, Mok SC, Yiu GK, Ho SM. Expression of gonadotropin receptor and growth responses to key reproductive hormones in normal and malignant human ovarian surface epithelial cells. *Cancer Res* 2001, **61**, 6768–76.

Tomassetti P, Migliori M, Campana D, Brocchi E, Piscitelli L, Salomone T, Corinaldesi R. Basis for treatment of functioning neuroendocrine tumours. *Dig Liver Dis* 2004, **36**, S35–41.

Trendle MC, Moertel CG & Kvols LK, Incidence and morbidity of cholelithiasis in patients receiving chronic octreotide for metastatic carcinoid and malignant islet cell tumours. *Cancer* 1997, **79**, 830–4.

Venkatesh S, Ordonez NG, Ajani J, *et al*. Islet cell carcinoma of the pancreas. A study of 98 patients. *Cancer* 1990, **65**, 354–7.

Webb P, Nguyen P, Valentine C, *et al*. The oestrogen receptor enhances AP-1 activity by two distinct mechanisms with different requirements for receptor transactivation functions. *Mol Endocrinol* 1999, **13**, 1672–85.

Welin SV, Janson ET, Sundin A, *et al*. High-dose treatment with a long-acting somatostatin analogue in patients with advanced midgut carcinoid tumours. *Eur J Endocrinol* 2004, **151**, 107–12.

Westerberg H, Nordenskjold B, de Schryver A, Notter G. Anti-ooestrogen therapy of advanced mammary carcinoma. *Acta Radiol Ther Phys Biol* 1976, **15**, 513–8.

Wise PM, Dubal DB, Wilson ME, Rau SW, Bottner M. Minireview. Neuroprotective effects of oestrogen: new insights into mechanisms of action. *Endocrinology* 2001, **142**, 969–73.

11

Photodynamic Therapy of Cancer

K. Eszter Borbas and Dorothée Lahaye

11.1 Introduction

This chapter aims to give an overview on the use of photodynamic therapy (PDT) in cancer. After presenting the chemical and biological background of PDT, the photosensitizers[1] (PSs) currently used in the clinic, or undergoing clinical trials are described (Figure 11.1). The availability/chemical synthesis of the compounds is presented, followed by a summary of the results of the clinical trials (if any). Also presented are compounds that have not entered the clinical trial phase yet, but have shown promise in *in vitro* experiments or animal models, or are expected to make an impact based on their photophysical properties. Due to the limitations of a book chapter, some subjects are only touched upon briefly. There have been a large number of excellent reviews published on PDT. For a more detailed discussion the reader is encouraged to turn to these references: Bonnett, 1995; Sternberg *et al.*, 1998; Pandey *et al.*, 2000; Osterloh and Vicente, 2002; Pushpan *et al.*, 2002; Dolmans *et al.*, 2003; Detty *et al.*, 2004; Lang *et al.*, 2004; Nyman and Hynninen, 2004; Castano *et al.*, 2006; Gorman *et al.*, 2006; Schleyer and Szeimies, 2006; Schneider *et al.*, 2006; Verma *et al.*, 2007.

Since the first report on photodynamic therapy (in 1903, using eosin as the PS) (Jesionek and von Tappenier, 1903), more than 8000 articles have been published on the topic (SciFinder Scholar search, September 2007). The attractiveness of PDT stems from its non-invasiveness, the lack of systemic toxicity (at least in principle), the good to excellent cosmetic outcome of the treatment as opposed to surgery, for example, and the inability of the tumour cells to develop resistance to PDT (*vide infra*). It has been suggested that PDT mobilizes the immune system against the tumour, thus eliminating metastases otherwise undetected or unreachable (See Biological background). In addition to being an exciting new treatment for a range of malignant and premalignant lesions, PDT has also been proposed for the treatment of bacterial, parasitic and viral infections, heart disease, and for the healing of wounds and burns.

---

[1] Throughout the text 'photosensitizer' and 'dye' are used interchangeably.

Anticancer Therapeutics Edited by Sotiris Missailidis
© 2008 John Wiley & Sons, Ltd

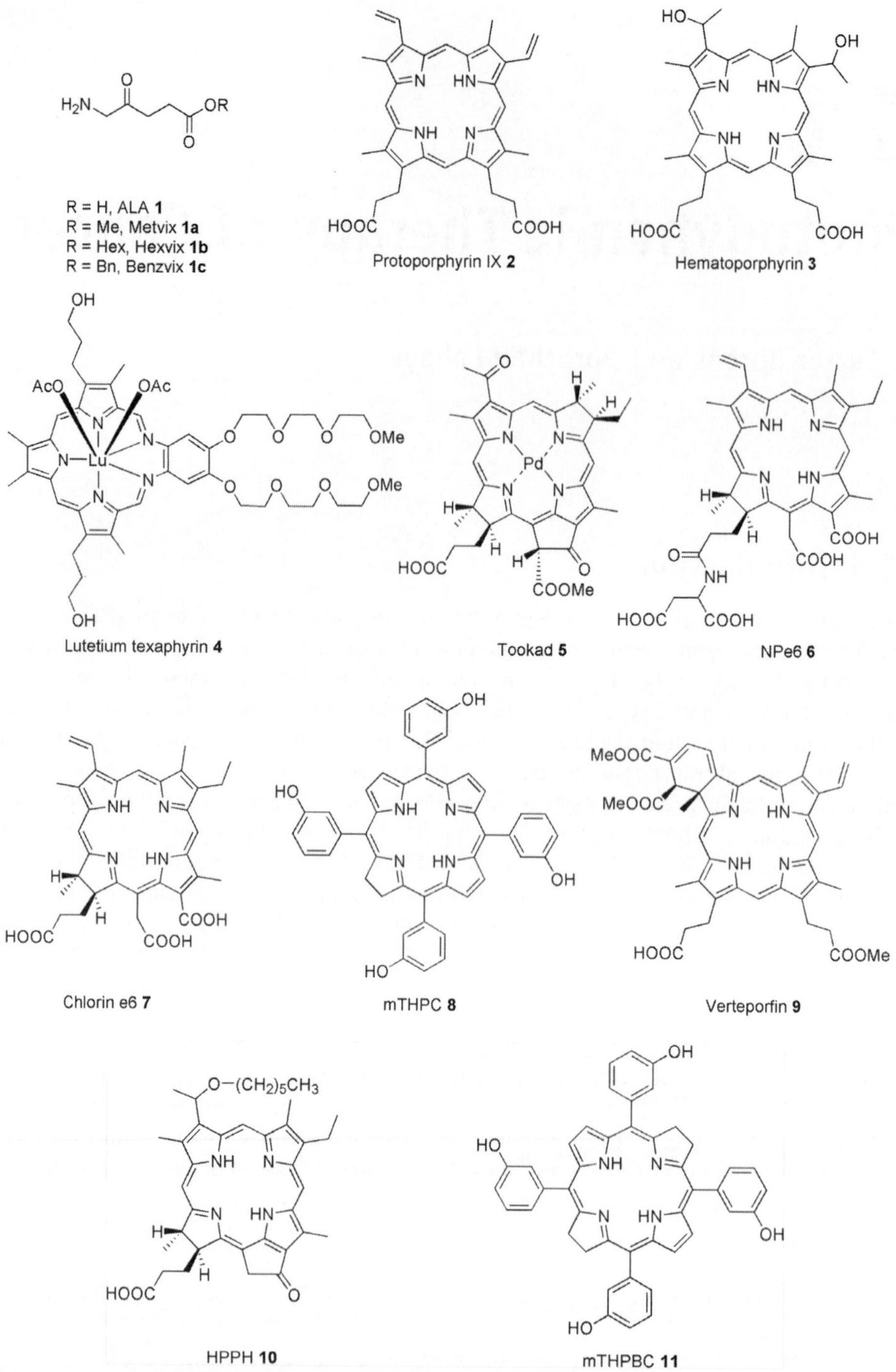

Figure 11.1 PSs used in clinical practice, or in various stages of clinical trials

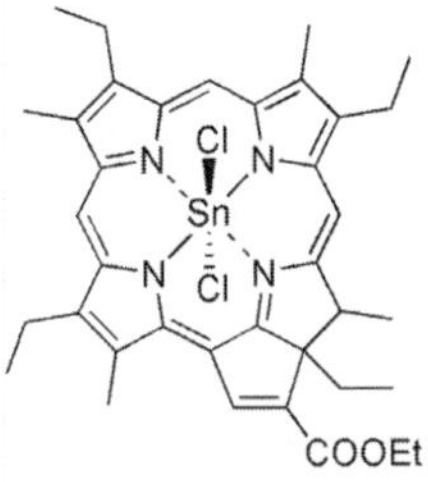

SnET2 **12** 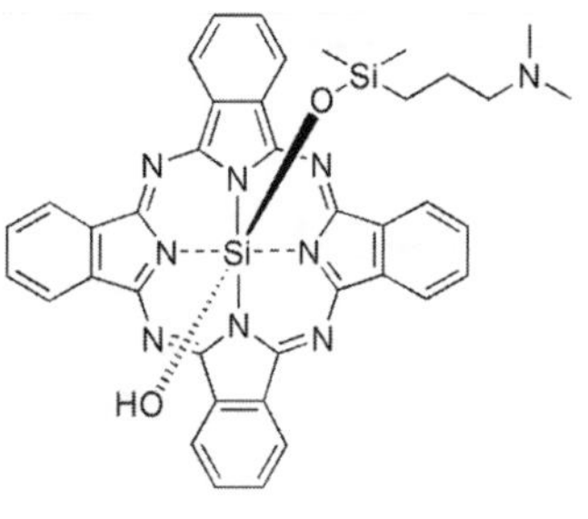   Pc4 **13**

Figure 11.1 *(continued)*

The side effects of PDT are generally mild and morbidity tends to be low. There are a handful of conditions that exclude the use of PDT, such as known photosensitivity disorders, like porphyria or lupus erythematosus.

11.1.1 Chemical background

There are two essential components of PDT: (1) a photosensitizer (drug) that preferentially accumulates in the diseased tissue, and (2) light. Only areas which contain the PS *and* are irradiated are affected. Ideal features for the design of an efficient photosensitizer are the following: (1) strong absorbance in the body's phototherapeutic window (620–850 nm) where skin permeation is optimal and light absorption by blood is low; (2) generation of the first triplet state T_1 of appropriate energy with a high quantum yield. The triplet state energy should match, for good efficiency, the first excitation energy of molecular oxygen, 0.98 eV; (3) non-toxicity in the dark; (4) accumulation of the PS in the selected target tissue; (5) rapid excretion from the body; (6) high purity; (7) easy chemical accessibility; (8) photostability in the presence of reactive oxygen species; (9) should act as a catalyst. Some of these conditions are more easily fulfilled than others, but a PS that conforms to all of the above is yet to be made.

The photophysical processes involved in photodynamic therapy are shown in Figure 11.2, and comprise of: (1) irradiation externally or internally through the use of optical fibres of the targeted tissue containing the photosensitizer; (2) absorption of a photon followed by internal conversion to generate the photosensitizer's first excited state (S_1); (3) conversion to the first excited triplet state through intersystem crossing (ISC) with high quantum yield ($S_1 \rightarrow T_1 +$ heat) [with minimal fluorescence ($S_1 \rightarrow S_0 + h\nu$) or internal conversion ($S_1 \rightarrow S_0 +$ heat)]; (4) the long-lived (µs to ms) triplet state generated can take part in two types of reactions: (i) the triplet state can either generate radicals through electron or hydrogen transfer (Type I) or (ii) transfer energy to a suitable substrate, most often triplet molecular oxygen, to generate cytotoxic singlet oxygen (Type II). In most cases, Type II mechanisms are predominant. Both pathways ultimately lead to the oxidative degradation of biomolecules (e.g. cholesterol, unsaturated fatty acids, amino acid residues, and DNA bases (Ali and van Lier, 1999). Photosensitizers in current PDT clinical practice rely on producing reactive oxygen species. The lifetime of singlet oxygen and superoxide radical is short (1–3 µs), and their effective radius

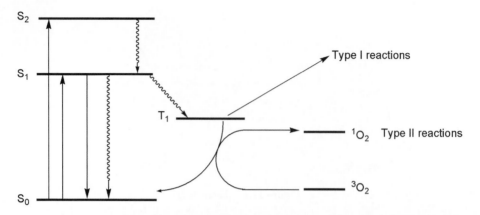

Figure 11.2 Simplified Jablonski diagram. S_0: ground state, S_1: first singlet excited state, S_2: second singlet excited state, T_1: first triplet excited state. Type I reactions generate radicals by electron or hydrogen transfer; Type II reactions give rise to singlet oxygen *via* energy transfer

is typically <10 nm. Thus, damage is usually localized at the site of the photochemical reaction. In hypoxic tissue alternative photochemical reactions may become important, and PSs that can undergo such reactions may be preferred.

Investigations and calculations leading to the theoretical determination of electronic spectra of potential compounds are being undertaken as they may help in the design of efficient photosensitizers (Petit *et al.*, 2006; Quartarolo *et al.*, 2007).

11.1.2 Biological background

Localization in tumours

A number of pathways have been proposed for the uptake of photosensitizers by cancer cells. Porphyrin-based compounds (porphyrins, hydroporphyrins, phthalocyanines, expanded and isomeric porphyrins), which make up the majority of PSs in current clinical practice, are taken up by several major routes. The preference for one or the other of these pathways is governed by factors such as hydrophobicity/hydrophilicity of the compound (structure), formulation, attachment to receptor ligands, and pH. For a detailed review of the topic, see Osterloh and Vicente (2002).

Structural considerations It has been suggested that hydrophobic porphyrins form complexes with lipoproteins, and it is in this form that they are transported to the tumour. Polar porphyrins localize in the outer shell of the lipoprotein, while lipophilic ones are nested in the hydrophobic inside. The actual uptake mechanism of such complexes is not clear, and probably involves endocytosis *via* the apo B/E receptor, which is overexpressed on the surface of rapidly proliferating cancer cells (Hamblin and Newman, 1994a). Phagocytosis of oxidatively modified low density lipoprotein (LDL)-PS conjugates is possible, as well as uptake by scavenger receptors on macrophage-like cells (Hamblin and Newman, 1994b). Charged porphyrins can be transported by association with serum proteins. Covalent conjugates of porphyrin-type PSs and serum proteins have also been described, and such conjugates were taken up by macrophages through the scavenger receptors (Hamblin and Newman, 1994c).

The positions of the peripheral substituents in porphyrinic PSs, and the balance between hydrophilicity/lipophilicity affect tumour selectivity and PDT efficacy (Zheng *et al.*, 2001a, b; Gryshuk *et al.*, 2007).

Formulation It has been shown that incorporation of PSs into liposomes, micelles formed with amphiphilic compounds, in the hydrophobic core of dendrimers (Kojima *et al.*, 2007; Nishiyama *et al.*, 2003, 2007), nanoparticles (Konan-Kouakou *et al.*, 2005) or in cyclodextrins facilitates tumour uptake, increases phototoxicity and tumour selectivity. The size of the delivery vehicle enables PS accumulation in the tumours due to the enhanced permeability and retention effect (EPR) (Maeda, 2001; Maeda *et al.*, 2001, 2006). Photosensitizers (Bengal Rose, protoporphyrin IX (PpIX), a zinc porphyrin and a zinc phthalocyanine) have been incorporated into dendrimers for cancer PDT (Kojima *et al.*, 2007; Nishiyama *et al.*, 2003, 2007). Dendrimers are attractive drug delivery vehicles, as they are non-toxic, non-immunogenic and can be produced in a rational manner (i.e. reproducibly). The dendrimer core is ideal for the incorporation of non-hydrophilic PSs, as PS aggregation and quenching are efficiently prevented, retaining the photosensitizing properties of the parent PS (Nishiyama *et al.*, 2007). The dendrimer shells can be tailored to be ionic (cationic or anionic), or receptor-specific (bearing peptides or sugars). Cationic and anionic dendrimers incorporating a zinc porphyrin or a zinc phthalocyanine were found to display excellent photosensitizing efficiency (Nishiyama *et al.*, 2007). The target cell component of such compounds is still under debate, although it is likely that they accumulate in membrane-bound organelles.

Receptor-mediated targeting Conjugation to receptor-specific biomolecules has been explored for the selective delivery to tumour sites and receptor-mediated uptake by cancer cells. A more detailed discussion of the subject is presented in Section 1.2.3.

The role of pH Charge-neutral, hydrophobic porphyrins can be taken up by passive diffusion through the cell membrane. Carboxylate-bearing porphyrins are negatively charged at normal tissue pH ($\sim$7.5) due to deprotonation, but their carboxylic acid units are protonated at the lower pH of tumour tissue (6.5–7.4). Once inside the tumour cell (which is essentially pH 7.5), the porphyrin gets deprotonated, and is retained inside the cell. For compounds that have pH-independent absorption spectra (i.e. hydroxyl- or sulfate-functionalized chlorins and porphyrins), the lower pH of tumour tissue does not facilitate cellular uptake. Even for such compounds, phototoxicity can be higher at lower pH, presumably due to a decrease in the repair of PDT-damage (Ma *et al.*, 1999).

Tumour-associated macrophage (TAM) uptake TAMs can play dual roles in tumour progression, as they can differentiate into either M1 TAMs (initiating tumour rejection), or M2 TAMs (stimulating metastasis, angiogenesis and tumour growth) (Lamagna *et al.*, 2006). Tumour infiltration by macrophages has been associated with poor prognosis and the occurrence of distant metastases. TAMs secrete mediators that stimulate angiogenesis, and thus facilitate tumour growth. Photosensitizers aimed at a tumour are taken up to a great extent by TAMs. All these factors point to TAMs being viable targets for PDT, and the idea has been put to practice by several workers (Hamblin and Newman, 1994a). Large porphyrin aggregates are probably taken up by phagocytosis, and LDL- and high density lipoprotein (HDL)-associated PSs *via* TAM apo B/E and scavenger receptors (Hamblin and Newman, 1994c). Damage to the PS carrier lipid (oxidation, peroxide formation, surface charge modifications), in particular, facilitates the scavenger receptor pathway. Scavenger receptors can be actively targeted by suitable ligands. Chlorin e_6 conjugated to maleylated serum albumin (a TAM scavenger receptor ligand) was selectively taken up by J774 cells, which express the receptor, but was not by the control OVCAR-5 human ovarian cancer cells (Hamblin *et al.*, 2000). Additionally, macrophages can take up non-aggregated porphyrin derivatives, usually to a much higher extent than tumour cells themselves. Photosensitized macrophages can respond to light by producing tumour necrosis factor (TNF), prostaglandin E_2 (PGE_2), and by affecting the tumour vasculature (Hamblin and Newman, 1994a).

Target organelles

Organelles such as cellular membranes, nuclei, and mitochondria can be affected by photo-generated reactive species resulting in apoptosis, necrosis, or a combination thereof (Castano *et al.*, 2006; Szacilowski *et al.*, 2005). The primary target of the PS depends on a range of factors, e.g. hydrophobicity/hydrophilicity, charge, and uptake mechanism (Osterloh and Vicente, 2002). Thus, for example, the primary target of Photofrin after short incubation times is the plasma membrane and after long incubation the mitochondria. In the case of Victoria Blue, the mitochondria are targeted, while NPe6 and Nile Blue affect lysosomes (Chan *et al.*, 2005; Morgan *et al.*, 2000).

Strategies to improve tumour selectivity

Targeted delivery Tumour selectivity of porphyrin-based photosensitizers is well documented, but, as shown in the previous sections, not completely understood. Tumour selectivity is achieved either by rapid clearance of the photosensitizer from healthy tissue and prolonged retention in the tumour, or selective accumulation in the tumour. As these processes involve different time scales, the optimal light application time and the period of skin photosensitivity varies from dye to dye, and can range from a few minutes to several days. Examples are provided with the detailed discussion of the respective photosensitizers.

To achieve more selective destruction of the tumour with minimal damage to healthy tissue, the selectivity of PS delivery needs to be improved. There are a number of potential targets available, (1) killing of the tumour cells directly, (2) damage to the endothelial cells causing collapse of the tumour vasculature, and (3) PDT-induced immune response, which involves macrophage-mediated destruction through release of toxic compounds (e.g. TNF), followed by tumour infiltration by neutrophils, mast cells, and monocytes (Hamblin and Newman, 1994a; Bown *et al.*, 2002; Castano *et al.*, 2006). The latter pathway is believed to play an important role in the elimination of distant metastases, as the inflammation-like response gives rise to antigen-presenting cells presenting tumour-peptides. The antigens are recognized by helper T lymphocytes, which have been recovered from distant lymphoid tissues. Thus, although PDT is localized to the treatment site, the elicited immune response can be systemic (Hamblin, 2002; Castano *et al.*, 2006).

Targeting of both cancer cells and the surrounding vascular epithelial cells has been reported, and several examples are given below. To date there have been no clinical trials using actively targeted photosensitizers. However, there is a large body of data concerning targeted PDT *in vitro* and in animal models, employing a variety of strategies. The synthesis of covalent conjugates requires the presence of suitable functional groups in the PS and the biomolecule. Ideally, only one such functional group per PS should be present to avoid the formation of cross-linked oligomers, although it is usually possible to suppress oligomer formation by a judicious choice of reaction conditions. In the most common PSs, either multiple reactive sites exist (haematoporphyrin, chlorin e_6, meso-tetrahydroxyphenylchlorin (mTHPC)), or none (aluminium phthalocyanines, zinc phthalocyanines). Access to suitably functionalized synthetic PSs (e.g. porphyrins, chlorins, bacteriochlorins, purpurins) enables the incorporation of one bioconjugatable site with the desired functionality (e.g. carboxylic acid, formyl group, iodoacetamide). In this regard, standard bioconjugation techniques (Hermanson, 1996) should be applicable for the synthesis of future PS-biomolecule conjugates. A large PS-carrier ratio is beneficial, as long as tumour recognition is maintained. Thus conjugates with as many as ten PS molecules per antibody (Ab) have been described (Rancan et al., 2007). To preserve the antigen binding ability of the Ab, the photosensitizers can be incorporated into a polymeric scaffold, instead of directly attaching them to the Ab.

HpD (haematoporphyrin derivative), chlorin e_6 and phthalocyanines (Pcs) have been attached to antibodies (van Dongen et al., 2004) and antibody fragments (Staneloudi et al., 2007) to achieve highly selective and tumour-specific delivery. The conjugates could be modified to obtain cationic or anionic macromolecules, which had a profound impact on the photosensitizing efficiency, although not on the selectivity of the conjugates (Hamblin, 2002). As with antibody-radioisotope, or antibody-toxin conjugates, a significant proportion of the drug ended up in the liver and kidney, although the non-toxicity of the photoimmunoconjugates in the absence of light makes this less of a problem. Antibodies are very specific and have high affinities for their targets, but there are also complications arising from tumour heterogeneity, immunogenicity, and inefficient internalization of the conjugates. PS-conjugates of other proteins, for which cancer cells overexpress receptors (e.g. transferrin, LDL), have also been investigated in vitro (Hamblin and Newman, 1994b). Highly hydrophilic PSs are less phototoxic in their free form than hydrophobic ones. Conjugation of water-soluble cationic dimethylaminophenyl-porphyrin (Malatesti et al., 2006) and trimethylpyridinium-porphyrin or anionic aluminium(III) phthalocyanine tetrasulfonate chloride to internalizing antibodies yielded conjugates with high photodynamic efficiency in vitro (Vrouenraets et al., 2000, 2001). The Pc-conjugates were more cytotoxic than the corresponding mTHPC-conjugates, presumably because of the different effects the PS had on the antibody structure and binding ability (Vrouenraets et al., 2002). Conjugation of porphyrinic PSs or Rose Bengal to anti-OGG1-mAb induced a non-singlet-oxygen-dependent photodegradation of the DNA repair enzyme 8-oxoguanine DNA glycosylase 1 (Conlon and Berrios, 2007). The destruction of a key DNA repair enzyme could boost PDT efficiency by inducing cell death.

Other targeting moieties have also been employed, such as oligonucleotides (Yamayoshi et al., 2007), nanoparticles (Konan-Kouakou et al., 2005; Oba et al., 2006), polyamines (Garcia et al., 2006), carbohydrates (Driaf, 1993; Oulmi et al., 1995; Sol et al., 1999; Chen and Drain, 2004; Pandey et al., 2007), drugs (Gacio et al., 2006) and steroids (Sharman et al., 2004; El-Akra et al., 2006). In general, such conjugates proved comparable to unmodified dyes as far as singlet oxygen quantum yield was concerned. Given the abundance of data available, only a few examples are mentioned for each type of conjugate.

Peptides are attractive alternatives to monoclonal antibodies, due to their small size, low cytotoxicity and antigenicity, good tissue permeability and rapid tumour localization (Schneider et al., 2006). Furthermore, they can be synthesized and manipulated (e.g. labelled) in reproducible and scalable chemical reactions, yielding well-defined entities. Due to their rapid clearance, peptides are not ideal for targeted PS delivery, although their stability can be increased by the incorporation of D-amino acids, amino acid analogues, and cyclization. PS-peptide conjugates have been prepared to target the tumour vasculature epithelial cells (Frochot *et al.*, 2007) or receptors overexpressed on tumour cells (Rahimipour *et al.*, 2003). To maximize photodynamic activity, PSs have been attached to nuclear localization signals to ensure delivery to the nucleus (Akhlynina *et al.*, 1999; Bisland *et al.*, 1999). 5-(4-Carboxyphenyl)-10,15,20-triphenylchlorin was attached to $\alpha_v\beta3$-specific tripeptides RGD (H-Arg-Gly-Asp-OH). The conjugate targets the neovasculature, indirectly affecting tumour eradication. The conjugate was incorporated up to 80–98-fold more than the free dye in $\alpha_v\beta3$-overexpressing human unbilical vein endothelial cells (Frochot *et al.*, 2007). The conjugates had molar absorption coefficients and singlet oxygen quantum yields only slightly lower than the unconjugated PS. In an attempt to selectively target cells that express gonadotropin-releasing hormone (GnRH) receptors, Protoporphyrin IX was conjugated to GnRH agonist and antagonist peptide sequences. The conjugates were more phototoxic, and ~10-fold more selective against their target cells than unconjugated PpIX. The conjugates act by two mechanisms, as GnRH agonists or antagonists and as photosensitizers (Rahimipour *et al.*, 2003). Chlorin e_6 was targeted to the cell nucleus by conjugation to a nucleus-directed linear peptide, or an octameric, branched derivative thereof (Bisland *et al.*, 1999). The conjugates were more phototoxic than the untargeted dye. A nucleus localization signal-chlorin e_6-insulin construct was prepared to confer receptor-specificity (insulin), and intracellular targeted delivery potential (NLS) onto the conjugate (Akhlynina *et al.*, 1999), and had a 2400-fold reduced EC_{50} compared to free chlorin e_6.

An antisense oligonucleotide equipped with a pyrimidine-specific psoralene photo-cross-linker has been found to induce apoptotic death in human papillomavirus (HPV)-positive cervical cancer cells upon ultraviolet (UV)-irradiation (Yamayoshi *et al.*, 2007). Enzymatically cleavable solubilizing phosphate moieties were incorporated into PpIX. Following cellular uptake, the phosphate groups were cleaved by phosphatases overexpressed in tumour cells, and the dye was retained in the cell. The porphyrin-phosphates displayed higher photocytotoxicity than the parent (unphosphorylated) PpIX, but also an appreciable dark toxicity, presumably due to the phenolic units in the molecule that can undergo oxidation to quinones (Xu *et al.*, 2007).

A range of glycosylated porphyrins (Driaf *et al.*, 1993; Oulmi *et al.*, 1995; Sol *et al.*, 1999; Chen *et al.*, 2004) and chlorins (Zheng *et al.*, 2001a; Pandey *et al.*, 2007) have been reported. Variations in the number, nature and position of the carbohydrate units gave rise to a series of PSs with varying levels of tumour selectivity and phototoxicity. It was found that globular porphyrins with ortho-glycosyl-phenyl substituents are poor photosensitizers, while flat, para-substituted ones are significantly better (Oulmi *et al.*, 1995; Sol *et al.*, 1999; Chen *et al.*, 2004). The nature of the carbohydrate unit, the linker, and the position of the substitution in purpurin PSs were crucial in determining the efficacy of purpurin-carbohydrate conjugates (Pandey *et al.*, 2007).

Tumour-specific singlet oxygen generation A more recent approach to selective tumour destruction is the tumour-specific production of singlet oxygen, whereby the PS becomes active upon a tumour-specific chemical reaction. Tumour-specific singlet oxygen production is far less common than targeted delivery, but the idea has nevertheless been shown feasible, and a handful of model systems have been put forward. In a particularly elegant approach (Zheng *et al.*, 2007), the photosensitizer is only 'turned on' by a tumour-specific stimulus, thus reducing damage to normal tissue. A matrix metalloproteinase-7 cleavable peptide (GPLGLARK) was capped on one end with Pyro (photosensitizer and FRET donor), and black hole quencher 3 ($^1O_2$ quencher and FRET acceptor, BHQ3), forming an

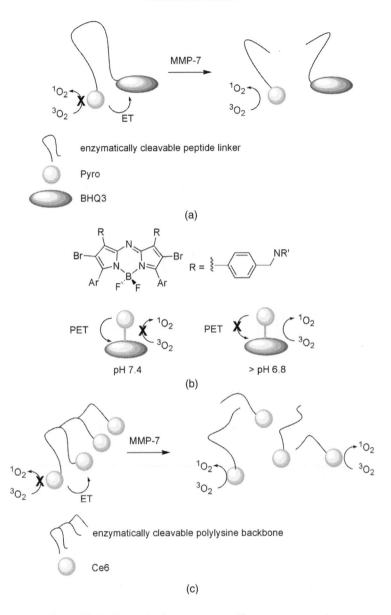

Figure 11.3 Strategies for tumour-specific oxygen generation

MMP-7-activable photosensitizing molecular beacon. Cleavage of the peptide removed BHQ3 from the proximity of the PS, enabling $^1O_2$ production (Figure 11.3a). The principle was demonstrated both *in vitro* and *in vivo*. The same PS-quencher pair was placed on two complementary DNA strands. Upon pairing of one strand with the target sequence, the PS and the quencher were separated, resulting in singlet oxygen generation upon illumination (Cló *et al.*, 2006). Azadipyrromethenes, substituted with various benzylic amines exploit the differences between normal tissue pH and that of the tumour interstitial fluid (6.4–6.8) (McDonnell *et al.*, 2005). Photoinduced electron transfer at pH 7.4 quenched fluorescence of the azadipyrromethene unit, and also singlet oxygen production (Figure 11.3b). In a poly-L-lysine loaded with chlorin e_6 PS molecules, self-quenching of the dye molecules prevented singlet oxygen generation. Photosensitizing activity could be recovered upon hydrolysis of the polylysine backbone by a tumour-specific enzyme, resulting in monomerization of the PS units, and a reduction in self-quenching (Figure 11.3c) (Choi *et al.*, 2006).

Resistance to PDT As there are numerous examples of repeated treatment sessions for a given patient, the issue of the development of resistance to PDT is of some relevance. *In vitro* results show that it is possible to culture cells resistant to a specific type of photosensitizer. Resistant lines can provide useful information on the modes of action of the PS, the defense mechanisms of the cell, and the possibilities of combination therapies.

Cells resistant to Nile Blue, Photofrin and aluminum phthalocyanine tetrasulfonate have been reported (Singh *et al.*, 2001). The levels of resistance found were in the 1.5–2.62-fold, which is a clinically relevant level, but not very large. It was found that the degree of resistance was not only dependent on the cell line, but also on the type of sensitizer used, which provides further support to the theory that the photodynamic action is sensitizer specific.

While Nile Blue, Photofrin and aluminium phthalocyanine tetrasulfonate are exogeneous photosensitizers, ALA-derived PpIX is produced endogeneously. Resistance has been induced in LM3 cells by exposure to increasing amounts of light in an ALA-containing medium (Casas *et al.*, 2006). Compared to the non-resistant parent cells, the resistant clones were twice the size, and had twice the protein content. Similar results were obtained for Photofrin-induced PDT-resistant mouse tumour cells (Luna and Gomer, 1991). Neither upregulated MDR mRNA levels, nor large differences in glutathione and antioxidant enzyme levels were seen in the PDT-resistant RIF cell lines (Luna and Gomer, 1991). It appears that resistance is caused not as much by a decreased production of PpIX, or major alterations in defence mechanisms, but more by the increase of target proteins to be destroyed per porphyrin. Changes in porphyrin distribution were also noted, with a slightly higher proportion of hydrophilic uroporphyrin and coproporphyrin being produced compared to non-resistant cells. As the photosensitizing efficiency of PpIX seems to be intimately tied up with its hydrophobicity, an increase in hydrophilic porphyrins may contribute to the resistance. It was found that the cells thus produced had some cross-resistance to exogeneous PpIX, but were sensitive to other photosensitizers, and to hyperthermia, which may help design combination therapies with PDT. Two RIF cell lines resistant to a Photofrin analogue and a phthalocyanine did not show cross-resistance to the PS, suggesting that there are several resistance mechanisms involved in PDT-resistant RIF cells, that are governed by factors other than simply PS structure or charge (Mayhew *et al.*, 2001). The authors also suggested, based on data from short and extended PS incubation experiments, that Photofrin resistance may be a consequence of membrane alteration. In this study, no differences in protein content, cell size, or enzyme levels were found in resistant lines compared to parent lines.

11.2 Photosensitizers

This section introduces the photosensitizers (Figure 11.1) currently used in clinical practice or undergoing clinical trials, and a select group of compounds that have shown promise *in vitro* or *in vivo*. Table 11.1 gives an overview of the most relevant properties of common PSs.

Table 11.1 Photosensitizers in the clinic, or in clinical trials. Under condition indications for which the PS has been registered or has been proposed are listed

Chemical name	Other names	Origin	λ_{abs} (ε/M/cm)	Condition
HpD	Photosan	Natural, chemically modified	630 nm (1170)	Cervical, oesophageal, bladder, gastric, endobronchial, brain
Photofrin		Partially purified HpD	630 nm (1170)	as HpD
δ-Aminolaevulinic acid	ALA, Levulan	Natural	635 nm (therapy)	BCC, AK, gynaecological tumours, head and neck tumours (therapy)
			375–400 nm (diagnosis)	Bladder, brain, head and neck tumours (diagnosis)
Aminolaevulinic acid methyl ester, MAL	Metvix	Synthetic	635 nm	BCC
Aminolaevulinic acid hexyl ester	Hexvix	Synthetic	375–400 nm	Bladder cancer diagnosis
Aminolaevulinic acid benzyl ester	Benzvix	Synthetic	635 nm	Gastrointestinal cancer
Protoporphyrin IX	PpIX	Natural	635 nm	AK, BCC
Meso-tetrahydroxyphenyl chlorin	mTHPC, Foscan	Synthetic	652 nm (30 000)	BCC, SCC
L-Aspartyl chlorin e_6	NPe6, MACE	Semisynthetic	645 nm (40 000) (Chlorin e_6)	Skin cancers including melanoma
Lutetium(III)-texaphyrin	Lu-Tex, Lutrin, Motexafin lutetium	Synthetic	732 nm (42 000)	Cervical, breast, and prostate cancer, brain tumours
Sn(II)-etiopurpurin	SnET2, Purlytin	Semisynthetic	660 nm (28 000)	BCC, Kaposi's sarcoma, prostate cancer, cutaneous metastatic breast cancer
Benzoporphyrin derivative monoacid A ring	Verteporfin, BPD, BPD-MA	Semisynthetic	689 nm (35 000)	Non-melanoma skin cancers, papillary capillary haemangioma, circumscribed choroidal haemangioma, conjunctival SCC
2-(1-Hexyloxyethyl)-2-devinyl pyropheophorbide-a	Photochlor, HPPH			BCC
Silicon phthalocyanine 4	Pc 4	Synthetic	670 nm (100 000)	Cutaneous and subcutaneous lesions of solid tumours

11.2.1 ALA and ALA derivatives, PpIX

e δ-Aminolaevulic acid (ALA, Figure 11.1, **1**) is a naturally occurring amino acid, and the biosynthetic precursor to protoporphyrin IX. In cells ALA is synthesized from glycine and succinyl coenzyme A. ALA production is under negative feedback control by heme, but the control can be bypassed by the administration of exogeneous ALA, which can go on forming protoporphyrin IX (Figure 11.1, **2**), an efficient PS ($\lambda_{abs} = 635$ nm, $\varepsilon < 5000/M/cm$). Eventually, PpIX gets metalated to its iron chelate (heme) through the action of ferrochelatase. The rate-determining step in heme synthesis is the metalation of PpIX. Thus, a large amount of ALA (from exogeneous administration) enables the synthesis and build-up of sufficient quantities of PpIX to be useful in PDT (Kennedy *et al.*, 1990; Pandey *et al.*, 2000). In a typical treatment session, irradiation follows ALA administration by 1–6 h, which allows for the build-up of practical quantities of PpIX. In addition to endogeneous PpIX, exogeneous PpIX (available from blood upon acidic demetalation) has been employed. There have been several attempts at improving the selectivity and phototoxicity of PpIX by targeted delivery (Rahimipour *et al.*, 2003; Yamayoshi *et al.*, 2007). In Europe, PpIX is approved for the PDT of basal cell carcinoma (BCC) and actinic keratosis (AK). Schleyer and Szeimies (2006) give an excellent overview of the results of the clinical trials for BCC and AK. In addition to AK and BCC, ALA has been employed as a PS for recurrent superficial bladder carcinoma (Skyrme *et al.*, 2005), SCC (Schleyer and Szeimies, 2006) and with great success in Bowen's disease (Schleyer and Szeimies, 2006). Side effects include stinging and burning during treatment, oedema, hyperpigmentation (Tschen *et al.*, 2006), and, in a small percentage of the cases, severe pain (Borelli *et al.*, 2007).

Unmodified ALA is hydrophilic, and has limited bioavailability. In particular, ALA cannot penetrate through normal keratin (but readily passes through abnormal keratin), which results in highly specific accumulation in abnormal epithelium, when applied topically (Schleyer and Szeimies, 2006). Modifications to the core structure of ALA, in particular esterification of the carboxylic acid moiety yielded Metvix, Hexvix, and Benzvix (Figure 11.1, **1a**, **1b** and **1c**). Metvix is approved for the treatment of gastrointestinal cancers, while Hexvix has found use in the diagnostics of bladder cancer, and Benzvix in the treatment of gastrointestinal tumours. Metvix (or MAL, methyl aminolaevulinate) has increased lipophilicity, and thus increased cell penetration. MAL-PDT has been employed in the treatment of AK (Angell-Peterson *et al.*, 2006; Caekelbergh *et al.*, 2006; Kasche *et al.*, 2006) BCC (Vinciullo *et al.*, 2005; Caekelbergh *et al.*, 2006), SCC (Morton *et al.*, 2006) and cutaneous T-cell lymphoma (Zane *et al.*, 2006). Topical application of ALA or MAL was proposed for the prevention of skin lesions (SCC and AK) in immunocompromised transplant patients. There did not seem to be a statistically significant difference in the occurrence of SCC between the treated and the control group (de Graaf *et al.*, 2006), although some beneficial effects were observed for the treatment group for AK (de Graaf *et al.*, 2006; Wulf *et al.*, 2006).

11.2.2 Haematoporphyrin derivatives

Haematoporphyrin derivative (HpD) is the most common photosensitizer. HpD is available from blood upon treatment with 5 % sulfuric acid in acetic acid, followed by treatment with a strong base (Figure 11.4) (Sternberg *et al.*, 1998; Pushpan *et al.*, 2002). The reaction yields a mixture of oligomeric species linked by ether and ester bonds. A purified form of HpD is Photofrin, which still contains up to 60 porphyrin derivatives. The lack of uniformity, along

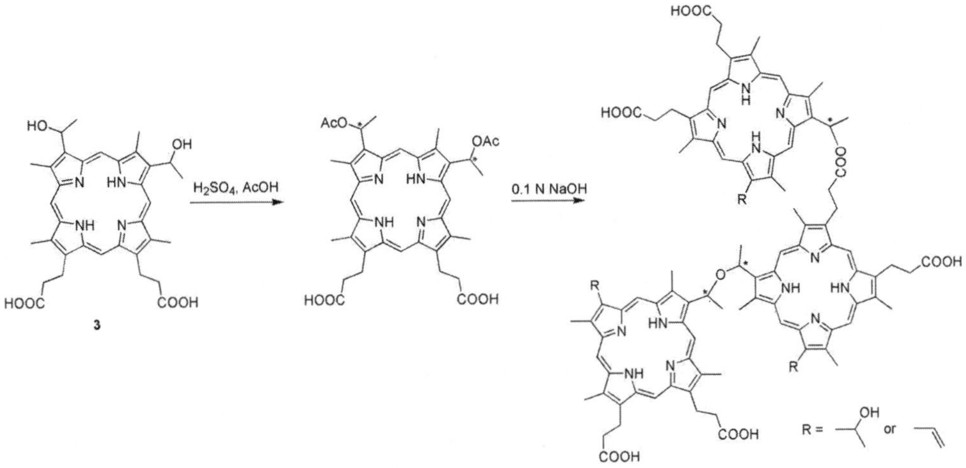

Figure 11.4 General structure and synthesis of HpD

with the varying quality of the PSs make, HpD a less than ideal PS. HpD and Photofrin have minimal absorption at 630 nm ($\varepsilon = 1170$/M/cm), prolonged *in vivo* retention in cutaneous tissue, and possess a terminal half-life of up to 452 h, causing photosensitivity for as long as 10 weeks (Ris *et al.*, 1996). Excessive drug doses or too much light result in increased morbidity, the main symptoms being pain at the treatment site, scarring, hyperpigmentation and fibrosis.

HpD or Photofrin are approved PDT agents in Canada, the USA, Europe and Japan for esophageal, endobronchial, cervical, bladder and gastric cancers (Detty *et al.*, 2004). They have been employed successfully in the palliation of non-resectable bile duct cancer (Berr *et al.*, 2000; Zoepf *et al.*, 2005), as an adjuvant treatment for high grade glioma (Stylli *et al.*, 2005), as an alternative to surgery for patients with cervical intraepithelial neoplasia (Yamaguchi *et al.*, 2004), and as an adjuvant therapy to surgery of recurrent pituitary adenomas (Marks *et al.*, 2000). There are a number of reports describing the use of Photofrin PDT in combination with debulking surgery for disseminated intraperitoneal tumours, although no significant improvements were recorded compared to previously existing methods (Hahn *et al.*, 2006; Hendren *et al.*, 2001; Menon *et al.*, 2001; Bauer *et al.*, 2001). PDT with HpD is an attractive alternative treatment to surgery for BCC, except for pigmented lesions (Kaviani *et al.*, 2005). Treatments usually produce a high response rate, with good to excellent cosmetic outcomes. (Kaviani *et al.*, 2005; Oseroff *et al.*, 2006) Treatments can be repeated, as there is no cumulative toxicity and multiple lesions can be treated in a short timeframe. A reasonable tumour selectivity (i.e. moderate skin photosensitivity) can be obtained with a low (1 mg/kg) dose of Photofrin and high doses of light, while maintaining the good response rate observed with high (2 mg/kg) doses (Oseroff *et al.*, 2006).

Photofrin-mediated PDT has shown excellent results for the treatment of chest wall progression of breast carcinoma (Allison *et al.*, 2001; Cuenca *et al.*, 2004). Once salvage surgery, radiation, or chemohormonal therapy fail for such patients, quality of life significantly worsens. PDT is an attractive option, as in the trial population appreciable rates of tumour responses were seen, combined with regression, excellent cosmetics, and reduced pain. The dose dependence of the therapy was investigated, and it was found that low dose

(0.8 mg/kg *vs.* 2 mg/kg Photofrin) therapy, combined with larger irradiation surfaces helped eradicate tumours better, and rim recurrence rates dropped compared to previous results. The authors ascribed this finding to photobleaching of the surrounding healthy tissue, resulting in decreased photosensitivity.

In the treatment of Barrett's oesophagus, the results obtained with Photofrin PDT were comparable to those of argon plasma coagulation (APC) (Ragunath *et al.*, 2005). However, PDT was significantly more costly. The major factor was a 2-day patient hospitalization (between Photofrin injection and light delivery), and the cost of the PS. With better patient education, the cost of hospitalization can be reduced. With PDT maturing into an accepted form of cancer therapy, cheaper PSs should become available. PDT was more effective at eradicating dysplasia than APC. The frequency and severity of negative events were comparable for the two modalities, and were usually limited to strictures in a small percentage of patients, with additional photosensitivity in some patients undergoing PDT. Attempts at reducing the incidence of strictures with the adminstration of oral steroids were not successful (Panjehpour *et al.*, 2000). In a phase III trial (Overholt *et al.*, 2005), PDT proved superior to omeprazole in ablating high grade dysplasia in Barrett's oesophagus patients (77 % complete ablation, *vs.* 39 % with omeprazole).

Photofrin- or HpD-mediated treatment of early superficial or advanced invasive carcinoma of the esophagus resulted in complete remission in 4 out of the 6 patients treated, and the mean survival rate was 27 months, with most of the patients still alive at the time of the report (13–47 months after PDT) (Okunaka *et al.*, 1990). The same study encompassed the palliative treatment of 14 patients with advanced invasive carcinoma, for whom alleviation of dysphagia was possible, improving the quality of life.

Photofrin was used in a clinical trial for the palliative treatment of acquired immune deficiency syndrome (AIDS)-related Kaposi's sarcoma (Bernstein *et al.*, 1999). Good tumour selectivity was achieved with 1.0 mg/kg Photofrin and up to 300 J/cm^2 light doses. However, the authors cautioned against using such high light doses for large tumours to avoid scarring, ulceration and infection.

PDT of early stage lung cancer has been employed with great success, and a large proportion of patients showed complete recovery from the illness (Ono *et al.*, 1992). In a study encompassing 100 patients with stage III or IV lung cancer (both SCLC and NSCLC), the major use of Photofrin PDT was palliation (Moghissi *et al.*, 1999). In this regard, there was a major difference between patients whose World Health Organization performance status was ≤2, and those >2. In the first group, there was an additional survival benefit associated with PDT, with 53 % of patients surviving 2 years or more. It is interesting to note that the sole determinant of survival rate seemed to be the performance status of the patients, and neither the histology, nor the stage of the tumour appeared to have an influence. Hundreds of patients have been treated for bladder cancer in various stages after the initial report in 1976 by Kelly and Snell. For those with carcinoma *in situ* (D'Hallewin and Baert, 1995), Photofrin PDT offered a similar response rate as intravesical Bacille Calmette-Guérin (BCG; 60–70 %); however, some patients non-responsive to BCG have responded to PDT. A careful adjustment of light dose is essential, as high doses result in vesicoureteral reflux with severe bladder irritability, loss of bladder volume, and ureteral stricture (Walther *et al.*, 1997). Total light energy as opposed to energy density seems to be a better measure of light dose, as it can account for the large variations in bladder volume.

In advanced gastrointestinal tumours, PDT with HpD proved a viable treatment option, with a complete response rate of ~10 %, of whom 86 % survived until the end of the follow-up time (1–5 years) (Jin *et al.*, 1990). When three treatment modalities were compared (PDT, chemotherapy, and a combination of the two), for patients with advanced cardiac cancer, PDT (HpD) in combination with chemotherapy (UFT with mitomycin C) was preferred to chemotherapy alone (58.5 % *vs.* 50 % response, 19.5 % *vs.* 8.3 % complete response) (Jin *et al.*, 1992). The median survival time for the PDT + chemotherapy group was significantly higher (12 months, with 1 patient alive seven years after treatment) than the chemotherapy alone group (mean survival rate 7 months, all patients died within two years). In early gastric cancer, PDT is recommended for patients who are at high risk at surgery because of other existing conditions (Mimura *et al.*, 1996). In such cases, PDT with HpD or Photofrin gave excellent response rates, with minimal recurrence for small lesions (67–100 % CR). For larger lesions or difficult-to-see tumours a major problem was the lack of uniform irradiation.

11.2.3 Synthetic porphyrins, and porphyrin isomers and analogues

Porphyrin derivatives have played an important role in the PDT of cancer. Apart from the naturally derived compounds described in this chapter, many synthetic compounds have also been put forward. The traditional (Adler-Longo) method of constructing tetraarylporphyrins was usually low-yielding, but satisfactory for the synthesis of *meso*-tetraarylporphyrins bearing identical substituents in the meso positions. Thus, *meso*-tetraphenylporphyrin (TPP) has been investigated as an alternative to HpD, but its insolubility in aqueous solvents was a distinct disadvantage. Sulfonation of the phenyl rings yielded water-soluble derivatives, which proved excellent PSs, but were neurotoxic (Detty *et al.*, 2004). The recent advances in porphyrin chemistry (Zaidi *et al.*, 2006; Smith, 2000) enabled the introduction of a vast variety of different functionalities, and such porphyrins can sometimes be prepared in multigram quantities (Zaidi *et al.*, 2006). Modifications of the procedure (macrocycle formation in the absence of an acid catalyst) allowed for the introduction of acid-sensitive functional groups, further expanding the scope of the methodology (Fan *et al.*, 2005; Taniguchi *et al.*, 2005; Dogutan *et al.*, 2007). However, most known porphyrins have absorption maxima at or at not much longer wavelengths than Photofrin. As the substitution pattern of a porphyrin has only a small effect on the position and intensity of the red absorption, it is likely that in the long run the importance of porphyrins in the PDT of cancer will decrease, although ALA-PDT should retain its importance for the treatment of superficial conditions (e.g. acne).

Porphyrin isomers which have one of the inner nitrogen atoms and a β-pyrrolic carbon atom formally switched (*N*-confused or carbaporphyrins) display red absorptions at longer wavelengths (> 730 nm) than the parent compounds (Furuta *et al.*, 1994) (Figure 11.5, **14**). Doubly *N*-confused porphyrins (Figure 11.5, **15**) have also been put forward as potential PSs, as their lowest energy absorption maxima appear at 710 and 665 nm (free base and metalated species, respectively). A singlet oxygen quantum yield of 0.81 was observed for the silver complex of the doubly *N*-confused *meso*-tetraphenylporphyrin derivative (Araki *et al.*, 2003). A major obstacle to their widespread use in PDT is their low-yielding and tedious synthesis. *N*-confused porphyrins are available *via* a modified Adler-Longo-method (5–7 % yield) (Smith, 2000). Stepwise procedures, employing either hydroxyfulvenedialdehyde and a tripyrrin (3+1 condensation), or a dipyrromethane and a *N*-confused dipyrromethane (2+2 condensation) are known, but are neither flexible, nor convenient. There is a single report on the synthesis of *N*-confused TPP with a 35 % isolated yield (providing 800 mg of product) (Geier *et al.*, 1999), employing methanesulfonic acid as the catalyst.

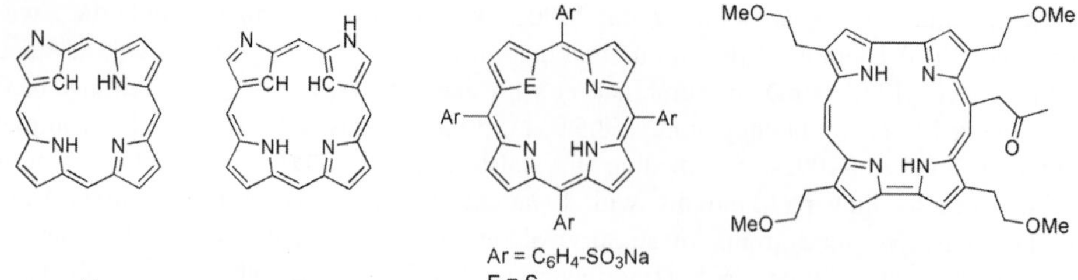

14   **15**   Ar = C₆H₄-SO₃Na   **17**
                               E = S **16**

Figure 11.5 *N*-confused, doubly *N*-confused porphyrins, core-modified porphyrin and porphycene

Core-modified porphyrins (e.g. Figure 11.5, **16**) (where one or more of the inner nitrogen atoms have been replaced by a sulfur or a selenium) have intense red absorptions (630–717 nm) (Stilts *et al.*, 2000; Hilmey *et al.*, 2002). The compounds are available in low yield (⩽5 %) by the acid-catalyzed condensation of pyrrole with 2,5-bis-(phenylhydroxymethylthiophene) or 2,5-bis-(phenylhydroxymethylselenothiophene) (Stilts *et al.*, 2000). The tetrasulfonated dithia-analogue of **16** was tested *in vivo* in BALB/c mice bearing Colo-26 tumours and was shown to be an effective PS (Stilts *et al.*, 2000). Carboxylated derivatives were even more effective than the tetrasulfonated analogues.

Porphycene (Figure 11.5, **17**) displays a strong absorption band ($\varepsilon = 51\,900$/M/cm) around 630 nm, making **17** a promising photosensitizer (Vogel *et al.*, 1986; Braslavsky *et al.*, 1997). Porphycene is obtained by McMurry coupling of 5,5′-diformyl-2,2′-dipyrrole (Vogel *et al.*, 1986). Despite the low yield (2–3 %), derivatives of **17** have been studied as PSs *in vitro* against cell lines derived from human SCC of the skin (Fickweiler *et al.*, 1999; Scherer *et al.*, 2004), and as polylysine conjugates for antimicrobial PDT (Polo *et al.*, 2000).

11.2.4 Lutetium texaphyrin

Texaphyrins are pentadentate expanded metalloporphyrins, composed of a tripyrrolyl-dimethene unit linked to a phenylenediamine by two imine-type bonds. This aromatic pentaaza monoanionic ligand absorbs strongly in the far-red region (720–780 nm), the result of a 22-π electron aromatic delocalization system (Sessler *et al.*, 1994).

A large number of metallotexaphyrins have been synthesized directly from nonaromatic precursors (Sessler *et al.*, 1993, 1994; Young *et al.*, 1996; Guldi *et al.*, 2000; Hannah *et al.*, 2002), and metal free texaphyrins have been rather difficult to obtain (Sessler *et al.*, 1994; Hannah *et al.*, 2001). Notably, stable lanthanide complexes can be accessed owing to a core cavity 20 % larger than that of porphyrins (Sessler *et al.*, 1993; Young *et al.*, 1996). Diamagnetic texaphyrins (Cd-, Y-, and Lu-chelates) are especially attractive as they display long-lived triplet states (μs range) and high singlet oxygen formation quantum yields. Among them, the water-soluble lutetium texaphyrin (Lu-Tex, Motexafin lutetium, Lutrin®, Figure 11.1, **4**) that bears alcohol and poly(ethylene glycol) substituents around the periphery of the macrocycle is undergoing clinical trials as a PS for a range of malignancies (*vide infra*).

Lu-Tex is available in two steps using generalized methods (Young *et al.*, 1996) (Figure 11.6). Step 1 entails acid-catalyzed condensation between a tripyrran (2,5-bis((5-formyl-3-(3-hydroxypropyl)-4-methylpyrrol-2-yl)methyl)-3,4-diethylpyrrole), and a substituted aromatic 1,2-diamine (1,2-diamino-4,5-bis(2-(2-(2-methoxyethoxy)ethoxy)ethoxy)

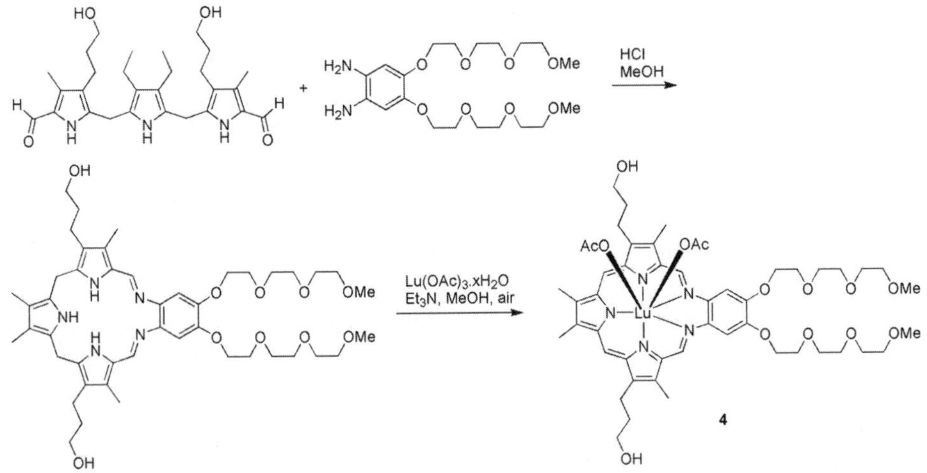

Figure 11.6   Synthesis of Lu-Tex **4**

benzene), yielding a non-aromatic texaphyrin precursor, also called 'sp^3-texaphyrin' in 92 % yield. The tripyrran unit itself can be acccessed in a high-yielding 4-step synthesis (Sessler *et al.*, 1993). Step 2 is the oxidative metalation of the macrocycle in the presence of Lu(OAc)$_3$.xH$_2$O and Et$_3$N, which provides the lanthanide chelate in moderate (~63 %) yield.

The lowest-energy transition of the lutetium complex appears at 732 nm ($\varepsilon = 42\,000/$ M/cm) in MeOH, the far red end of the visible spectrum, allowing for deep tissue penetration without interference from hemoglobin (Sessler and Miller, 2000). As it contains a heavy atom, this complex undergoes intersystem crossing very efficiently, and singlet oxygen quantum yields up to 70 % have been observed.

Lu-Tex has a number of appealing properties that give it an edge over the compounds presented so far. It is a chemically produced, pure compound, as opposed to HpD. Lu-tex remains monomeric in aqueous solutions up to 100 µM concentrations (Sessler *et al.*, 1997), and is thus easily administered *via* i.v. injection. It is retained in cancerous tissue approximately ten times more than in surrounding normal tissue. It displays favorable pharmacokinetics, and clears rapidly from the plasma (hours *vs.* months for Photofrin), which results in low dark toxicity (Sessler and Miller, 2000).

Lu-Tex-PDT has been proposed for the treatment of prostate cancer. Preclinical results on canines (Hargus *et al.*, 2007), and phase I trials have shown that interstitial Lu-Tex-mediated PDT is a viable treatment option for locally recurrent prostate adenocarcinoma (Du *et al.*, 2006; Verigos *et al.*, 2006). Preliminary results did not show a dose-limiting toxicity. Side effects were mild, and generally much better tolerated than those of other treatment modalities. However, drug distribution was found to be very uneven, and photobleaching during treatment was noted. Further studies should be undertaken to explore the full possibilities in the PDT of prostate cancer. Lutetium texaphyrin has also been the subject of phase I clinical trials in the treatment of cutaneous and subcutaneous tumours and phase II clinical trials for treatment of recurrent breast cancer to the chest wall (Sessler and Miller, 2000). *In vivo* light dosimetry on five patients involved in the later trial showed large variations in optical properties in patients (Dimofte *et al.*, 2002).

11.2.5 Hydroporphyrin photosensitizers

Chlorins and bacteriochlorins can be formally derived from porphyrins by reduction of one and two pyrrolic double bonds, respectively (Figure 11.7). The reduction results in an increase in the intensity of the Q_y band for chlorins, and a pronounced red shift for bacteriochlorins (λ_{abs} = 600–650 nm for chlorins, >700 nm for bacteriochlorins), which make these compounds better suited for PDT. The majority of clinical trials were conducted on natural or naturally-derived hydroporphyrins.

Chlorins and bacteriochlorins have been prepared by osmium-mediated dihydroxylation of porphyrins and chlorins, cyclopropanation of pyrrolic double bonds, and by extensions of the conjugated π-system to produce benzochlorins, naphthalochlorins and spirochlorins (Sternberg *et al.*, 1998; Pandey *et al.*, 2000). In some cases significant bathochromic shifts were observed compared to the parent molecules, along with respectable singlet oxygen quantum yields. However, these results did not translate into high *in vivo* activities. Naturally occurring bacteriochlorins are prone to oxidation when exposed to light and/or air, thus their potential in PDT is severely limited, although Tookad (Figure 11.1, **5**), a bacteriochlorophyll-derived second generation photosensitizer is rather stable, and has received a lot of attention recently (Chen *et al.*, 2002; Schreiber *et al.*, 2002; Huang *et al.*, 2005; Mazor *et al.*, 2005). Synthetic bacteriochlorins, which bear substituents that lock-in the oxidation state of the reduced pyrrole units are stable under ambient conditions, and should be considered promising PSs (Kim and Lindsey, 2005; Fan *et al.*, 2007). Although to date only organic-soluble derivatives have been prepared, it is expected that the methodology will enable the introduction of functional groups relevant for PDT.

After several decades of tetrapyrrole synthetic chemistry (Pandey *et al.*, 1994; Robinson, 2000; Jacobi *et al.*, 2001; Laha *et al.*, 2006; O'Neal *et al.*, 2006; Muthiah *et al.*, 2007; Taniguchi *et al.*, 2007), virtually any combination of substituents can be introduced into tetrapyrrole complexes of various oxidation states. Thus, *de novo* synthesis of porphyrins, chlorins, bacteriochlorins and purpurins is possible, with tailored lipo/hydrophilicities, and added bioconjugatable sites. Tailored, synthetic chlorins are available in a convergent manner (O'Neal *et al.*, 2005; Kim *et al.*, 2006) (Figure 11.8). Extensive work on synthetic and natural chlorin derivatives has revealed profound relationships between the type and position of peripheral substituents and photophysical properties, which should be exploited in the design of new PS candidates (Muthiah *et al.*, 2007; Taniguchi *et al.*, 2007; Laha *et al.*, 2006).

One of the major obstacles is the availability of the desired compounds, as some of the reported syntheses consist of a large number (more than five) of steps, some of which proceed

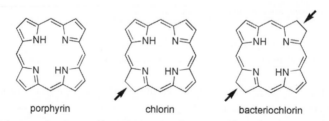

porphyrin   chlorin   bacteriochlorin

Figure 11.7 The relationship between porphyrins, chlorins and bacteriochlorins

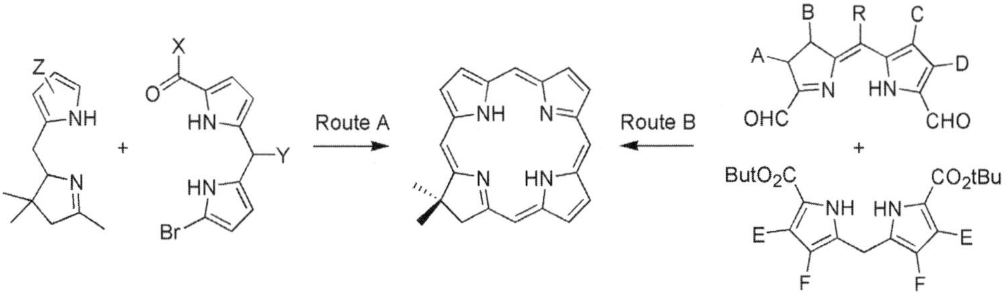

Figure 11.8 Chlorin synthesis from dihydrodipyrrin and dipyrromethane by the Lindsey (Route A) and Jacobi (Route B) methods

with only moderate or poor yields. A path of PDT research may thus be the optimization of existing syntheses, and the development of novel routes to access the PSs. Incorporation of the new PSs into conjugates for targeted delivery, or site-specific singlet oxygen generation would be a further goal.

NPe6

NPe6 (mono-(L)-aspartyl chlorin-e_6, Talaporfin, MACE, LS-11, Figure 11.1, **6**) is formed in the coupling reaction of the chlorophyll α-derivative, chlorin e_6 (Figure 11.1, **7**), and aspartic acid. The reaction is reported to yield one regioisomer in >95 % purity (Hargus et al., 2007), despite the availability of three carboxylic acids in chlorin e_6.

A phase I trial involving 14 cancer patients with primary and secondary skin cancers evaluated the pharmacokinetics, efficacy and safety of NPe6 (Chan et al., 2005). Five different doses of NPe6 were administered, while the light dose was varied according to tumour shape and size. Two patients had mild to moderate adverse effects (oedema, erythema, burning pain at the tumour site) related to the treatment. Photosensitivity caused by NPe6 ceased after only about 1 week in most patients. At a four-week follow-up, out of 22 BCC lesions, 20 (91 %) showed complete response, 1 (5 %) partial response, and 1 (5 %) no response. Of the 12 SCC lesions, 9 (75 %) showed complete response, and 8 % partial response. Similar results were obtained for papillary breast carcinoma, while metastatic papillary carcinomas of the thyroid were not responsive. Fewer non-responsive lesions were observed at higher NPe6 doses (1.65 mg/kg, or greater). Topical application of NPe6 for PDT has been proposed after experiments with BALB/c mice models (Wong et al., 2003). The dose dependence of the effectiveness was evaluated with NPe6 doses ranging from 0.5–3.5 mg/kg (Taber et al., 1998). It was found that doses below 1.65 mg/kg produced short-lived results, and tumour regression occurred within 12 weeks of treatment, while selectivity was lost when at least 2.5 mg/kg NPe6 was administered.

In early stage lung cancer, easily accessible, small lesions (up to 2 cm in diameter) were treated with a combination of 40 mg/m^2 NPe6 and 100 J/cm^2 laser irradiation (Kato et al., 2003). A complete response rate of >80 % was registered. The study also found that a new, portable diode laser system (AlGaInP) was a suitable light source. Given the small size (20 kg) of the equipment, it is likely to become a useful addition to the instruments currently in use.

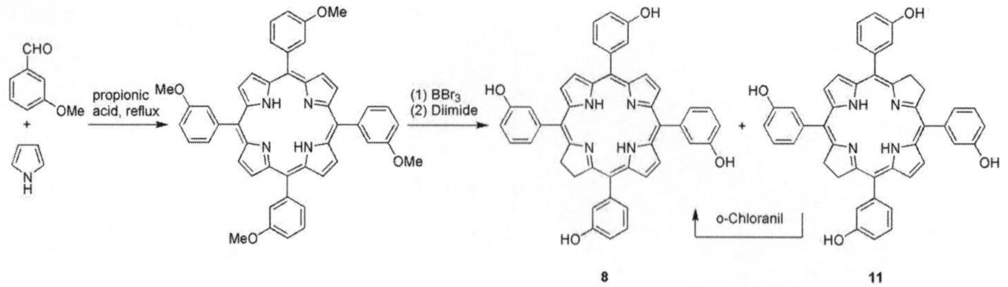

Figure 11.9 Synthesis of *meso*-tetrahydroxyphenylchlorin **8**

mTHPC

mTHPC (Foscan, Figure 11.1, **8**), a fully synthetic systemic photosensitizer is available from the corresponding porphyrin by diimide-reduction (Figure 11.9). The reaction yields a mixture of chlorin and bacteriochlorin, the latter of which can be oxidized back to the chlorin upon treatment with DDQ or *o*-chloranil (Gorman *et al.*, 2006). The long activation wavelength of mTHPC (652 nm, $\varepsilon = 30\,000$/M/cm) enables deeper tissue penetration compared to porphyrin derivatives (up to 10 mm), thus allowing for the treatment of nodular skin tumours, for example. It also has a relatively short treatment time (100–200 s), so multiple lesions can be treated at the same time. In one instance (Kübler *et al.*, 1999), this meant the treatment of 83 lesions in one patient in a single session. Foscan photosensitivity lasts only 1–2 weeks (although there are reports of up to 6 weeks (Detty *et al.*, 2004)), compared to up to several months with Photofrin. Regioselectively glycosylated derivatives of mTHPC have been described (Varamo *et al.*, 2007), with the aim of altering the tumour uptake, localization and phototoxicity of the parent compound, and increasing photoefficiency. mTHPC has been approved for the palliative treatment of head and neck cancer in Europe (Detty *et al.*, 2004).

mTHPC-PDT in early-stage squamous cell carcinoma of the head and neck resulted in 95 % complete remission for T1 lesions and 57 % for T2 lesions. Recurrent cancers, as well as metastases could be managed by conventional therapies (radiotherapy or surgery) (Copper *et al.*, 2003). For patients with advanced stage, incurable head and neck cancers, mTHPC-mediated PDT offered an improved quality of life and an increase in survival (Lou *et al.*, 2004; D'Cruz *et al.*, 2004). In particular, when 100 % tumour mass reduction was observed, the patients' mean survival time increased considerably (16 months *vs.* 2 months), compared to non-responsive tumours (Lou *et al.*, 2004). The best results were obtained for lesions <10 mm depth and receiving full illumination across the area, while non-responders were bulky tumours in the proximity of vital structures, which were difficult to illuminate.

PDT of BCC with mTHPC was carried out on five patients presenting a total of 187 lesions. Illumination on days 1 and 2 with 10 or 15 J/cm^2 yielded the best results. Under such conditions, 70–86 % of treated lesions gave complete response at a 12-month follow-up (Baas *et al.*, 2001). Further optimization of the procedure (Triesscheijn *et al.*, 2006) revealed that the ideal illumination time was 24 h after i.v. administration of 0.1 mg/kg mTHPC. Good results were obtained with treatment times as low as 100 s. PDT of non-melanomous skin cancers on the head and neck in 18 patients, with altogether 97 tumours (88 BCC, 9 SCC), was carried out (Kübler *et al.*, 1999). The patients were injected with either 0.15 mg/kg or 0.10 mg/kg Foscan, followed by irradiation at 652 nm, 96 h post injection. The response rate,

(92.7 % complete, 7.3 % partial), as well as the cosmetic outcomes were excellent. The latter depended on the light dose, and better results were obtained with 10 J/cm^2 than with 15 J/cm^2. The authors observed that tumour necrosis was not uniform even when similar lesions were treated on different patients, and suggested that the discrepancies may stem from the different distribution of the drug caused by variations in body weight.

A phase I study was conducted to evaluate the effectiveness of mTHPC in early prostate cancer (Moore *et al.*, 2006). No attempts were made at that point to eradicate the whole tumour, but the preliminary results suggested that mTHPC could be useful for patients for whom a radical treatment is not a viable option. A further advantage of PDT is that the side effects (incontinence, erectile dysfunction) associated with a radical treatment can be avoided, significantly improving the quality of life. Curative resection of pancreatic cancer is possible in approximately 10 % of the cases, but numbers as low as 2.6 % have been suggested. A large proportion of the tumours do not respond to either chemotherapy, or radiotherapy. In such cases, mTHPC-mediated PDT was proposed as a promising treatment modality. In a phase I study, a median survival time of 9.5 months was obtained, and the 1-year survival rate was 44 %, compared with 19 % with chemotherapy, and 41 % with a combination of radiotherapy and chemotherapy (Bown *et al.*, 2002).

A preliminary study on Barrett's oesophagus showed that mTHPC-mediated PDT can yield complete responses in the majority of the cases (Javaid *et al.*, 2002). In this study, the authors also evaluated the use of an alternative light source to eliminate the need for lasers. For mTHPC, a xenon arc lamp (Paterson lamp), equipped with suitable filters, was appropriate. Irradiation at 514 nm (green light) was also feasible, although the general applicability of green light is not obvious, given the limited tissue penetration of such short-wavelength light (Etienne *et al.*, 2004).

Diffuse malignant mesothelioma is a usually fatal disease that has no standard therapy yet, and attempts at improving survival have been generally unsuccessful. In a preliminary trial, intraoperative PDT with mTHPC was attempted to achieve local tumour control and pain relief (Ris *et al.*, 1996). Good selectivity of mTHPC for tumour tissue was observed, which generally spared non-malignant areas during treatment. During follow-up, the treated site proved disease free, and effectively painless, but distant metastases could not be avoided. It was possible to combine PDT with photodynamic diagnosis (PDD, fluorescence detection) using mTHPC (Zimmermann *et al.*, 2001). The red fluorescence of mTHPC could be detected either with the naked eye, or by a video camera. Additional sensitivity could be achieved by filters. The technique was applied to the surgery of malignant brain tumours. Following resection, residual cancer cells could be detected because of the mTHPC fluorescence, and the remaining tumour tissue could be treated by PDT.

In recurrent breast cancer to the chest wall, PDT with low (0.1–0.15 mg/kg) doses of mTHPC proved a useful modality (Wyss *et al.*, 2001). In this study, all (89) lesions gave complete responses. As a comparison, radiotherapy of such tumours was only successful in ~65 % of the cases (Wyss *et al.*, 2001). Contrary to surgical restriction, PDT does not result in scarring, and offers the same high response rate.

Verteporfin

Verteporfin (Figure 11.1, **9**), is synthesized by the Diels-Alder addition of dimethyl acetylenedicarboxylate to PpIX dimethyl ester, and exists as a mixture of isomers (Figure 11.10) (Morgan *et al.*, 1990). The monoesters have proved superior photosensitizers

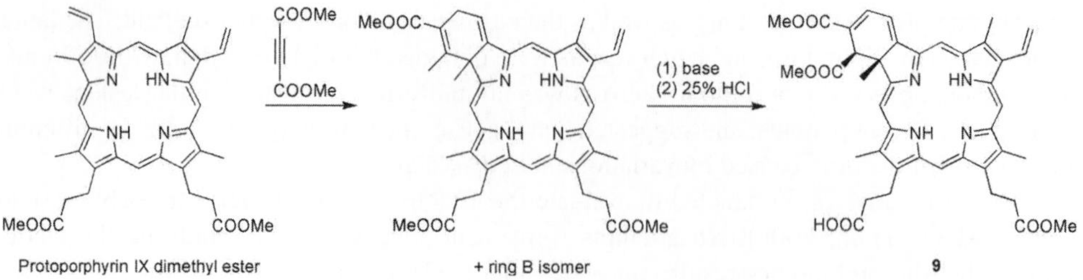

COOMe

MeOOC

COOMe

MeOOC   COOMe

Protoporphyrin IX dimethyl ester

(1) base
(2) 25% HCl

MeOOC

MeOOC

MeOOC   COOMe

+ ring B isomer

MeOOC

MeOOC

HOOC   COOMe

9

Figure 11.10   Synthesis of Verteporfin **9**

in vivo to both the diacid and the diester. It is an FDA-approved treatment for age-related macular degeneration. Clinical trials have been conducted to assess its use in the treatment of nonmelanoma skin cancers (Lui *et al.*, 2004). Verteporfin combines the long excitation wavelength (689 nm) (hence deep tissue penetration) of traditional porphyrin-based photosensitizers, with the rapid skin clearance and reduced light sensitivity of ALA, making it a near-ideal PS, although further experiments are required to fully establish its potential.

In a series of clinical trials, the groups of Barbazetto, Schmidt-Erfurth and Jurklies have explored the possibility of Verteporfin-mediated treatment of various eye conditions, among them conjunctival SCC (Barbazetto *et al.*, 2004; Jurklies *et al.*, 2003). In all cases, PDT offered the possibility of tumour regression without significant risk to ocular structures. The results of all studies were promising. Verteporfin acts primarily on the vasculature, and the authors noted that it may be less well suited for early and poorly vascularized tumours (Barbazetto *et al.*, 2004). Larger tumours could be treated by applying the laser irradiation to several spots. Despite overlaps between irradiated regions, an increase in adverse effects was not observed, compared to single-spot treatments. When patients with circumscribed choroidal haemangioma were treated, PDT only had a beneficial effect on the visual acuity for patients with a significant tumour distance from the fovea, a pretreatment visual acuity of >0.1, and who had had the symptoms for less than 30 months (Jurklies *et al.*, 2003).

Other hydroporphyrin photosensitizers in clinical trials

PDT has shown promise in a number of different types of cancers, but only a very small number of trials have assessed it for the treatment of melanoma. One of the reasons for the lack of results may be the intense absorption of melanin-containing cells, which requires photosensitization at the near infrared region of the spectrum. Most approved photosensitizers (ALA, Photofrin, HpD) simply do not have an appropriate absorption band for such applications. Chlorins, phorbides and bacteriochlorins absorb strongly in the red region of the visible spectrum (600–800 nm), making them promising candidates for the PDT of melanoma. In a phase I study, chlorin e_6 (Figure 11.1, **7**) was evaluated for safety and efficacy. Fourteen patients with skin metastases from melanoma were treated with 5 mg/kg chlorin e_6 and 80–120 J/cm^2 light doses. In eight cases, a complete response was observed after a single treatment, and in the remaining six patients, after several rounds of PDT (Sheleg *et al.*, 2004).

Photosensitivity caused by Photochlor (HPPH, 2-(1-hexyloxyethyl)-2-devinyl pyropheophorbide-a, Figure 11.1, **10**) was found to be significantly milder and shorter

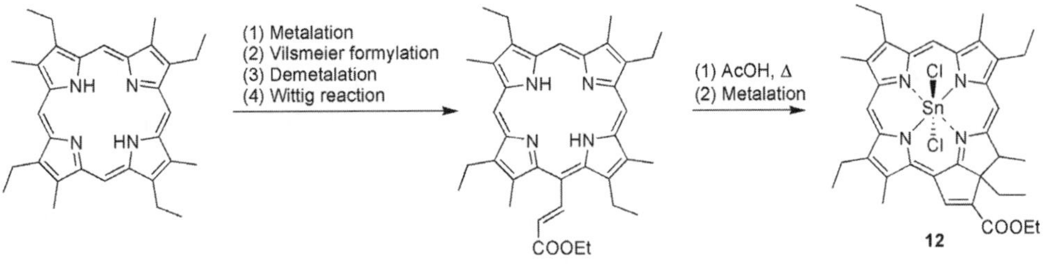

Figure 11.11 Synthesis of SnET2 **12** from etioporphyrin

lived than that caused by Photofrin (Bellnier *et al.*, 2006). Photochlor entered phase II clinical trials for lung cancer in August 2007.

The bacteriochlorin analogue of mTHPC, (5,10,15,20-tetrakis(*m*-hydroxyphenyl) bacteriochlorin, mTHPBC, Figure 11.1, **11**) was evaluated for interstitial PDT of non-resectable, deeply seated colorectal liver metastases (van Duijnhoven *et al.*, 2005). Tumour progression was stopped in 84 % of the cases. Side effects included PDT damage to the pancreas, mild pain, and hyperpigmentation at the administration site.

Purpurins are chlorin derivatives, and have originally been obtained by chlorophyll degradation. They can be synthesized (Figure 11.11) from *N*-protected porphyrins (e.g. a stable metal chelate, such as Ni) by Vilsmeier formylation of one of the *meso* positions (Morgan *et al.*, 1989). Wittig reaction with (carbethoxymethylene)-triphenylphosphorane yields the *meso*-(b-(ethoxycarbonyl)vinyl)porphyrin, which after demetalation can cyclize to the purpurin when refluxed in glacial acetic acid under an inert atmosphere. The cyclization proceeds with excellent yields, and when the starting porphyrin is readily available, this is a simple and straightforward route to purpurins. The compounds thus obtained had absorption maxima in the 638–690 nm region, depending on the nature of the peripheral substituents, and the metalation state (Garbo, 1996). The purpurins were efficient generators of singlet oxygen ($\Phi(^1O_2)$ = 0.6–1.0). *In vitro*, the most active derivative was a tin-metalated etioporphyrin-derived compound (tin etiopurpurin, SnET2, Figure 11.1, **12**), which underwent clinical trials for certain skin malignancies and AIDS-related Kaposi's sarcoma (Garbo, 1996). The response rates were excellent for patients with basal cell carcinoma, metastatic adenoma carcinoma, and with basal cell nevus syndrome (100 %, 96 % and 86 % complete response, respectively). The major pathway for tumour destruction seems to be vascular damage.

11.2.6 Phthalocyanines

Phthalocyanines (Pcs) are synthetic tetrapyrrolic macrocycles, extensively used in industry as colorants, redox catalysts, and photoconducting agents (Sharman *et al.*, 2003; McKeown, 2003; Rodríguez-Morgade *et al.*, 2003). The core structure is an 18-π-electron aromatic macrocycle, wherein four isoindoles are held together by aza bridges. Pcs display an absorption maximum around 670 nm with a high extinction coefficient ($\varepsilon = 10^5$/M/cm), which makes them attractive for photodynamic therapy. They also benefit from chemical stability and are easily synthesized and modified. Unsubstituted or symmetrically substituted Pcs are readily available (albeit in only moderate yields) *via* cyclotetramerization of suitable precursors, such as phthalonitriles (Sharman *et al.*, 2003; Tomoda *et al.*, 1980; Hu *et al.*, 1998; Choi *et al.*, 2004; Lo *et al.*, 2007), phthalimide (Sharman *et al.*, 2003), phthalic

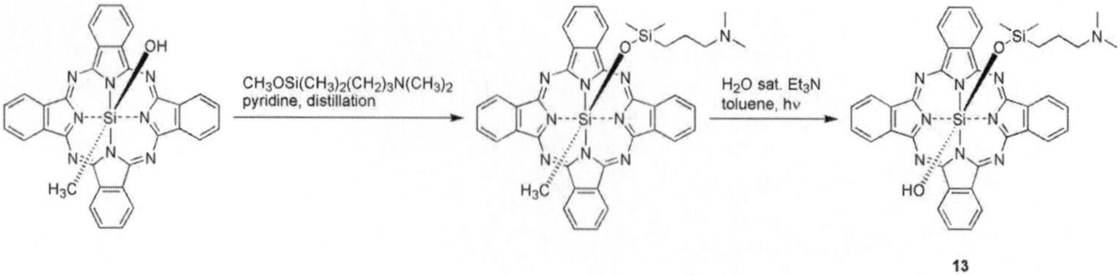

Figure 11.12 Synthesis of Pc4 **13**

acid (Sharman *et al.*, 2003) or other phthalic acid derivatives (Sharman *et al.*, 2003; de la Torre *et al.*, 2000). Metal Pcs are available either directly from tetramerization when performed in the presence of a metal template (McKeown, 2003; Aoudia *et al.*, 1997); through transmetalation (McKeown, 2003); or metalation of free base Pc (McKeown, 2003; Aoudia *et al.*, 1997; Thompson *et al.*, 1993; McKeown *et al.*, 1990). Unsymmetrically substituted Pcs can be prepared either *via* statistical reaction, followed by rather tedious purification of the desired component (McKeown, 2003), or by various directed approaches (McKeown, 2003; de la Torre *et al.*, 2000). There are selective approaches towards the synthesis of *cis*-disubstituted Pcs, which have shown more promising PDT-activity than their *trans* counterparts (Sharman *et al.*, 2003). Unsubstituted Pcs are poorly soluble either in aqueous or most organic solvents. Recent work on Pc-mediated PDT has focused on improving PS bioavailability by increasing their solubility in biological media e.g. *via* sulfonation, carboxylation (Choi *et al.*, 2004), hydroxylation (Hu *et al.*, 1998), the attachment of cationic groups to the macrocycle (Lo *et al.*, 2007), or by incorporation of penta- or hexacoordinate metals (e.g. Si), bearing axial ligands (Lee *et al.*, 2005), which can facilitate both aqueous solubility and improve the tumour targeting properties. Tumour targeting upon conjugation to monoclonal antibodies has also been described (Ogura *et al.*, 2006). Closed-shell diamagnetic phthalocyanines (Zn, Al, Si), which can efficiently generate singlet oxygen with yields in the range of 0.2–0.5, have especially received a lot of attention as PSs.

Silicon phthalocyanine 4 (Pc4, Figure 11.1, **13**), a silicon-inserted phthalocyanine axially ligated to hydroxyl and silyloxy moieties, is currently in phase I clinical trial for the treatment of patients with premalignant and malignant skin conditions (actinic keratosis, Bowen's disease, skin cancer, or stage I or stage II mycosis fungoides) through topical application (http://www.cancer.gov/search/ViewClinicalTrials.aspx?cdrid=410675&version=HealthProfessional&protocolsearchid=3854835). It previously underwent phase I evaluation in the treatment of patients with cutaneous malignancies (http://www.cancer.gov/search/ViewClinicalTrials.aspx?cd rid=68862&version=HealthProfessional&protocolsearchid=3854839).

This complex can be synthesized in a two-step procedure, by ligand substitution of the hydroxyl substituent followed by photolysis of a methyl,hydroxyl-ligated silicon phthalocyanine with an overall yield of 73 % (Figure 11.12) (Oleinick *et al.*, 1993). This complex displays a maximum absorption at 669 nm (CH$_2$Cl$_2$) (Oleinick *et al.*, 1993).

11.2.7 Non-porphyrinic photosensitizers

Acridine orange (AO, Figure 11.13), a cationic fluorescent dye, has been employed in combination with surgery as a new limb salvage modality (Kusuzaki *et al.*, 2005). AO

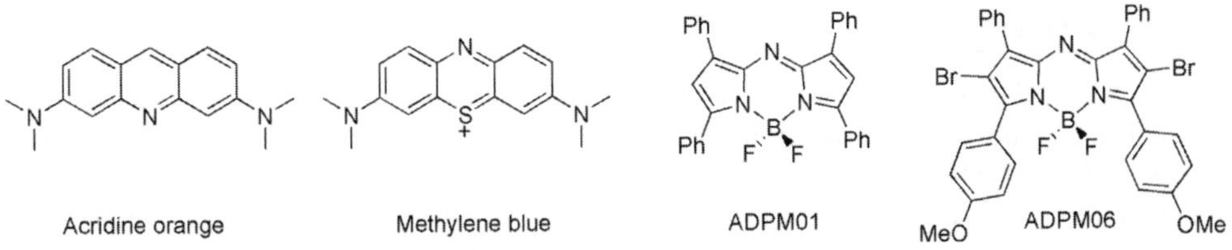

Figure 11.13 Structures of some non-porphyrinic photosensitisers

has reported antitumour activity, photosensitizing activity, and antifungal-, antiviral- and antiparasitic activity (Kusuzaki *et al.*, 2005). AO-sensitive tumour cells emit in the green upon blue-light irradiation, thus, the sensitivity of the cells can be confirmed by biopsy. A phase I clinical trial utilized a unique combination of AO-PDT treatment with surgical removal of musculoskeletal sarcomas, either with or without an additional 5 Gy irradiation. Ten patients underwent AO-PDT and surgery, half of whom received the additional X-ray irradiation. Of the five AO-PDT only patients, one showed local recurrence of the tumour. All ten patients were alive, and regained full use of their limb in the two years of follow-up.

The phenothiazine dye, methylene blue, has antimicrobial activity in combination with light (Kömerik *et al.*, 2002). It has been proposed for systemic PDT, but proved ineffective, presumably due to the high aqueous solubility of the compound. As more hydrophobic derivatives are readily available through the introduction of lipophilic side chains, this group of PSs may eventually be useful (Figure 11.13) (Bonnett, 1995).

Two azadipyrromethanes (Figure 11.13, ADPM01 and ADPM06) were tested against various cell types for anticancer activity. The compounds had no dark toxicity, and were effective in the micro- to nanomolar range when illuminated, with ADPM06 retaining activity under hypoxic conditions (Gallagher *et al.*, 2005).

Semiconductor quantum dots (QD) are a group of nanomaterials that have been increasingly employed in the biomedical sciences. Their tunable absorption and emission characteristics make them very popular in imaging. Recently, it has been suggested that QD could be employed in PDT, either as fluorescence resonance energy transfer (FRET) donors, that can be matched to sensitize known organic photosensitizers, or for direct singlet oxygen generation (Samia *et al.*, 2006). The preliminary results have been encouraging; for example, a CdSe QD linked with a Pc acceptor participated in FRET, and singlet oxygen was generated upon illumination of the QD. There has been only one example of singlet oxygen production by a QD, with a rather low 5 % yield. Nevertheless, it is expected that given their photophysical flexibility, QD could become an interesting class of PSs. There are a number of issues that need addressing for *in vivo* applications, such as the potential toxicity of the heavy metal core of QD, and the effect of surface modifications on tumour selectivity, localization, and singlet oxygen generation.

Finally, fullerenes have been investigated by some research groups as potential PSs (Mroz *et al.*, 2007; and references therein). It was found that fullerenes (unlike the majority of PSs) initiate Type I photochemical reactions, in particular under reducing conditions (e.g. aqueous NADPH solutions). The major disadvantage of this class of compounds is that their absorption lies in the blue region of the spectrum, and the rather cumbersome derivatization and/or formulation necessary for tumour delivery. In addition to cancer therapy, fullerenes have been explored in the photodynamic killing of bacteria and viruses.

11.3 Outlook

The search for the ideal photosensitizer continues. A promising line of research is the synthetic analogues of currently used naturally derived photosensitizers. Rational chemical synthesis enables the fine tuning of the dye properties, be it the absorption maximum, the singlet oxygen quantum yield, or the aqueous solubility. Better tumour selectivity will be a major goal for the next generation of PSs. Whether the new PSs will exploit receptor-mediated delivery, incorporation into nanoparticles, or a fine-tuning of the lipophilicities/aqueous solubilities of the PS, or some other mechanism, remains to be seen. The immune response elicited by PDT is an exciting field of cancer PDT therapy, and certainly has relevance to the PDT of other diseases. The possibility to eliminate distant metastases using the body's own defense mechanism, and without systemic damage would give PDT a definite edge over both surgery and chemotherapy. PDT is expected to become a widely accepted and useful treatment modality, possibly in combination with other treatments not only for cancer, but also for a range of other diseases; the next decade will be crucial in defining the areas where PDT will have the greatest impact.

11.4 Acknowledgement

This chapter was written while the authors were postdoctoral researchers at North Carolina State University in the group of Prof. Jonathan S. Lindsey.

References

Akhlynina TV, Jans DA, Statsyuk NV, *et al*. Adenoviruses synergize with nuclear localization signals to enhance nuclear delivery and photodynamic action of internalizable conjugates containing chlorin e$_6$. *Int J Cancer* 1999, **81**, 734–40.

Ali H, van Lier JE. Metal complexes as photo- and radiosensitizers. *Chem Rev* 1999, **99**, 2379–450.

Allison R, Mang T, Hewson G, Snider W, Dougherty D. Photodynamic therapy for chest wall progression from breast carcinoma is an underutilized treatment modality. *Cancer* 2001, **91**, 1–8.

Angell-Petersen E, Sørensen R, Warloe T, Soler AM, Moan J, Peng Q, Giercksky K-E. Porphyrin formation in actinic keratosis and basal cell carcinoma after topical application of methyl 5-aminolevulinate. *J Invest Dermatol* 2006, **126**, 265–71.

Aoudia M, Cheng G, Kennedy VO, Kenney ME, Rodgers MAJ. Synthesis of a series of octabutoxy- and octabutoxybenzophthalocyanines and photophysical properties of two members of the series. *J Am Chem Soc* 1997, **119**, 6029–39.

Araki K, Engelmann FM, Mayer I, Toma HE, Baptista MS, Maeda H, Osuka A, Furuta H. Doubly N-confused porphyrins as efficient sensitizers for singlet oxygen generation. *Chem Lett* 2003, **32**, 244–5.

Baas P, Saarnak AE, Oppelaar H, Neering H, Stewart FA. Photodynamic therapy with metatetrahydroxyphenylchlorin for basal cell carcinoma: a phase I/II study. *Br J Dermatol* 2001, **145**, 75–8.

Barbazetto IA, Lee TC, Abramson DH. Treatment of conjunctival squamous cell carcinoma with photodynamic therapy. *Am J Ophthalmol* 2004, **138**, 183–9.

Bauer TW, Hahn SM, Spitz FR, Kachur A, Glatstein E, Fraker DL. Preliminary report of photodynamic therapy for intraperitoneal sarcomatosis. *Ann Surg Oncol* 2001, **8**, 254–9.

Bellnier DA, Greco WR, Nava H, Loewen GM, Oseroff AR, Dougherty TJ. Mild skin photosensitivity in cancer patients following injection of Photochlor (2-(1-hexyloxyethyl)-2-devinyl pyropheophorbide-a; HPPH) for photodynamic therapy. *Cancer Chemother Pharmacol* 2006, **57**, 40–5.

Bernstein ZP, Wilson BD, Oseroff AR, *et al*. Photofrin photodynamic therapy for treatment of AIDS-related cutaneous Kaposi's sarcoma. *AIDS* 1999, **13**, 1697–704.

Berr F, Wiedmann M, Tannapfel A, *et al*. Photodynamic therapy for advanced bile duct cancer: evidence for imrpoved palliation and extended survival. *Hepatology* 2000, **31**, 291–8.

Bisland SK, Singh D, Gariépy J. Potentiation of chlorin e_6 photodynamic activity *in vitro* with peptide-based intracellular vehicles. *Bioconjugate Chem* 1999, **10**, 982–92.

Bonnett R. Photosensitizers of the porphyrin and phthalocyanine series for photodynamic therapy. *Chem Soc Rev* 1995, **24**, 19–33.

Borelli C, Herzinger T, Merk K, Berking C, Kunte C, Plewig G, Degitz K. Effect of subcutaneous infiltration anesthesia on pain in photodynamic therapy: a controlled open pilot trial. *Dermatol Surg* 2007, **33**, 314–18.

Bown SG, Rogowska AZ, Whitelaw DE, *et al*. Photodynamic therapy for cancer of the pancreas. *Gut* 2002, **50**, 549–57.

Braslavsky SE, Müller M, Mártire DO, *et al*. Photophysical properties of porphycene derivatives (18π porphyrinoids). *J Photochem Photobiol B: Biol* 1997, **40**, 191–8.

Caekelbergh K, Annemans L, Lambert J, Roelandts R. Economic evaluation of methyl aminolaevulinate-based photodynamic therapy in the management of actinic keratosis and basal cell carcinoma. *Br J Dermatol* 2006, **155**, 784–90.

Casas A, Perotti C, Ortel B, di Venosa G, Saccoliti M, Batlle A, Hasan T. Tumour cell lines resistant to ALA-mediated photodynamic therapy and possible tools to target surviving cells. *Int J Oncol* 2006, **29**, 397–405.

Castano AP, Mroz P, Hamblin MR. Photodynamic therapy and anti-tumour immunity. *Nature Rev Cancer* 2006, **6**, 535–45.

Chan AL, Juarez M, Allen R, Volz W, Albertson T. Pharmacokinetics and clinical effects of mono-L-aspartyl chlorin e_6 (NPe6) photodynamic therapy in adult patients with primary or secondary cancer of the skin and mucosal surfaces. *Photodermatol Photoimmunol Photomed* 2005, **21**, 72–8.

Chen Q, Huang Z, Luck D, *et al*. WST09 (TOOKAD) mediated photodynamic therapy as an alternative modality in the treatment of prostate cancer. *Proc SPIE* 2002, **4612**, 29–39.

Chen X, Drain CM. Photodynamic therapy using carbohydrate conjugated porphyrins. *Drug Des Rev* 2004, **1**, 215–34.

Choi C-F, Tsang P-T, Huang J-D, Chan EYM, Ko W-H, Fong W-P, Ng DKP. Synthesis and *in vitro* photodynamic activity of new hexadeca-carboxy phthalocyanines. *Chem Commun* 2004, **19**, 2236–2237.

Choi Y, Weissleder R, Tung C-H. Protease-mediated phototoxicity of a polylysine-chlorin e_6 conjugate. *ChemMedChem* 2006, **1**, 698–701.

Cló E, Snyder JW, Voigt NV, Ogilby PR, Gothelf KV. DNA-programmed control of photosensitized singlet oxygen production. *J Am Chem Soc* 2006, **128**, 4200–1.

Conlon KA, Berrios M. Site-directed photoproteolysis of 8-oxoguanine DNA glycosylase 1 (OGG1) by specific porphyrin-protein probe conjugates: a strategy to improve the effectiveness of photodynamic therapy for cancer. *J Photochem Photobiol B: Biol* 2007, **87**, 9–17.

Copper MP, Tan IB, Oppelaar H, Ruevekamp MC, Stewart FA. Meta-tetra(hydroxyphenyl)chlorin photodynamic therapy in early-stage squamous cell carcinoma of the head and neck. *Arch Otolaryngol Head Neck Surg* 2003, **129**, 709–11.

Cuenca RE, Allison RR, Sibata C, Downie GH. Breast cancer with chest wall progression: treatment with photodynamic therapy. *Ann Surg Oncol* 2004, **11**, 322–7.

D'Cruz AK, Robinson MH, Biel MA. mTHPC-mediated photodynamic therapy in patients with advanced, incurable head and neck cancer: a multicenter study of 128 patients. *Head Neck* 2004, **26**, 232–40.

D'Hallewin M-A, Baert L. Long-term results of whole bladder wall photodynamic therapy for carcinoma *in situ* of the bladder. *Urology* 1995, **45**, 763–7.

de Graaf YGL, Kennedy C, Wolterbeek R, Collen AFS, Willemze R, Bouwes Bavinck JN. Photodynamic therapy does not prevent cutaneous squamous-cell carcinoma in organ-transplant recipients: results of a randomized-controlled trial. *J Invest Dermatol* 2006, **126**, 569–74.

de la Torre G, Claessens CG, Torres T. Phthalocyanines: the need for selective synthetic approaches. *Eur J Org Chem* 2000, 2821–30.

Detty MR, Gibson SL, Wagner SJ. Current clinical and preclinical photosensitizers for use in photodynamic therapy. *J Med Chem* 2004, **47**, 3897–915.

Dimofte A, Zhu TC, Hahn SM, Lustig RA. *In vivo* light dosimetry for motexafin lutetium-mediated PDT of recurrent breast cancer. *Lasers Surg Med* 2002, **31**, 305–12.

Dogutan DK, Ptaszek M, Lindsey JS. Direct synthesis of magnesium porphine via 1-formyl-dipyrromethane. *J Org Chem* 2007, **72**, 5008–11.

Dolmans DEJGJ, Fukumura D, Jain RK. Photodynamic therapy for cancer. *Nat Rev Cancer* 2003, **3**, 380–7.

Driaf K, Krausz P, Verneuil B, Spiro M, Blais JC, Bolbach G. Glycosylated cationic porphyrins as potential agents in cancer phototherapy. *Tetrahedron Lett* 1993, **34**, 1027–30.

Du KL, Mick R, Busch TM, *et al*. Preliminary results of interstitial motexafin lutetium-mediated PDT for prostate cancer. *Lasers Surg Med* 2006, **38**, 427–34.

El-Akra N, Noirot A, Faye J-C, Souchard J-P. Synthesis of estradiol-pheophorbide a conjugates: evidence of nuclear targeting, DNA damage and improved photodynamic activity in human breast cancer and vascular endothelial cells. *Photochem Photobiol Sci* 2006, **5**, 996–9.

Etienne J, Dorme N, Bourg-Haeckly G, Raimbert P, Flijou JF. Photodynamic therapy with green light and m-tetrahydroxyphenyl chlorin for intramucosal adenocarcinoma and high-grade dysplasia in Barrett's esophagus. *Gastrointest Endosc* 2004, **59**, 880–9.

Fan D, Taniguchi M, Yao Z, Dhanalekshmi S, Lindsey JS. 1,9-Bis(N, N-dimethylaminomethyl) dipyrromethanes in the synthesis of porphyrins bearing one or two meso substituents. *Tetrahedron* 2005, **61**, 10291–302.

Fan D, Taniguchi M, Lindsey JS. Regioselective 15-bromination and functionalization of a stable synthetic bacteriochlorin. *J Org Chem* 2007, **72**, 5350–57.

Fickweiler S, Abels C, Karrer S, Bäumler W, Landthaler M, Hofstädter F, Szeimies R-M. Photosensitization of human skin cell lines by ATMPn (9-acetoxy-2,7,12,17-tetrakis-(β-methoxyethyl)-porphycene) *in vitro*: mechanism of action. *J Photochem Photobiol B: Biol* 1999, **48**, 27–35.

Frochot C, di Stasio B, Vanderesse R, Belgy M-J, Dodeller M, Guillemin F, Viriot M-L, Barberi-Heyob M. Interest of RGD-containing linear or cyclic peptide targeted tetraphenylchlorin as novel photosensitizers for selective photodynamic therapy. *Bioorg Chem* 2007, **35**, 205–20.

Furuta H, Asano T, Ogawa T. 'N-confused porphyrin': a new isomer of tetraphenylporphyrin. *J Am Chem Soc* 1994, **116**, 767–8.

Gacio AF, Fernandez-Marcos C, Swamy N, Dunn D, Ray R. Photodynamic cell-kill analysis of breast tumour cells with a tamoxifen-pyropheophorbide conjugate. *J Cell Biochem* 2006, **99**, 665–70.

Gallagher WM, Allen LT, O'Shea C, Kenna T, Hall M, Gorman A, Killoran J, O'Shea DF. A potent nonporphyrin class of photodynamic therapeutic agent: cellular localisation, cytotoxic potential and influence of hypoxia. *Br J Cancer* 2005, **92**, 1702–10.

Garbo GM. Purpurins and benzochlorins as sensitizers for photodynamic therapy. *J Photochem Photobiol B: Biol* 1996, **34**, 109–116.

Garcia G, Sol V, Lamarche F, Granet R, Guilloton M, Champavier Y, Krausz P. Synthesis and photocytotoxic activity of new chlorin–polyamine conjugates. *Bioorg Med Chem Lett* 2006, **16**, 3188–92.

Geier GR, Haynes DM, Lindsey JS. An efficient one-flask synthesis of N-confused tetraphenylporphyrin. *Org Lett* 1999, **1**, 1455–58.

Gorman SA, Brown SB, Griffiths J. An overview of synthetic approaches to porphyrin, phthalocyanine, and phenothiazine photosensitizers for photodynamic therapy. *J Environ Pathol Toxicol Oncol* 2006, **25**, 79–108.

Gryshuk A, Chen Y, Goswami LN, *et al.* Structure-activity relationship among purpurinimides and bacteriopurpurinimides: Trifluoromethyl substituent enhanced the photosensitizing efficacy. *J Med Chem* 2007, **50**, 1754–67.

Guldi DM, Mody TD, Gerasimchuk NN, Magda D, Sessler JL. Influence of large metal cations on the photophysical properties of texaphyrin, a rigid aromatic chromophore. *J Am Chem Soc* 2000, **122**, 8289–98.

Hahn SM, Fraker DL, Mick R, *et al.* A phase II trial of intraperitoneal photodynamic therapy for patients with peritoneal carcinomatosis and sarcomatosis. *Clin Cancer Res* 2006, **12**, 2517–25.

Hamblin MR. Covalent photosensitizer conjugates for targeted photodynamic therapy. *Trends Photochem Photobiol* 2002, **9**, 1–24.

Hamblin MR, Miller JL, Ortel B. Scavenger-receptor targeted photodynamic therapy. *Photochem Photobiol* 2000, **72**, 533–40.

Hamblin MR, Newman EL. On the mechanism of the tumour-localising effect in photodynamic therapy. *J Photochem Photobiol B: Biol* 1994a, **23**, 3–8.

Hamblin MR, Newman EL. Photosensitizer targeting in photodynamic therapy I. Conjugates of haematoporphyrin with albumin and transferrin. *J Photochem Photobiol B: Biol* 1994b, **26**, 45–55.

Hamblin MR, Newman EL. Photosensitizer targeting in photodynamic therapy II. Conjugates of haematoporphyrin with serum lipoproteins. *J Photochem Photobiol B: Biol* 1994c, **26**, 147–57.

Hannah S, Lynch VM, Gerasimchuk N, Magda D, Sessler JL. Synthesis of a metal-free texaphyrin. *Org Lett* 2001, **3**, 3911–14.

Hannah S, Lynch V, Guldi DM, Gerasimchuk N, MacDonald CLB, Magda D, Sessler JL. Late first-row transition-metal complexes of texaphyrin. *J Am Chem Soc* 2002, **124**, 8416–27.

Hargus JA, Fronczek FR, Vicente MGH, Smith KM. Mono-(L)-aspartylchlorin-e_6. *Photochem Photobiol* 2007, **83**, 1006–15.

Hendren SK, Hahn SM, Spitz FR, Bauer TW, Rubin SC, Zhu T, Glatstein E, Fraker DL. Phase II trial of debulking surgery and photodynamic therapy for disseminated intraperitoneal tumours. *Ann Surg Oncol* 2001, **8**, 65–71.

Hermanson GT. *Bioconjugate Techniques.* Academic Press, San Diego, 1996.

Hilmey DG, Abe M, Nelen MI, *et al.* Water-soluble, core-modified porphyrins as novel, longer-wavelength-absorbing sensitizers for photodynamic therapy. II. Effects of core heteroatoms and *meso*-substituents on biological activity. *J Med Chem* 2002, **45**, 449–61.

http://www.cancer.gov/search/ViewClinicalTrials.aspx?cdrid=410675&version=HealthProfessional& protocolsearchid =3854835

http://www.cancer.gov/search/ViewClinicalTrials.aspx?cdrid=68862&version=HealthProfessional&pro tocolsearchid=3854839

Hu M, Brasseur N, Yildiz SZ, van Lier JE, Leznoff CC. Hydroxyphthalocyanines as potential photodynamic agents for cancer therapy. *J Med Chem* 1998, **41**, 1789–802.

Huang Z, Chen Q, Luck D, *et al.* Studies of a vascular-acting photosensitizer, Pd-Bacteriopheophorbide (Tookad), in normal canine prostate and spontaneous canine prostate cancer. *Lasers Surg Med* 2005, **36**, 390–7.

Jacobi PA, Lanz S, Ghosh I, Leung SH, Löwer F, Pippin D. A new synthesis of chlorins. *Org Lett* 2001, **3**, 831–4.

Javaid B, Watt P, Krasner N. Photodynamic therapy (PDT) for oesophageal dysplasia and early carcinoma with mTHPC (m-tetrahydroxyphenyl chlorin): a preliminary study. *Lasers Med Sci* 2002, **17**, 51–6.

Jesionek A, von Tappenier H. Zur Behandlung der Hautcarcinomit mit fluorescierenden Stoffen. *Muench Med Wochneshschr* 1903, **47**, 2042.

Jin ML, Yang BQ, Zhang W, Ren P. Review of photodynamic therapy for gastrointestinal tumours in the past 6 years in China. *J Photochem Photobiol B: Biol* 1990, **7**, 87–92.

Jin ML, Yang BQ, Zhang W, Ren P. Combined treatment with photodynamic therapy and chemotherapy for advanced cardiac cancers. *J Photochem Photobiol B: Biol* 1992, **12**, 101–6.

Jurklies B, Anastassiou G, Ortmans S, Schüler A, Schilling H, Schmidt-Erfurth U, Bornfeld N. Photodynamic therapy using verteporfin in circumscribed choroidal haemangioma. *Br J Ophthalmol* 2003, **87**, 84–9.

Kasche A, Luderschmidt S, Ring J, Hein R. Photodynamic therapy induces less pain in patients treated with methyl aminolevulinate compared to aminolevulinic acid. *J Drugs Dermatol* 2006, **5**, 353–6.

Kato H, Furukawa K, Sato M, *et al.* Phase II clinical study of photodynamic therapy using mono-L-aspartyl chlorin e_6 and diode laser for early superficial squamous cell carcinoma of the lung. *Lung Cancer* 2003, **42**, 103–11.

Kaviani A, Ataie-Fashtami L, Fateh M, Sheikhbahaee N, Ghodsi M, Zand N, Djavid GE. Photodynamic therapy of head and neck basal cell carcinoma according to different clinicopathologic features. *Lasers Surg Med* 2005, **36**, 377–82.

Kelly JF, Snell ME. Hematoporphyrin derivative: a possible aid in the diagnosis and therapy of carcinoma of the bladder. *J Urol* 1976, **115**, 150–1.

Kennedy JC, Pottier RH, Pross DC. Photodynamic therapy with endogeneous protoporphyrin IX: basic principles and present clinical experience. *J Photochem Photobiol B: Biol* 1990, **6**, 143–8.

Kim H-J, Lindsey JS. De novo synthesis of stable tetrahydroporphyrinic macrocycles: bacteriochlorins and a tetradehydrocorrin. *J Org Chem* 2005, **70**, 5475–86.

Kim H-J, Dogutan DK, Ptaszek M, Lindsey JS. Synthesis of hydrodipyrrins tailored for reactivity at the 1- and 9-positions. *Tetrahedron* 2006, **63**, 37–55.

Kojima C, Toi Y, Harada A, Kono K. Preparation of Poly(ethylene glycol)-attached dendrimers encapsulating photosensitizers for application in photodynamic therapy. *Bioconjugate Chem* 2007, **18**, 663–70.

Kömerik N, Curnow A, MacRobert AJ, Hopper C, Speight PM, Wilson M. Fluorescence biodistribution and photosensitising activity of toluidine blue O on rat buccal mucosa. *Lasers Med Sci* 2002, **17**, 86–92.

Konan-Kouakou YN, Boch R, Gurny R, Allémann E. *In vitro* and *in vivo* activities of verteporfin-loaded nanoparticles. *J Controlled Rel* 2005, **103**, 83–91.

Kübler AC, Haase T, Staff C, Kahle B, Rheinwald M, Mühling J. Photodynamic therapy of primary non-melanomatous skin tumours of the head and neck. *Lasers Surg Med* 1999, **25**, 60–8.

Kusuzaki K, Murata H, Matsubara T, *et al.* Clinical trial of photodynamic therapy using acridine orange with/without low dose radiation as a new limb salvage modality in musculoskeletal sarcomas. *Anticancer Res* 2005, **25**, 1225–36.

Laha JK, Muthiah C, Taniguchi M, Lindsey JS. A new route for installing the isocyclic ring on chlorins yielding 13^1-oxophorbines. *J Org Chem* 2006, **71**, 7049–52.

Lamagna C, Aurrand-Lions M, Imhof BA. Dual role of macrophages in tumour growth and angiogenesis. *J Leukoc Biol* 2006, **80**, 705–13.

Lang K, Mosinger J, Wagnerová DM. Photophysical properties of porphyrinoid sensitizers noncovalently bound to host molecules; models for photodynamic therapy. *Coord Chem Rev* 2004, **248**, 321–50.

Lee PPS, Lo P-C, Chan EYM, Fong W-P, Ko W-H, Ng DKP. Synthesis and *in vitro* photodynamic activity of novel galactose-containing phthalocyanines. *Tetrahedron Lett* 2005, **46**, 1551–54.

Lo P-C, Zhao B, Duan W, Fong W-P, Ko W-H, Ng DKP. Synthesis and *in vitro* photodynamic activity of mono-substituted amphiphilic zinc(II) phthalocyanines. *Bioorg Med Chem Lett* 2007, **17**(4), 1073–77.

Lou P-J, Jäger HR, Jones L, Theodossy T, Bown SG, Hopper C. Interstitial photodynamic therapy as salvage treatment for recurrent head and neck cancer. *Br J Cancer* 2004, **91**, 441–6.

Lui H, Hobbs L, Tope WD, *et al.* Photodynamic therapy of multiple nonmelanoma skin cancers with verteporfin and red light-emitting diodes: two-years results evaluating tumor response and cosmetic outcomes. *Arch Dermatol* 2004, **140**, 26–32.

Luna MC, Gomer CJ. Isolation and initial characterization of mouse tumour cells resistant to porphyrin-mediated photodynamic therapy. *Cancer Res* 1991, **51**, 4243–9.

Ma LW, Bjørklund E, Moan J. Photochemotherapy of tumours with mesotetrahydroxyphenyl chlorin is pH dependent. *Cancer Lett* 1999, **138**, 197–201.

Maeda H. The enhanced permeability and retention (EPR) effect in tumour vasculature: the key role of tumour-selective macromolecular drug targeting. *Advan Enzyme Regul* 2001, **41**, 189–207.

Maeda H, Sawa T, Konno T. Mechanism of tumour-targeted delivery of macromolecular drugs, including the EPR effect in solid tumour and clinical overview of the prototype polymeric drug SMANCS *J Controlled Rel* 2001, **74**, 47–61.

Maeda H, Greish K, Fang J. The EPR effect and polymeric drugs: a paradigm shift for cancer chemotherapy in the 21st century. *Adv Polym Sci* 2006, **193**, 103–21.

Malatesti N, Smith K, Savoie H, Greenman J, Boyle RW. Synthesis and *in vitro* investigation of cationic 5,15-diphenyl porphyrin-monoclonal antibody conjugates as targeted photodynamic therapy sensitizers. *Int J Oncol* 2006, **28**, 1561–9.

Marks PV, Belchetz PE, Saxena A, *et al.* Effect of photodynamic therapy on recurrent pituitary adenomas: clinical phase I/II trial – and early report. *Br J Neurosurg* 2000, **14**, 317–25.

Mayhew S, Vernon DI, Schofield J, Griffiths J, Brown SB. Investigation of cross-resistance to a range of photosensitizers, hyperthermia and UV-light in two radiation-induced fibrosarcoma cell strains resistant to photodynamic therapy *in vitro*. *Photochem Photobiol* 2001, **73**, 39–46.

Mazor O, Brandis A, Plaks V, Neumark E, Rosenbach-Belkin V, Salomon Y, Scherz A. WST11, a novel water-soluble bacteriochlorophyll derivative; Cellular uptake, pharmacokinetics, biodistribution and vascular-targeted photodynamic activity using melanoma tumours as a model. *Photochem Photobiol* 2005, **81**, 342–51.

McDonnell SO, Hall MJ, Allen LT, Byrne A, Gallagher WM, O'Shea DF. Supramolecular photonic therapeutic agents. *J Am Chem Soc* 2005, **127**, 16360–61.

McKeown NB. The synthesis of symmetrical phthalocyanines. In: *The Porphyrin Handbook*, edited by KM Kadish, KM Smith and R Guilard. Academic Press, San Diego, 2003, Vol. 15, Chapter 98.

McKeown NB, Chambrier I, Cook MJ. Synthesis and characterization of some 1,4,8,11,15,18,22,25-octa-alkyl- and 1,4,8,11,15,18-hexa-alkyl-22,25-bis(carboxypropyl)phthalocyanines. *J Chem Soc Perkin Trans* 1990, **1**, 1169–77.

Menon C, Kutney SN, Lehr SC, Hendren SK, Busch TM, Hahn SM, Fraker DL. Vascularity and uptake of photosensitizer in small human tumour nodules: implications for intraperitoneal photodynamic therapy. *Clin Cancer Res* 2001, **7**, 3904–11.

Mimura S, Ito Y, Nagayo T, *et al.* Cooperative clinical trial of photodynamic therapy with Photofrin II and excimer dye laser for early gastric cancer. *Lasers Surg Med* 1996, **19**, 168–72.

Moghissi K, Dixon K, Stringer M, Freeman T, Thorpe A, Brown S. The place of bronchoscopic photodynamic therapy in advanced unresectable lung cancer: experience of 100 cases. *Eur J Cardiothorac Surg* 1999, **15**, 1–6.

Moore CM, Nathan TR, Lees WR, Mosse CA, Freeman A, Emberton M, Bown SG. Photodynamic therapy using meso tetra hydroxyl phenyl chlorin (mTHPC) in early prostate cancer. *Lasers Surg Med* 2006, **38**, 356–63.

Morgan AR, Rampersaud A, Garbo GM, Keck RW, Selman SH. New sensitizers for photodynamic therapy: controlled synthesis of purpurins and their effect on normal tissue. *J Med Chem* 1989, **32**, 904–8.

Morgan AR, Garbo GM, Keck RW, Miller RA, Selman, SH, Skalkos D. Diels-Alder adducts of vinyl porphyrins: synthesis and *in vivo* photodynamic effect against a rat bladder tumour. *J Med Chem* 1990, **33**, 1258–62.

Morgan J, Potter WR, Oseroff AR. Comparison of photodynamic targets in a carcinoma cell line and its mitochondrial DNA-deficient derivative. *Photochem Photobiol* 2000, **71**, 747–57.

Morton C, Horn M, Leman J, Tack B, Bedane C, Tjioe M, Ibbotson S, Khemis A. Comparison of topical methyl aminolevulinate photodynamic therapy with cryotherapy or fluorouracil for treatment of squamous cell carcinoma *in situ*. *Arch Dermatol* 2006, **142**, 729–35.

Mroz P, Tegos GP, Gali H, Wharton T, Sarna T, Hamblin MR. Photodynamic therapy with fullerenes. *Photochem Photobiol Sci* 2007, **6**, 1139–49.

Muthiah C, Bhaumik J, Lindsey JS. Rational routes to formyl-substituted chlorins. *J Org Chem* 2007, **72**, 5839–42.

Nishiyama N, Stapert HR, Zhang G-D, Takasu D, Jiang D-L, Nagano T, Aida T, Kataoka K. Light-harvesting ionic dendrimer porphyrins as new photosensitizers for photodynamic therapy. *Bioconjugate Chem* 2003, **14**, 58–66.

Nishiyama N, Jang W-D, Kataoka K. Supramolecular nanocarriers integrated with dendrimers encapsulating photosensitizers for effective photodynamic therapy and photochemical gene delivery. *New J Chem* 2007, **31**, 1074–82.

Nyman ES, Hynninen PH. Research advances in the use of tetrapyrrolic photosensitizers for photodynamic therapy. *J Photochem Photobiol B: Biol* 2004, **73**, 1–28.

Oba T, Shishikura M, Ugajin A, Ito S, Hiratani K. Development of novel peptide-photosensitizer conjugates for photodynamic therapy. *Peptide Sci* 2006, **43**, 303.

Ogura S-I, Tabata K, Fukushima K, Kamachi T, Okura I. Development of phthalocyanines for photodynamic therapy. *J Porphyrins Phthalocyanines* 2006, **10**, 1116–24.

Okunaka T, Kato H, Conaka C, Yamamoto H, Bonamino A, Eckhauser ML. Photodynamic therapy of esophageal carcinoma. *Surg Endosc* 1990, **4**, 150–3.

Oleinick NL, Antunez AR, Clay ME, Rihter BD, Kenney ME. New phthalocyanine photosensitizers for photodynamic therapy. *Photochem Photobiol* 1993, **57**, 242–7.

O'Neal WG, Roberts WP, Ghosh I, Jacobi PA. Studies in chlorin chemistry. II. A versatile synthesis of dihydrodipyrrins. *J Org Chem* 2005, **70**, 7243–51.

O'Neal WG, Roberts WP, Ghosh I, Wang H, Jacobi PA. Studies in chlorin chemistry. 3. A practical synthesis of C,D-ring symmetric chlorins of potential utility in photodynamic therapy. *J Org Chem* 2006, **71**, 3472–80.

Ono R, Ikeda S, Suemasu K. Hematoporphyrin derivative photodynamic therapy in roentgenographically occult carcinoma of the tracheobronchial tree. *Cancer* 1992, **69**, 1696–701.

Oseroff AR, Blumenson LR, Wilson BD, *et al.* A dose ranging study of photodynamic therapy with porfimer sodium (Photofrin®) for treatment of basal cell carcinoma. *Lasers Surg Med* 2006, **38**, 417–26.

Osterloh J, Vicente MGH. Mechanism of porphyrinoid localization in tumours. *J Porphyrins Phthalocyanines* 2002, **6**, 305–24.

Oulmi D, Maillard P, Guerquin-Kern J-L, Huel C, Momenteau D. Glycoconjugated porphyrins. 3. Synthesis of flat amphiphilic mixed meso-(glycosylated aryl)arylporphyrins and mixed meso-(glycosylated aryl)alkylporphyrins bearing some mono- and disaccharide groups. *J Org Chem* 1995, **60**, 1554–64.

Overholt BF, Lightdale CJ, Wang KK, *et al.* Photodynamic therapy with porfimer sodium for ablation of high-grade dysplasia in Barrett's esophagus: international, partially blinded, randomized phase III trial. *Gastrointest Endosc* 2005, **63**, 488–98.

Pandey RK, Shiau F-Y, Sumlin AB, Dougherty TJ, Smith KM. Syntheses of new bacteriochlorins and their antitumour activity. *Bioorg Med Chem Lett* 1994, **4**, 1263–1267.

Pandey RK, Zheng G. Porphyrins as photosensitizers in photodynamic therapy. In: *The Porphyrin Handbook*, edited by KM Kadish, KM Smith and R Guillard. Academic Press, New York, 2000, Vol. 6, pp 157–230.

Pandey SK, Zheng X, Morgan J, *et al.* Purpurinimide carbohydrate conjugates: effect of the position of the carbohydrate moiety in photosensitizing efficacy. *Mol Pharmaceutics* 2007, **4**, 448–464.

Panjehpour M, Bergein F, Overholt F, Haydek JM, Lee SG. Results of photodynamic therapy for ablation of dysplasia and early cancer in Barrett's esophagus and effect of oral steroids on stricture formation. *Am J Gastroenterol* 2000, **95**, 2177–84.

Petit L, Quartarolo A, Adamo C, Russo N. Spectroscopic properties of porphyrin-like sensitizers: insights from theory. *J Phys Chem B* 2006, **110**, 2398–404.

Polo L, Segalla A, Bertoloni G, Jori G, Schaffner K, Reddi E. Polylysine-porphycene conjugates as efficient photosensitizers for the inactivation of microbial pathogens. *J Photochem Photobiol B: Biol* 2000, **59**, 152–8.

Pushpan SK, Venkatraman S, Anand VG, Sankar J, Parmeswaran D, Ganesan S, Chandrashekar TK. Porphyrins in photodynamic therapy – a search for ideal photosensitizers. *Curr Med Chem – Anti-Cancer Agents* 2002, **2**, 187–207.

Quartarolo AD, Russo N, Sicilia E, Lelj F. Absorption spectra of the potential photodynamic therapy photosensitizers texaphyrins complexes: a theoretical analysis. *J Chem Theory Comput* 2007, **3**, 860–9.

Ragunath K, Krasner N, Raman VS, Haqqani MT, Phillips CJ, Cheung I. Endoscopic ablation of dysplastic Barrett's oesophagus comparing argon plasma coagulation and photodynamic therapy: a randomized prospective trial assessing efficacy and cost-effectiveness. *Scand J Gastroenterol* 2005, **40**, 750–8.

Rahimipour S, Ben-Aroya N, Ziv K, Chen A, Fridkin M, Koch Y. Receptor-mediated targeting of a photosensitizer by its conjugation to gonadotropin-releasing hormone analogues. *J Med Chem* 2003, **46**, 3965–74.

Rancan F, Helmreich M, Mölich A, Ermilov EA, Jux N, Röder B, Hirsch A, Böhm F. Synthesis and *in vivo* testing of a pyropheophorbide-a-fullerene hexakis adduct immunoconjugate for photodynamic therapy. *Bioconjugate Chem* 2007, **18**, 1078–86.

Ris H-B, Altermatt HJ, Nachbur B, Stewart CM, Wang Q, Lim CK, Bonnett R, Althaus U. Intraoperative photodynamic therapy with m-tetrahydroxyphenylchlorin for chest malignancies *Lasers Surg Med* 1996, **18**, 39–45.

Robinson BC. Bacteriopurpurins: synthesis from meso-diacrylate substituted porphyrins. *Tetrahedron* 2000, **56**, 6005–14.

Rodríguez-Morgade MS, de la Torre G, Torres T. Design and synthesis of low-symmetry phthalocyanines and related systems. In: *The Porphyrin Handbook*, edited by KM Kadish, KM Smith and R Guilard. Academic Press, San Diego, 2003, Vol. 15, Chapter 99.

Samia ACS, Dayal S, Burda C. Quantum dot-based energy transfer: perspectives and potential for applications in photodynamic therapy. *Photochem Photobiol* 2006, **82**, 617–25.

Scherer K, Abels C, Bäumler W, Ackermann G, Szeimies R-M. Structure-activity relationships of three differently substituted 2,7,12,17-tetrakis-(β-methoxyethyl) porphycene derivatives *in vitro*. *Arch Dermatol Res* 2004, **295**, 535–41.

Schleyer V, Szeimies R-M. ALA/MAL-PDT in dermatology. In: *Photodynamic Therapy with ALA: A Clinical Handbook*, edited by R Baumgartner, B Krammer, R Pottier and H Stepp. Royal Society of Chemistry, London, 2006, Vol. 7, Chapter 3.

Schneider R, Tirand L, Frochot C, Vanderesse R, Thomas N, Gravier J, Guillemin F, Barberi-Heyob M. Recent improvements in the use of synthetic peptides for a selective photodynamic therapy. *Anticancer Agents Med Chem* 2006, **6**, 469–88.

Schreiber S, Gross S, Brandis A, Harmelin A, Rosenbach-Belkin V, Scherz A, Salomon Y. Local photodynamic therapy (PDT) of rat C6 glioma xenografts with Pd-Bacteriopheophorbide leads to decreased metastases and increase of animal cure compared with surgery. *Int J Cancer* 2002, **99**, 279–85.

Sessler JL, Miller RA. Texaphyrins. New Drugs with diverse clinical applications in radiation and photodynamic therapy. *Biochem Pharmacol* 2000, **59**, 733–9.

Sessler JL, Hemmi G, Mody TD, Murai T, Burrell A, Young SW. Texaphyrins: synthesis and applications. *Acc Chem Res* 1994, **27**, 43–50.

Sessler JL, Dow WC, O'Connor D, *et al*. Biomedical applications of lanthanide(III) texaphyrins. Lutetium(III) texaphyrins as potential photodynamic therapy photosensitizers. *J Alloy Cmpd* 1997, **249**, 146–52.

Sessler JL, Mody TD, Hemmi GW, Lynch V. Synthesis and structural characterization of lanthanide(III) texaphyrins. *Inorg Chem* 2003, **32**, 3175–87.

Sharman WM, van Lier JE. Synthesis of phthalocyanine precursors. In: *The Porphyrin Handbook*, edited by KM Kadish, KM Smith and R Guilard. Academic Press, San Diego, 2003, Vol. 15, Chapter 97.

Sharman WM, van Lier JE, Allen CM. Targeted photodynamic therapy via receptor mediated delivery systems *Adv Drug Deliv* 2004, **56**, 53–76.

Sheleg SV, Zhavrid EA, Khodina TV, Kochubeev GA, Istomin YP, Chalov VN, Zhuravkin IN. Photo-dynamic therapy with chlorin e_6 for skin metastases of melanoma. *Photodermatol Photoimmunol Photomed* 2004, **20**, 21–6.

Singh G, Espiritu M, Shen XY, Hanlon JG, Rainbow AJ. *In vitro* induction of PDT resistance in HT29, HT1367 and SK-N-MC cells by various photosensitizers. *Photochem Photobiol* 2001, **73**, 651–6.

Skyrme RJ, French AJ, Datta SN, Allman R, Mason MD, Matthews PN. A phase-1 study of sequen-tial mitomycin C and 5-aminolevulinic acid-mediated photodynamic therapy in recurrent superficial bladder carcinoma. *BJU Int* 2005, **95**, 1206–10.

Smith KM. Syntheses and chemistry of porphyrins. *J Porphyrins Phthalocyanines* 2000, **4**, 319–24.

Sol V, Blais JC, Carré V, Granet R, Guilloton M, Spiro M, Krausz P. Synthesis, spectroscopy, and pho-totoxicity of glycosylated amino acid porphyrin derivatives as promising molecules for cancer pho-totherapy. *J Org Chem* 1999, **64**, 4431–44.

Staneloudi C, Smith KA, Hudson R, Malatesi N, Savoie H, Boyle RW, Greenman J. Development and characterization of novel photosensitizer: ScFv conjugates for use in photodynamic therapy of cancer. *Immunology* 2007, **120**, 512–17.

Sternberg ED, Dolphin D, Büchner C. Porphyrin-based photosensitizers for use in photodynamic ther-apy. *Tetrahedron* 1998, **54**, 4151–202.

Stilts CE, Nelen MI, Hilmey DG, *et al*. Water-soluble, core-modified porphyrins as novel, longer-wavelength-absorbing sensitizers for photodynamic therapy *J Med Chem* 2000, **43**, 2403–10.

Stylli SS, Kaye AH, MacGregor L, Hoews M, Rajendra P. Photodynamic therapy of high grade glioma – long term survival. *J Clin Neurosci* 2005, **12**, 389–98.

Szacilowski K, Macyk W, Drzewiecka-Matuszek A, Brindell M, Stochel G. Bioinorganic photochem-istry: frontiers and mechanisms. *Chem Rev* 2005, **105**, 2647–94.

Taber SW, Fingar VH, Coots CT, Wieman TJ. Photodynamic therapy using mono-L-aspartyl chlorin e_6 (NPe6) for the treatment of cutaneous disease: a phase I clinical study. *Clin Cancer Res* 1998, **4**, 2741–6.

Taniguchi M, Balakumar A, Fan D, McDowell BE, Lindsey JS. Imine-substituted dipyrromethanes in the synthesis of porphyrins bearing one or two meso substituents. *J Porphyrins Phthalocyanines* 2005, **9**, 554–74.

Taniguchi M, Ptaszek M, McDowell BE, Boyle PD, Lindsey JS. Sparsely substituted chlorins as core contructs in chlorophyll analogue chemistry. Part 3: spectral and structural properties. *Tetrahedron* 2007, **63**, 3850–63.

Thompson JA, Murata K, Miller DC, Stanton JL, Broderick WE, Hoffmann BM, Ibers JA. Synthesis of high-purity phthalocyanines (pc): High intrinsic conductivities in the molecular conductors $H_2(pc)I$ and $Ni(pc)I$. *Inorg Chem* 1993, **32**, 3546–53.

Tomoda H, Saito S, Ogawa S, Shiraishi S. Synthesis of phthalocyanines from phthalonitrile with organic strong bases. *Chem Lett* 1980, **10**, 1277–1280.

Triesscheijn M, Ruevekamp M, Antonini N, Neering H, Stewart FA, Baas P. Optimizing meso-tetra-hydroxyphenyl-chlorin-mediated photodynamic therapy for basal cell carcinoma. *Photochem Photo-biol* 2006, **82**, 1686–90.

Tschen EH, Wong DS, Pariser DM, Dunlap FE, Houlihan A, Ferdon MB. Photodynamic therapy using aminolevulinic acid for patients with nonhyperkeratotic actinic keratosis of the face and scalp: phase IV multicentre trial with 12-month follow-up. *Br J Dermatol* 2006, **155**, 1262–9.

van Dongen GAMS, Visser GWM, Vrouenraets MB. Photosensitizer-antibody conjugates for detection and therapy of cancer. *Adv Drug Deliv* 2004, **56**, 31–52.

van Duijnhoven FH, Rovers JP, Engelmann K, Krajina Z, Purkiss SF, Zoetmulder FAN, Vogl TJ, Terp-stra OT. Photodynamic therapy with 5,10,15,20-tetrakis(m-hydroxyphenyl) bacteriochlorin for col-orectal liver metastases is safe and feasible: results from a phase I study. *Ann Surg Oncol* 2005, **12**, 808–16.

Varamo M, Loock B, Maillard P, Grierson DS. Development of strategies for the regiocontrolled synthesis of meso-5,10,20-triaryl-2,3-chlorins. *Org Lett* 2007, **9**, 4689–92.

Verigos K, Hsiung Stripp DC, Mick R, *et al*. Updated results of a phase I trial of Motexafin lutetium-mediated interstitial photodynamic therapy in patients with locally recurrent prostate cancer. *J Environ Pathol* 2006, **25**, 373–8.

Verma S, Watt GM, Mai Z, Hasan T. Strategies for enhanced photodynamic therapy effects. *Photochem Photobiol* 2007, **83**, 996–1005.

Vinciullo C, Elliott T, Francis D, *et al*. Photodynamic therapy with topical methyl aminolevulinate for 'difficult-to-treat' basal cell carcinoma. *Br J Dermatol* 2005, **152**, 765–72.

Vogel E, Köcher M, Schmickler H, Lex J. Porphycen – Ein neuartiges Porphin-Isomer. *Angew Chem* 1986, **98**, 262–3.

Vrouenraets MB, Visser GWM, Loup C, *et al*. Targeting of a hydrophilic photosensitizer by use of internalizing monoclonal antibodies: A new possibility for use in photodynamic therapy. *Int J Cancer* 2000, **88**, 108–14.

Vrouenraets MB, Visser GWM, Stigter M, Oppelaar H, Snow GB, van Dongen GAMS. Targeting of aluminium (III) phthalocyanine tetrasulfonate by use of internalizing monoclonal antibodies: Improved efficacy in photodynamic therapy. *Cancer Res* 2001, **61**, 1970–5.

Vrouenraets MB, Visser GWM, Stigter M, Oppelaar H, Snow GB, van Dongen GAMS. Comparison of aluminium (III) phthalocyanine tetrasulfonate- and meta-tetrahydroxyphenylchlorin-monoclonal antibody conjugates for their efficacy in photodynamic therapy *in vitro*. *Int J Cancer* 2002, **98**, 793–8.

Walther MM, Delaney TF, Smith PD, *et al*. Phase I trial of photodynamic therapy in the treatment of recurrent superficial transitional cell carcinoma of the bladder. *Urology* 1997, **50**, 199–206.

Wong T-W, Aizawa K, Sheyedin I, Wushur C, Kato H. Pilot study of topical delivery of mono-L-aspartyl chlorine e6 (NPe6): implication of topical NPe6-photodynamic therapy. *J Pharmacol Sci* 2003, **93**, 136–42.

Wulf HC, Pavel S, Stender I, Ahb Bakker-Wensveen C. Topical photodynamic therapy for prevention of new skin lesions in renal transplant recipients. *Acta Derm Venereol* 2006, **86**, 25–8.

Wyss P, Schwarz V, Dobler-Girdziunaite D, Hornung R, Walt H, Degen A, Fehr MK. Photodynamic therapy of locoregional breast cancer recurrences using a chlorin-type photosensitizer. *Int J Cancer* 2001, **93**, 720–4.

Xu B, Liang G, Wang L, Yang Z, Chan K, Chang CK. Enhancing PDT drug delivery by enzymatic cleavage of porphyrin phosphates. *Proc of SPIE* 2007, **6427**, 64271A.

Yamaguchi S, Tsuda H, Takemori M, *et al*. Photodynamic therapy for cervical intraepithelial neoplasia. *Oncology* 2004, **69**, 110–16.

Yamayoshi A, Kato K, Suga S, *et al*. Specific apoptosis induction in human papillomavirus-positive cervical carcinoma cells by photodynamic antisense regulation. *Oligonucleotides* 2007, **17**, 66–79.

Young SW, Woodburn KW, Wright M, Mody TD, Fan Q, Sessler JL, Dow WC, Miller RA. Lutetium texaphyrin (PCI-0123): Near-infrared, water-soluble photosensitizer. *Photochem Photobiol* 1996, **63**, 892–7.

Zaidi SHH, Loewe RS, Clark BA, Jacob MJ, Lindsey JS. Nearly chromatography-free synthesis of the A₃B-porphyrin 5-(4-hydroxymethylphenyl)-10,15,20-tri-*p*-tolylporphyrinatozinc(II). *Org Proc Res Dev* 2006, **10**, 304–14.

Zane C, Venturini M, Sala R, Calzavara-Pinton P. Photodynamic therapy with methylaminolevulinate as a valuable treatment option for unilesional cutaneous T-cell lymphoma. *Photodermatol Photoimmunol Photomed* 2006, **22**, 254–8.

Zheng G, Chen K, Stefflova K, Jarvi M, Li H, Wilson BC. Photodynamic molecular beacon as an activable photosensitizer based on protease-controlled singlet oxygen quenching and activation. *Proc Natl Acad Sci USA* 2007, **104**, 8989–94.

Zheng G, Graham A, Shibata M, Missert JR, Oseroff AR, Dougherty TJ, Pandey RK. Synthesis of b-galactose-conjugated chlorins derived by enyne metathesis as galectin-specific photosensitizers for photodynamic therapy. *J Org Chem* 2001a, **66**, 8709–16.

Zheng G, Potter WR, Camacho SH, *et al.* Synthesis, photophysical properties, tumour uptake, and preliminary *in vivo* photosensitizing efficacy of a homologous series of 3-(1'-alkyloxy)ethyl-3-devinylpurpurin-18-*N*-alkylimides with variable lipophilicity. *J Med Chem* 2001b, **44**, 1540–59.

Zimmermann A, Ritsch-Marte M, Kostron H. mTHPC-mediated photodynamic diagnosis of malignant brain tumours. *Photochem Photobiol* 2001, **74**, 611–16.

Zoepf T, Jakobs R, Arnold JC, Apel D, Riemann JF. Palliation of nonresectable bile duct cancer: improved survival after photodynamic therapy. *Am J Gastroenterol* 2005, **31**, 291–8.

12
Target-directed Drug Discovery

Tracey D. Bradshaw

12.1 Introduction

Cancer will affect one in three of the UK population, and be responsible for one in four deaths; 284K new cases were diagnosed in the UK in 2004 (www.cancerresearchuk.org). There are more than 200 different types of cancer, but four of them – lung, breast, colorectal and prostate – account for over half of new cases. Although cancer has been termed a disease of aging, 65 % of cases occur in those over 65; this is not exclusively so; in children, leukaemia is the most common cancer; in young men (20–39 years), testicular cancer occurs most frequently.

Tumourigenesis was first identified as a multistage process in 1938 (Kidd and Rous, 1938) comprising initiation with a suboptimal dose of carcinogen, promotion, progression and metastasis. A single genetic mutation is thought to be sufficient to render the damaged cell in an initiated state. A controversial discussion point is whether this event occurs in a stem, progenitor or somatic cell. Each day our DNA is bombarded by thousands of DNA-damaging events including UV radiation, exposure to or ingestion of carcinogens, cellular metabolism and replication errors during mitoses (Figure 12.1). To counter such damage, complex surveillance and repair mechanisms have evolved to maintain DNA integrity.

Initiation confers upon the cell a survival/proliferative advantage. It is able to self renew and clonal expansion of the mutated, initiated cell follows. Tumour promotion and progression may take decades during which time tumour cells acquire increased genetic (and epigenetic) instability. The age-related incidence of prostate cancer exemplifies this point. In addition, approximately two decades separate the two cumulative graphs comparing smoking and lung cancer incidence (Figure 12.2) (www.nih.org). The final stage allows metastatic spread of tumour cells *via* circulatory and lymphatic systems to distant organ sites and is responsible for 95 % of all cancer deaths.

It is now established beyond doubt that smoking is a (if not the) major cause of many cancers. As early as 1950, epidemiological work demonstrated a clear relationship between

Anticancer Therapeutics   Edited by Sotiris Missailidis
© 2008 John Wiley & Sons, Ltd

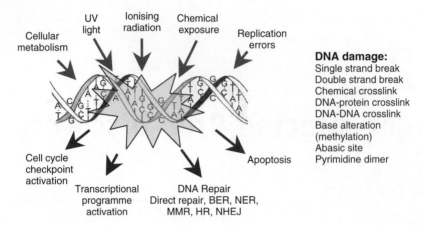

DNA damage:
Single strand break
Double strand break
Chemical crosslink
DNA-protein crosslink
DNA-DNA crosslink
Base alteration
(methylation)
Abasic site
Pyrimidine dimer

Figure 12.1   DNA faces many events which cause deleterious and carcinogenic mutations and complex mechanisms have evolved to detect and repair DNA damage

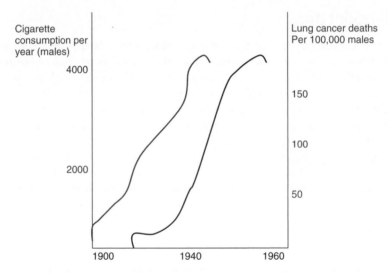

Figure 12.2   Cumulative graphs illustrating approximately the number of cigarettes smoked per person per year during the 20th century in the US and ensuing lung cancer incidence (www.nih.org)

smoking and lung cancer deaths (Doll and Hill, 1950). Inhaled cigarette smoke contains a toxic cocktail of >4000 chemicals harbouring 43 known carcinogens. Indeed, tobacco exposure is now a known risk factor for many cancer types, accounting for an estimated 30 %of all cancer mortality.

Cancer cells acquire defects in genes which encode proteins whose roles in regulatory circuits govern normal cell proliferation and homeostasis. Such genes include proto-oncogenes whose mutant oncogenes drive proliferation of genetically aberrant cells, and mutant/inactive tumour suppressor genes which allow unrepaired genetically aberrant cells to proliferate and evade apoptosis. Mutations in 'caretaker' genes encoding DNA damage recognition and repair proteins allow accumulation of genetic instability, and predispose individuals with inherited mutations within these critical repair pathways to cancer.

In a seminal paper, six principles of tumourigenesis were described by Hanahan and Weinberg (2000).

1. Self-sufficiency in growth signals

2. Insensitivity to growth-inhibitory signals

3. Evasion of programmed cell death

4. Limitless replicative potential

5. Sustained angiogenesis

6. Tissue invasion and metastasis.

An additional important property of tumour cells is their ability to respire anaerobically. Cancer, therefore, comprises many different and vastly complex diseases. During the past decade however, tremendous advances in molecular biology have led to enhanced understanding of cancer cell biology; aberrant protein expression and signal transduction networks are being elucidated. This in turn enables design and development of specific agents which target these proteins and pathways perturbed in, and important for cancer cell survival.

The rationale of target-directed drug discovery is the development of agents that:

• possess selective activity against a validated target critical for tumour cell proliferation and survival

• allow patient selection

• evoke limited, tolerated side effects and therefore possess better toxicity profiles than chemotherapeutic agents, resulting in improved quality of life

• improve long term prognoses for cancer patients

• augment the armamentarium against cancer.

Procedures adopted during the discovery programme typically may include:

• identification of a validated biological target (normally a protein)

• a 'molecule hunt', typically involving a high-throughput screening

• hit-to-lead development to identify (a) lead compound(s)

• lead development (involving *in vivo* testing and of absorption, distribution, metabolism, excretion and toxicity (ADMET) properties)

• clinical trials with appropriate biomarker or pharmacodynamic (PD) endpoint assays to monitor the efficacy of the therapy under evaluation (Workman and Kaye, 2002; Workman, 2005).

It is recoganised that many anomalous proteins and signalling cascades are implicated in the aetiology and pathophysiology of malignancy. Aberrant activation of the phosphatidyl inositol 3-kinase-like serine/threonine protein kinase (PIKK) family offer novel targets for intervention (Abraham, 2004; Liang *et al.*, 2003); for example enhanced activation of AKT has been detected during progression of Barrett's oesophagus to oesophageal adenocarcinoma (Sagatys *et al.*, 2007). Certain serine/threonine protein kinase inhibitors that target such pathways, impacting cell cycle progression and apoptotic processes are considered. However, initially, within this chapter, it is oncogenic mutations in specific tyrosine kinase proteins conferring constitutive signal transduction pathway activation which will be discussed.

12.2 Tyrosine Kinases – Role and Significance in Cancer

Tyrosine kinases, enzymes that catalyse the transfer of phosphate groups onto tyrosine residues of substrate proteins, are important mediators of many signalling cascades. They determine key roles in diverse biological processes including growth, differentiation, metabolism and apoptosis in response to external and internal stimuli, and have been implicated in the pathophysiology of many cancers (Paul and Mukhopadhyay, 2004). In normal cells, tyrosine kinase activity is tightly regulated; however, transforming functions may result from mutations, overexpression or autocrine paracrine stimulation driving unchecked proliferation and malignancy. It has been shown that constitutive oncogenic activation in cancer cells can be blocked by selective tyrosine kinase inhibitors, thus offering a promising approach for intervention and therapy. Mechanisms of oncogenic activation and examples of two different approaches for tyrosine kinase inhibition, small molecule inhibitors and monoclonal antibodies, will be considered.

As angiogenesis is critical for cancer growth and metastasis (Hanahan and Weinberg, 2000), the development of tyrosine kinase inhibitors that target the angiogenic process as a new mode of cancer therapy will be discussed.

12.3 Targeted Therapy for the Treatment of Non-small Cell Lung Cancer (NSCLC)

NSCLC is the most common form of lung cancer and the leading cause of cancer mortality. In approximately 50 % of patients, NSCLC is too advanced to be operable, and first-line treatment with platinum agents give only a modest increase in survival. Thus, the limited efficacy and intolerable side effects of cytotoxic chemotherapy for NSCLC expose the urgency for new therapies for this disease. The desire for novel therapies is to achieve cancer-specific effects which minimise adverse effects by targeting aberrant molecular pathways underlying NSCLC growth to improve treatment efficacy. Significant in the development and progression of NSCLC is the aberrant epidermal growth factor receptor (EGFR) signalling. Transduction through EGFR cascades (Figure 12.3) leads to inhibition of apoptosis, and promotion of tumour cell proliferation, angiogenesis and metastasis. EGFR is expressed in up to 93 % and overexpressed in approximately 45 % of NSCLC cases. Moreover, the level of EGFR expression correlates with poor prognosis and reduced survival. It is convincingly argued therefore that EGFR represents a good target (Hirsch *et al.*, 2003; Scaglioti *et al.*, 2004).

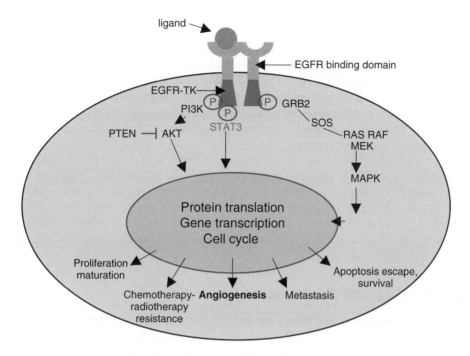

Figure 12.3   Epidermal growth factor signal transduction cascades

Two treatment strategies have been developed to thwart activation of EGFR:

- Monoclonal antibodies directed against the extracellular ligand binding domain prevent ligand binding. An example of such an antibody licensed for therapy is cetuximab (Erbitux a registered trademark of ImClone Systems Incorporated).

- Small molecule selective inhibitors of the intracellular tyrosine kinase domain of EGFR. Licensed examples of such molecules designed to inhibit signalling, following inhibition of receptor autophosphorylation include gefitinib (Iressa) and erlotinib (Tarceva; Figure 12.4) (Herbst *et al*., 2004; Ng *et al*., 2002).

We shall consider each small molecule EGFR inhibitor in turn.

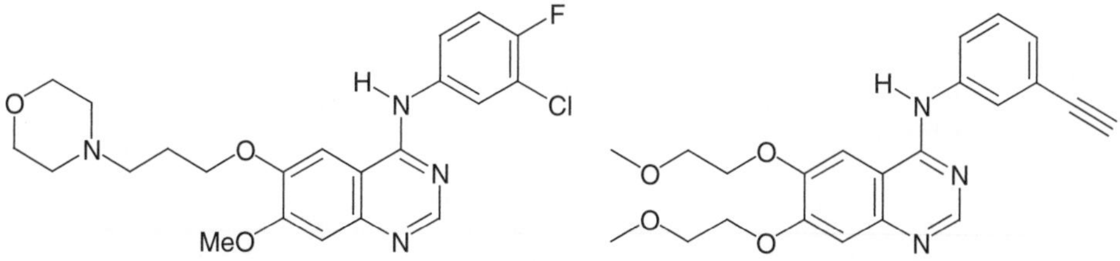

ZD1839 (Iressa)   **OSI-774 (Tarceva)**

Figure 12.4   EGFR inhibitors: Iressa and Tarceva

Gefitinib (Iressa; Astra Zeneca)

Gefinitib is a small molecule inhibitor of epidermal growth factor receptor (EGFR) tyrosine kinase which targets the ATP binding pocket (Hirsch *et al.*, 2003). Oral bioavailability is good and preclinically, the growth of a broad range of human solid tumour xenografts was inhibited dose dependently, with marked regressions observed in some tumours. The observed long half-life in humans is compatible with once-daily oral dosing and single agent activity was reported in a phase II clinical trial. However, a large phase III study in combination with cytotoxic agents was disappointing.

On 17 December 2004, the 'Iressa survival evaluation in lung cancer (ISEL)' trial results were reported. No overall survival advantage was revealed in a highly refractory cancer patient population. A statistically significant improvement in tumour shrinkage failed to translate into statistically significant survival benefit. Prospective subgroup analyses suggested survival benefits in patients of oriental origin and in patients who never smoked (www.astrazeneca.com). Tragically, in Japan, Iressa has been linked to 588 deaths (www.timesonline.co.uk).

Erlotinib (Tarceva; OSI Pharmaceuticals)

A synthetic anilinoquinazoline, this orally available small molecule receptor tyrosine kinase inhibitor is structurally and mechanistically very similar to Iressa, also competing with ATP, and binding to the intracellular TK domain of EGFR (Scagliotti *et al.*, 2004). Receptor autophosphorylation is inhibited and downstream signal transduction blocked.

In a single-agent NSCLC phase III trial Tarceva significantly improved overall survival, the primary endpoint, by 37 % in patients with relapsed stage IIIB/IV NSCLC ($P < 0.001$; Figure 12.5); (OSI Pharmaceuticals Inc, 2004); www.tarceva/professional/survival/results.jsp).

In November 2004 approval was granted for second line therapy in the treatment of NSCLC.

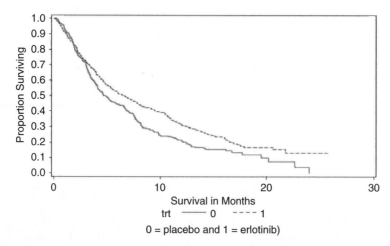

Figure 12.5 Tarceva significantly prolongs progression-free survival (www.tarceva.com/professional/survival/results.jsp)

Tarceva has shown antitumour activity and given symptom relief (improved quality of life) in patients where standard chemotherapy has failed; only mild adverse reactions (skin rash, diarrhoea) were experienced; no benefit was derived when Tarceva was administered in combination with platinum-based agents. It was concluded that there is potential therapeutic use for Tarceva in patients who have previously experienced severe co-morbidities and who are intolerant of cytotoxic chemotherapy.

Somatic mutations in the tyrosine kinase domain of EGFR correlate with a subset of patients whose tumours are exquisitely sensitive to Tarceva (and Iressa) (Krejsa *et al.*, 2006). Such patients are typically female, have never smoked, are of East Asian descent and have bronchoalveolar adenocarcinoma. Trials indicate that certain patients without EGFR mutations may still derive treatment benefit. The underlying molecular reason for this sensitivity is thought to be EGFR gene amplification.

Comprehensive reviews by Dancey and Sausville (2003), and Workman (2005) describe further protein kinase inhibitors in clinical development.

12.4 Targeted Therapy for the Treatment of Chronic Myeloid Leukaemia

Unusual in cancer aetiology and crucial to its pathogenesis, and successful treatment, a single genetic mutagenic translocation drives chronic myeloid (or myelogenous) leukaemia (CML). A reciprocal chromosomal translocation between chromosomes 9 and 22 gives rise to the distinct 'Philadelphia chromosome' (Ph) found in 95% cases of CML (t(9;22)(q34;q11); Figure 12.6). Chromosome 22, severed at the breakpoint cluster region (bcr), fuses with the short section of chromosome 9, on which c-abl proto-oncogene is encoded. Crucially, this mutagenic translocation gives rise to the *bcr-abl* fusion oncogene, the protein product of which possesses constitutively active tyrosine kinase activity. This event is critical and sufficient to initiate CML and promote myeloid cell proliferation.

Thus, the *bcr-abl* gene, formed by juxtaposition of *c-abl* oncogene on chromosome 9 with the *bcr* oncogene on chromosome 22, encodes an oncoprotein distinct in activity between normal and leukaemic cells and represents a discrete drug target.

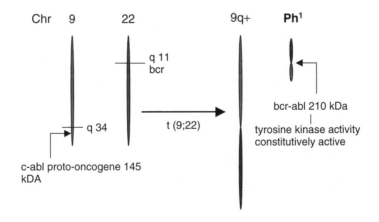

Figure 12.6 Pathogenesis of chronic myeloid leukaemia

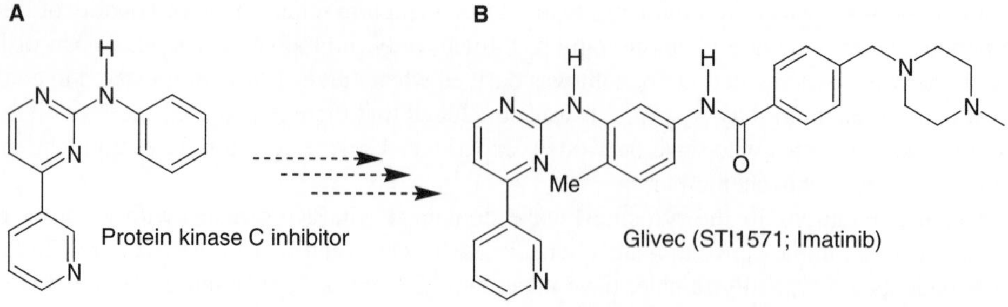

Figure 12.7   Chemical optimization of glivec

12.4.1 Chemical optimisation of glivec

The small molecule inhibitor of bcr-abl tyrosine kinase activity (4-[(4-methyl-1-piperazinyl)methyl]-N-[4-methyl-3-[{4-(3-pyridinyl)-2-pyrimidinyl}amino]phenyl] benzamide methanesulfonate) (glivec; STI1571; imatinib; Gleevec) has been developed by Novartis (Moen *et al.*, 2007). The initial compound (Figure 12.7A) was identified in a high throughput screen and is a protein kinase C (serine/threonine kinase) inhibitor. Crucially, it had promising 'lead-like' properties and high potential for chemical diversity. Structure activity (SAR) studies aided by molecular modelling led to the synthesis of Glivec (Figure 12.7B). Glivec binds in the ATP pocket, inhibiting downstream signal transduction inducing apoptosis in the bcr-abl-'addicted' CML cells. Preclinically, dose-dependent inhibition of CML tumour cell growth was observed in mouse xenograft models.

Outstanding clinical activity has been obtained in CML patients, particularly patients whose disease was in the chronic phase. Unprecedented in phase II human trials, a 95 %response rate was observed. For patients in blast crisis, the response rate was significantly lower (29 %) and high relapse rates were recorded for these patients in the later stages of disease.

It was found that glivec targets other receptor tyrosine kinases including c-kit – constitutively activated in gastrointestinal stromal tumours (GIST) and platelet-derived growth factor receptor (PDGFR), activated in a variety of human tumours. Drug resistance, intrinsic or acquired, is a major problem faced by patients, and clinicians, severely impeding successful treatment. In the late stages of CML, resistance to glivec emerges as a consequence of mutations in the ATP binding pocket.

New lead agents are reaching the clinic which demonstrate activity in Gleevec-resistant CML models, examples include nilotinib and dasatinib (Weisberg *et al.*, 2007).

12.5 Targeted Therapy for the Treatment of Breast Cancer

Many human cancers express cell surface molecules that are specific to the cancer cell. The role of these antigens in tumourigenesis and cancer progression is often not fully elucidated, what is understood is that they impart proliferative and/or survival advantage. These antigens provide attractive targets for diagnostic and therapeutic applications. Antibody binding to tumour cells often leads to their destruction: antigen-dependent cellular cytotoxicity. The antibody, targeted to the cancer cells, thus sparing normal cells, may also be conjugated to a

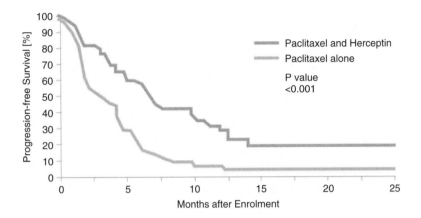

Figure 12.8 Herceptin increases progression-free survival in HER2+ metastatic breast cancer

toxin or radionuclide. Antibody based cancer therapy has been pioneered by the breakthrough antibody medicine Herceptin.

In 25–30 % of metastatic breast cancer patients the human epidermal growth factor receptor 2 – (HER2) protein is overexpressed (10–100-fold) through amplification of the *HER2* gene. HER2 signalling, particularly activation of downstream effector pathways including Ras/Raf/MAPK and PI3K/Akt signal transduction, has been implicated in the pathogenesis of breast cancer. HER2 overexpressing (HER2+) breast cancer is an especially aggressive disease, associated with significantly decreased disease-free survival periods than HER2 negative tumours. These cancers are less responsive to conventional chemotherapy and hormonal therapy.

12.5.1 Herceptin

Herceptin, developed by Genentech is a humanised monoclonal antibody that selectively targets the extracellular domain of the HER2 protein, specifically blocking the HER2 receptor on the surface of breast cancer cells, thus, debilitating the permanently 'on' signal for cell division. Herceptin is licensed for treatment of HER2-positive metastatic breast cancer and significantly increases survival time of women with advanced disease. Sustained improvement in progression-free survival was observed in the pivotal clinical trial of Herceptin administered with combination chemotherapy (doxorubicin or epirubicin plus cyclophosphamide or paclitaxel (Figure 12.8) (Genentech Inc, 2006; Slamon *et al.*, 2001; Mass *et al.*, 2005).

12.6 Angiogenesis

Capillaries extend into virtually all the tissues of the body, replenishing nutrients and removing waste products. Under normal conditions the endothelial cells which line these narrow tubes do not divide, except for example when tissue is damaged, then the vessels grow rapidly. The proliferation of new capillaries is known as angiogenesis (or neovascualrisation) and is typically short-lived (1 or 2 weeks), tumour cells however 'switch on' angiogenesis, indeed, angiogenesis is essential for tumour growth (promotion of a small, cluster of mutated cells to a large malignant growth) and metastasis (Hanahan and Weinberg, 2000; Plank and Sleeman, 2004). When the distance between a tumour cell and a blood vessel exceeds 1 mm, the supply of oxygen and nutrients diminish (and waste products accumulate). As hypoxia develops, tumour cells begin to respire anaerobically and the hypoxia-inducible response is triggered. Hypoxia-inducible factor 1-alpha (HIF-1α) protein stabilises and dimerises with constitutive HIF-1β, the heterodimer binds to the hypoxia response element (HRE) of DNA initiating transcription of a plethora of proteins, activating signalling cascades that control environmental pH, glucose metabolism, vascular endothelial cell growth and angiogenesis. Hypoxic tumour cells secrete vascular endothelial growth factor (VEGF) essential for endothelial cell recruitment, angiogenesis and hence tumour progression. VEGF gene expression can, however, be upregulated by various stimuli (not only hypoxia) including the genetic mutations associated with malignant transformation for example: loss of p53; activation of ras, v-src, HER2; PDGF, tumour necrosis factor-α (TNF-α), epidermal growth factor (EGF), transforming growth factor-beta (TGF-β), interleukin-6 (IL-6), IL-1β, insulin-like growth factor-1 (IGF-1); nitric oxide (NO) and oestrogen (Figure 12.9) (Ferrar, 2004).

The actions of VEGF are mediated through binding to receptor tyrosine kinases (VEGFR-1, VEGFR-2 and VEGFR-3). Receptor binding triggers phosphorylation of a multitude of proteins activating signal transduction cascades including the RAS/RAF/MEK/ERK pathway driving proliferation as well as angiogenesis.

Angiogenesis is a multistep process, and VEGF acts at several stages:

- It is a potent mitogen for vascular endothelial cells

- It mediates the secretion and activation of enzymes that degrade the extracellular matrix, e.g. matrix metalloproteinases, collagenase

- Through inhibition of apoptosis, it is a survival factor for endothelial cells

- It modulates endothelial cell migration to sites of angiogenesis

- It is important for vascular maintenance.

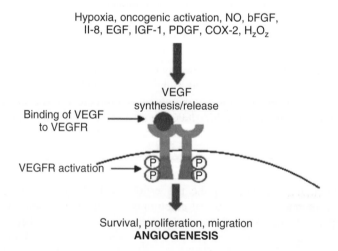

Survival, proliferation, migration
ANGIOGENESIS

Figure 12.9 Activation of VEGFR signal transduction (Ferrar, 2004)

In addition, VEGF can promote lymphangiogenes, and act directly on tumour cells, as in addition to paracrine stimulation of vascularisation, many tumour cells themselves express VEGF receptors.

Increased VEGF expression has been detected in the majority of cancers examined to date and overexpression correlates with enhanced risk of recurrence and poor prognosis. Intervention of VEGF signal transduction therefore represents a validated target.

Drugs targeting endothelial cell VEGF activity at endothelial cells do not need to penetrate tumours. This is constructive as the aberrant tumour vasculature and raised interstitial pressure within tumours renders tumour penetration an obstacle to drug treatment. Angiogenesis inhibition may improve the outcome of co-administered chemotherapy/radiotherapy by reducing treatment-induced elevation in VEGF which may contribute to tumour resistance to apoptosis, and by preventing rapid tumour cell repopulation after cytotoxic chemotherapy (Kerbel, 2006). It has been proposed that anti-VEGF therapy may normalise 'tumour vasculature, reducing intratumoural pressure and allowing better delivery of drugs. Importantly, drug resistance may be considered less likely than with traditional chemotherapeutics, a consequence of greater stability within the endothelial cell genome compared to that of the cancer cell.

12.6.1 Targeting VEGF receptor tyrosine kinase activity

Monoclonal antibodies directed at VEGF have been developed. Bevacizumab, marketed by Genentech (and its parent company Roche), as Avastin is a humanised neutralising IgG monoclonal antibody directed against VEGF. Avastin binds to all VEGF isoforms with high affinity and blocks their ability to bind to receptors. It was first approved by the United States Food and Drug Administration (US FDA) in 2004 for use in combination with standard chemotherapy for first-line treatment of metastatic colorectal cancer (mCRC). A pivotal phase III clinical trial revealed that addition of Avastin to 5-fluorouracil, leukovorin, irinotecan (IFL) chemotherapy increased progression-free survival time by 4.4 months; 45 %of patients receiving Avastin plus chemotherapy responded to treatment compared with 35 %of patients receiving chemotherapy alone. Avastin continues to improve clinical outcome for patients with mCRC.

In a large randomised trial, sponsored by the National Cancer Institute (NCI), Avastin plus chemotherapy also prolonged survival in metastatic NSCLC (www.cancer.gov). Additional late stage clinical trials were initiated to determine safety and efficacy in patients with metastatic breast, ovarian or prostate cancer, renal cell carcinoma, glioblastoma multiforme and metastatic or locally advanced pancreatic carcinoma. Data is now available revealing the benefit to patients with metastatic clear cell carcinoma of the kidney of exposure to antibody-targeted inhibition of VEGF signalling pathways (Figure 12.10) (Yang *et al.*, 2003; Yang, 2004).

Small molecule tyrosine kinase inhibitors have also been developed to inhibit VEGF receptor signalling. $N-[2-(diethylamino)ethy]-5-[(Z)-(5-fluoro-1,2-dihydro-2-oxo-3-indol-3-ylidine)methyl]-2,4-dimethyl-1H-pyrrole-3-carboxamide$, or sunitinib (Figure 12.11) is marketed as Sutent by Pfizer and inhibits tyrosine kinase autophosphorylation of VEGFR, thereby preventing downstream signal transduction.

Two independent multicentre phase II trials of sunitinib in cytokine-refractory metastatic renal cell carcinoma (RCC) have been completed. In the first trial objective tumour responses (partial response PR) were observed in 40 %of patients and stable disease was achieved in

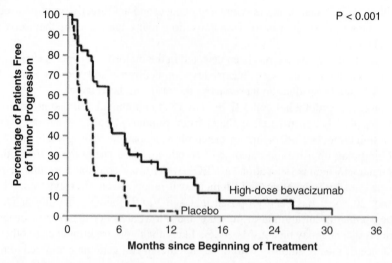

Figure 12.10 Statistically significant but modest improvement in median time to progression in metastatic clear cell RCC patients receiving high dose Avastin

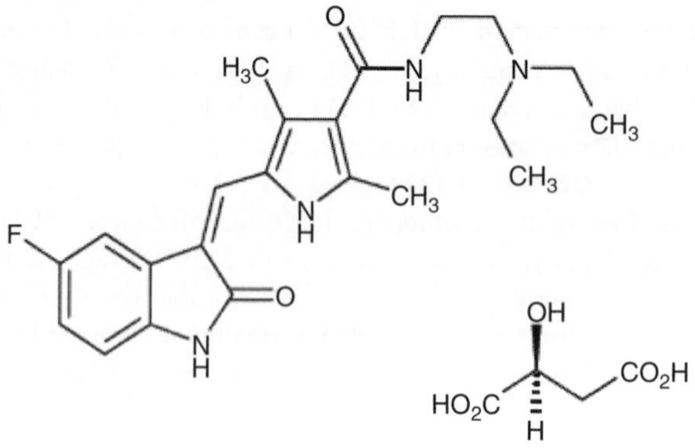

Figure 12.11 Structure of sunitinib

27 %of patients. In the second trial of 83 assessable patients treated with sunitinib, 29 %of patients had >30 %decrease in tumour size, 1 complete response (CR), 16 confirmed PR and 7 unconfirmed PR were reported. Such high response rates, coupled with manageable toxicity prompted a phase III trial of sunitinib versus interferon-α as first-line therapy for metastatic RCC (Larkin and Eisen, 2006). The results of this trial confirmed the superior antitumour efficacy of sunititnib which more than doubled progression-free survival times (Motzer *et al.*, 2007). Benefit was derived by all patient subgroups receiving sunitinib including poor-risk patients, again 1 CR was observed. The investigators concluded from this study that sunitinib should be the new standard of care for first line treatment of metastatic clear cell RCC.

Sunitinib, as many small molecule agents, possesses a rich and diverse pharmacology and is in fact a multi-targeted receptor tyrosine kinase inhibitor. Simultaneous inhibiton of

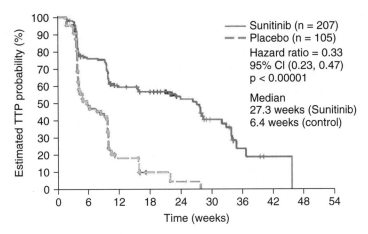

Figure 12.12   Results of phase III trial of sunitinib in glivec refractory GIST patients. (Demetri 2006). Time to tumour progression is shown

VEGFR and PDGFR (targeting angiogenesis and tumour cell proliferation) leads to reduced tumour, cancer cell death, and ultimately tumour shrinkage. Sunitinib also inhibits flt3, constitutively activated in acute myeloid leukaemia (AML) (Au *et al.*, 2004) and c-KIT, the receptor tyrosine kinase which drives the majority of GISTs. In 2006, following impressive trial results sunitinib was approved by the FDA for treatment of glivec-refractory GIST (Figure 12.12) (Demetri *et al.*, 2006).

12.7 Targeting Cell Cycling

Mutations in tumour suppressor genes (e.g. *p*53) that negatively regulate cell growth can also generate cancer by allowing cells to escape important growth checkpoints. Typically, tumour suppressor genes are involved in regulating the cell cycle, the process that a cell undergoes during its division into two daughter cells (see below). Cyclin-dependent kinases (CDKs) are serine/threonine kinases that are involved in the control of the cell cycle, cellular checkpoints and apoptosis.

Flavopiridol is a synthetic flavonoid inhibitor, which is under development by Sanofi Aventis in collaboration with the National Cancer Institute (NCI). Flavopiridol is a small molecule derivative of the alkaloid rohitukine, which inhibits cyclin-dependent kinases (CDK) causing a halt to the cell cycle. At the same time it decreases the expression of a number of proteins that regulate the cell cycle.

Ispinesib is a novel small molecule drug candidate that inhibits cell proliferation and promotes cancer cell death by specifically disrupting the function of a cytoskeletal protein known as kinesin spindle protein, or KSP. KSP is essential for cell proliferation. Ispinesib is being studied in a broad clinical trials program that consists of nine phase II trials and five phase I/Ib trials. The phase II trials are evaluating ispinesib as monotherapy in each of NSCLC, ovarian cancer, breast cancer, colorectal

Flavopiridol

cancer, hepatocellular cancer, melanoma, head and neck cancer, prostate cancer and renal cell cancer. The phase I trials are evaluating ispinesib in solid tumours or acute leukaemia, CML and advanced myelodysplastic syndromes. Additionally, three phase Ib trials are being conducted to study ispinesib in combination with each of carboplatin, capecitabine or docetaxel.

Ispinesib

P276-00 is another rohitukine-derived small molecule drug that acts as a selective CDK4-D1 and CDK1-B inhibitor and is being developed by Nicholas Piramal India Limited. It is currently in phase I/II trials in Canada and India for the treatment of advanced refractory neoplasms.

BI 2536 is a novel inhibitor of the serine/threonine kinase Polo-like kinase 1 (Plk-1), a cell cycle switch essential for cell proliferation and is being developed by Boehringer Ingelheim. BI 2536 inhibits mitosis, resulting in cancer cell death by apoptosis. It is the first Plk-1 inhibitor being tested in clinical trials and is at phase I in the treatment of advanced non-Hodgkin's lymphoma and phase II in both advanced NSCLC and SCLC.

SNS-032 is a novel aminothiazole small molecule cell-cycle modulator that targets CDK2, CDK7 and CDK9 that is being developed by Sunesis Pharmaceuticals. Preclinical studies have shown that SNS-032 induces cell-cycle arrest and apoptosis (or cell death) across multiple cell lines. SNS-032 is a small molecule that is currently administered by intravenous infusion, but also has the potential to be developed as an oral formulation. It is currently undergoing phase I clinical trials for the treatment of a variety of advanced solid tumours. Sunesis are also developing SNS-595 is a novel naphthyridine analogue that induces a G_2 cell cycle arrest. This drug is currently in phase I trials for the treatment of a variety of hematological malignancies and phase II clinical trials for both ovarian and small cell lung cancers.

AT7519 is Astex Therapeutic's most progressed drug candidate and is currently in phase I clinical trials for advanced solid tumours and for refractory non-Hodgkin's lymphoma. AT7519 is a potent cell cycle inhibitor that targets the cyclin dependent kinases.

MKC-1 is a novel, orally active cell cycle inhibitor that is being developed by EntreMed and is currently in phase II clinical trials for the treatment of breast cancer and NSCLC. MKC-1 inhibits mitotic spindle formation, thus preventing chromosome segregation during mitosis, and inducing apoptosis. It is thought that this arises due to the binding of MKC-1 to tubulin and the importin β proteins.

12.8 Targeting Apoptosis

Defects in the apoptotic process can lead to the onset of cancer by allowing cells to grow unchecked when an oncogenic signal is present; therefore, activating the apoptosis machinery is an ideal way in which to reduce tumour volume. The figure below illustrates the large number of proteins that are involved in apoptosis and that are a potential target for candidate small molecular drugs.

Obatoclax is currently the only small molecule inhibitor that targets apoptosis-regulating proteins in clinical trials and is being developed by GeminX Biotechnologies. It is designed to restore apoptosis through inhibition of the Bcl-2 family of proteins, thereby reinstating the natural process of cell death that is often inhibited in cancer cells. It currently in phase I clinical trials for use against mantle-cell lymphoma and phase II trials for the treatment of myelodysplastic syndromes, NSCLC, follicular lymphoma, myelofibrosis and Hodgkin's lymphoma.

Obatoclax

12.9 Targeting mTOR

Rapamycin (Sirolimus), a macrolide antiobiotic first isolated from the bacterium *Streptomyces hygroscopicus*, possesses potent immunosuppressive and antiproliferative properties. The mammalian target of rapamycin (mTOR) is a serine threonine kinase which regulates biogenesis impacting cell growth, proliferation, survival, protein translation and angiogenesis. mTOR integrates input from multiple signal transduction pathways (Hay and Sonenberg, 2004). The two most well characterised mTOR substrates are components of protein translational machinery; p70S6 kinase 1 (S6K1), and translational repressor, eukaryotic initiation factor 4E (eIF4E) binding protein (4EBP), thus activating eIF4E. Rapamycin therefore, critically perturbs translational regulation, cell cycle progression, modulating cell size and adversely impacting on cell survival. mTOR is a direct target of AKT (protein kinase B). AKT is activated or overexpressed in a number of human malignancies (including colon, pancreas and breast), and implicated in mediation of cell growth, proliferation and survival. Its activity is negatively regulated by the tumour suppressor phosphatase PTEN (phosphatase and tensin homologue), which is mutated and inactivated in multiple cancers, allowing malignant cells uncontrolled proliferative capacity and evasion of apoptosis.

Temsirolimus is an intravenous drug for the treatment of renal cell carcinoma (RCC), developed by Wyeth Pharmaceuticals and approved by the FDA in late May 2007 (www.wyeth.com/news?nav=display&navTo=/wyeth_html/home/news/pressreleases/2007/1180576865144.html) also being approved by the European Medicines Agency (EMEA) in November 2007. It is a derivative of rapamycin and is sold as Torisel.

Temsirolimus

The drug shows promise for RCC patients. A phase III clinical study demonstrated a 49 % increase in patients' median overall survival time (10.9 months). The drug was administered to patients who had received no prior systemic therapy; however, this study only included patients with a poor prognosis. The benefits of temsirolimus in patients with favorable or intermediate prognostic factors remain to be elucidated (http://www.wyeth.com/news?nav=display&navTo=/wyeth_html/home/news/pressreleases/ 200 7/1 180 576 865 144.html)

12.10 The Future of Molecularly Targeted Therapy

A single genetic mutation (bcr-abl) underpins the pathogenesis of CML, explaining the spectacular success of glivec administered in the chronic disease stage. However, at the time of diagnoses, common solid tumours have acquired multiple genetic mutations (Figure 12.13). Cancer is a clinically, histopathologically and molecularly heterogeneous disease.

Consider Vogelstein's model for solid tumourigenesis (Figure 12.13; Fearon and Vogelstein, 1990).

Indeed, it has been reported that sporadic colorectal cancers contain at least 11 000 genomic alterations per cell (Boland and Ricciardiello, 1999; Stoler et al., 1999). Recently, Vogelstein reports that the genomic landscapes of human breast and colorectal cancers is 'composed of a handful of commonly mutated gene 'mountains' and a much larger number of gene 'hills' mutated at low frequency' (Wood et al., 2007).

Targeting such solid tumours may seem an insurmountable challenge. However, there are 'driver' and 'passenger' mutations and tumour cells become 'addicted' to pathways activated by dominant oncogenically transformed proteins.

One therapeutic approach, already introduced, is to target multiple pathways. Indeed, small molecules have been synthesised that possess dual, or multiple tyrosine kinase inhibitory action. The therapeutic benefit of one such molecule, sunitinib, has been discussed. The number of small molecule kinase inhibitors introduced into the clinic increases, and evaluation of efficacy is underway. Sorafenib (developed by Bayer) is a bi-aryl urea, designed as an inhibitor of the non-receptor serine threonine kinases b-RAF and c-RAF (members of the RAF/MEK/ERK pathway) (Zhong and Bowen, 2007). In addition to b-RAF and c-RAF,

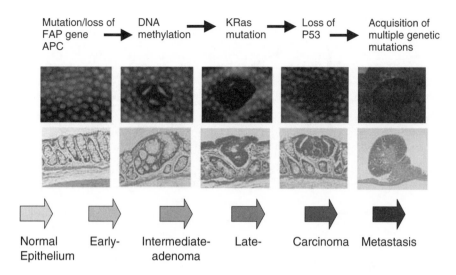

Figure 12.13   A genetic model for colorectal tumourigenesis

sorafenib also inhibits the receptor tyrosine kinases VEGFR2 and 3, FLT-3, c-KIT and PDGFR. In preclinical studies, sorafenib demonstrated antitumour activity in NSCLC, breast and colon xenograft models, significantly inhibiting angiogenesis (Wilhelm *et al.*, 2004). A multicentre phase III randomised double blind trial revealed that sorafenib doubled the median progression-free survival time in advanced RCC (Escudier *et al.*, 2005).

In the wake of Iressa, Astra Zeneca is developing the orally active dual EGFR/VEGFR2 inhibitor vandetanib (Zactima). Preclinical and early clinical data were promising and trials with Zactima have progressed to phase III. In a randomised double blind multicentre phase III clinical trial, antitumour efficacy of Zactima in combination with docetaxel is being compared with docetaxel treatment alone in patients with locally advanced or metastatic NSCLC. Zactima has been granted orphan drug designation for the treatment of patients with follicular, medullary, anaplastic and locally advanced and metastatic papillary thyroid cancer. As well as targeting VEGF and EGF signalling pathways, Zactima also inhibits RET kinase which drives the growth and survival of certain tumours and is believed to be important in medullary thyroid cancer. As observed, combination of 'molecularly-targeted' agents with conventional cytotoxic agents is a strategy adopted. Examples of such combination therapy include cetuximab (Erbitux; monoclonal antibody targeting EGFR) and irinotecan for the treatment of advanced colorectal cancer. Herceptin (trastuzumab), targeting the HER2 protein is administered in combination with paclitaxel to patients with advanced metastatic HER2+ breast cancer; and the monoclonal antibody Avastin has been combined with paclitaxel or carboplatin for NSCLC treatment.

Excellent trial results followed combination of the orally active small molecule quinazoline kinase inhibitor lapatinib with cytotoxic chemotherapy (Figure 12.14) (Geyer *et al.*, 2006); (Geyer *et al.*, 2007). Lapitanib targets epidermal growth factor receptor members, reversibly blocking phosphorylation of EGFR and HER2. In addition, the activity of ERK-1, ERK-2 and AKT kinases is inhibited. In 2007, the FDA approved lapatinib (Tyverb) in combination with capecitibine for treatment of advanced or metastatic HER2+ Herceptin refractory breast cancer.

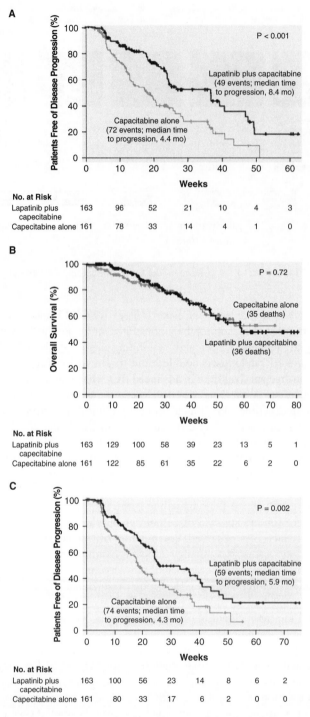

Figure 12.14 Increased time to progression in patients receiving lapiatinib and cytotoxic combined therapy (Geyer *et al.*, 2007)

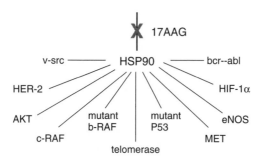

Figure 12.15   Inhibition of HSP90 results in multiple signal transduction disruption

The effect of combinations of the mTOR inhibitor temsirolimus and an oestrogen receptor-α (ERα) antagonist, ERA-923 on breast carcinoma models *in vitro* and *in vivo* has been studied (Sadler *et al.*, 2006). Synergy was achieved *in vitro* following combination of non-inhibitory doses of temsirolimus with suboptimal concentration of ERA-923. Moreover, *in vivo*, complete tumour growth inhibition was achieved. Synergy was also encountered combining temsirolimus with raloxifene or 4-hydroxytamoxifen. The authors conclude that combination of temsirolimus and a pure anti-oestrogen has excellent anticancer activity in preclinical models, and may have clinical implications in the treatment of hormone-dependent tumours.

It can be argued that by inhibiting mTOR, rapamycin analogues target proteins whose functions are fundamental to a number of pathways pertinent in tumourigenesis. Another such protein is heat-shock protein 90 (HSP90), overproduced in several types of cancer, orchestrates the folding, intracellular disposition and proteolytic turnover of key regulators of signal transduction pathways subverted in malignancy, Geldanamycin analogues 17-(allylamino)-17-demethoxygeldanamycin (17AAG) and 17-(dimethylaminoethylamino)-17-demethoxygeldanamycin (17DMAG), which occupy the ATP binding pocket of thus ubiquitous molecular chaperone are being evaluated clinically. Protein folding and assembly of immature kinases is inhibited; consequently and degradation of client proteins mutated in tumour cells is observed (eg v-src, bcr-abl) resulting in cellular apoptosis. Multiple signal transduction pathways harbouring mutant or overexpressed proteins that promote growth or survival of tumour cells are disrupted by HSP 90 inhibition: e.g. HER2, AKT, c-RAF, mutant b-RAF, mutant p53 (Figure 12.15) (Witesell and Lindquist, 2005; Rowan, 2005).

A final word should be assigned to chemistry-driven drug discovery (Westwell and Stevens, 2004), often inspired by nature, which yields structurally diverse compound libraries and experimental agents with antitumour activity *in vitro* and efficacy *in vivo*. As target-driven drug discovery necessitates a 'molecule hunt', chemistry-driven drug discovery necessitates target deconvolution. Both strategies are vital to cancer research and inspirational drug discovery.

References

Abraham RT. PI3-kinase related kinases: 'big' players instress-induced signalling pathways. *DNA Repair* 2004, **3**, 883–7.

Au WY, Fung A, Chim CS, Lie AK, *et al*. FLT-3 aberrations in acute promyelocytic leukaemia: clinico-pathological associations and prognostic impact. *Br J Haematol* 2004, **125**, 463–9.

Boland C, Ricciardiello L. How many mutations does it take to make a tumor? *Proc Natl Acad Sci U S A* 1999, **96**, 14675–7.

Dancey J, Sausville EA. Issues and progress with protein kinase inhibitors. *Nat Rev Drug Discov* 2003, **2**, 296–313.

Demetri GD, van Oosterom AT, Garrett CR, *et al.* Efficacy and safety of sunitinib in patients with advanced gastrointestinal stromal tumour after failure of imatinib: a randomised controlled trial. *Lancet* 2006, **368**(9544), 1329–38.

Doll R, Hill AB. Smoking and carcinoma of the lung: preliminary report. *Br Med J* 1950, **221**, 739–48.

Escudier B, Szczylik C, Eisen T, Stadler WM, Schwartz B, Shan M, Bukowski RM. Randomized phase III trial of the Raf kinase and VEGFR inhibitor sorafenib (BAY 43–9006) in patients with advanced renal cell carcinoma (RCC). *J Clin Oncol* ASCO annual meeting proceedings 2005, **23**, 4510.

Fearon ER, Vogelstein B. A genetic model for colorectal tumorigenesis. *Cell* 1990, **61**(5), 759–67.

Ferrar N. Vascular endothelial growth factor as a target for anticancer therapy. *Oncologist* 2004, **9**, 2–10.

Geyer CE, Foster J, Lindquist D, *et al.* Lapiatinib plus capecitabine in HER2-Positive Advanced Breast Cancer. *New England Journal of Medicine*, **355**, 2733–43 2006.

Geyer CE, Martin A, Newstat B, *et al.* Lapiatinib plus capecitabine in HER2+ advanced breast cancer: genomic and updated efficacy data. *J Clin Oncol* ASCO annual meeting proceedings 2007, **25**, 1035.

Hanahan D, Weinberg RA. The hallmarks of cancer. *Cell* 2000, **100**, 57–70.

Hay N, Sonenberg N. Upstream and downstream of mTOR. *Genes Dev* 2004, **18**, 1926–1945.

Herbst RS, Fukuoka M, Baselga J. Gefitinib – a novel targeted approach to treating cancer. *Nat Rev Cancer* 2004, **4**, 956–65.

Herceptin (Trastuzumab) Prescribing Information. Genentech, Inc. November 2006.

Hirsch F, Scagliotti GV, Langer CJ, Varella-Garcia M, Franklin WA Epidermal growth factor family of receptors in preneoplasia and lung cancer. perspectives for targeted therapies. *Lung Cancer* 2003, **41**, S29–41.

Kerbel RS. Antiangiogenic therapy: a universal chemosensitization strategy for cancer? *Science* 2006, **312**, 1171–5.

Kidd JG, Rous P. The carcinogenic effect a papilloma virus on the tarred skin of rabbits: Major factors determining the phenomenon: The manifold effects of tarring. *J Exp Med* 1938, **68**, 529–62.

Krejsa C, Rogge M, Sadee W. New applications for pharmacogenetics. *Nat Rev Drug Discov* 2006, **5**, 507–21.

Larkin JMG, Eisen T. Kinase inhibitors in the treatment of renal cell carcinoma. *Crit Rev Oncol/Hematol* 2006, **60**, 216–26.

Liang K, Jin W, Knuefermann C, Schmidt M, Mills GM, Ang K, Milas L, Fan Z. Targeting the phosphatidylinositol 3-kinase/Akt pathway for enhancing breast cancer cells to rafiotherapy. *Mol Cancer Ther* 2003, **2**, 353–60.

Mass RD, Press MF, Anderson S, *et al.* Evaluation of clinical outcomes according to HER2 detection by fluorescence *in situ*hybridization in women with metastatic breast cancer treated with trastuzumab. *Clin Breast Cancer* 2005, **6**, 240–6.

Moen MD, McKeage K, Plosker GL, Siddiqui MAA. Imatinib: A review of its use in chronic myeloid leukaemia. *Drugs* 2007, **67**, 299–320.

Motzer RJ, Huston TE, Tomczak P, Michaelson RM. Sunitinb versus interferon-alfa in renal-cell carcinoma. *N Engl J Med* 2007, **356**, 115–24

Ng SSW, Tsao M-S, Nicklee T, Hedley DW. Effecys of the epidermal growth factor receptor inhibitor OSI-774, Tarceva on downstream signalling pathways and apoptosis in human pancreatic adenocarcinoma. *Mol Cancer Ther* 2002, **1**, 777–83.

OSI Pharmaceuticals Inc, Tarceva (erlotinib) full prescribing information, 2004.

Paul MK, Mukhopadhyay AK. Tyrosine kinase – role and significance in cancer. *Int J Med Sci* 2004, **1**, 101–15).

Plank MJ, Sleeman BD. Tumour-induced angiogenesis: a review. *J Theoret Med* 2004, **5**, 137–53.

Rowan A. Lead optimisation: Improving natural strength. *Nat Rev Drug Discov* 2005, **4**, 1041.

Sadler TM, Gavriil M, Annable T, Frost P, Greenberger LM, Zhang Y. Combination therapy for treating breast cancer using antiestrogen, ERA-923, nad the mammalian target of rapamycin inhibitor, temsirolimus. *Endocr Rel Cancer* 2006, **13**, 863–73.

Sagatys E, Garrett CR, Boulware D, Kelley S, Malafa M, Cheng JQ, Sebti S, Coppola D. Activation of the serine/threonine protein kinase at during the progression of Barrett neoplasia. *Hum Pathol* 2007, **38**, 1526–31.

Scagliotti GV, Selvaggi G, Novello S, Hirsch FR. The biology of epidermal growth factor in lung cancer. *Clin Cancer Res* 2004, **10**, 4227S–232S.

Slamon DJ, Leyland-Jones B, Shak S, Fuchs H, Paton V. Use of chemotherapy plus a monoclonal antibody against HER2 for metastatic breast cancer that overexpresses HER2. *N Engl J Med* 2001, **344**, 783–92.

Stoler DL, Chen N, Basik M, Kahlenberg MS, Rodriguez-Bigas MA, Petrelli NJ, Anderson GR. The onset and extent of genomic instability in sporadic colorectal tumor progression. *Proc Natl Acad Sci U S A* 1999, **96**, 15121–6.

Weisberg E, Manley PW, Cowan-Jacob SW, Hochaus A, Griffin JD. Second generation inhibitors of BCR-ABL for the treatment of imatinib-resistant chronic myeloid leukaemia. *Nat Rev Cancer* 2007, **7**, 345–56.

Westwell AD, Stevens MFG. Hitting the chemotherapy jackpot. *Drug Discovery Today* 2004, **9**, 625–7.

Whitesell L, Lindquist SL. HSP90 and the chaperoning of cancer. *Nat Rev Cancer* 2005, **5**, 761–72.

Wilhelm SM, Carte Cr, Tang LY, *et al*. BAY 43–9006 exhibits broad spectrum oral antitumor activity and targets the RAF/MEK/ERK pathway and receptor tyrosine kinases involved in tumor progression and angiogenesis. *Cancer Res* 2004, **64**, 7099–109.

Wood LD, Williams Parsons D, Jones S, *et al*. Genomic landscapes of human breast and colorectal cancers. *Science* 2007; Oct 11 online.

Workman P, Kaye SB. Translating basic cancer research into new cancer therapeutics. *Trends Mol Med* 2002; **8**, S1–S9.

Workman P. Genomica and the second golden era of cancer drug development. *Molec Biosystems* 2005, **1**, 17–26.

www.astrazeneca.com [accessed 29 April 2008]

www.cancer.gov [accessed 29 April 2008]

www.cancerresearchuk.org [accessed 29 April 2008]

www.nih.org [accessed 29 April 2008]

www.tarceva.com/professional/survival/results.jsp

www.timesonline.co.uk [accessed 29 April 2008]

www.wyeth.com/news?nav=display&navTo=/wyeth_html/home/news/pressreleases/2007/1180576865144.html

Yang JC, Haworth L, Sherry RM, *et al*. A randomized trial of bevacizumab, an anti-vascular endothelial growth factor antibody, for metastatic renal cancer. *N Engl J Med* 2003, **349**, 427–34.

Yang JC. Bevacizubab for patients with metastatic renal cell cancer: an update. *Clin Cancer Res* 2004, **10**, 6367s–70s.

Zhong HZ, Bowen JP. Molecular design and clinical development of VEGFR kinase inhibitors. *Curr Top Med Chem* 2007, **7**, 1379–93.

13
Tumour Hypoxia: Malignant Mediator

Jill L. O'Donnell, Aoife M. Shannon, David Bouchier-Hayes

13.1 Introduction

Hypoxia occurs in tissue when there is an inadequate supply of O_2, which compromises normal biological processes in the cell. Solid tumours contain regions of hypoxia, a finding which is a prognostic indicator and determinant of malignant progression, metastatic development and chemoradio-resistance (Shannon *et al.*, 2003). Tumour tissue pO_2 is dependent on O_2 supply and consumption parameters and the diffusion properties of the tissue (Figure 13.1) (Degner and Sutherland, 1998). Tumour hypoxia arises from inadequate perfusion, as a result of severe structural and functional abnormalities of the tumour microcirculation, and can also be caused by an increase in diffusion distances as cells distant ($>70\,\mu m$) from the blood vessel receive less O_2 than needed (Hockel and Vaupel, 2001). Tumour-associated or therapy-induced anaemia can lead to a reduction in the O_2 transport capacity of the blood, further contributing to the development of hypoxia (Kelleher *et al.*, 1996; Shannon *et al.*, 2003; Vaupel *et al.*, 2003, 2005). When the tissue pO_2 falls below a critical value (median $pO_2 <10\,\text{mmHg}$) a progressive decrease in O_2 consumption and adenosine triphosphate (ATP) production occurs; tumour cells switch from oxidative phosphorylation to glycolysis which results in a reduction in energy supply and severe extracellular acidosis in the tumour microenvironment (Galarraga *et al.*, 1986). The high rate of glycolysis in tumours is known as the Warburg effect (Warburg, 1956). Areas of hypoxia are a hallmark of solid tumours (Teicher, 1995), with nearly 50 % of locally advanced breast cancers containing areas of hypoxic tissue (Vaupel *et al.*, 2002). A comparison of pO_2 in cervical cancer tissues with the corresponding normal tissues, measured using polarographic needle electrodes, illustrated a reduction in the median pO_2 in tumour tissues from 42 mmHg to 8 mmHg, with hypoxic pO_2

Anticancer Therapeutics   Edited by Sotiris Missailidis
© 2008 John Wiley & Sons, Ltd

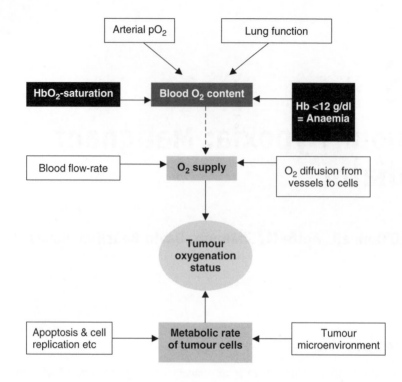

Figure 13.1   Factors contributing to tumour hypoxia (Hb = haemoglobin)

values $\leqslant 2.5$ mmHg occurring in approximately 60% of lesions investigated (Vaupel *et al.*, 2001).

13.2 Hypoxia Inducible Factor-1 and Hypoxia

Hypoxia induces a wide range of responses in cells and tissues (Table 13.1). The degree of this intra-tumoural hypoxia is positively correlated with the expression of the transcription factor hypoxia-inducible factor 1 (HIF-1). HIF 1 was initially discovered as the mediator of erythropoietin (Epo) production by renal cells in response to hypoxia (Semenza and Wang, 1992) but its malignant influence is now appreciated. Biological pathways that are regulated by hypoxia inducible genes, largely under the influence of HIF-1, include apoptosis, cell cycle arrest, angiogenesis, glycolysis and pH regulation.

HIF-1 is composed of two sub-units, 120 and 80 kDa, respectively HIF 1 alpha (HIF-1α) and HIF-1beta (HIF-1β). Both subunits contain a basic-helix-loop-helix (bHLH) motif and a PAS (Per, ARNT, Sim) protein–protein interaction domain (Wang *et al.*, 1995). Production of the transcription factor HIF-1α is however the key element in allowing cells to adapt and survive in a hostile hypoxic environment (Bradbury, 2001). The PAS and the bHLH domain are required for heterodimerization and DNA binding (Jiang *et al.*, 1996). HIF-1β is also known as aryl hydrocarbon receptor nuclear translocator (ARNT) and is constitutively expressed (Wood *et al.*, 1996).

Table 13.1 Gene targets for HIF-1

Process	Gene target for HIF-1
Red cell production	Erythropoietin
Iron metabolism	Transferrin, transferrin receptor, ceruloplasmin
Angiogenesis	VEGF, PAI-1, adrenomedullin, endothelin-1, nitric oxide synthase, haemoxygenase-1, α1β-adrenoreceptor
pH control	Carbonic anhydrases 9 & 12
Matrix metabolism	Prolyl-4-hydroxylase-α1, collagen v-α1, urokinase plasminogen activator receptor, matrix metalloproteinase 2
Glucose metabolism	Glucose transporter 1 & 3, hexokinase 1 & 2, phosphofructokinase L, aldolase A & C, glyceraldehyde-3-phosphate dehydrogenase, phosphoglycerate kinase 1, enolase 1, pyruvate kinase M, lactate dehyrogenase A, adenylate kinase 3
Cellular growth	Insulin-like binding factors 1, 2 & 3, insulin like growth factor II, transforming growth factor β3, p21, Nip3, cyclin G2, differentiated chondrocyte 1
Negative feedback	p53

HIF-1α is stabilised by hypoxia at the protein level and is rapidly degraded in normoxia (Wang *et al.*, 1995). The tyrosine kinases HER2neu, SRC, IGF and EGF receptors stimulate phosphoinositol-3K-AKT-FRAP signal transduction, which leads to increased translation of HIF-1α mRNA into protein (Semenza, 2002).

13.3 HIF-1α Post-translational Changes

HIF-1α functions as a physiological mediator of cellular response to hypoxia in both normal and malignant tissues as it determines the activity of HIF-1 and transactivates the hypoxia inducible genes, such as vascular endothelial growth factor (VEGF). Full transcriptional activity of HIF-1 requires the N-terminal and C-terminal *trans*-activation domains in the HIF-1α subunit. Its N-terminal is essential for DNA binding and dimerisation (Jiang *et al.*, 1996) and its C-terminal contains two trans-activation domains and a nuclear localisation signal (Jiang *et al.*, 1997). The central portion of HIF-1α contains an ODDD (oxygen dependent degradation domain), which determines the stability of HIF-1α protein (Huang *et al.*, 1998). In normoxia, proyl-hydroxylases (PHD1, PHD2, PHD3) modify two of the proline residues at either end of the ODDD. These proyl-hydroxylases are oxygen sensors (Mazure *et al.*, 2004). Hydoxylated HIF-1α binds to the Von Hippel–Lindau protein (pVHL), which targets it for proteasomal degradation by polyubiquination.

A second hydroxylation by the asparagine hydroxylase, FIH-1 (factor inhibiting HIF-1) also occurs. Importantly, FIH-1 transcription is completely independent of oxygen concentration and is mainly located in the cytoplasm, even under hypoxic conditions. FIH-1 inactivates the C-terminal transactivation domain of HIF-1α and, therefore, C-terminal binding to the transcriptional co-activator p300/CBP is inhibited (Lando *et al.*, 2002; Hewitson *et al.*, 2002). Recruitment of p300/CBP is necessary for transcription and hence transactivation of HIF-1 target genes. Mazure *et al.* (2004) propose that if HIF-1α escapes the degradation process mediated by the proyl-hydroxylases/VHL pathway, then hydroxylation of asparagine by FIH-1 should abolish residual transactivation of HIF-1α in normoxia (Mazure *et al.*, 2004).

SUMO (small ubiquitin related modifier) is a member of the ubiquitin-like protein family. Unlike ubiquitin, it does not signal proteins for degradation however, but instead may protect proteins from ubiquination and thereby influence cellular localisation and protein–protein interactions (Seeler and Dejean, 2003). Under hypoxic conditions, stabilised HIF-1α transfers to the nucleus and is SUMO modified, as it initially passes through the nuclear pore complex, where it comes into contact with the SUMO E3 ligase RanBP2 (Sondergaard *et al.*, 2002). Down-regulation of HIF-1β and HIF-1α has been demonstrated by SUMO post-translational modification (Tojo *et al.*, 2002; Berta *et al.*, 2004).

Recently published work shows that for HIF-1α to become active in transcription it needs the protease SENP1. SENP1 (sentrin/SUMO-specific protease-1) snips SUMO from SUMO-modified HIF-1α. If this does not occur, SUMO will act like ubiquitin, targeting destruction of HIF-1α. Work with SENP1-deficient mice showed that hypoxia-induced transcription of HIF-1α-dependent genes, such as VEGF and glucose transporter 1 (GLUT1), were markedly reduced (Cheng *et al.*, 2007). Inhibiting SENP1 would allow SUMO target HIF-1α for destruction and potentially arrest tumour growth. Work is underway to target this newly recognised pathway. The arrest defective protein 1 (ARD1), like the proyl-hydroxylases, makes HIF-1α unstable and acts thus as another negative regulator of HIF-1. Acetylation of a lysine residue within the ODDD by ARD1 results in pVHL and, hence, ubiqination and degradation of HIF-1α. ARD1 is not thought to be dependent on oxygen but ARD1 mRNA does decrease with hypoxia (Jeong *et al.*, 2002) and acetylation of HIF-1α appears to be reduced under hypoxic conditions.

When HIF-1α is stabilised under hypoxic conditions, p42/44 mitogen-activated kinase (MAPK) can phosphorylate HIF-1α. MAPK activation is associated with increased transcriptional activity of HIF-1 but does not affect its stability or DNA-binding of HIF-1α (Richard *et al.*, 1999). Suzuki *et al.* (2001) propose that HIF-1β binds preferentially to phosphorylated HIF-1α, increasing transcriptional activity of HIF-1. Another study showed that phosphorylation of the HIF 1α co-activator p300 by MAPK increased the interaction between the HIF-1α C terminal and p300 (Sang *et al.*, 2003). This, in turn, leads to *trans*-activation of the HIF-1 target genes.

CITED 2 (CBP/p300 interacting *trans*-activator with ED-rich tail 2) is a negative regulator of HIF-1α. It is a nuclear protein and binds with high affinity to the same domain of HIF-1α as p300/CBP (Freedman *et al.*, 2003). It then recruits the accessory co-activators such as histone, acetyltransferases SRC-1, TIF-2, and redox factor Ref-1. By competing with p300/CBP for binding to HIF-1α, CITED2 represses transcription of the HIF-1 target genes.

13.4 How Genetics Can Modify HIF

HIF-1α expression may be modulated by genetic alterations in oncogenes, tumour suppressor genes and stimulation of receptor tyrosine kinases.

Mutations within the tumour suppressor p53 gene lead to an environment where damaged and proliferating cells can survive and progress through the cell cycle. This has a potential knock-on effect of mutations occurring within other genes and the activation of proto-oncogenes. The expression of HIF-1α and p53 are closely linked, with loss of p53 expression leading to an augmentation of HIF 1α expression and activity in p53$^{-/-}$ knock-out colon carcinoma cells as compared with parental p53$^{+/+}$ cells (Hanahan and Folkman, 1996). The nuclear protein p53 confers survival advantage to cancer cells and is one of the genes most commonly mutated in cancer (Steele *et al.*, 1998). The combination of p53 mutation and over-expression of HIF-1α may result in clonal selection of cancer cells resistant to apoptosis, which can survive and grow despite hypoxic conditions, thus facilitating a more malignant phenotype. Graeber *et al.* (1994) demonstrated this by showing that small numbers of p53 deficient tumour cells could outgrow large numbers of p53 wild type tumour cells when treated with hypoxia.

In normoxia pVHL recognises HIF-1α and targets it for proteasomal degradation by HPTF (HIF-1α proteasome targeting factor). In response to low oxygen tension, its basal expression is maintained, allowing protein expression of its target genes. Loss of pVHL function as seen in von Hippel–Lindau syndrome, a familial cancer syndrome, is associated with high levels of HIF-1α (Maynard *et al.*, 2004).

Loss of function mutations in tumour suppressor genes or activating mutations in onco-genes have been shown to deregulate growth factor signal transduction pathways such as epidermal growth factor, insulin-like growth factor, insulin and interleukin 1 and their cognate tyrosine kinase receptors. Increased signalling *via* tyrosine kinase receptors activates the mitogen-activated kinase kinase (MEK) and phosphinositol-3-kinase (PI3K) pathways. This in turn activates AKT (protein kinase B) and among others effects, induces HIF-1α expression in tumours (Fukuda *et al.*, 2002; Treins *et al.*, 2002; Shi *et al.*, 2004; Skinner *et al.*, 2004). Loss of PTEN (phosphatase and tensin homologue deleted on chromosome 10) results in increased HIF-1α expression (Wu *et al.*, 1998). PTEN is a tumour suppressor and antag-onizes the PI3K pathway. Loss of PTEN function in prostate cancer cells correlates with increased angiogenesis and appears to be critical for progression to hormone refractory dis-ease (McMenamin *et al.*, 1999).

Increased activity of the HER2 receptor tyrosine kinase, a member of the epidermal growth factor receptor family, which occurs in one-third of breast tumours, is associated with increased tumour grade, chemotherapy resistance, and decreased patient survival. HER2's tyrosine kinase activity occurs in the absence of any known ligand. It heterodimerises with HER3 and HER4 and then binds the ligand heregulin (Tzahar *et al.*, 1994). In breast cancer cells heregulin activates AKT and upregulates the PI3K pathway (Liu *et al.*, 1999). However, the AKT/PI3 pathway needs FRAP (FKBP-rapamycin associated protein) kinase activity to achieve induced increased expression of HIF-1α (Zhong *et al.*, 2000).

HIF-1α over-expression is associated with HER2 and VEGF expression and with microvas-cular density in human ductal carcinoma *in situ* and invasive breast cancer and has been shown to be an independent prognostic indicator in lymph node negative patients (Bos *et al.*, 2001, 2003). HER2 signalling increases HIF-1 activity in non-hypoxic breast cancer cells, not by ubiquination and proteasomal degradation, but by increasing HIF-1α protein synthe-sis (Laughner *et al.*, 2001). Hypoxia increases the stability of HIF-1α protein and its specific transcriptional activity, such that a combination of HER2 over-expression and hypoxia has a synergistic effect on VEGF gene expression.

13.5 How Tumours Overcome Hypoxia with HIF-1

Biological pathways that are regulated by HIF-1α include angiogenesis, glycolysis, apoptosis and cell cycle. Some of these pathways, i.e. angiogenesis, decreased apoptosis, increased cell cycling, loss of cell cycle arrest etc, lead to a more aggressive cancer phenotype. This can render the cells more resistant to the cytotoxic effects of chemoradiotherapy.

Hypoxia results in nuclear accumulation of HIF-1α and allows dimerisation with HIF-1β. HIF-1 recognises HREs (hypoxia-responsive elements) in target genes such as VEGF and will recruit p300/cbp *via* the C-terminal of HIF-1α. HIF-1 plays a critical role in angiogenesis through activation of pro-angiogenic agents (i.e. VEGF and thymidine phosphorylase). Moreover, HIF-1 also targets the gene encoding inducible nitric oxide synthase (iNOS), which influences tumour vascularisation. In the absence of new vessel formation solid tumours are incapable of growing beyond a 2–3 mm^3 owing to the limited supply of oxygen, glucose and other nutrients (Folkmann and D'Amore, 1996). In response to hypoxia tumour cells and host cells within the tumour express VEGF *via* gene transcription. This production of VEGF in turn initiates and sustains new blood vessel development. This angiogenesis is essential for cellular proliferation and tumour development.

Tumour cells characteristically display a high glycolytic rate, even when growing in the presence of oxygen. Under normoxic conditions, cells generate ATP *via* oxidative phosphorylation. However in an expanding tumour mass, characterised by low levels of oxygen and a high glucose consumption rate, anaerobic glycolysis can become the predominant pathway of ATP generation (Vaupel *et al.*, 1989). This metabolic shift appears to be regulated by HIF-1. Glycolytic enzymes including aldolase A, phosphoglycerate kinase 1 and pyruvate kinase M are induced by HIF-1 *in vitro*, and lactate dehydrogenase is induced in breast carcinoma lines (Semenza *et al.*, 1996). The efficiency of the glycolytic response is enhanced by overexpression of other proteins, including glucose transporters i.e. GLUT1 which facilitates glucose uptake by the cells, and hexokinase, which then converts the intra-cellular glucose into glucose-6-phosphate; a 'priming' step of glycolysis (Semenza *et al.*, 1994). HIF-1-induced adaptive responses ensures that the energy requirements of the cell are met; thereby allowing their survival in a hostile environment (Vaupel, 2004).

Chemokines are small proteins that stimulate and attract leukocytes to sites of inflammation (Moser and Loetscher, 2001). Stromal derived factor 1 (SDF-1) or CXCL12, is a chemokine shown to enhance breast cancer growth, migration and invasion (Hall and Korach, 2003; Muller *et al.*, 2001). CXCR4, the receptor for SDF-1/CXCL12 plays an important role in lymphocyte trafficking (Gupta *et al.*, 1998). Breast and other cancer cells express the receptor CXCR4, and frequently metastasise to organs that produce its ligand SDF-1/CXCL12 (Muller *et al.*, 2001; Samara *et al.*, 2004). SDF-1 enhances ischaemic vasculogenesis and angiogenesis *via* increased AKTt endothelial nitric oxide synthase (eNOS) activity in a murine model of ischaemia (Hiasa *et al.*, 2004). HIF-1α had not been known to directly regulate the expression of members of the chemokine family, such as SDF1. Ceradini *et al.* (2004), however, established that the recruitment of CXCR4 positive progenitor cells to regenerating tissues is mediated by hypoxic gradients *via* HIF-1 induced expression of SDF1. The tumour microenvironment may continuously recruit circulating stem and progenitor cells, effectively hijacking the body's capacity for tissue regeneration to support tumour growth. Ceradini *et al.* (2004) implied that efforts to decrease tumour vascularity (anti-angiogenic approaches) may be counter-productive because they increase tumour hypoxia, thereby potentially enhancing the recruitment of circulating stem and progenitor cells and enlisting host mechanisms for survival and growth.

Cancer cells are characterised by uncontrolled growth. The first barrier to cell proliferation is senescence. Telomeres from normal cells undergo shortening after repeated rounds of DNA replication. This shortening continues until the cells reach a second proliferative block known as end-to-end fusions, and cell senescence. To become immortal a cancer cell must overcome this. Telomerase plays an important role in cell growth, cell immortalisation and cancer progression (Holt *et al.*, 1996). This can be achieved by telomere DNA stabilisation through telomerase activation where telomeres are added back on to the chromosome allowing the cell to continue proliferating (Stewart and Weinberg, 2000). Hypoxia extends the lifespan of vascular smooth muscle cells through telomerase activation (Minamino *et al.*, 2001). In hypoxia, HIF-1 induces hTERT (human telomerase catalytic sub-unit) by binding to the hTERT promoter and hence increases telomerase expression (Nishi *et al.*, 2004). Nishi *et al.* (2004) proposed two potential mechanisms whereby hypoxia induces telomerase activity. Firstly hypoxia can induce a DNA-damage response by causing telomere damage. HIF-1α may then induce telomerase in order to heal the damaged chromosome ends. Alternatively, the hypoxic induction of telomerase can trigger an anti-apoptotic response (Zhang *et al.*, 2003).

Apoptosis is regulated by a balance of pro- and anti-apoptotic proteins. Cells undergoing environmental stress can undergo apoptosis, which is mediated by the caspase proteins. Anti-apoptotic proteins include Bcl2 and Bcl-xL and pro-apoptotic proteins include Bax, Bad, Bak and bid. HIF-1α can induce apoptosis *via* BNIP3 (BCLA/adenovirus E1B 19 kDa interacting protein 3). The BNIP3 promoter contains an HRE so that HIF-1 can induce expression of the gene (Kothari *et al.*, 2003). BNIP3, and its human homolog BNIP3α, bind to and inhibit the anti-apoptotic proteins Bcl-2 and Bcl-xL in breast cancer cells (Yasuda *et al.*, 1999). BNIP3 may represent an important protein for the elimination of damaged cells by triggering rapid cell death (Greijer and Van der Wall, 2004). HIF-1 may also have an anti-apoptotic function, which may in part explain why tumours with a high HIF-1 expression are often chemoradio-resistant tumours (Unruh *et al.*, 2003). There is also conflicting evidence between HIF-1 and the anti-apoptotic protein Bcl2. In breast carcinoma there is a strong positive correlation between tumours expressing both HIF-1 and Bcl2, and a worse prognosis (Costa *et al.*, 2001). Pidgeon *et al.* (2001) showed that in human mammary adenocarcinoma cells that VEGF up-regulates Bcl2 and down-regulates tumour cell apoptosis. There is a HRE in the promoter of the VEGF gene which is a target of HIF-1. However, in a study of non-small cell carcinoma patients HIF 1 expression showing a significant inverse association with Bcl2 expression (Giatromanolaki *et al.*, 2001). Increased Bcl2 expression in lung carcinoma is correlated with a better survival (Koukourakis *et al.*, 1997). Greijer and colleagues propose that these differences may be explained by tissue specific regulation of hypoxia induced apoptosis (Greijer and Van der Wall, 2004).

In 1953 Gray *et al.* established that radiation resistance is conferred by hypoxia (Gray *et al.*, 1953). The presence of low oxygen partial pressure (pO_2) levels can protect cells from ionising radiation, and it has been shown that approximately two to three times higher radiation dose is necessary to kill hypoxic cells versus well-oxygenated cells (Gray *et al.*, 1994). Also a number of factors associated either directly or indirectly with tumour hypoxia have been shown to contribute to resistance to standard chemotherapy (Shannon *et al.*, 2003). For example, many drugs are dependent on cellular oxygenation for maximal efficacy, i.e. bleomycin, actinomycin D, vincristine (Teicher *et al.*, 1981). Hypoxia causes cells to cycle more slowly, rendering them less sensitive to cytotoxic drugs that preferentially kill rapidly proliferating cells (Amellem and Peterson, 1991). DNA-damaging chemotherapeutic agents such as platinum compounds may have compromised function due to increased activity of DNA repair enzymes under hypoxic conditions (Walker *et al.*, 1994). Hypoxia increases interstitial fluid pressure, which interferes with drug delivery and distribution. In addition hypoxia compromises the cytotoxic functions of immune cells that infiltrate a tumour (Jain, 2005).

13.6 HIF-1 Therapeutics

Differences between hypoxic cancer cells and normal cells give researchers a basis upon which to design rational drugs. As HIF-1α plays a major role in activating gene transcription specifically for maintaining homeostasis under hypoxic conditions, it is an obvious target for the development of novel cancer therapeutics.

13.6.1 Gene therapy

Once HIF-1α has been stabilised in response to hypoxia it must translocate to the nucleus and dimerise with HIF-1β to form HIF-1. HIF-1 then binds to HRE sequences and activates hypoxia-inducible genes. Chun and colleagues developed two HIF-1α variant cDNAs that compete with HIF-1α for dimerisation with HIF-1β, hence inhibiting the hypoxic activation of HIF-1 and reducing the mRNA expression of HIF-1 targeted genes (Chun *et al.*, 2001a, 2002; Yeo *et al.*, 2004). The inhibitory PAS (Per-ARNT-Sim) domain protein (IPAS), whilst not a HIF-1α variant, may also inhibit HIF1 dimerisation. It has a similar structure to HIF-1α and the two aforementioned DNA variants. IPAS functions as a dominant-negative regulator of HIF-1 mediated gene induction. In hepatoma cells IPAS is expressed and associated with impaired induction of hypoxia inducible genes regulated by HIF-1, resulting in retarded tumour growth and low vascular density *in vivo*. To date these anti HIF-1 approaches have not demonstrated high therapeutic effect in patients. Advanced gene transfers techniques utilising these antisense plasmids and dominant negative isoforms may provide future anticancer strategies (Yeo *et al.*, 2004). RNA interference is the process of sequence specific post-translational gene silencing (Hunter *et al.*, 1975). Small interfering RNAs exhibit gene specific RNA inhibition, thereby increasing their suppressive effects and avoiding nonspecific gene silencing (Miyagashi *et al.*, 2003). In a recent cell model studying pancreatic and hepatobiliary carcinomas, the gene transfer of HIF-1α siRNA expression vectors reduced HIF-1α mRNA levels, VEGF, GLUT-1 and aldolase A expression. Their use also significantly inhibited cancer cell colony formation *in vitro* and the development of solid tumours *in vivo* (Mizuno *et al.*, 2006).

13.6.2 mTOR/rapamycin

Growth factors bind to their receptor and activate receptor tyrosine kinases, which in turn activate the PI3K/AKT/mTOR pathway. In prostate cancer cells, growth factor-mediated activation of PI3K leads to an increased expression of HIF-1α (Zhong *et al.*, 2000). Loss of PTEN function can increase stabilisation of HIF-1α *via* activation of PI3K signalling and oncogenic Akt. Loss of PTEN can sensitise cells to mTOR inhibition and clinical trials are now testing whether rapamycin (an mTOR inhibitor) is useful in PTEN-null cancers (Huang and Houghton, 2002). The mechanism leading to mTOR-dependent elevated HIF-1α activity remains unclear. In a murine prostate cancer model elevated HIF-1α activity was reversed with mTOR inhibition (Majumder *et al.*, 2004). Majumder *et al.* (2004) also demonstrated that response to mTOR inhibition was mediated through independent apoptotic and HIF-1α pathways with transgenic AKT and Bcl2 mice developing prostatic intraepithelial neoplasia (PIN) resistant to rapamycin-induced apoptosis. Both PTEN loss and BCL2 over-expression have been linked to increased grade and progression of prostatic cancer (McMenamin *et al.*, 1999; Baltaci *et al.*, 2000). Therefore it is suggested that mTOR inhibition alone may be less effective in advanced cancer characterised by BCL2 overexpression, and that it should perhaps be used in combination with BCL2 inhibitors (Baltaci *et al.*, 2000).

13.6.3 2-methoxyoestradiol (2ME2)

Head and neck squamous cancers over-express HIF-1α. 2ME2 is a natural compound derived from oestradiol, with HIF-1α inhibitory activity that is currently being evaluated in phase I and II clinical trials for advanced solid tumours and multiple myeloma. Usage of 2ME2 on head and neck carcinoma lines results in decreased nuclear HIF 1α binding activity and affected the downstream transcriptional activation of genes, i.e. bid upregulation (pro-apoptotic bcl2 family member) and VEGF down-regulation (Ricker *et al.*, 2004). 2ME2 inhibits tumour growth and angiogenesis at concentration that disrupt tumour microtubules *in vivo*. Inhibition of HIF-1 occurs downstream of the 2ME2/tubulin interaction, as disruption of interphase tumour microtubules for HIF-1α down-regulation (Mabjeesh *et al.*, 2003). It has also been demonstrated that 2-methoxyestradiol has anti-angiogenic and anti-cancer effects on xenoplanted human breast tumours and melanoma in immune deficient mice (Klauber *et al.*, 1997; Ireson *et al.*, 2004).

13.6.4 HSP90/geldanamycin

HSP90 (heat-shock protein-90) is involved in the folding of HIF 1α and many other proteins (Minet *et al.*, 2002). HIF 1α can be destabilised when HSP90 binding is inhibited with geldanamycin in prostate cancer cells. Geldanamycin induces degradation of HIF-1α *via* the proteosome pathway, in a dose and time dependent manner under both normoxia and hypoxia. This is accompanied by a respective inhibition of HIF-1α functional transcription activity of VEGF.

13.6.5 YC-1

The synthetic compound YC-1 (3′(5′hydroxymethyl-2′-furyl)-1-benzyl indazole) has inhibitory effects on platelet aggregation and smooth muscle cell proliferation (Tulis *et al.*, 2002), vascular endothelial cell (Hsu *et al.*, 2003) and hepatocellular carcinoma cell proliferation (Wang *et al.*, 2004). Although its effects are independent of nitric oxide (NO), it mimics some of the biological actions of NO. Effects were thought to be solely linked to the activation of soluble guanyl cyclase (sGC) and the elevation of cyclic guanine monophosphate (cGMP) (Ko *et al.*, 1994). Soluble guanyl-cyclase is a heterodimeric enzyme needed for NO activity. YC-1 diminishes the hypoxic induction of Epo and VEGF mRNA, and inhibits the DNA binding activity of HIF-1. However, sGC inhibitors failed to block the actions of YC-1. YC-1 suppressed hypoxic accumulation of HIF-1α, but not its mRNA level. This suggests that the HIF-1 inhibitory effect of YC-1 is unlikely to be related to sGC/cGMP signal transduction but to post-translational inhibition of HIF-1α accumulation (Chun *et al.*, 2001). Recent *in vitro* work on prostatic cancer cell lines has shown that YC-1 suppresses PI3K/Akt/mTOR/4E-BP pathways, which regulate HIF-1α at the translational level (Sun *et al.*, 2007). Tumours from intraperitoneal YC-1-treated immunodeficient mice demonstrated significantly smaller tumours, lower levels of HIF-1α, less vascularisation and lower levels of HIF-1 inducible genes: i.e. VEGF, enolase, aldolase (Yeo *et al.*, 2003). The

YC-1 treatment was also well tolerated and has been proposed as a new anti-angiogenic treatment (Yeo *et al.*, 2004).

13.6.6 Topoisomerase 1 inhibition and MSC

Irinotecan and topotecan are both inhibitors of topoisomerase 1. Topoisomerase 1 is an enzyme that produces reversible single strand breaks in DNA during DNA replication. These single strand breaks relieve torsional strain and allow DNA replication to proceed.

Topotecan is a semisynthetic, water soluble derivative of camptothecin. Topotecan binds to the topoisomerase 1-DNA complex and prevents religation of the DNA strand, resulting in double strand DNA breakage and cell death (USP DI, 2000). Unlike irinotecan, topotecan is found predominantly in the inactive carboxylate form at neutral pH and it is not a pro-drug (Cersosimo, 1998). As a result, topotecan has different antitumour and toxicity profiles than irinotecan. *In vitro* topotecan blocks HIF-1α and VEGF induced by insulin-like growth factor-1 in neuroblastoma cells (Bernier-Chastagner *et al.*, 2005). For this reason topotecan was proposed as a radiosensitiser in advanced brain tumours but phase II trials have been disappointing in both the adult and paediatric populations (Bernier-Chastagner *et al.*, 2005; Chintagumpala *et al.*, 2005; Pipas *et al.*, 2005). Multicentre trials have demonstrated the utility of topotecan in recurrent ovarian and lung cancer (Michener and Belinson, 2005; Huber *et al.*, 2006). New studies would also suggest that its use with thalidomide shows promise in patients with high risk myelodysplastic syndromes (Raza *et al.*, 2006).

The active metabolite of irinotecan, SN38, selectively inhibits endothelial cell proliferation and decreases the HIF-1α and VEGF expression of glioma cells in a dose-dependent and time-dependent manner under both hypoxic and normoxic conditions *in vitro* (Kamiyama *et al.*, 2004). It has dual angiosuppressive actions, including both the inhibition of endothelial proliferation and tube formation, and the inhibition of the angiogenic cascade in glioma cells. Clinical trials have, however, showed conflicting evidence of the drugs efficacy in recurrent malignant glioma (Prados *et al.*, 2006; Reardon *et al.*, 2005). New work suggests that Se-methylselenocysteine (MSC) is a selective modulator of the antitumour activity and selectivity of irinotecan. In an *in vivo* model of squamous carcinoma of the head & neck, curative rate increased from 30 to 100 % when MSC was given in conjunction with irinotecan. Observed therapeutic synergy correlated with inhibition of neoangiogenesis through down-regulation of HIF-1α, iNOS and cyclo-oxygenase-2 (Yin *et al.*, 2006).

13.6.7 Increasing tumour oxygenation using Epo

Erythropoietin is a 30.4-kDa glycoprotein hormone that is synthesised and secreted primarily by interstitial renal fibroblasts in response to hypoxic stimulus by HIF-1 (Krantz, 1991). Epo plays a vital role in the regulation of erythropoiesis and is responsible for the maturation of erythroid progenitor cells in the bone marrow that are converted to red blood cell precursors, and subsequently to erythrocytes (Jelkmann, 1992). Recombinant human Epo (rHuEpo) is an important clinical therapeutic agent in the treatment of cancer or renal failure patients suffering from anaemia (Erslev, 2000).

There is evidence to support a direct relationship correlating lower blood haemoglobin and a higher level of hypoxia (Becker *et al.*, 2000). Degner and colleagues showed that increasing blood haemoglobin concentration has a strong influence on hypoxic tissue volume (Degner and Sutherland, 1988). The rationale therefore exists that as the chemotherapy response of some tumours is dependent on the oxygenation status of the tumour tissue, increasing low haemoglobin concentrations using Epo will improve therapy response. rHuEpo has been shown to improve tumour radio-chemosensitivity following the correction of anaemia (Silver and Piver, 1999; Thews *et al.*, 1998, 2001). The investigators hypothesized that these results reflected improved oxygenation of the hypoxic tumour tissue, as a result of the increase in oxygen availability due to Epo. Preventing tumour hypoxia in this manner increases the therapeutic index of most chemotherapy drugs but also has the added benefit of slowing down the process of hypoxia-induced malignant progression. The implication of this is that chemotherapy regimens can be potentially scaled down from 'maximum tolerated dose' when combined with Epo treatment, to the benefit of cancer patients. Several cancer cell lines do not have functional erythropoietin receptors. Concerns about propagating tumour growth with erythropoietin are reasonable but *in vitro* evidence would suggest this is dependent on the individual cell type and its associated receptors (Laugsch *et al.*, 2008).

13.7 Conclusion

Scientific evidence highlights the importance of hypoxia in tumour progression and resistance to current therapeutic regimes. The development of cancer is a complex multistep process with external carcinogenic influences combined with sporadic or inherited genetic defects contriving to disturb the normal pattern of cellular homeostasis. Cancer cells adapt to hypoxia by producing numerous angiogenic factors such as VEGF, thymidine phosphorylase (TP) that initiate and sustain angiogenesis. HIF-1α is thought to play a critical role in this process. Studies in patients with advanced cancers of the uterine cervix showed that patients with low pO_2 tumours (median pO_2 < 10 mmHg) exhibited a higher rate of malignant progression prior to treatment and significantly worse disease-free survival following surgery or radiotherapy, compared to patients with well oxygenated tumours (Hockel *et al.*, 1996). In patients with oropharyngeal tumours expressing high levels of HIF-1α, radiotherapy was less likely to result in remission than patients with tumours expressing low levels of HIF-1α (Aebersold *et al.*, 2001) and HIF-1α overexpression is also associated with an increased risk of mortality in early-stage cervical cancer (Birner *et al.*, 2000).

Several papers have reported that over-expression of HIF-1α may correlate with tumours that are poor responders to established chemoradiotherapy regimes. This represents an area of potential focused therapeutic intervention. Anti-HIF 1 agents have been used in conjunction with established regimes (Ricker *et al.*, 2004). Irinotecan, an anti-HIF-1α and anti-angiogenic agent, has recently shown some encouraging results for recurrent malignant glioma, when used in conjunction with celecoxib in phase II trials (Reardon *et al.*, 2005). However, evidence for paradoxical effects of HIF-1 on tumour growth, such as growth acceleration of HIF-1$\alpha^{-/-}$ ES-(oestradiol) derived tumours has been observed (Carmeliet *et al.*, 1998). Cell type, oncogenic mutations and microenvironment will undoubtedly influence HIF-1 behaviour. As more light is shown on the activation-stabilisation process of HIF-1 i.e. the recent discovery of SENP1, the greater the chance to develop more efficacious therapeutic modalities. It is an area that demands further study, particularly for non-responders to established chemoradiotherapeutic regimes.

References

Aebersold DM, Burri P, Beer KT, Laissue J, Djoniv V, Greiner RH, Semenza GL. Expression of hypoxia inducible factor 1alpha: a novel predicitive and prognostic parameter in the radiotherapy of oroopharyngeal cancer. *Cancer Res* 2001, **61**(7), 2911–16.

Amellem O, Peterson EO. Cell inactivation and cell cycle inhibition as induced by extreme hypoxia: the role of cell cycle arrest as a protection against hypoxia-induced lethal damage. *Cell Prolif* 1991, **244**, 127–141.

Baltaci S, Orhan D, Ozer G, Tolunay O, Gogous O. Bcl-2 proto-oncogene expression in low and high grade prostatic intra-epithelial neoplasia. *BJU Int* 2000, **85**, 155–59.

Becker A, Stadler P, Lavey RS, *et al.* Severe anaemia is associated with poor tumour oxygenation in head and neck squamous carcinomas. *Int J Radiat Oncol Biol Phys* 2000, **46**, 459–66.

Bernier-Chastagner V, Grill J, Doz F, *et al.* Topotecan as a radiosensitiser in the treatment of children with malignant diffuse brain-stem glioms: results of a French Society of Pediatric Oncology Phase II study. *Cancer* 2005, **104**, 2792–97.

Berta M, Brahimini-Horn C, Pouyssegur J. Regulation of the Hypoxia Inducible Factor 1 alpha: A breath of fresh air in hypoxia research. *J Soc Biol* 2004, **198**(2), 113–20.

Birner P, Schindl M, Obermair A, Plank C, Breitenecker G, Oberhuber G. Overexpression of hypoxia inducible factor 1alpha is a marker for an unfavourable prognosis in early stage invasive cervical cancer. *Cancer Res* 2000, **60**(17), 4693–96.

Bos R, van der Groep P, Greijer AE, *et al.* Levels of hypoxia inducible factor 1 alpha independently predict prognosis in patients with lymph node negative breast carcinoma. *Cancer* 2003, **97**(6), 1573–81.

Bos R, Zhong C, Hanrahan F, *et al.* Levels of hypoxia inducible factor 1 during breast carcinogenesis. *J Natl Cancer Inst* 2001, **93**(4), 309–14.

Bradbury J. Breathing hard to keep up with HIF-1. *Lancet* 2001, **358**, 9294.

Carmeliet P, Dor Y. Herbert JM, *et al.* Role of HIF 1 alpha in hypoxia mediated apoptosis, cell proliferation and tumour angiogenesis. *Nature* 1998, **394**(6692), 485–90.

Ceradini DJ, Kulkarni AR, Callaghan MJ, *et al.* Progenitor cell trafficking is regulated by hypoxic gradients through HIF-1 induction of SDF-1. *Nat Med* 2004, **10**(8), 858–64.

Cersosimo RJ. Topotecan: a new topoisomerase 1 inhibiting anti-neoplastic agent. *Ann Pharmacother* 1998, **32**, 1334–43.

Cheng J, Kang X, Zhang S, Yeh ET. SUMO specific protease-1 is essential for stabilisation of HIF-1α during hypoxia. *Cell* 2007, **131**(3), 584–95.

Chintagumpala MM, Friedman HS, Stewart CF, *et al.* A phase II window trial of procarbazine and topotecan in children with high-grade glioma: a report from the childrens oncology group. *J Neurooncol* 2005, **Nov 29**, 1–6.

Chun YS, Choi E, Yeo EJ, Lee JH, Kin MS, Park JW. A new HIF-1 variant induced by zinc ion suppresses HIF-1 mediated hypoxic responses. *J Cell Sci* 2001a, **114**(22), 4051–61.

Chun YS, Yeo EJ, Choi E, *et al.* Inhibitory effect of YC-1 on the hypoxic induction of erythropoietin and vascular endothelial growth factor in Hep3B cells. *Biochem Pharmacol* 2001b, **61**, 947–54.

Chun YS, Choi E, Kim TY, Kim MS, Park JW. A dominant-negative isoform lacking exons 11 and 12 of the human hypoxia inducible factor 1 alpha gene. *Biochem J* 2002, **362**(1), 71–9.

Costa A, Coradini D, Carassi A, Erdas R, Sardella A, Daidone MG. Re: Levels of hypoxia inducible factor 1 alpha during breast carcinogenesis. *J Natl Cancer Inst* 2001, **93**, 1175–77.

Degner FL, Sutherland RM. Mathematical modelling of oxygen supply and oxygenation in tumour tissues: prognostic, therapeutic, and experimental implications. *Int J Radiat Oncol Biol Phys* 1988, **15**, 391–97.

Erslev AJ. Erythropoeitin and anemia of cancer. *Eur J Haematol* 2000; **64**, 353–58.

Folkmann J, D'Amore PA. Blood vessel formation: what is its molecular basis? *Cell* 1996, **87**(7), 1153–55.

Freedman SJ, Sun ZY, Kung AL, France DS, Wagner G, Eck MJ. Structural basis for the negative regulation of hypoxia inducible factor-1 alpha by CITED2. *Nat Struct Biol* 2003, **10**(7), 504–12.

Fukuda R, Hirota K, Fan F, Jung YD, Ellis LM, Semenza GL. Insulin-like growth factor 1 and induces hypoxia-inducible factor 1 mediated vascular endothelial growth factor expression, which is dependent on MAP kinase and phosphatidyl 3 kinase signalling in colon cancer cells. *J Biol Chem* 2002, **277**(41), 38205–11.

Galarraga J, Loreck DJ, Grahan JF, DeLaPaz RL, Smith BH, Hallgren D, Cummins CJ. Glucose metabolism in human gliomas – correspondence of *in situ* and *in vitro* metabolic rates and altered energy metabolism. *Metab Brain Dis* 1986, **1**(4), 279–91.

Giatromanolaki A, Koukourakis MI, Sivridis E, Turley H, Tlaks K, Pezella F, Gatter KC, Harris AL. Relation of hypoxia inducible factor 1 alpha and 2 alpha in non-operable non-small cell lung cancer to angiogenic/molecular profile of tumours and survival. *Br J Cancer* 2001, **85**(6), 881–90.

Graeber TG, Peterson JF, Tsai M, Minoca K, Fornace AJ Jnr, Giaccia AJA. Hypoxia induces accumulation of p53 protein, but activation of a G1-phase checkpoint by low oxygen conditions is independent of p53 status. *Mol Cell Biol* 1994, **14**, 6263–77.

Gray LH, Conger AD, Evert M, *et al.* The concentration of oxygen in tissues at the time of irradiation as a factor in radiotherapy. *Br J Radiol* 1953, **26**, 638–48.

Gray LH, Cygler J, Klassen NV, *et al.* The survival of aerobic and anoxic human glioma and melanoma cells after irradiation at ultra-high and clinical doses rates. *Radiat Res* 1994, **140**, 79–84.

Greijer AE, van der Wall E. The role of hypoxia inducible factor 1 in hypoxia induced apoptosis. *J Clin Pathol* 2004, **57**(10), 1009–14.

Gupta SK, Lysko PG, Pillariesetti K, Ohlstein E, Stadel JM. Chemokine receptors in human endothelial cells. Functional expression of CXCR4 and its transcriptional regulation by inflammatory cytokines. *J Biol Chem* 1998, **273**(7), 4282–87.

Hall JM, Korach KS. Stromal derived growth factor 1, a novel target of estrogen receptor action, mediates the mitogenic effects of estradiol in ovarian and breast cancer cell lines. *Mol Endocrinol* 2003, **17**(5), 792–803.

Hanahan D, Folkman J. Patterns and emerging mechanisms of the angiogenic switch during tumorigenesis. *Cell* 1996, **86**, 353–64.

Hewitson KS, McNeill LA, Riordab, MV, *et al.* Hypoxia inducible factor (HIF) asparagines hydroxylase is identical to factor inhibiting HIF (FIH) and is related to the cupin structural family. *J Biol Chem* 2002, **277**(29), 26351–55.

Hiasa K, Ishibashi M, Ohtani K, *et al.* Gene transfer of stromal cell-derived factor-1 alpha enhances ischaemic vasculogenesis and angiogenesis via vascular endothelial growth factor/endothelial nitric oxide synthase-related pathway: next generation chemokine therapy for therapeutic neovascularisation. *Circulation* 2004, **109**(25), 2454–61.

Hockel M, Schlenger K, Aral B, Mitze M, Schafer U, Vaupel P. Association between tumour hypoxia and malignant progression in advanced cancer of the uterine cervix. *Cancer Res* 1996, **56**(19), 4509–15.

Hockel M, Vaupel P. Tumour hypoxia: definitions & current clinical, biologic and molecular aspects. *J Natl Cancer Inst* 2001, **93**(4), 266–76.

Holt SE, Shay JW, Wright WE. Refining the telomere-telomerase hypothesis of aging and cancer. *Nat Biotechnol* 1996, **14**(7), 836–39.

Hsu HK, Juan SH, Ho PY, Liang YC, Lin CH, Teng CM, Lee WS. YC-1 inhibits the proliferation of human vascular endothelial cells through a cyclic GMP-independent pathway. *Biochem Pharmacol* 2003, **66**(2), 263–71.

Huang LE, Gu J, Schau M, Bunn HF. Regulation of hypoxia inducible factor 1 alpha is mediated by an O_2 dependent domain via the ubiquitin-proteasome pathway. *Proc Natl Acad Sci U S A* 1998, **95**(14), 7987–92.

Huang S, Houghton PJ. Inhibitors of mammalian target of rapamycin as novel anti-tumour agents: from bench to clinic. *Curr Opin Invest Drugs* 2002, **3**, 295–304.

Huber RM, Reck M, Gosse H, VonPawel J, Mezger J, Saal JG, Steppert C, Steppling H. Efficacy of toxicity-adjusted topotecan therapy in recurrent small-cell lung cancer. *Eur Respir J* 2006, **6**, 1183–9

Hunter T, Hunt T, Jackson RJ, Robertson HD. The characteristics of inhibition of protein synthesis by double stranded ribonucleic acid in reticulocyte lysates. *J Biol Chem* 1975, **250**, 409–17.

Ireson CR, Chander SK, Purohit A, *et al*. Pharmacokinetics and efficacy of 2-methoxyoestradiol and 2-methoxyestradiol-bis-sulphunate *in vivo* in rodents. *Br J Cancer* 2004, **90**(4), 932–7.

Jain RK. Normalisation of tumour vasculature: An emerging concept in anti-angiogenic therapy. *Science* 2005, **307**, 58–62.

Jelkmann W. Erythropoietin: structure, control of production and function. *Physiol Rev* 1992, **72**, 449–89.

Jeong JW, Bae MK, Ahn MY, *et al*. Regulation and destabilisation of HIF-1 alpha by ARD-1 mediated acetylation. *Cell* 2002, **111**, 709–20.

Jiang BH, Rue E, Wang GL, Roe R, Semenza GL. Dimerisation, DNA binding and transactivation properties of hypoxia inducible factor 1. *J Biol Chem* 1996, **271**, 17771–78.

Jiang BH, Zheng JZ, Leung SW, Roe R, Semenza GL. Transactivation and inhibitory domains of hypoxia inducible factor 1 alpha. Modulation of transcriptional activity by oxygen tension *J Biol Chem* 1997, **272**(31), 19253–60.

Kamiyama H, Takano S, Tsuboi K, Matsumura A. Anti-angiogenic effects of SN38(active metabolite of irinotecan): inhibition of hypoxia inducible factor 1 alpha (HIF 1 alpha)/vascular endothelial growth factor (VEGF) expression of glioma and growth of endothelial cells. *J Cancer Res Clin Oncol* 2005, **131**(4), 205–13.

Kelleher DK, Mattheinsen U, Thews O, Vaupel P. Blood flow oxygenation, and bioenergic status of tumours after erythropoietin treatment in normal and anaemic rats. *Cancer Res* 1996, **56**(20), 4728–34.

Klauber N, Parangi S, Flynn E, Hamel E, D'amato RJ. Inhibition of angiogenesis and breast cancer in mice by the microtubule inhibitors 2-methoxystrdaiol and taxol. *Cancer Res* 1997, **57**(1), 81–6.

Ko FN, Wu CC, Kuo SC, Lee FY, Teng CM. YC-1, a novel activator of platelet guanylate cyclase. *Blood* 1994, **84**, 4226–33.

Kothari s, Cizeau J, McMillan-Ward E, Israels SJ, Bailes M, Ens K, Kirshenbaum LA, Gibson SB. BNIP3 plays a role in hypoxic cell death in human epithelial cells that is inhibited by growth factors EGF and IGF. *Oncogene* 2003, **22**(30), 4734–44.

Koukourakis MI, Giatromanolaki A, O'Byrne KJ, Whitehouse RM, Talbot DC, Gatter KC, Harris AL. Potential role of bcl-2 as a suppressor of tumour angiogenesis in non-small cell cancer. *Int J Cancer* 1997, **74**(6), 565–70.

Krantz SB. Erythropoietin. *Blood* 1991, **77**, 419–34.

Lando D, Peet DJ, Gorman JJ, Whelan DA, Whitelaw ML, Bruick RK. FIH-1 is an asparginyl hyroxylase enzyme that regulates the transcriptional activity of hypoxia inducibe factor. *Genes Dev* 2002, **16**, 1466–71.

Laughner E, Taghavi P, Chiles K, Mahon PC, Semenza GL. Her2(neu) signalling increases the rate of hypoxia-inducible factor 1 alpha (HIF-1 alpha) synthesis: mechanisms for HIF-1 mediated vascular endothelial growth factor expression. *Mol Cell Biol* 2001, **21**(12), 3995–4004.

Laugsch M, Metzen E, Svensson T, Depping R. Lack of functional erythropoietin receptors in cancer cell lines. *Int J Cancer* 2008, **122**(5), 1005–11

Liu W, Li J, Roth RA. Heregulin regulation of Akt/protein kinase B in breast cancer cells. *Biochem Biophy Res Commun* 1999, **261**(3), 897–903.

Mabjeesh NJ, Escuin D, LaVallee TM, *et al*. 2ME2 inhibits tumour growth and angiogenesis by disrupting microtubules and dysregulating HIF. *Cancer Cell* 2003, **4**, 363–75.

Majumder PK, Febbo PG, Bikoff R, *et al*. mTOR inhibition reverses Akt-dependent prostate intra-epithelial neoplasia through regulation of apoptotic and HIF-1 dependent pathways. *Nat Med* 2004, **10**, 594–601.

Maynard MA, Ohh M. Von Hippel–Lindau tumour suppressor protein and hypoxia inducible factor in kidney cancer. *Am J Nephrol* 2004, **24**(1), 1–13.

Mazure NM, Brahimi-Horn MC, Berta MA. HIF-1: master and commander of the hypoxic world. A pharmacological approach to its regulation by siRNAs. *Biochem Pharm* 2004, **68**, 971–80.

McMenamin ME, Soung P, Perera S, Kaplan I, Loda M, Sellers WR. Loss of PTEN expression in paraffin-embedded primary prostate cancer correlates with high Gleason score and advanced stage. *Cancer Res* 1999, **59**(17), 4291–96.

Michener CM, Belinson JL. Modern management of recurrent ovarian carcinoma. A systematic approach to chronic disease. *Oncology* 2005, **10**, 1277–85.

Minamino T, Mitsialis SA, Kourembanas S. Hypoxia extends the lifespan of vascular smooth muscle cells through telomerase activation. *Mol Cell Biol* 2001, **21**(10), 3336–42.

Minet M, Mottet D, Michel G, *et al*. Hypoxia induced activation of HIF-1: role of HIF 1alpha-Hsp90 interaction. *FEBS Lett* 1999, **460**, 251–56.

Mabjeesh NJ, Post DE, Willard MT, Kaur B, Van Meir EG, Simons JW, Zhong H. Geldanamycin induces the degradation of hypoxia inducible factor 1 alpha protein via the proteasome pathway in prostate cancer cells. *Cancer Res* 2002, **62**(9), 2478–82.

Miyagashi M, Hayashi M, Taira K. Comparison of the suppressive effects of antisense oligonucleotides and siRNAs directed against the same targets in mammalian cells. *Antisense Nucleic Acid Drug Dev* 2003, **13**, 1–7.

Mizuno T, Nagao M, Yamada Y, Narikiyo M, Ueno M, Miyagashi M, Taira K, Nakajima Y. Small interfering RNA expression vector targeting hypoxia inducible factor – 1 alpha inhibits tumour growth in hepatobiliary and pancreatic cancers. *Cancer Gene Ther* 2006, **13**, 131–40.

Moser B, Loetscher P. Lymphocyte traffic control by chemokines. *Nat Immunol* 2001, **2**(2), 123–28.

Muller A, Homey B, Soto H, *et al*. Involvement of chemokine receptors in breast cancer metastasis. *Nature* 2001, **410**(6824), 50–6.

Nishi H, Nakada T, Kyo S, Inoue M, Shay JW, Isaka K. Hypoxia-inducible factor 1 mediates the upregulation of telomerase(hTERT). *Mol Cell Biol* 2004, **24**(13), 6076–83.

Pidgeon GP, Barr MP, Harmey JH, Foley DA, Bouchier-Hayes DJ. Vascular endothelial growth factor (VEGF) up-regulates Bcl-2 and inhibits apoptosis in human and murine mammary adenocarcinoma cells. *Br J Cancer* 2001, **85**(2), 272–78.

Pipas JM, Meyer LP, Rhodes CH, Cromwell LD, McDonnell CE, Kingman LS, Rigas JR, Fadul CE. A phase II trial of paclitaxel and topotecan with filgrastim in patients with recurrent or refractory glioblastoma multiforme or anaplastic astrocytomas. *J Neuro-oncol* 2005, **71**, 301–5.

Prados MD, Lamborn K, Yung WK, *et al*. A phase 2 trial of irinotecan in patients with recurrent malignant glioma; A North American Brain Tumour Consortium. *Neuro-oncol* 2006, **2**, 183–93.

Raza A, Lizak L, Billmeier J, Pervaiz H, Mumtaz M, Gohar S, Wahid K, Galili N. Phase II trial of topotecan and thalidomide in patients with high risk myelodysplastic syndromes. *Leuk Lymphoma* 2006, **47**, 433–40.

Reardon DA, Quinn JA, Rich JN, *et al*. Phase 1 trial of irinotecan plus temozolomide in adults with recurrent malignant glioma. *Cancer* 2005, **104**, 1478–86.

Reardon DA, Quinn JA, Vredenburgh J, *et al*. Phase 2 trial of irinotecan plus celecoxib in adults with recurrent malignant glioma. *Cancer* 2005, **103**(2), 329–38.

Richard DE, Berra E, Gothie E, Roux D, Pouyssegur J. p42/p44 mitogen activated protein kinases phosphorylate hypoxia inducible factor 1 alpha (HIF 1alpha) and enhance the transcriptional activity of HIF-1. *J Biol Chem* 1999, **274**(46), 32631–37.

Ricker JL, Chen Z, Yang XP, Pribluda VS, Swartz GM, Van Waes C. 2 methoxyestradiol inhibits hypoxia inducible factor 1 alpha, tumour growth, and angiogenesis and augments paclitaxel efficacy in head and neck squamous cell carcinoma. *Clin Cancer Res* 2004, **10**(24), 8665–73.

Samara GJ, Lawrence DM, Chiarelli CJ, Valentino MD, Lyubsky S, Zucker S, Vaday GG. CXCR4-mediated adhesion and MMP-9 secretion in head and neck squamous cell carcinoma. *Cancer Lett* 2004, **214**, 213–41.

Sang N, Steihl DP, Bohensky J, Leshchinsky I, Srinivas V, Caro J. MAPK signalling up-regulates the activity of hypoxia inducible factors by its effects on p300. *J Biol Chem* 2003, **278**(16), 14013–19.

Seeler JS, Dejean A. Nuclear and unclear functions of SUMO. *Nat Rev Mol Cell Biol* 2003, **4**(9), 690–99.

Semenza GL. HIF-1 and tumour progression: pathophysiology and therapeutics *Trends Mol Med* 2002, **8**(4 suppl), S62–7.

Semenza GL, Wang GL. A nuclear transcription factor induced by hypoxia via de novo protein synthesis binds to the human erythropoietin gene enhancer at a site required for transcriptional activation. *Mol Cell Biol* 1992, **12**(12), 5447–54.

Semenza GL, Roth PH, Fang HM, Wang GL. Transcriptional regulation of genes encoding glycolytic enzymes by hypoxia inducible factor 1. *J Biol Chem* 1994, **269**(38), 23757–63.

Semenza GL, Jiang BH, Leung SW, Passantino R, Concordet JP, Marie P, Giallongo A. Hypoxia response elements in the aldolase A, enolase 1, and lactate dehydrogenase A gene promoters contain essential binding sites for hypoxia inducible factor 1. *J Biol Chem* 1996, **271**(51), 32529–37.

Shannon AM, Bouchier-Hayes DJ, Condron CM, Toomey D. Tumour hypoxia, chemotherapeutic resistance and hypoxia related therapies. *Cancer Treat Rev* 2003, **29**, 297–307.

Shi YH, Wang YX, You JF, Heng WJ, Zhong H, Fang WG Activation of HIF 1 by bFGF in breast cancer: role of PI-3K and MEK1/ERK pathways. *Zhonghua Yi Xue Za Zi* 2004, **84**(22), 1899–903.

Silver DF, Piver MS. Effects of recombinant human erythropoietin on the anti-tumour effect of cisplatin in SCID mice bearing human ovarian cancer: A possible oxygen effect. *Gynecol Oncol* 1999, **73**, 280–84.

Skinner HD, Zheng JZ, Fang J, Agani F, Jiang BH. Vascular endothelial growth factor transcriptional activation is mediated by hypoxia-inducible factor 1α, HDM2, and p70S6K1 in response to phosphatidylinositol 3-kinase/AKT signalling. *J Biol Chem* 2004, **279**, 45643–51.

Sondergaard KL, Hilton DA, Penney M, Ollerenshaw M, Demaine AG. Expression of hypoxia-inducible factor-1α in tumors of patients with glioblastoma. *Neuropathol Appl Neurobiol* 2002, **28**, 210–17.

Steele RJC, Thompson AM, Hall PA, Lane DP. The p53 tumour suppressor protein. *BJS* 1998, **85**, 1460–67.

Stewart SA, Weinberg RA 2000. Telomerase and human tumorigenesis. *Semin Cancer Biol* 2000, **10**(6), 399–406.

Sun HL, Liu YN, Huang YT, *et al*. YC-1 inhibits HIF-1 expression in prostate cancer cells: contribution of Akt/NF-Kb signalling to HIF-1α. *Oncogene* 2007, **26**, 3941–51.

Suzuki H, Tomida A, Tsuruo T. Dephosphorylated hypoxia inducible factor 1 alpha as a mediator of p53-dependent apoptosis during hypoxia. *Oncogene* 2001, **20**(41), 5779–88.

Teicher BA. Physiologic mechanisms of therapeutic resistance. Blood flow and hypoxia. *Haematol Oncol Clin North Am* 1995, **9**(2), 475–506.

Teicher BA, Lazo JS, Sartorelli AC. Classification of anti-neoplastic agents by their selective toxicities toward oxygenated and hypoxic tumour cells. *Cancer Res* 1981, **41**, 73–81.

Thews O, Koenig R, Kelleher DK, Kutzner J, Vaupel P. Enhanced radiosensitivity in experimental tumours following erythropoietin treatment of chemotherapy-induced anaemia. *Br J Cancer* 1998, **78**, 752–56.

Thews O, Kelleher DK, Vaupel P. Erythropoietin restores the anaemia-induced reduction in cyclophosphamide cytotoxicity in rat tumours. *Cancer Res* 2001, **61**, 1358–61.

Tojo M Matsuzaki K, Minami T, *et al*. The aryl-hydrocarbon receptor nuclear transporter is modulated by the SUMO-1 conjugation system. *J Biol Chem* 2002, **277**(48), 46576–85.

Treins C, Giorgetti-Peraldi S, Murdaca J, Semenza GL, Van Obberghen E. Insulin stimulates hypoxia-inducible factor 1 through a phosphatidyl 3-kinase/target of rapamycin-dependent signalling pathway. *J Biol Chem* 2002, **277**(31), 27975–81.

Tulis DA, Bohl Masters KS, Lipke EA, *et al*. YC-1 mediated vascular protection though inhibition of smooth muscle cell proliferation and platelet function. *Biochem Biophys Res Commun* 2002, **291**, 1014–21.

Tzahar E, Levkowitz G, Kaunagaran D, *et al*. ErbB-3 and ErbB-4 function as the respective low and high affinity receptors of all Neu differentiation factor/heregulin isoforms. *J Biol Chem* 1994, **269**(40), 25226–33.

Unruh A, Ressel A, Mohamed HG, Johnson RS, Nadowitz R, Richter E, Katschinski DM, Wenger RH. The hypoxia inducible factor 1 alpha is a negative factor for tumour therapy. *Oncogene* 2003, **22**(21), 3213–20.

USP DI. Volume 1. Drug information for health professional. Topotecan. Update monographs. Englewood, Colorado, Micromedex, Inc, 2000.

Vaupel P. The role of hypoxia-induced factors in tumour progression. *Oncologist* 2004, **9**(Suppl 5), 1017.

Vaupel P, Kallinowski F, Okunieff P. Blood flow, oxygen and nutrient supply, and metabolic microenvironment of human tumours: a review. *Cancer Res* 1989, **49**(23), 6449–65.

Vaupel P, Thews O, Hockel M. Treatment resistance of solid tumours: role of hypoxia and anaemia. *Med Oncol* 2001, **18**(4), 243–59.

Vaupel P, Briest S, Hockel M. Hypoxia in breast cancer: pathogenesis, characterisation and biological/therapeutic implications. *Wien Med Wochenschr* 2002, **152**(13–14), 334–42.

Vaupel P, Mayer A, Briest S, Hockel M. Oxygenation gain factor: a novel parameter characterising the association between haemoglobin level and oxygenation status of breast cancers. *Cancer Res* 2003, **63**(22), 7634–37.

Vaupel P, Dunst J, Engert A, Fandey J, Freyer P, Freund M, Jelkmann W. Effects of recombinant human erythropoietin on tumour control in patients with cancer induced anaemia. *Onkologie* 2005, **28**(4), 216–21.

Walker LJ, Craig RB, Harris AL, Hickson ID. A role for the human DNA repair enzyme HAP1 in cellular protection against DNA damaging agents and hypoxic stress. *Nucl Acids Res* 1994, **22**, 4884–89.

Wang GL, Jiang BH, Rue EA, Semenza GL. Hypoxia inducible factor 1 is a basic loop helix loop PAS heterodimer regulated by cellular O2 tension. *Proc Natl Acad Sci U S A* 1995, **92**(12), 5510–14.

Wang SW, Pan SL, Guh JH, *et al.* YC-1 exhibits a novel anti-proliferative effect and arrests the cell cycle in G0-G1 in human hepatocellular carcinoma cells. *J Pharmacol Exp Ther* 2005, **312**(3): 917–25

Warburg O. On the origin of cancer cells. *Science* 1956, **123**(3191), 309–14.

Wood SM, Gleadle JM, Pugh CW, Hankinson O, Ratcliffe PJ. The roel of the acryl hydrocarbon receptor nuclear translocator (ARNT) in hypoxic induction of gene expression. Studies in ARNT deficient cells. *J Biol Chem* 1996, **271**, 15117–23.

Wu X, Senechal K, Neschat MS, Whang YE, Sawyers Cl. The PTEN/MMAC1 tumour suppressor phosphatase functions as a negative regulator of the phosphoinositide 3-kinase/Akt pathway. *Proc Natl Acad Sci U S A* 1998, **95**(26), 15587–91.

Yasuda M, Han JW, Dionne CA, Boyd JM, Chinnadurai G. BNIP3alpha: a human homolog of mitochondrial pro-apoptotic protein BNIP3. *Cancer Res* 1999, **59**(3), 533–37.

Yeo EJ, Chun YS, Cho YS, *et al.* YC-1: a novel potential anti-cancer drug targeting hypoxia inducible factor 1. *J Natl Cancer Inst* 2003, **5**, 516–25.

Yeo EJ, Chun YS, Park JW. New anti-cancer strategies targeting HIF-1. *Biochem Pharmacol* 2004, **68**(6), 1061–69.

Yin MB, Li ZR, Toth K, *et al.* Potentiation of irinotecan sensitivity by SMC in an *in vivo* tumour model is associated with down-regulation of COX2, iNOS and HIF-1α. *Oncogene* 2006, **25**, 2509–19.

Zhang P, Chan SL, Fu W, Mendoza M, Mattson MP. TERT suppresses apoptosis at a pre-mitochondrial step by a mechanism requiring reverse transcription activity and 14-3-3 protein binding ability. *FASEB J* 2003, **17**(6), 767–69.

Zhong H, Chiles K, Feldser D, Laughner E, Hanrahan C, Georgescu MM, Simons JW, Semenza GL. Modulation of hypoxia-inducible factor 1 alpha expression by the epidermal growth factor/phosphatidylinositol 3-kinase/PTEN/AKT/FRAP pathway in human prostate cancer cells: implications for tumour angiogenesis and therapeutics. *Cancer Res* 2000, **60**(6), 1541–45.

14
Resistance to Chemotherapy Drugs

Robert O'Connor and Laura Breen

14.1 Introduction

The use of chemotherapeutic drugs, whether alone or in combination with other modalities of cancer treatment, is a mainstay in the management of many forms of cancer, especially advanced and/or metastatic disease. However, although these agents are typically very toxic, many tumours and tumour types either fail to respond to the actions of such drugs or, after a period of initial response, the tumour mass spreads or expands, ultimately leading to patient death. This 'response failure' is typically termed *resistance* and advancing our understanding of the identification, management and treatment of resistance is key to improving the utility and applicability of chemotherapy in the treatment of many of the most common forms of cancer.

14.2 What are the Factors Limiting the Efficacy of Cancer Chemotherapy Treatment?

An examination of the pharmacology of cancer treatment highlights a collection of fundamental challenges which make drug-based treatment of cancer uniquely difficult. These can be divided under a number of headings.

14.2.1 Heterogeneity

Cancer is a very heterogeneous collection of cellular diseases that give rise to body cells which, by malfunctioning in their growth/death and/or motility characteristics, threaten the

Anticancer Therapeutics   Edited by Sotiris Missailidis
© 2008 John Wiley & Sons, Ltd

existence of the whole organism. To avoid inaccurate generalizations, one must generally limit observations pertaining to the classification of any aspect of cancer (e.g. resistance) to the specific forms of the disease where experimental, epidemiological and clinical evidence have shown these factors to have significance. For example, cancer is generally a chronic disease affecting people as they get older, but conversely, childhood malignancies are well documented. In the case of leukaemia, childhood forms are generally treatable with over-all cure rates of 75% and cure rates for some specific forms are often now in excess of 95% (Ravindranath, 2003), while many adult forms of leukaemia do not show nearly the same rate of cure (Burnett, 2002). Therefore, depending on the perspective taken, leukaemia might be regarded as either a sensitive or resistant cancer; on the whole, then, leukaemia is a heterogeneous collection of haematological malignancies with difference response and treatment characteristics.

Much of our understanding of cancer resistance mechanisms comes from detailed exami-nations of specific, often *in vitro*, models of the disease but the characteristics of an individual patient's cancer can vary hugely from such models and from the features other patients with cancer of similar origin (Vargo-Gogola and Rosen, 2007). This aspect of heterogeneity means that even looking at resistance in a specific form of cancer, such as cancer of the breast, we are generalizing about how we think drugs usually fail rather than how we know they are actually failing in all forms and instances of the disease. Indeed, there is clear evidence that charac-terizing tumours on the basis of their tissue of origin confounds attempts to overcome resis-tance and molecular characterization and profiling may give a more accurate picture (Schaner *et al.*, 2003; Sorlie *et al.*, 2003; Vargo-Gogola and Rosen, 2007). This heterogeneity presents challenges in developing effective therapeutic strategies that deal with even the majority of pharmacological eventualities and resistance mechanisms that might occur in, for example, breast cancer.

14.2.2 Therapeutic index

Therapeutic index can be described as the ratio of the amount of drug that causes the toxic, or in the case of cancer, sometimes fatal effect, to that which generates a useful therapeutic action. With most pharmaceuticals, e.g. antimicrobials, this ratio is typically a significant multiple of 1 and therefore the dose that causes undue toxicity might be several times the therapeutic dose; while with cancer chemotherapy drugs, significant, and in some cases life-threatening, toxicity can be evident (although generally managed) in the majority of patients on standard regimens (Davies, 2006; Minchinton and Tannock, 2006).

Cancer chemotherapy drugs are generally regarded as being almost uniquely non-specific in their actions. Through a variety of means, most chemotherapy drugs appear to be effective by being generally cytotoxic or cytostatic (Davies, 2006). Those who challenge conventional models of chemotherapy action have shown that some chemotherapy drugs may in fact be much more molecularly targeted than we once thought (e.g. anti-angiogenic actions of taxanes for example) (Zelnak and O'Regan, 2007); however, the point still remains that the dose at which drugs show anti-cancer action is very close to, if not the same as that where we see unacceptable toxicity. The low therapeutic index, therefore, means that quite small alterations in anyone of a large number of factors affecting tumour sensitivity can tip the balance of tumour growth away from death and towards survival (Davies, 2006).

14.2.3 Common nature of tumour and normal cells

Allied to the low therapeutic index (and indeed the major explanation for this phenomenon) is the fact that cancer cells are facsimiles of our own normal body cells, albeit with generally minor alterations which make them dangerous to the host. As a result it is extremely difficult, outside of certain classified oncogenic transgene products, e.g. BCR-Abl, to identify targetable alterations unique to the cancer cell (Kummar *et al.*, 2006; Sharifi and Steinman, 2002). Commensurate with the low therapeutic index of chemotherapeutic agents, this similarity means that very slight molecular alterations reducing sensitivity can lead to ineffective tumour cell treatment (resistance) (Dean *et al.*, 2005).

14.2.4 Defence mechanisms

Cancer drugs, by their actions, tend to send cells in the body, including cancer cells, into what might best be termed 'survival mode' (Tortora *et al.*, 2007). Recognizing that cancer cells are inherently similar to normal cells, exposure to toxic xenobiotic chemotherapy drugs can, and usually will, illicit activation of a wide set of cellular defence mechanisms, which can include many of the best characterized resistance mechanism, such as drug efflux pumps, metabolic enzymes and cellular repair mechanisms (Leonessa and Clarke, 2003). The fact that chemotherapy drugs work at all suggests, though, that the loss of growth coordination, common in cancer cells, may make such defence mechanisms perhaps less effective than in normal, fully coordinated and integrated body cells.

14.3 A Classification of the Important Chemotherapy Resistance Mechanisms

Broadly speaking, resistance might best be broken down by two primary distinctions (Dean *et al.*, 2005; Fojo and Coley, 2007; Leonessa and Clarke, 2003; Matheny *et al.*, 2001):

- Pharmacokinetic resistance – where factors construe to reduce the concentration of anti-cancer agent at its target, and;

- Pharmacodynamic resistance – where conventionally effective concentrations of drug are generated at the target, but there is a failure to generate the anticipated response e.g. cellular apoptosis.

14.3.1 Context of resistance mechanisms

While these factors are discussed in isolation in the following text, there is ample evidence in the literature to demonstrate that it is common for multiple mechanism of resistance to act in concert to reduce the efficacy of cancer drug treatments (Leonessa and Clarke, 2003).

It should be borne in mind that some forms of resistance, e.g. reduced oxygenation, may be precipitated by the basic inherent growth characteristics of a tumour (Brown, 2007), while many others, such as drug efflux pump over-expression, represent a combination of selection and/or adaptive responses on the part of the tumour cell (Dean *et al.*, 2005).

Some mechanisms may also have knock-on effects across a spectrum of discrete actions, for example, alteration in the pH of a tumour, which can be caused by rapid tumour cell growth, can affect the pharmacokinetics of how a cancer therapeutic gets into a tumour cell and its subsequent subcellular distribution as well a pharmacodynamic aspects, such as reduced propagation of pro-apoptotic signals (Tredan *et al.*, 2007).

Finally, as mentioned in the opening comments, tumours by their nature can often be very heterogeneous, with cells in different stages of the cell cycle and different phenotypic profiles evident throughout the tumour. This will often mean that some cells in a tumour will be sensitive to drug action, while others will be resistant. This observation is particularly important in the context of the theory of stem cell populations existing in tumour cells (Dean *et al.*, 2005; Kvinlaug and Huntly, 2007). Whether these cells are altered stem cells giving rise to the tumour cell population or they represent a de-differentiation of a tumour cell to produce a stem cell-like population is unclear (both possibilities may be existing in different tumours). Stem cells represent a collection of cells uniquely designed to protect themselves through the intrinsic expression of multiple protective mechanisms (e.g. drug efflux pumps). Antitumour drug therapy may therefore eradicate the vast bulk of conventional tumour cells while leaving the more resistant stem cell populations untouched (Dean *et al.*, 2005; Kvinlaug and Huntly, 2007).

14.3.2 Summary of some of the important contributory mechanisms to cancer resistance

Pharmacokinetic resistance mechanisms

Table 14.1 outlines some of the common mechanisms of pharmacokinetic resistance; i.e. actions that alter delivery of the drug to the target, which have been associated with tumour cell resistance.

Table 14.1 A summary of some of the major forms of resistance to anticancer treatment which might be classed as pharmacokinetic resistance

Mechanism	Reference
Supracellular	
Blood perfusion alterations – reducing delivery of drug	Au *et al.*, 2002; Feldmann *et al.*, 1999
Tumour encapsulation	Jang *et al.*, 2003
Reduction in tumour permeation	Lu *et al.*, 2007
Alterations in tumour and cellular pH	Zhou *et al.*, 2006
Alterations in whole body metabolism, distribution and/or elimination	Leonessa and Clarke, 2003; Michael and Doherty, 2005
Intracellular	
Reduced cellular drug uptake (active or passive mechanisms)	Kapp *et al.*, 2006; Kim, 2003
Increased expression/activity of drug efflux pumps	Juliano and Ling, 1976; Leonard *et al.*, 2003
Vesicular drug localization/sequestration	Cleary *et al.*, 1997; Duvvuri and Krise, 2005
Increased expression of metabolic enzymes	Rodriguez-Antona *et al.*, 2006
Decreased expression of metabolic enzymes (for prodrugs)	Daly, 2007; Rodriguez-Antona *et al.*, 2006

Table 14.2   A summary of some of the major forms of resistance to anticancer treatment that might be classed as pharmacodynamic resistance

Mechanism	Reference
Intratumoural alterations with pharmacodynamic consequences	
Alterations in oxygen tension	Tredan *et al.*, 2007
Alterations in redox potential	Martinez-Sanchez and Giuliani, 2007; Trueba *et al.*, 2004
Intracellular pharmacodynamic alterations	
Decreased expression of target enzyme/receptor	Assaraf, 2007
Increased expression of target molecules	Kantarjian *et al.*, 2006; Leonessa and Clarke, 2003
Mutation of target	Noguchi, 2006
Alterations in apoptotic cascade	Fulda, 2007; Tsuruo *et al.*, 2003
Alterations in tumour cell growth rate (inc. senescence)	Rebbaa, 2005
Alterations in cellular repair mechanism	Zaremba and Curtin, 2007
Stromal/autocrine/paracrine factors secreted which protect tumour cells from drug treatment	Dalton, 2003

Pharmacodynamic resistance mechanisms

Table 14.2 outlines several of the dominant mechanisms of pharmacodynamic resistance where near similar concentrations of drug are generated at the cellular target but with significantly reduced action.

A thorough discussion of the causes of drug resistance would necessitate multiple volumes of text, and indeed several books and reviews are available that provide more detail on mechanisms of drug resistance. This section will therefore attempt to illustrate selected specific mechanisms which contribute to the resistance phenomenon.

14.4 Illustrative Mechanisms of Pharmacokinetic Resistance

Despite the diverse array of molecular changes that can occur in various cellular drug targets and response intermediates, the fundamentals of whether a drug can reach a relevant target concentration, i.e. cellular drug pharmacokinetics, remains of key importance to the life and death of a tumour cell (Leonard *et al.*, 2003).

14.4.1 Alterations in drug uptake mechanisms

Early pharmacology texts assumed that drug uptake was governed primarily by the process of diffusion and, while this tenet appears to still hold for some drugs, it has become apparent that the processes controlling drug entry into cells are generally more complex. A number of pumps/transporters have been identified, which appear to play a significant role in uptake of specific cancer drugs in cells (Kim, 2003; Smith *et al.*, 2005). These uptake proteins belong to the solute carrier family (SLCs) and include a number of organic anion transporter proteins (OATPs). Proteins such as OATP1B3 (OATP8) have been shown to play an important role in the uptake of taxane chemotherapy drugs (Smith *et al.*, 2005). In the case of cisplatin, a copper transporting pump has been demonstrated to play a significant role in the uptake of this drug in cells, with down-regulation of pump expression leading to reduced drug toxicity in resistant cancer cells (Kapp *et al.*, 2006). Tissue expression of these proteins is known to be varied; however, the broader contribution of cellular uptake mechanisms to resistance remains to be more thoroughly investigated.

14.4.2 Alterations in drug efflux protein activity

The significance of drug efflux mechanisms in drug resistance first emerged in 1976 with the description of the multidrug resistance protein-1 (MDR-1; P-gp, ABCB1) protein and its characterization as a cellular efflux pump (Juliano and Ling, 1976). The identification of P-gp subsequently led to the characterization of a vast protein family of protein transporters, which play a pivotal role in homeostasis throughout the various genera of life. Cancer researchers have, in particular, spent a lot of time examining the function of these transporters and their contribution to therapeutic resistance. P-gp has the ability to pump a large number of drugs, including many of the most broadly active cytotoxic agents (Dean *et al.*, 2005). Therefore, cells overexpressing this pump are often resistant to a broad swathe of cancer drugs with little similarity in properties or mechanisms of action, a phenomenon termed, multiple drug resistance. Two other pump proteins, MRP-1 (multiple drug resistance protein-1, ABCC1) and BCRP (breast cancer resistance protein, ABCG2) have also been demonstrated to be commonly overexpressed in various resistant cancers and be capable of transporting an often overlapping collection of anticancer agents. Several members from the same immediate protein family as MRP-1 (e.g. MRP-2, MRP-5, etc.) have also been shown to be capable of generating resistance to a narrower and individually distinct collection of drugs (Plasschaert *et al.*, 2005).

Although the exact mechanism by which each pump transports drug varies, and in some cases is not well understood, the net effect of overexpression of such pumps is to reduce the amount of drug which gets to its cellular target. In the case of anthracyclines, overexpression of P-gp can be shown to greatly reduce the concentration of drug in the nucleus of resistant cells, with inhibition of P-gp significantly (as illustrated in Figure 14.1) increasing cellular anthracycline levels.

The potential contribution of such mechanisms is therefore easy to conceptualize. In reality, the clinical picture is much more complicated. While P-gp can, undoubtedly, contribute to tumour MDR, the pump has important physiological roles, especially for drug absorption and elimination, and we now recognize that pumps often work together in a coordinated manner to move drug through tissues, a process termed *drug vectoring* (Nobili *et al.*, 2006; Su *et al.*, 2004). The complexity of drug vectoring and difficulties in correlating P-gp expression with the extent of resistance, cloud our understanding of the specific role of MDR-1 and other drug efflux mechanisms to cancer resistance. Nonetheless, emerging theories of cancer biology have shone a new spotlight on pump-mediated drug resistance. It is now recognized that so-called cancer stem cells, populations of cancer-progenitor cells giving rise to the tumour mass, have many if not all of the properties of normal stem cells, including the increased expression of drug efflux pumps such as P-gp. In many forms of cancer it may very well be difficult to eradicate the full tumour mass without having a strategy to deal with the inherent pharmacokinetic resistance of the stem cell sub-population (Dean *et al.*, 2005; Fojo and Coley, 2007; Kvinlaug and Huntly, 2007).

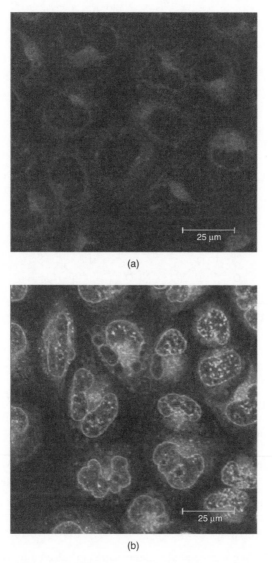

(a)

(b)

Figure 14.1   Laser scanning confocal microscopic view of the anthracycline epirubicin in live P-g-p-overexpressing, DLKP-A cells. In (a), cells have been exposed to 2 μM epirubicin for 2 h. In (b), the same cells have been exposed to epirubicin 2 μM for 2 h plus 0.25 μM P-gp inhibitor, elacridar, greatly increasing the concentration of the anthracycline. Image courtesy of Dr Finbarr O'Sullivan and Dr. Denis Collins, NICB, Dublin City University

Multidrug resistance inhibitors

With a clear role for drug efflux pumps in chemotherapeutic drug resistance established experimentally and agents identified which could inhibit these pumps, it was inevitable that attempts would be made to inhibit drug pumps, especially P-gp, in clinical cancer treatment. Initially, existing pharmaceuticals, including verapamil, quinine, megace and cyclosporin, were used in

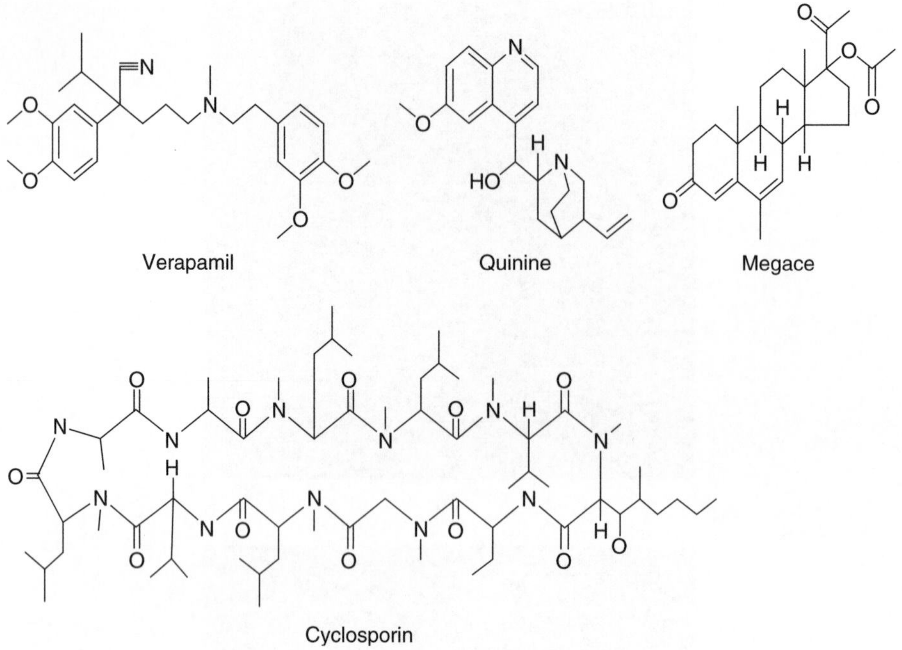

Verapamil   Quinine   Megace

Cyclosporin

Figure 14.2 Chemical structures of some of the first-generation P-gp inhibitors

conjunction with P-gp substrate drugs (see Figure 14.2 for structures). Unfortunately, efficacy trials failed to show any benefit from such combinations. It appears likely that these agents failed for the following reasons (Thomas and Coley, 2003):

- Insufficient P-gp inhibitory activity – the concentrations of putative inhibitor needed to inhibit P-gp *in vitro* were at or above the achievable peak plasma concentrations.

- Pharmaceutical effects – each agent had distinct drug activity, e.g. cyclosporin is a potent inhibitor of immune function which only compounded the inherent toxicity of concomitant chemotherapeutic drug usage.

- Lack of patient selection – in general patients were not selected for on the basis of demonstrated P-gp activity. As a result the impact of P-gp inhibition would always be difficult to correlate with resistance status.

Further synthetic iterations produced competitive P-gp inhibitors which were much more specific, e.g. PSC833, dexverapamil, dexniguldipine and biricodar (see structures in Figure 14.3); however, clinical efficacy trials of combinations were also disappointing and a number of explanations have emerged (Leonard *et al.*, 2003):

- Many of these agents are metabolized by the same common metabolic systems responsible for chemotherapy drug inactivation.

- P-gp has an important protective role in parts of the body and, in particular, is responsible for the elimination of several important groups of chemotherapy drugs from the body.

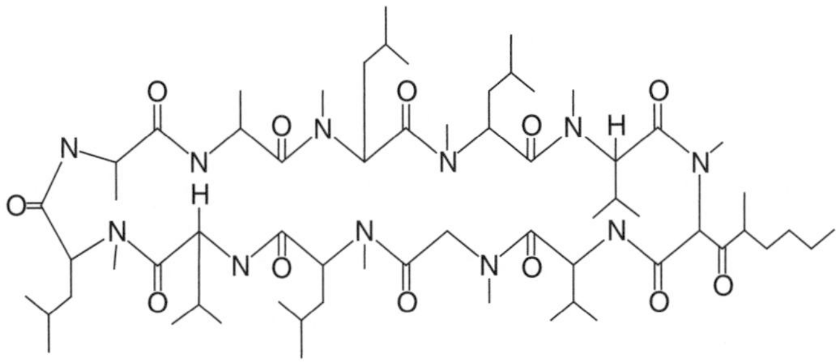

PSC-833

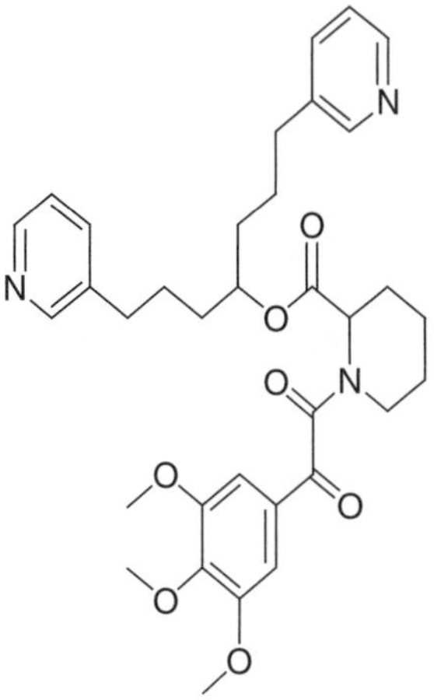

Biricodar

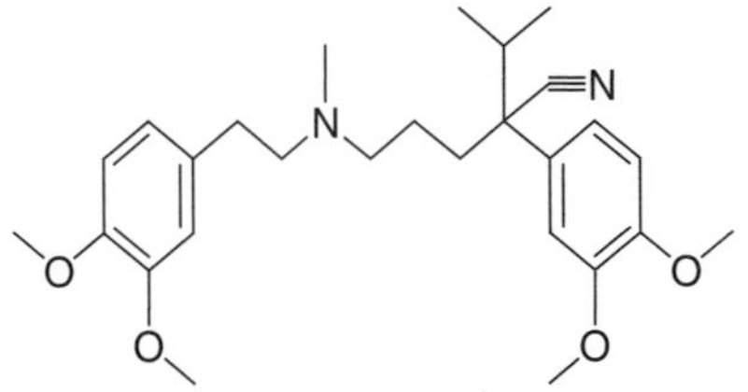

Dexverapamail (R-enantiomer of verapamil)

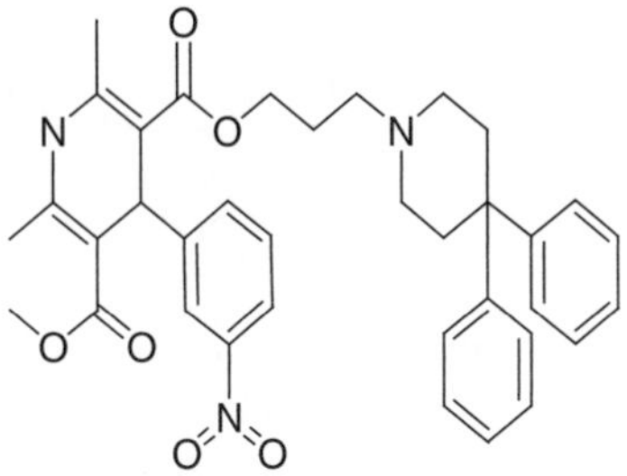

Dexniguldipine

Figure 14.3 Chemical structures of some of the second-generation P-gp inhibitors

- A combination of the two issues above meant that concurrent use of second-generation inhibitors with P-gp substrate chemotherapeutic drugs often produced increased circulating levels of the anticancer agent giving rise to increases in toxicity and side effects and a necessity to reduce the dose of the cytotoxic.

Although the majority of solid tumours where P-gp inhibition has been examined have recorded, at best, no increase in treatment efficacy, there have been some indications of positive findings in a small number of the trials conducted (Leonard *et al.*, 2003).

More recently, newer, non-competitive P-gp inhibitors such as tariquidar, zosuquidar, laniquidar, ONT-093 and elacridar (see Figure 14.4 for structures) have been examined for therapeutic efficacy in combination with P-gp substrates. Some reports have indicated these agents do not appear to affect the pharmacokinetics of P-gp substrate cytotoxics, while others have shown significant interactions.

In particular, later stage clinical trials have demonstrated higher incidences of toxicity and side effects, necessitating dose reduction, presumably associated with elevations in circulating chemotherapeutic drug levels, and clinical development of these strategies seems to have largely stalled (Nobili *et al.*, 2006).

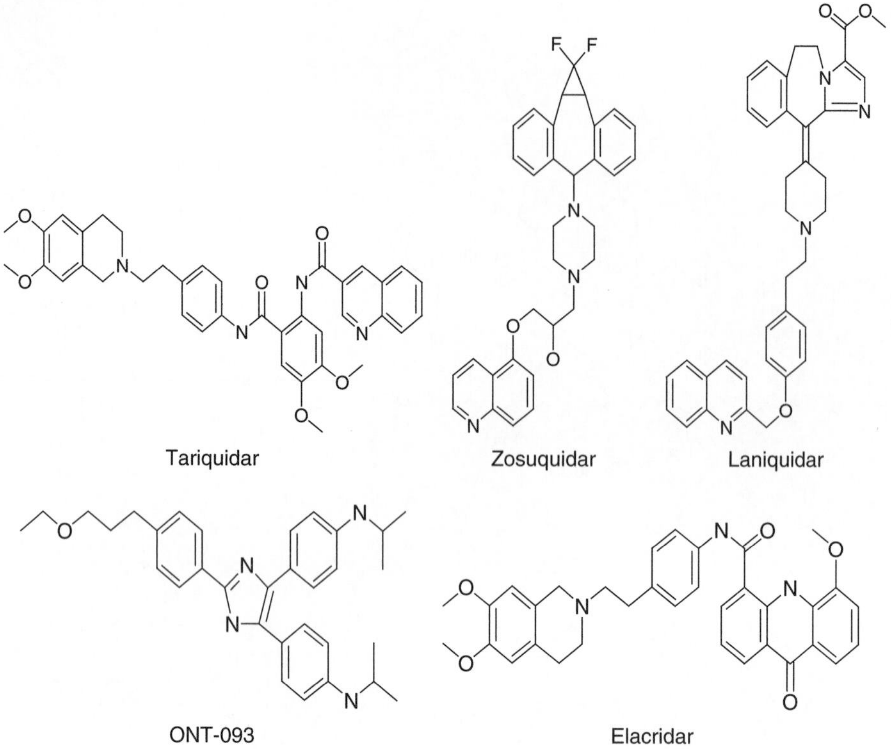

Figure 14.4 Chemical structures of some of the third-generation P-gp inhibitors

Inhibition of other MDR pumps

Research published by our group in 1998 indicated that the MRP-1 drug resistance pump could be competitively inhibited, at pharmacologically relevant concentrations, by specific non-steroidal anti-inflammatory drugs (NSAIDs), in particular, sulindac (see Figure 14.5 for structure) (Duffy *et al.*, 1998). A phase I trial illustrated that this agent could be successfully used in combination with the MRP-1 substrate agent epirubicin (O'Connor *et al.*, 2006). We are currently examining the efficacy of this combination in the treatment of advanced melanoma, a highly resistant form of cancer that has a high expression rate for MRP-1.

Inhibitors of other drug efflux pumps, such as BCRP have also been characterized; however, to date no efficacy-based trials have been conducted to evaluate the efficacy of inhibition of these resistance mechanisms.

Other approaches to modulation of drug efflux pump-mediated resistance

Other strategies which have been employed to overcome pump-mediated drug resistance include alterations in the formulation of the cytotoxic drug (Mayer and Shabbits, 2001) and engineering of cancer drugs to make them poor substrates for pumps (Lee and Swain, 2005). However, in the case of P-gp, examination of the genetic apparatus leading to overexpression of the transporter has lead to new inhibitor strategies based on downregulation of the MDR-1

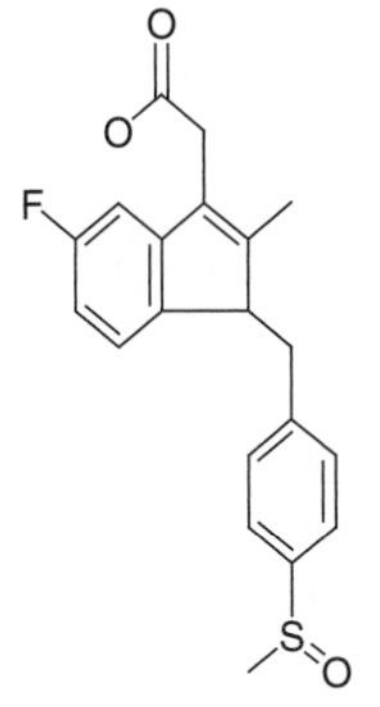

Sulindac

Figure 14.5   Chemical structures of sulindac, a competitive inhibitor of MRP-1

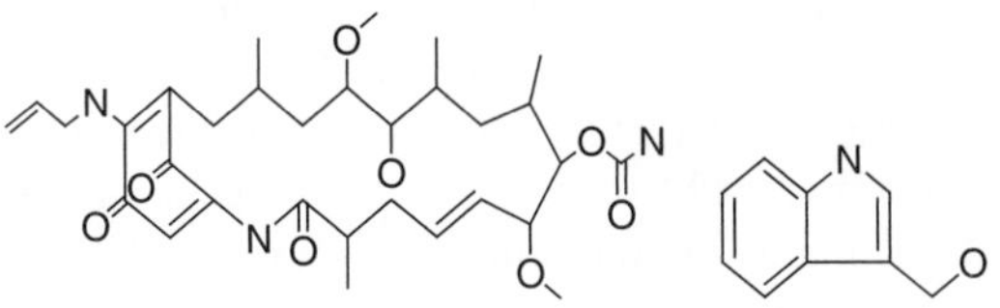

17-AAG (tanespimycin)   Indole-3-carbinol

Figure 14.6   Chemical structures of some of the agents which have been demonstrated to downregulate P-gp overexpression

gene overexpression. This has been successful accomplished *in vitro* using gene silencing technology, e.g. siRNA, ribozyme, etc. (Daly *et al.*, 1996; Xing *et al.*, 2007). However, it has also been shown that treatment with certain drugs can reduce transcription of the MDR-1 gene product, effectively reversing P-gp-mediated MDR, at least *in vitro* (Arora *et al.*, 2005). Some of the agents that reduce MDR-1 transcription include drugs such as indole-3-carbinol and 17-AAG (see Figure 14.6 for structures) (Katayama *et al.*, 2007), which are already being investigated in cancer treatment and it seems likely that clinical evaluation of MDR-1gene downregulation as a treatment for P-gp-mediated resistance will be undertaken shortly.

14.5 Illustrative Mechanisms of Pharmacodynamic Resistance

The number of chemotherapy drug resistance mechanisms, which could be classified under the theme of pharmacodynamic resistance mechanisms, is indeed myriad and the area becomes particularly complex when one looks at more than one form of cancer. For sake of brevity, this section of the chapter will focus on resistance associated with altered responses to DNA damage.

14.5.1 DNA repair inhibition

A significant factor contributing to treatment failure in chemotherapy with DNA-damaging agents is the natural cellular system of DNA repair. Many cellular stresses can induce DNA damage. The ability of a cell to sense and repair this damage is crucial to maintaining genome integrity and preventing passing on of potentially lethal mutations. Many chemotherapeutic agents cause DNA damage, so the cells ability to deal with this injury plays a major role in the response of a tumour to a particular agent. DNA repair mechanisms are an important cellular defence system; however, if the cellular response to DNA damage caused by chemotherapeutic agents results in the cells surviving, this may lead to treatment failure or drug resistance. The response to this damage can have several possible outcomes: induction of apoptosis; cell-cycle arrest and/or DNA-repair (Madhusudan and Hickson, 2005). DNA repair pathways include direct repair, base excision repair (BER), nucleotide excision repair (NER), homologous recombination (HR), non-homologous end joining (NHEJ), DNA inter-strand crosslink repair and DNA mismatch repair (MMR). These repair pathways generally occur during cell cycle arrest, allowing any damage to be repaired before the cell continues to replicate. Inhibitors of these pathways are attractive therapeutic adjuncts because they have the potential to influence the way the cell responds to drug treatment.

Many inhibitors of DNA repair have been discovered to be useful in cancer therapy. Traditionally, inhibitors of DNA repair have been used in combination with DNA-targeting drugs and typically induce a response from that particular DNA repair target. For example, several known inhibitors of HR are available, such as wortmannin and caffeine (Figure 14.7), which prevent activation of the ATM tumour suppressor gene, a key protein kinase which controls response to DNA double strand damage. A number of DNA–protein kinase (PK) inhibitors have been found to sensitize tumour cells to the effects of cytotoxic drugs and radiation (Bentle *et al.*, 2006). The DNA–PK complex regulates NHEJ DNA repair. These and other areas of the DNA repair machinery have been targeted for inhibition in cancer therapy but we will focus on two targets in particular, PARP (poly(ADP-ribosylation) of DNA-binding protein) and MGMT (O^6-methylguanine-DNA-methyltransferase).

PARP inhibitors

PARP are a family of protein enzymes found in eukaryotes that catalyse poly(ADP-ribosylation) of DNA-binding proteins. These enzymes play an important part in the cellular response to DNA damage (Cepeda *et al.*, 2006). PARP enzymes have an important role, particularly in base excision repair (Plummer, 2006). PARP-1 is the most studied member of the PARP family and has been labelled a 'molecular nick sensor' (Plummer, 2006). It is activated by sensing single and double stranded DNA breaks and binding to them. Homodimers formed by the DNA binding lead to the cleavage of NAD^+ into nicotinamide

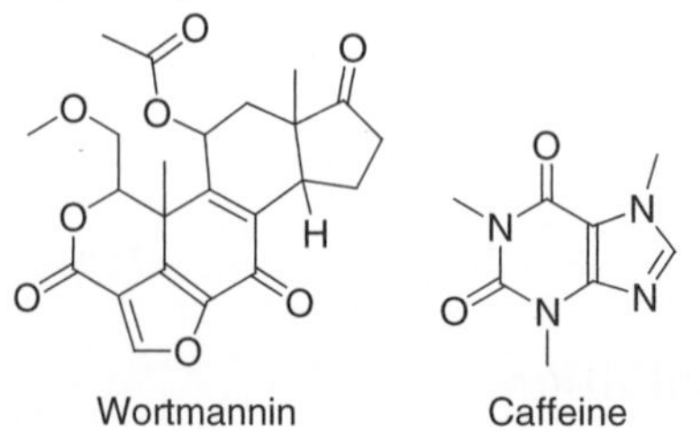

Wortmannin   Caffeine

Figure 14.7 Chemical structures of two 'classical' HR repair inhibitors

and ADP-ribose (Virag and Szabo, 2002). After DNA repair, the inactivated PARP-1 is released to bind to further DNA damage.

The promise of PARP inhibitors in anticancer therapy lies in the fact that increased PARP activity is a common mechanism used by tumour cells to avoid apoptosis, which may be caused by DNA-damaging agents (Ratnam and Low, 2007). Therefore, targeting PARP should cause damaged cells to die by apoptosis rather than allowing them to survive with carcinogenic lesions.

PARP inhibitors first began clinical examination in 2003 in combination with temozolomide and with phase I results showing tolerability, phase II results are keenly awaited. There are also some data to suggest that PARP inhibitors have promise as single-agents in the treatment of cancer (Plummer, 2006). It has been suggested that PARP inhibitors may be most potent as single-agents in tumours with defects in *BRCA1* and *BRCA2* (Bryant *et al.*, 2005; Farmer *et al.*, 2005).

The first generation PARP inhibitors, benzamides, were mostly analogues of nicotinamide (Curtin, 2005). One of the most studied of this type of inhibitor is 3-aminobenzamide (3-AB) (see Figure 14.8 for structure) which has been found to delay DNA strand-break repair and increase the cytotoxicity of some alkylating agents. These PARP inhibitors have activity by virtue of being competitive inhibitors of nicotinamide adenine dinucleotide (NAD^+, Ratnam and Low, 2007). The PARP inhibitor, ANI (4-amino-1,8-naphthalimide) (Figure 14.8), has been shown to potentiate the effect of the chemotherapeutic agent doxorubicin in human breast cancer cell lines (Munoz-Gamez *et al.*, 2005). The PARP inhibitors GPI 15427 (10-(4-methyl-piperazin-1-ylmethyl)-2H-7-oxa-1,2-diaza-benzo[*de*]anthracen-3-one), AG14361, NU1025 and NU1085 (see Figure 14.8 for structures) have been shown to enhance the cytotoxic effects of temozolomide *in vitro* (Curtin *et al.*, 2004; Delaney *et al.*, 2000; Tentori *et al.*, 2003). NU1025 has also been shown to potentiate the cytotoxicity of camptothecin, but to have no effect on the cytotoxicity of etoposide (Bowman *et al.*, 2001). The PARP inhibitor CEP-6800 can enhance the cytotoxicity of the chemotherapeutic agents cisplatin, temozolomide and irinotecan (Miknyoczki *et al.*, 2003).

A number of newly developed PARP inhibitors are currently in clinical trial. Agents such as AG14447 (Pfizer) are all being tested in combination with temozolomide. Other agents (ABT-888, Abbott Laboratories (Figure 14.8) and KU59436 (AstraZeneca and BSI-201, BiPar) are being tested as single agent therapeutics (these agents are so new that information on their chemical structures has not yet been published) (Ratnam and Low, 2007).

MGMT inhibitors

Another DNA repair protein that has been identified as having a role in cancer therapy response is the product of O^6-methylguanine-DNA-methyltransferase (MGMT) gene, O^6-alkylguanine-DNA alkyltransferase (AGT). MGMT can initiate direct repair leading to removal of methyl adducts caused by numerous DNA-damaging agents. This is achieved through transfer of an alkyl group from the target base to a cysteine residue within the active site of MGMT. AGT can repair O^6-AG lesions, which can result from treatment with many chemotherapeutic agents. Repair of these lesions is essential for maintenance of cellular integrity (Gerson, 2002). Aberrant expression of AGT has been noted in many different cancers. AGT-expressing tumours may be up to 10-fold more resistant to methylating agents than tumours lacking MGMT expression (Gerson, 2004).

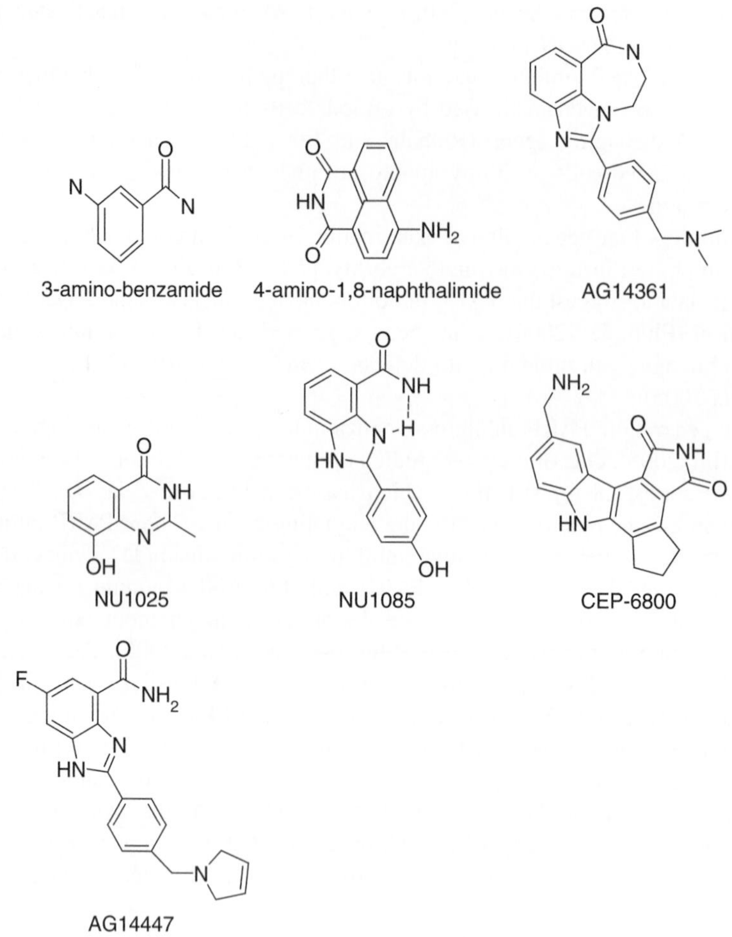

Figure 14.8   Chemical structures of some of the inhibitors of PARP

Alkylating agents, which mainly cause the type of lesions repaired by AGT, fall into two major groups: methylating and chloroethylating agents. Methylating agents, like temozolomide and dacarbazine, are commonly used in the treatment of many cancers, including gliomas and melanomas, and form O^6-methylguanine (O^6-MG) DNA adducts. Chloroethylating agents, including carmustine and lomustine commonly used in the treatment of lymphomas, lead to the creation of O^6-chloroethylguanine adducts (Sabharwal and Middleton, 2006). Resistance to the chemotherapeutic agent temozolomide is usually associated with increased MGMT activity as well as impaired MMR (Alvino *et al.*, 2006).

The importance of MGMT/AGT has lead to interest in developing combination therapies of methylating and chloroethylating chemotherapeutic agents with inhibitors of this repair mechanism. O^6-benzylguanine (O^6-BG) is a non-toxic inhibitor of AGT which was discovered in 1990 (see Figure 14.9 for structure) (Gerson, 2004). It exerts its effect by covalent transfer of the benzyl group to the active site cysteine. *In vitro* studies have shown that O^6-BG can enhance the toxicity of many alkylating chemotherapeutic agents (Friedman, 2002; Dolan,

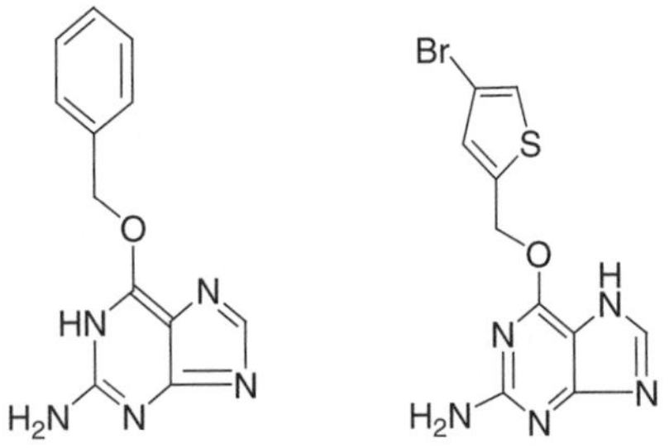

O^6-benzylguanine O^6-(4-bromothenyl) guanine

Figure 14.9 Chemical structures of some of the inhibitors of MGMT/AGT

1991). This agent is currently under investigation as a possible tumour-sensitizing treatment. Several phase I clinical studies of the combination of O^6-BG with BCNU have been carried out in patients with cancer (Gerson, 2002; Schilsky *et al.*, 2000). A phase II trial, which was focused on patients with melanoma (Gajewski *et al.*, 2005), concluded that tumour regression in patients treated with the combination of O^6-BG with BCNU was not significantly different to what would be expected with BCNU alone. Other clinical trials combining O^6-BG with different agents and different cancer types are ongoing.

A related compound, O^6-(4-bromothenyl) guanine (also known as lomeguatrib or PaTrin-2) (Figure 14.9), is also under clinical development. This newer agent has several advantages since it shows improved activity against human MGMT and is orally bioavailable (Woolford *et al.*, 2006).

Treatment with DNA repair inhibitors, alone and/or in combination with DNA-damaging cytotoxics, including those discussed here, represents a promising angle to improving current chemotherapeutic options for cancer patients, however, we are currently too early in the clinical evolution of such agents to have a clear picture of their likely therapeutic benefit.

14.6 Conclusion

Successful pharmaceutical management and treatment of the various forms of cancer that afflict humans is made complex by the ability of tumour cells to mount an effective resistance to the anticancer actions of these agents. A vast array of different physiological, pharmacological and biochemical mechanisms can negate the therapeutic effects of conventional cytotoxic agents, as well as newer molecularly-targeted agents. The net result being that a cancer drug fails to reach sufficient concentration at the tumour cell target and/or despite reaching sufficient concentration, the agent does not stimulate the key death/apoptotic pathways required for the cell to die.

An understanding of the molecular causes of resistance has allowed the development of a number of resistance modulation strategies, but much research remains to be conducted to identify the most effective means to overcome resistance and use these in the most appropriate way for a given patient.

Further progress is emerging as we identify and better characterize the dominant mechanisms of resistance in particular forms of cancer, and move away from traditional strategies based purely on tumour origin. A significant arsenal of agents is emerging to modulate certain common mechanisms of cancer drug resistance. In addition to these resistance circumvention strategies, experimental evidence now suggests that it may be possible to pharmacologically intervene in the emergence of certain forms of efflux pump-mediated resistance and thus stop this form of resistance from appearing, rather than inhibiting it after it has emerged.

While generations of resistance research may have to date largely 'only' managed to aid our understanding of the causes of resistance, the field may now be sufficiently mature to deliver real therapeutic advances for patients.

References

Alvino E, Castiglia D, Caporali S, *et al*. A single cycle of treatment with temozolomide, alone or combined with O(6)-benzylguanine, induces strong chemoresistance in melanoma cell clones *in vitro*: role of *O*(6)-methylguanine-DNA methyltransferase and the mismatch repair system. *Int J Oncol* 2006, **29**, 785–97.

Arora A, Seth K, Kalra N, Shukla Y. Modulation of P-glycoprotein-mediated multidrug resistance in K562 leukemic cells by indole-3-carbinol. *Toxicol Appl Pharmacol* 2005, **202**, 237–43.

Assaraf YG. Molecular basis of antifolate resistance. *Cancer Metastasis Rev* 2007, **26**, 153–81.

Au JLS, Jang SH, Wientjes MG. Clinical aspects of drug delivery to tumors. *J Controlled Release* 2002, **78**, 81–95.

Bentle M, Bey E, Dong Y, Reinicke K, Boothman D. New tricks for old drugs: the anticarcinogenic potential of DNA repair inhibitors. *J Mol Histol* 2006, **37**, 203–18.

Bowman K, Newell D, Calvert A, Curtin N. Differential effects of the poly (ADP-ribose) polymerase (PARP) inhibitor NU1025 on topoisomerase I and II inhibitor cytotoxicity in L1210 cells *in vitro*. *Br J Cancer* 2001, **84**, 106–12.

Brown JM. Tumor hypoxia in cancer therapy. *Methods Enzymol* 2007, **435**, 295–321.

Bryant H, Schultz N, Thomas H, *et al*. Specific killing of BRCA2-deficient tumours with inhibitors of poly(ADP-ribose) polymerase. *Nature* 2005, **434**, 913–17.

Burnett AK. Acute myeloid leukemia: treatment of adults under 60 years. *Rev Clin Exp Hematol* 2002, **6**, 26–45; discussion 86–7.

Cepeda V, Fuertes M, Castilla J, Alonso C, Quevedo C, Soto M, Perez J. Poly(ADP-ribose) polymerase-1 (PARP-1) inhibitors in cancer chemotherapy. *Recent Patents on Anti-Cancer Drug Discovery* 2006, **1**, 39–53.

Cleary I, Doherty G, Moran E, Clynes M. The multidrug-resistant human lung tumour cell line, DLKP-A10, expresses novel drug accumulation and sequestration systems. *Biochem Pharmacol* 1997, **53**, 1493–502.

Curtin N. PARP inhibitors for cancer therapy. *Expert Rev Mol Med* 2005, **7**, 1–20.

Curtin N, Wang L, Yiakouvaki A, *et al*. Novel poly(ADP-ribose) polymerase-1 inhibitor, AG14361, restores sensitivity to temozolomide in mismatch repair-deficient cells. *Clin Cancer Res* 2004, **10**, 881–89.

Dalton WS. The tumor microenvironment: focus on myeloma. *Cancer Treat Rev* 2003, **29**(Suppl 1), 11–9.

Daly AK. Individualized drug therapy. *Curr Opin Drug Discov Devel* 2007, **10**, 29–36.

Daly C, Coyle S, McBride S, O'Driscoll L, Daly N, Scanlon K, Clynes M. mdr1 ribozyme mediated reversal of the multi-drug resistant phenotype in human lung cell lines. *Cytotechnology* 1996, **19**, 199–205.

Davies SM. Pharmacogenetics, pharmacogenomics and personalized medicine: are we there yet? *Hematology Am Soc Hematol Educ Program* 2006, 111–7.

Dean M, Fojo T, Bates S. Tumour stem cells and drug resistance. *Nat Rev Cancer* 2005, **5**, 275–84.

Delaney C, Wang L, Kyle S, *et al*. Potentiation of temozolomide and topotecan growth inhibition and cytotoxicity by novel poly(adenosine diphosphoribose) polymerase inhibitors in a panel of human tumor cell lines. *Clin Cancer Res* 2000, **6**, 2860–67.

Dolan M, Mitchell R, Mummert C, RC M, Pegg A. Effect of O6-benzylguanine analogs on sensitivity of human tumor-cells to the cytotoxic effects of alkylating-agents. *Cancer Res* 1991, **51**, 3367–72.

Duffy C, Elliott C, O'Connor R, *et al*. Enhancement of chemotherapeutic drug toxicity to human tumour cells *in vitro* by a subset of non-steroidal anti-inflammatory drugs (NSAIDs). *Eur J Cancer* 1998, **34**, 1250–59.

Duvvuri M, Krise JP. Intracellular drug sequestration events associated with the emergence of multidrug resistance: A mechanistic review. *Frontiers Biosci* 2005, **10**, 1499–509.

Farmer H, McCabe N, Lord C, *et al*. Targeting the DNA repair defect in BRCA mutant cells as a therapeutic strategy. *Nature* 2005, **434**, 917–21.

Feldmann HJ, Molls M, Vaupel P. Blood flow and oxygenation status of human tumors - Clinical investigations. *Strahlentherapie Und Onkologie* 1999, **175**, 1–9.

Fojo T, Coley HM. The role of efflux pumps in drug-resistant metastatic breast cancer: new insights and treatment strategies. *Clin Breast Cancer* 2007, **7**, 749–56.

Friedman H, Keir S, Pegg A, *et al*. O-6-benzylguanine-mediated enhancement of chemotherapy. *Mol Cancer Ther* 2002, **1**, 943–48.

Fulda S. Inhibitor of apoptosis proteins as targets for anticancer therapy. *Expert Rev Anticancer Ther* 2007, **7**, 1255–64.

Gajewski T, Sosman J, Gerson S, Liu L, Dolan E, Lin S, Vokes E. Phase II trial of the O-6-alkylguanine DNA alkyltransferase inhibitor O-6-benzylguanine and 1,3-bis(2-chloroethyl)-1-nitrosourea in advanced melanoma. *Clin Cancer Res* 2005, **11**, 7861–65.

Gerson S. Clinical relevance of MGMT in the treatment of cancer. *J Clin Oncol* 2002, **20**, 2388–99.

Gerson S. MGMT: its role in cancer aetiology and cancer therapeutics. *Nat Rev Cancer* 2004, **4**, 296–307.

Jang SH, Wientjes MG, Lu D, Au JLS. Drug delivery and transport to solid tumors. *Pharm Res* 2003, **20**, 1337–50.

Juliano RL, Ling V. A surface glycoprotein modulating drug permeability in Chinese hamster ovary cell mutants. *Biochim Biophys Acta* 1976, **455**, 152–62.

Kantarjian HM, Talpaz M, Giles F, O'Brien S, Cortes J. New insights into the pathophysiology of chronic myeloid leukemia and imatinib resistance. *Ann Intern Med* 2006, **145**, 913–23.

Kapp T, Muller S, Gust R. Dinuclear alkylamine platinum(II) complexes of [1,2-bis(4-fluorophenyl)ethylenediamine]platinum(II): influence of endocytosis and copper and organic cation transport systems on cellular uptake. *ChemMedChem* 2006, **1**, 560–4.

Katayama K, Yoshioka S, Tsukahara S, Mitsuhashi J, Sugimoto Y. Inhibition of the mitogen-activated protein kinase pathway results in the down-regulation of P-glycoprotein. *Mol Cancer Ther* 2007, **6**, 2092–102.

Kim RB. Organic anion-transporting polypeptide (OATP) transporter family and drug disposition. *Eur J Clin Invest* 2003, **33**(Suppl 2), 1–5.

Kummar S, Gutierrez M, Doroshow JH, Murgo AJ. Drug development in oncology: classical cytotoxics and molecularly targeted agents. *Br J Clin Pharmacol* 2006, **62**, 15–26.

Kvinlaug BT, Huntly BJ. Targeting cancer stem cells. *Expert Opin Ther Targets* 2007, **11**, 915–27.

Lee JJ, Swain SM. Development of novel chemotherapeutic agents to evade the mechanisms of multidrug resistance (MDR). *Semin Oncol* 2005, **32**, S22–6.

Leonard GD, Fojo T, Bates SE. The role of ABC transporters in clinical practice. *Oncologist* 2003, **8**, 411–24.

Leonessa F, Clarke R. ATP binding cassette transporters and drug resistance in breast cancer. *Endocr Relat Cancer* 2003, **10**, 43–73.

Lu D, Wientjes MG, Lu Z, Au JLS. Tumor priming enhances delivery and efficacy of nanomedicines. *J Pharmacol Exp Ther* 2007, **322**, 80–8.

Madhusudan S, Hickson I. DNA repair inhibition: a selective tumour targeting strategy. *Trends Mol Med* 2005, **11**, 503–11.

Martinez-Sanchez G, Giuliani A. Cellular redox status regulates hypoxia inducible factor-1 activity. Role in tumour development. *J Exp Clin Cancer Res* 2007, **26**, 39–50.

Matheny CJ, Lamb MW, Brouwer KR, Pollack GM. Pharmacokinetic and pharmacodynamic implications of P-glycoprotein modulation. *Pharmacotherapy* 2001, **21**, 778–96.

Mayer LD, Shabbits JA. The role for liposomal drug delivery in molecular and pharmacological strategies to overcome multidrug resistance. *Cancer Metastasis Rev* 2001, **20**, 87–93.

Michael M, Doherty MM. Tumoral drug metabolism: Overview and its implications for cancer therapy. *J Clin Oncol* 2005, **23**, 205–29.

Miknyoczki S, Jones-Bolin S, Pritchard S, *et al.* Chemopotentiation of temozolomide, irinotecan, and cisplatin activity by CEP-6800, a poly(ADP-ribose) polymerase inhibitor. *Mol Cancer Ther* 2003, **2**, 371–82.

Minchinton AI, Tannock IF. Drug penetration in solid tumours. *Nat Rev Cancer* 2006, **6**, 583–92.

Munoz-Gamez J, Martin-Oliva D, Aguilar-Quesada R, *et al.* PARP inhibition sensitizes p53-deficient breast cancer cells to doxorubicin-induced apoptosis. *Biochem J* 2005, **386**, 119–25.

Nobili S, Landini I, Giglioni B, Mini E. Pharmacological strategies for overcoming multidrug resistance. *Curr Drug Targets* 2006, **7**, 861–79.

Noguchi S. Predictive factors for response to docetaxel in human breast cancers. *Cancer Sci* 2006, **97**, 813–20.

O'Connor R, O'Leary M, Ballot J, *et al.* A phase I clinical and pharmacokinetic study of the multidrug resistance protein-1 (MRP-1) inhibitor sulindac, in combination with epirubicin in patients with advanced cancer. *Cancer Chemother Pharmacol* 2007, **59**, 79–87.

Plasschaert SL, de Bont ES, Boezen M, *et al.* Expression of multidrug resistance-associated proteins predicts prognosis in childhood and adult acute lymphoblastic leukemia. *Clin Cancer Res* 2005, **11**, 8661–8.

Plummer E. Inhibition of poly(ADP-ribose) pollyrnerase in cancer. *Curr Opin Pharmacol* 2006, **6**, 364–68.

Ratnam K, Low J. Current development of clinical inhibitors of poly(ADP-ribose) polymerase in oncology. *Clin Cancer Res* 2007, **13**, 1383–8.

Ravindranath Y. Recent advances in pediatric acute lymphoblastic and myeloid leukemia. *Curr Opin Oncol* 2003, **15**, 23–35.

Rebbaa A. Targeting senescence pathways to reverse drug resistance in cancer. *Cancer Lett* 2005, **219**, 1–13.

Rodriguez-Antona C, Ingelman-Sundberg M. Cytochrome P450 pharmacogenetics and cancer. *Oncogene* 2006, **25**, 1679–91.

Sabharwal A, Middleton M. Exploiting the role of O6-methylguanine-DNA-methyltransferase (MGMT) in cancer therapy. *Curr Opin Pharmacol* 2006, **6**, 355–63.

Schaner ME, Ross DT, Ciaravino G, *et al.* Gene expression patterns in ovarian carcinomas. *Mol Biol Cell* 2003, **14**, 4376–86.

Schilsky R, Dolan M, Bertucci D, Ewesuedo R, Vogelzang N, Mani S, Wilson L, Ratain M. Phase I clinical and pharmacological study of O-6-benzylguanine followed by carmustine in patients with advanced cancer. *Clin Cancer Res* 2000, **6**, 3025–31.

Sharifi N, Steinman RA. Targeted chemotherapy: chronic myelogenous leukemia as a model. *J Mol Med* 2002, **80**, 219–32.

Smith NF, Acharya MR, Desai N, Figg WD, Sparreboom A. Identification of OATP1B3 as a high-affinity hepatocellular transporter of paclitaxel. *Cancer Biol Ther* 2005, **4**, 815–8.

Sorlie T, Tibshirani R, Parker J, *et al.* Repeated observation of breast tumor subtypes in independent gene expression data sets. *Proc Natl Acad Sci U S A* 2003, **100**, 8418–23.

Su Y, Zhang X, Sinko PJ. Human organic anion-transporting polypeptide OATP-A (SLC21A3) acts in concert with P-glycoprotein and multidrug resistance protein 2 in the vectorial transport of Saquinavir in Hep G2 cells. *Mol Pharm* 2004, **1**, 49–56.

Tentori L, Leonetti C, Scarsella M, *et al.* Systemic administration of GPI 15427, a novel poly(ADP-ribose) polymerase-1 inhibitor, increases the antitumor activity of temozolomide against intracranial melanoma, glioma, lymphoma. *Clin Cancer Res* 2003, **9**, 5370–9.

Thomas H, Coley HM. Overcoming multidrug resistance in cancer: an update on the clinical strategy of inhibiting p-glycoprotein. *Cancer Control* 2003, **10**, 159–65.

Tortora G, Bianco R, Daniele G, *et al*. Overcoming resistance to molecularly targeted anticancer therapies: Rational drug combinations based on EGFR and MAPK inhibition for solid tumours and haematologic malignancies. *Drug Resist Updat* 2007, **10**, 81–100.

Tredan O, Galmarini CM, Patel K, Tannock IF. Drug resistance and the solid tumor microenvironment. *J Natl Cancer Inst* 2007, **99**, 1441–54.

Trueba GP, Sanchez GM, Giuliani A. Oxygen free radical and antioxidant defense mechanism in cancer. *Front Biosci* 2004, **9**, 2029–44.

Tsuruo T, Naito M, Tomida A, Fujita N, Mashima T, Sakamoto H, Haga N. Molecular targeting therapy of cancer: drug resistance, apoptosis and survival signal. *Cancer Sci* 2003, **94**, 15–21.

Vargo-Gogola T, Rosen JM. Modelling breast cancer: one size does not fit all. *Nat Rev Cancer* 2007, **7**, 659–72.

Virag L, Szabo C. The therapeutic potential of poly(ADP-ribose) polymerase inhibitors. *Pharmacol Rev* 2002, **54**, 375–429.

Woolford L, Southgate T, Margison G, Milsom M, Fairbairn L. The P140K mutant of human O(6)-methylguanine-DNA-methyltransferase (MGMT) confers resistance *in vitro* and *in vivo* to temozolomide in combination with the novel MGMT inactivator O(6)-(4-bromothenyl)guanine. *J Gene Med* 2006, **8**, 29–34.

Xing H, Wang S, Weng D, *et al*. Knock-down of P-glycoprotein reverses taxol resistance in ovarian cancer multicellular spheroids. *Oncol Rep* 2007, **17**, 117–22.

Zaremba T, Curtin NJ. PARP inhibitor development for systemic cancer targeting. *Anticancer Agents Med Chem* 2007, **7**, 515–23.

Zelnak AB, O'Regan RM. Targeting angiogenesis in advanced breast cancer. *BioDrugs* 2007, **21**, 209–14.

Zhou J, Schmid T, Schnitzer S, Brune B. Tumor hypoxia and cancer progression. *Cancer Lett* 2006, **237**, 10–21.

15

Cancer Immunotherapy

Maria Belimezi

15.1 The Molecular Basis of Cancer Immunotherapy and Gene Immunotherapy of Cancer

Classical surgery, radiotherapy and chemotherapy are, without doubt, the leading efficient therapeutic options for the treatment of cancer. The need to develop new therapeutic protocols is urgent, in order to handle aggressive malignancies that do not respond to the anticancer approaches currently used in the clinic, and to improve on the lifespan and the quality of life of cancer patients. Triggering the host's antitumour immunity using gene therapy strategies in combination with cytokines, costimulatory molecules, sensitized lymphocytes and antibodies is a very promising approach according to the outcome of many clinical trials (Qian and Prieto, 2004).

Immune surveillance is one of the major defence mechanisms against cancer. Immuno-suppressed individuals are more prone to cancer and at the same time many human tumours are either non-immunogenic or weakly immunogenic. The immune system, evolved to rid the body of unwanted intruders, could be instructed and reinforced to eliminate cancer cells.

Cancer immunotherapy aims to elicit an immune response using either passive or active approaches. Passive immunotherapy includes the use of highly specific antibodies against cancer antigens or adoptive transfer of lymphocytes to the patient. Active immunotherapy increases the immunogenicity of tumours through the induction of local cytokine production and gene expression. Therapeutic protocols using the *ex-vivo* activation of T cells with cytokines or the use of genetically engineered T cells expressing antitumour agents, followed by reintroduction to the patient, are a very promising immunotherapeutic approach. Finally, the use of DNA vaccines is also a major area of focus in cancer immunotherapy.

15.2 Recombinant Monoclonal Antibodies

The efficiency of antibodies in the treatment of cancer resulted in the approval by the US Food and Drug Administration (FDA) of more than nine antibodies for therapeutic use since 1995. Among them are the humanized murine monoclonal antibody (mAb) trastuzumab

Anticancer Therapeutics   Edited by Sotiris Missailidis
© 2008 John Wiley & Sons, Ltd

(Herceptin) that targets Her2/neu receptor and the chimeric monoclonal antibody cetuximab (Erbitux) against EGFR (ErbB1) receptor (Hynes and Lane, 2005, Astsaturov *et al.*, 2006). Herceptin is the first antibody that received FDA approval for the treatment of solid tumours that overexpress Her2/neu, while another humanized mAb, pertuzumab (Omnitarg) that inhibits Her2/neu heterodimerization is now found in phase II clinical trials (Albanell *et al.*, 2003). More than 40 monoclonal antibodies for cancer treatment are currently at the clinical trial stage. Apart from the ErbB proteins, some of the most common cancer antigens used as targets are: the carcinoembryonic antigen (CEA), involved in cancers of the intestine, MUC1 that is related with breast and lung cancers and is involved in the ErbB network, and CD20 in B cells, a biological marker in non-Hodgkin' s lymphoma and the vascular endothelial growth factor (VEGF).

The use of antibodies as anticancer agents became possible with the development of novel technologies that were created in order to overcome the restrictions imposed by the application of murine antibodies in humans. The hybridoma technology, using cells from immunized mice (Kohler and Milstein, 1975), constituted a decisive step for the development of mAbs, even though the outcome of clinical trials was disappointing (Larrick and Fry, 1991).

In this technique, B lymphocytes extracted from the spleen of immunized mice are fused with myeloma cells. The hybrid cells that result from the fusion are called hybridomas. Each hybridoma has the ability to produce a specific antibody (a property of B cells) and at the same time can be maintained and propagated in culture infinitely (a property of cancer cells). The hybridomas can be safely stored in liquid nitrogen ($-120°C$) for long periods of time. With this method, mAbs can be easily produced at large scale. Murine mAbs have short half-lives in patient serum and are unable to activate effector molecules in humans; the biggest problem, though, was the sensitization of the immune system and the production of human anti-mouse antibodies (HAMA response, Khazaeli *et al.*, 1994). The first attempt to improve the immunogenicity of murine antibodies was the creation of chimeric structures, using genetic engineering techniques. In these engineered antibodies the variable regions were of murine origin while the constant regions originated from human immunoglobulins (Boulianne *et al.*, 1984; Morrison *et al.*, 1984). Although less immunogenic, these chimeric antibodies were also recognized by the human immune system, which develops an antichimeric immune response (HACA) (Bell and Kamm, 2000). Next, followed the generation of 'humanized' antibodies, that contained only the hypervariable regions (CDR) of the murine mAb embedded between the framework regions of the human immunoglobulin (Jones *et al.*, 1986). The ability of transformation of antibodies to a human-like derivatives made feasible their application in the clinic. The chimeric and 'humanized' antibodies are 65–90 % and 95 % human, respectively (Figure 15.1).

Advancements in the study of immunoglobulins, the isolation of genes of variable domains as well as their successful expression in *Escherichia coli* (Better *et al.*, 1988; Skerra and Pluckthun, 1988) were the basis for the development of the phage display technology that allowed the expression of cDNA libraries in the surface of bacteriophages (McCafferty *et al.*, 1990). The construction of phage libraries that express monovalent fragments of human immunoglobulins (scFv or Fab) made feasible the isolation of antibodies that exclusively originate from humans.

In general, the advantage of *in vitro* selection of antibodies is the improved specificity and affinity towards a wide range of antigens that might include toxic agents that cannot be used

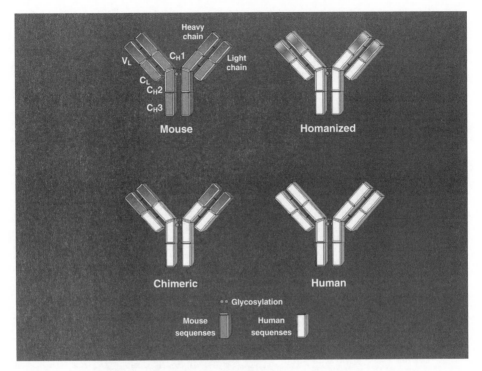

Figure 15.1 Monoclonal antibody optimization towards the development and isolation of fully human antibodies and minimization of unwanted responses during immunotherapeutic protocols

for animal immunization such as, toxins, various drugs, cytokines or other molecules. Human antibodies have also been isolated from transgenic mice that express one or even more genetic loci of human immunoglobulins, while at the same time the expression of their endogenous genes of immunoglobulins has been suppressed. These mice can produce human antibodies after their immunization with a specific antigen. The antibodies are subsequently isolated with the use of hybridoma technology, as discussed before (Mendez *et al.*, 1997; Green, 1999; Nicholson *et al.*, 1999).

15.2.1 Mechanisms of antibody action

The aim of using antibodies in cancer therapy is the direct destruction of cancer cells by specific targeting and killing tumour cells or indirectly by inhibiting angiogenesis and as a result restricting tumours from growing.

Herceptin and a number of other antibodies have been characterized as '*naked*' antibodies as long as they are not conjugated with any radioisotope or toxin. The mechanisms that are involved in their anti-oncogenic action mainly concern the activation of the host's immune system. The Fc fragments of immunoglobulins induce complement-dependent cytotoxicity (CDC) through the binding of the C1q molecule, a component of the

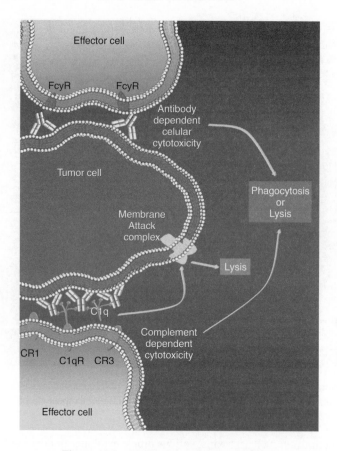

Figure 15.2   Mechanisms of antibody action

complement, and antibody-dependent cellular cytotoxicity (ADCC), through binding to the Fc receptors (Carter, 2001; Brekke and Sandlie, 2003) (Figure 15.2).

The action of antibodies that do not have the Fc fragment (Fab, scFv) is also significant since they are capable of inhibiting ligand binding to the receptor and subsequent activation of its signalling (Weir *et al.*, 2002). Although the bivalent nature of antibodies is considered to be an essential property for inducing endocytosis through receptor cross-linking (Srinivas *et al.*, 1993), it has been observed that scFv fragments can spontaneously form bivalent molecules, though at low frequency (Nielsen *et al.*, 2002).

One of the great advantages of using genetic engineering and the phage display approach is the possibility to create bispecific antibodies. This type of antibodies has been used for coupling cells molecules. Initially they were designed to direct the active molecules of the immune system towards cell-targets (Holliger *et al.*, 1993; Kontermann *et al.*, 1997), and consequently for the transport of radioactive molecules, drugs and toxins (Segal *et al.*, 1999; van Spriel *et al.*, 2000). Coupling of antibody fragments forms bivalent and trivalent structures and creates multivalent antibody fragments with stronger binding to the target molecules (avidity) and with the ability to promote receptor cross-linking (Iliades *et al.*, 1997; Kortt *et al.*, 2001).

Antibodies with double specificity against ErbB2 and FcγRIII for example, were systematically infused in severe combined immunodeficient mice with ovarian cancer and a significant improvement in the survival rate was observed without any toxicity (Weiner *et al.*, 1993). The above results opened the way for several clinical studies of new antibodies against the ErbB2, ErbB1 and FcγRI receptors for the treatment of several types of cancer (Curnow *et al.*, 1997; van Ojik *et al.*, 1997).

15.2.2 Induction of signalling pathways

Depending on the nature of cancer targets, the antibodies have the ability to couple the receptors and to trigger the activation of several signalling pathways that control tumour growth and results in apoptosis (Cragg *et al.*, 1999). In that way monoclonal antibodies behave as 'surrogate' linkers that bind to receptors and mimic or differentiate their signalling action. The first clinical indications on the role of membrane signalling tumour development came from a study by Vuist *et al.* (1994), which showed that the tumour reduction in 24 patients with non-Hodgkin's lymphoma was not due to the antibodies' isotope but was the result of the antibodies' ability to activate the intracellular signalling and to increase tyrosine phosphorylation.

15.2.3 Conjugated antibodies

Many antibodies have been used as carriers of therapeutic substances to the tumour site or into the cytoplasm, in the case that are endocytosed (Hudziak *et al.*, 1989; van Leeuwen *et al.*, 1990; Hurwitz *et al.*, 1995). The specific targeting of cancer cells allows the administration of high doses of the above complexes without the side effects that are observed when general cytotoxic agents are administered. Among several studies are those that used the antibody's property to endocytosed for the introduction of liposomes that carry a cytotoxic load (doxorubicin) into the cytoplasm of cancer cells (Park *et al.*, 1995; Goren *et al.*, 1996).

15.2.4 Monoclonal antibodies as anticancer drugs

The development and optimization of the characteristics and properties of antibodies resulted in their introduction to clinic for cancer treatment. A list of monoclonal antibodies approved in the treatment of cancer is presented in Table 15.1 followed by a short description.

Alemtuzumab: a monoclonal antibody against CD52

Campath (alemtuzumab) is a chimeric humanized monoclonal antibody (Campath-1H) against the 21–28 kD cell surface glycoprotein, CD52. It is an immunoglobulin G1 (IgG1) kappa with human variable framework and constant regions, and complementarity-determining regions from a murine (rat) monoclonal antibody (Campath-1G). CD52 is widely expressed on the surface of all mature (normal and malignant B and T) lymphocytes, and Campath is indicated for the treatment of B-cell chronic lymphocytic leukaemia (B-CLL). It has been approved by the FDA in 2001, after exhibiting a 33 % response in patients with relapse cases of chronic lymphocytic leukaemia (CLL), as a single agent in patients who have been treated with alkylating agents and who have failed fludarabine therapy (Keating

Table 15.1   FDA approved monoclonal antibodies for the treatment of cancer

Antibody generic/trade name	Type	Target	Company	Approval date	Cancer type
Alemtuzumab/ Campath-1H	Humanized	CD52	Millenium And ILEX Partners	2001	B cell chronic lymphocytic leukaemia
Bevacizumab/ Avastin	Humanized	VEGF	Roche/Genentech	2004	Metastatic colorectal cancer
Cetuximab/ Erbitux	Chimeric	EGFR	Merck KgGA	2004	Colorectal cancer
				2006	Head and neck
Gemtuzumab ozogamicin/ Mylotarg	Humanized conjugated to calicheamicin	CD33	Wyeth	2000	Acute myelogenous leukaemia
Ibritumomab tiuxe- tan/Zevalin	Murine conjugated to ^{90}Y	CD20	Biogen IDEC	2002	B-cell non-Hodgkin's lymphoma
Panitumumab/ Vectibix	Human	EGFR	Amgen/Abgenix	2006	Colorectal cancer
Rituximab/ Rituxan Mab Thera	Chimeric	CD20	Roche/Genentech	1997	non-Hodgkin's lymphoma
Tositumomab/ Bexxar	Murine conjugated to ^{131}I	CD20	Corixa	2003	non-Hodgkin's lymphoma
Trastuzumab/ Herceptin	Humanized	Her2/neu	Roche/Genentech	1998	Breast cancer

EGFR, epidermal growth factor; VEGF, vascular endothelial growth factor.

et al., 2002). Campath has also been used in combination with Rituximab and showed a 63 % response in patients with CLL that were resistant to conventional chemotherapy (Faderl *et al.*, 2003).

Bevacizumab: a monoclonal antibody against vascular endothelial growth factor (VEGF)

Angiogenesis is a critical step in tumour cell proliferation, growth, and metastasis. VEGF plays an integral role in tumour neovascularization and has been shown to be highly expressed in several cancer patients. Bevacizumab (Avastin), a humanized monoclonal antibody that targets the VEGF signalling pathway, was the first commercially available angiogenesis inhibitor. Bevacizumab stops tumour growth by preventing the formation of new blood vessels by targeting and inhibiting the function of VEGF that stimulates new blood vessel formation. The drug was first developed as a genetically engineered version of a mouse antibody that contains both human and mouse components. The FDA approved bevacizumab for use in colon cancer in 2004. It was developed by Genentech and is marketed, in the United States by Genentech, and elsewhere by Roche (Genentech's parent company). It is used in combination with standard chemotherapy drugs in patients with metastatic colorectal cancer.

Bevacizumab has yielded improved clinical outcomes when combined with other regimens in colon, lung and breast cancers (Sandler *et al.*, 2005). Avastin is the third top selling anticancer drug (sales $2.2 billion in 2005) after Rituximab and Imatinib (Glivec).

A phase II trial conducted by the Cancer and Leukaemia Group B (CALGB 90006) evaluated the efficacy of bevacizumab in combination with docetaxel/estramustine in patients with progressive metastatic hormone resistant prostate cancer (HRPC). The initial results of CALGB 90006 showed that bevacizumab in combination with docetaxel/estramustine yielded a PSA RR of 77 % and a median time to PSA progression of 10.3 months (Di Lorenzo *et al.*, 2008)

Cetuximab: a monoclonal antibody against epidermal growth factor (EGFR)

Cetuximab (Erbitux, Merck) is a chimeric monoclonal antibody against the ligand binding site in the extracellular domain of EGFR for treatment of metastatic colorectal cancer and head and neck cancer. EGFR activation results in phosphorylation of several downstream intracellular substrates that induce cell proliferation, angiogenesis, and inhibition of apoptosis. Cetuximab competes with ligands for receptor binding, causing receptor internalization and preventing ligand-mediated receptor tyrosine kinase phosphorylation (Gill *et al.*, 1984; Peng *et al.*,1996, Lilenbaum 2006). Cetuximab has 10-fold greater affinity for EGFR than the natural ligand. Clinical trials have demonstrated that cetuximab is synergistic with chemotherapy for patients with colorectal cancer. As a result, cetuximab was approved by the FDA for use in metastatic colorectal cancer in February 2004 (Saltz *et al.*, 2004; Cunningham *et al.*, 2004). It is the 13th best selling anticancer drug with sales of $686 million in 2005. Cetuximab was approved by the FDA in March 2006 for use in combination with radiation therapy for treating squamous cell carcinoma of the head and neck (SCCHN) or as a single agent in patients who have had prior platinum-based therapy. In locally advanced head and neck carcinoma, cetuximab in combination with radiotherapy significantly improved survival compared with radiotherapy alone. Patients treated with cetuximab occasionally develop a magnesium-wasting syndrome with aberrant urinary excretion.

Gemtuzumab ozogamicin: a monoclonal antibody against CD33 linked to a cytotoxic agent

Gemtuzumab ozogamicin (Mylotarg) is a monoclonal antibody, marketed by Wyeth, used to treat acute myelogenous leukaemia in older patients (>60 years) who can not undergo other cytotoxic chemotherapy (Bross *et al.*, 2001, Giles *et al.*, 2001, Wadleighet *et al.*, 2003). It is a monoclonal antibody to CD33, linked to a cytotoxic antitumour antibiotic agent, calicheamicin. CD33 is expressed in most leukaemic blast cells but is not found on normal haematopoietic stem cells. Upon binding to the receptor, the antibody–CD33 is internalized into the cytoplasm, delivering the cytotoxic load inside the cell and causing double-strand DNA breaks and inhibition of DNA synthesis due to calicheamicin action.

Ibritumomab tiuxetan: a monoclonal antibody against CD20 for radioimmunotherapy

Ibritumomab tiuxetan (Zevalin) is a monoclonal antibody radioimmunotherapy treatment for some forms of B cell non-Hodgkin's lymphoma and was the first radioimmuno-agent to be approved by the FDA. Zevalin is a monoclonal mouse IgG1 antibody (Ibritumomab) conjugated to the chelator tiuxetan, to which a radioactive isotope (either ^{90}Y or ^{111}In) is added. In the same mode as Rituximab (see below), Zevalin binds to the CD20 antigen found on the surface of normal and malignant B cells (but not B-cell precursors), and triggers cell death *via* ADCC, CDC and apoptosis (Wiseman *et al.*, 2001, 2002). Additionally, releases radiation from the attached isotope (mostly beta emission), killing the target cell as well as some neighbouring cells. The combination of the above properties eliminate B cells from the body, allowing a new population of healthy B cells to develop from lymphoid stem cells with higher efficiency than Rituximab ($\sim$24 %) (Witzig *et al.*, 2002).

Panitumumab: a fully human monoclonal antibody against EGFR

Panitumumab (ABX-EGF) (Vectibix) is the first fully human monoclonal antibody that has received its license for clinical use on humans by the FDA, for patients suffering with non-curable colorectal cancer in September 2006 (Cohenuram and Saif, 2007). It is an IgG2 antibody, produced by immunization of transgenic mice that carry human genes and produce human immunoglobulin light and heavy chains. The mice were immunized and a specific clone of B cells that produce an antibody against EGFR was selected and immortalized for the generation of the antibody, using the hybridoma technology. It targets the extracellular ligand-binding domain of the receptor and acts in the same manner as Cetuximab, resulting in blockade of essential downstream signalling pathways that control apoptosis, proliferation and differentiation. Panitumumab is efficient in the treatment of a wide variety of cancer types, including non-small-cell lung cancer, renal, and colorectal cancer as monotherapy and in combination with standard chemotherapeutic agents. Unlike Cetixumab, it does not induce any immune response and there was no observation of unfavourable drug–drug interactions or any effect on the pharmacokinetic properties of other cytotoxic chemotherapeutic agents with which it is combined. Panitumumab has a very high affinity for epidermal growth factor receptor and has been generally well tolerated (Cohenuram and Saif, 2007). A mild to moderate rash has been observed, which is a common reaction among drugs targeting the EGFR. Currently, several studies are undergoing using panitumumab in combination with irinotecan and fluorouracil or FOLFOX (fluorouracil/leucovorin/oxaliplatin) for the treatment of colorectal cancer (Saif *et al.*, 2006).

Rituximab: a monoclonal antibody against CD20

Rituximab, (Rituxan and MabThera), is a monoclonal chimeric antibody against the B-cell surface antigen CD20 (Binder at al., 2006), which is present in 95 % of non-Hodgkin's lymphoma (NHL) cells (Avivi *et al.*, 2003). Rituximab is the first monoclonal antibody approved by FDA in 1997 for relapsed or refractory low-grade or follicular CD20-positive, B-cell lymphomas. Marketed as Mabthera by Roche it was the top selling drug against cancer in 2005 with sales of 2.7 billion USD. Although originally developed against B-cell lymphoma, it has now found applications in other diseases such as autoimmune diseases, including rheumatoid arthritis and systemic lupus erythematosis (SLE) (Kazkaz and Isenber, 2004). The antibody binds to CD20 molecules on the B-cell membrane through its murine anti-human CD20 portion, and targeting them for destruction by the body's own immune system. The actual mechanism for Rituximab is to eliminate B cells inducing the ADCC, CDC, responses and apoptosis through the human IgG1 portion of the antibody and especially the Fc fragment.

This results in the elimination of B cells (including the cancerous ones) from the body, allowing a new population of healthy B cells to develop from lymphoid stem cells. Rituximab is efficiently used in bone marrow transplantation and in the management of renal transplant recipients due to its property to kill lymphoma cells that cause rejection of grafts to the recipient (Hale *et al.*, 1983). It is especially useful in transplants involving incompatible blood groups and is also used as induction therapy in highly sensitized patients going for renal transplantation.

The most common side effects include fever, chills and malaise even though premedication with acetaminophen (paracetamol) and diphenhydramine is being administered. Nausea, vomiting, flushing, urticaria, angioedema, hypotension, dyspnoea, bronchospasm, fatigue, headache, rhinitis and pain at disease sites are other infusion-related side effects. These symptoms are self-limited and are generally improved with longer infusions. Other uncommon symptoms include short-lived myelosuppression, abdominal pain, myalgia, arrhythmias and angina pectoris (Lamanna *et al.*, 2006; Hiddemann, 2005).

Tositumomab: a monoclonal antibody against CD20 covalently bound to [131]I

Tositumomab (Bexxar) is a monoclonal antibody derived from immortalized mouse cells. It is an IgG2a anti-CD20 antibody and is covalently bound to [131]I. It has been approved by the FDA in 2003 and was used to treat CD20 expressing lymphomas that were resistant to Rituximab (Wahl, 2005). Its mechanism of action is similar to Zevalin radioimmunotherapy.[131]I emits both beta and gamma radiation, and is broken down rapidly in the body. Clinical trials have established the efficacy of tositumomab in patients with relapsed follicular lymphoma in approximately 65 % (Kaminski *et al.*, 2001). Bexxar advantage over conventional radiotherapy is the targeted delivery of radiation to cancer cells due to the specific antibody binding that significantly reduces toxicity to other organs. Radioimmunoconjugates usually cause myelosuppression, which is reversible after treatment. In general, side effects are manageable.

Trastuzumab: a monoclonal antibody against HER2/neu

Trastuzumab (Herceptin) is a humanized mAb against the extracellular domain of HER2/neu (erbB2) receptor. The Her2/neu receptor is overexpressed in 25–30 % of breast cancers with metastatic aggressiveness and poor prognosis. Her2/neu encodes a transmembrane tyrosine kinase p185-erbB2 glycoprotein and is part of a complicated signalling cascade that involves the activation of PI3K/Akt and MAPK pathways and the promotion of invasion, survival and angiogenesis of cells (Nahta *et al.*, 2003). Herceptin is used against breast cancer in patients whose tumours overexpress HER2/neu with a response rate approximately 11 %, although it resulted in stabilization of tumours in many cases. It was the fourth top selling anticancer drug in 2005 with sales of ∼$2.2 billion.

Herceptin acts through its Fc domain and thus inducing an ADCC immune response and cancer cell lysis. It also inhibits the proteolytic cleavage of the Her2/neu extracellular domain which causes the constitutive activation of the receptor and loss of cell proliferation control. It is believed that Herceptin arrest cells in the G_1 phase of the cell cycle and has also been suggested that its antiproliferative effects are due to the downregulation of the Her2/neu receptor and the inhibition of its dimerization that leads to receptor activation (Baselga *et al.*, 2001; Molina *et al.*, 2001).

An international multicentre randomized trial showed that 1 year of treatment with trastuzumab after adjuvant chemotherapy has a significant overall survival benefit after a median follow-up of 2 years (Romond *et al.*, 2005, Smith *et al.*, 2007).

Herceptin is administered either once a week or once every three weeks intravenously for 30 to 90 min. Herceptin is administered in combination with chemotherapy such as Adriamycin, cisplatin and taxanes. Herceptin is associated with cardiac dysfunction in 2–7 % that can advance in severe cardiotoxicity and congestive heart failure when combined with Adriamycin and anthracycline substances, which should be avoided (Slamon *et al.*, 2001).

15.3 Cell Immunotherapy

Cell immunotherapy describes the adoptive transfer of autologous T cells or donor T cells after *ex vivo* activation and expanding with the use of specific cytokines and adjuvants. Additionally it can also embraces the whole-cell vaccine approach as well as the gene therapy field when employes genetic modidifications of T cells (Guinn *et al.*, 2007; June, 2007; Gattinoni *et al.*, 2006).

Cells undergoing malignant transformation are believed to be eliminated from the body by white blood cells including natural killer cells (NK), lymphokine-activated killer cells (LAK), cytotoxic T lymphocytes (CTL), tumour-infiltrating lymphocytes (TIL), and activated macrophages. Lymphocytes, upon encountering an antigen, initiate a cascade of events that leads to either a T helper 1 (Th1) or a Th2 immune response. Dendritic cells recognize various pathogens and foreign substances and present their antigens to the immune system to make the most appropriate type of response. Viruses, bacteria, and abnormal cells such as cancer cells, trigger a Th1 response, whereas detection of a parasite infection results in the initiation of a Th2 response. The Th1 response leads to the production of specific cytokines, including interferon-alpha (INF-α), interferon-gamma (INF-γ), and interleukins (ILs) as well as the generation of natural killer (NK) and specific protective IgG antibodies from B cells. Cytokines including IL-2, 4, 7, 9, 15 and 21 have been shown to program the effector cytotoxic T cells (CD8$^+$) and together these two factors comprise the body's most potent anti-infective weapons. A Th1 response to a specific antigen or allergen provides memory for subsequent encounter with the same antigen. Activation of the Th2 response results in the production of other cytokines, IL-4, IL-5 and IL-13 and IgE antibodies. These cytokines attract inflammatory cells such as eosinophils, basophils, and mast cells capable of destroying the invading organism.

15.3.1 Cytokines/adjuvants

Since established cancers in the human body may escape this potential defence mechanism of immunologic surveillance, cancer patients have been treated with various cytokines such as IL-2, IL-12, IL-7, INF-γ and other co-stimulatory molecules to induce their cellular immune mechanisms and kill cancer cells. Stable and complete remissions, however, were at a low rate and complications were encountered by the toxicity caused by the systemic administration of cytokines mainly to acute immune response. IL-2 and IL-12 mechanisms of action are described in the following paragraphs as an example.

15.3.2 IL-2

IL-2 had initially shown great promise in cancer immunotherapy. It promotes its signal through γ_c containing receptors and is widely used both *in vitro* and *iv vivo* to stimulate the differentiation of precursor lymphocytes into LAK cells and further promote T-cell proliferation (Waldmann, 2006). IL-2-based regimens shown to activate cellular antitumour immunity and are the basis of immunotherapies against melanoma and kidney. In contrast IL-2 can result in increased toxicity by inducing the production of cytokines and their elevated accumulation in the body. Recent finding show IL-2 can lead to T regulatory expansion and induction of self-tolerance and immunosuppression. Because IL-2 effects are so ambiguous other γ_c signaling cytokines such as IL-7, IL-15, and IL-21 are considered more beneficial (Colombo and Piconese, 2007).

15.3.3 IL-12

Increasing interest has been recently accumulated in the anti-tumour effects of IL-12, which mainly targets and activates lymphocytes (Trinchieri, 2003). Phagocytes and dendritic cells are the basic physiological producers of IL-12 in response to pathogens, inflammatory components or activated T cells (Figure 15.3). IL-12 induces the production of INF-γ resulting in increased cytotoxicity and anti-angiogenesis effect of the innate resistance machinery. It has also shown to favour a Th1 response and promote the production of specific class of immunoglobulins from B cells that generates a tumour antigen-specific adaptive immunity (Figure 15.4).

Preclinical studies showed great promise for the use of IL-12 as a powerful anticancer agent. *In vivo* treatment with IL-12 has been shown to have a significant antitumour effect on mouse tumours, by inhibiting establishment of tumours or by inducing regression of

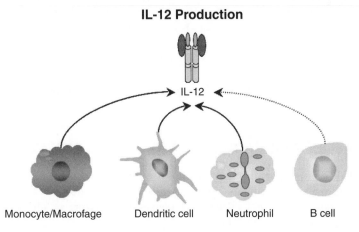

IL-12 Production

Monocyte/Macrofage   Dendritic cell   Neutrophil   B cell

Figure 15.3 IL-12 is a heterodimeric pro-inflammatory cytokine that induces the production of IFN-g and favours the Th1 response. The phagocytes (monocytes, macrophages and neutrophils) and dendritic cells are the main physiological producers of IL-12

Anti-tumour activity of IL-12

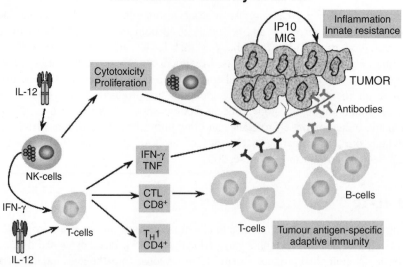

Figure 15.4 IFN-γ induced by IL-12 has a direct toxic effect on the tumour cells and/or might activate potent anti-angiogenic mechanisms. IL-12, by inducing Th1 responses, also augments the production of opsonizing and complement-fixing classes of IgG antibodies with antitumour activity

established tumours. As with IL-2 administration the induction of an inflammatory response by IL-12 can also result in toxicity resembling the response of acute infection.

Clinical utility of gene therapy systems (see Chapter 16) has been the main focus on expressing cytokines and exploiting their anti-cancer properties. LipoVIL12 (patented by Regulon Inc) is a liposomally encapsulated Semliki Forest Virus (SFV) carrying both p35 and p40 subunits of the human IL-12 gene and expressing a functional and secreted IL-12 with a potential in immunotherapy of cancer but also against viral and other infections. LipoVIL12 has been approved for phase I/II clinical trials in Greece against major human malignancies.

Cytokines are essential for the *ex vivo* approaches in immunotherapy in stimulating the isolated autologous T cells, in cell culture *(in vitro)*, before their transfer back to the patient. The ultimate goal is the activation of tumour-specific T lymphocytes capable of rejecting tumour cells from patients.

15.3.4 Cytolytic T cells (CTLs)

CTLs (isolated from peripheral blood lymphocytes/PBLs) transplantation has been used in clinical trials in melanoma patients with refractory metastatic disease, with an overall low antitumour response. Cancer cells develop antigen escape variants and this is a major drawback in CTL therapy that is soon disarmed.

15.3.5 Tumour-infiltrating lymphocytes (TILs)

TIL-based therapies have also been applied in patients with metastatic melanoma, with disappointing results for the most part. TIL isolation and expansion is a laborious and time demanding procedure and requires availability of tumour tissue, which has usually a low yield (30–40 %) of T cells, one of the main reasons TIL therapies have only been applied in melanoma patients, where biopsies are of a more accessible nature. On the other hand, TILs are still a promising immunological approach, since recent findings have correlated their presence with better survival outcomes in patients with ovarian and colorectal cancers. As we have already mentioned, adoptive cell transfer therapies achieve T-cell stimulation *ex vivo* by activating and expanding autologous tumour-reactive T-cell populations to large numbers of cells that are then transferred back to the patient. TILs are selected for their ability to secrete high levels of IFN-γ when cultured with autologous or allogeneic major histocompatibility complex (MHC)-matched tumour cell lines. Co-transfer with T helper cells and cytokines also enhances adoptive transfer effect and is a common strategy in many protocols.

15.3.6 Haematopoietic stem cell transplatation (HSCT)

HSCT from an autologous or allogeneic source is currently one of the most potent immunotherapeutic protocols especially for haematological malignancies. More interesting is a new perspective of the allogeneic non-myeloblative HSCT for the treatment of human solid tumours. A non-myeloblastive treatment describes the induction (usually through chemotherapy) of a transient leucopenia to patients that do not permanently destroy HSC, thus allowing the recovery of the hosts' haematological function. It has been observed that a graft-versus-tumour effect (GVT) can be generated in patients with metastatic renal cell cancer after HSCT and this response is investigated for other solid tumours (Renga *et al.*, 2003, Takahashi Y *et al.*, 2008). Preconditioning treatments also include HSCT.

15.3.7 Genetically engineered T cells

The development of *genetically engineered T cells* tried to overcome the problems of the antigen variant escape, low persistence of allogeneic T-cell or lethal graft-versus-host-disease (GVHD). Genetically modified T cells have been shown to persist for years in humans following adoptive transfer. Although several safety issues may arise from genetic modifications such as in HSCs that can result in cellular transformation, investigators further face these problems by an effort to control expression using for example tissue specific inducible promoters. Furthermore, through expression of suicide molecules, such as herpes simplex virus thymidine kinase (HSV-TK), fusion proteins contain FAS or caspase death domain, it is possible to control the transplanted T cell population and silence expression for long-term approaches, limiting the GVHD result. The suicide proteins themselves can be proved immunogenic to the patient, so the next step is to employ less immunogenic endogenous proteins.

Loss of tumour immunogenicity during T-cell therapy is possible to overcome by exploiting a broad range of tumour antigen-specific T-cell repertoire and increasing the efficiency of epitope spreading or utilizing CTL clones with multiple antigenic specificities. The engineering of novel chimeric receptors, that have antibody-based external receptor structures and cytosolic domains that encode signal transduction modules of the T-cell receptor, is another strategy to augment the efficacy of adoptively transferred T cells that have higher specificity for the target antigen and escape MHC restriction. The generation of T cells transduced to express selected T-cell receptors (TCR) of known specificity and avidity for tumour antigen is believed to hold a great potential in immunotherapy protocols, since they are powerful tools in targeting target cells and most importantly have not been edited by the host thymic tolerance mechanisms.

Tumour cells have also been used in clinical protocols. A recent example is the GVAX™ cancer immunotherapeutic that comprises tumour cells that are genetically modified to secrete a cytokine, granulocyte–macrophage colony-stimulating factor (GM-CSF), and are irradiated in order to deactivate their malignant potential. GVAX™ is currently in phase II clinical trials for NSCLC (www.cellgenesys.com).

Immune tolerance is a basic factor that can limit the outcome of cell immunotherapy. Preconditioning regimens that result in immunosuppression of the host improves the anti-tumour efficacy of adoptive TIL therapy. Lymphodepletion in patients, prior to the adoptive cell transfer, is known to significantly increase antitumour immunity. Moreover, specific lymphodepletion for the elimination of the immunosuppressive cells, such as regulatory T cells, can minimize the toxic effects of standard non-specific preconditioning protocols using chemotherapy or radiotherapy. In the past few years, researchers have identified that a population of cells, known as myeloid-derived suppressor cells (MDSCs), is markedly increased in cancer patients. MDSCs are potent suppressors of the immune system, with mechanisms that can block both the innate and adoptive pathways, and might play a role in the original growth of cancers and in the failure of immunotherapies.

Currently, there are no FDA-approved adoptive T-cell therapies for cancer and randomized clinical trials are especially difficult to design and perform, since they require the development of personalized vaccines, which are technically demanding, as well as needing preconditioning of the patient. Nevertheless, hope for T-cell immunotherapy has not been abandoned, but in contrast, the establishment of improved cell culture techniques, genetic engineering modifications and more specific preconditioning regimens, have evolved the protocols and reinforced the initial expectations. Data accumulation has also been significant for the determination of the doses, duration, and number of infusion of T cells, for accomplishment of the maximum immune response and lowest toxicity. An advantage in the combination approaches of vaccines and T-cell therapy for the designing of clinical protocols has been the positive outcome in improving boosting of the host's immune response.

Final issues for the feasibility of cell immunotherapy are the manufacturing of this laborious therapy on a large scale, under GMP standards, and at a permissive cost for its clinical use.

15.4 Cancer Vaccines

Anticancer vaccination embodies an ideal non-toxic treatment, capable of evoking tumour-specific immune responses that can ultimately recognize and kill cancer cells.

There are two main categories that cancer vaccines fit into: specific cancer vaccines and universal cancer vaccines. As the name already suggests, specific cancer vaccines are designed to treat specific types of cancers (Coico and Eli, 2003; Janeway et al., 2004; Goldsby and Orsborne, 2005). A more attractive cancer vaccine would be one that could fight cancer cells regardless of cancer type. This type of vaccine is called a universal cancer vaccine. Within these two categories there are more specific types of cancer vaccines. Currently, five types of cancer vaccines are being developed: antigen vaccines, anti-idiotypic vaccines, dendritic cell vaccines, DNA vaccines and tumour cell vaccines. Whichever the type, a cancer vaccine works on the same basic principle: the vaccine, which contains tumour cells or antigens, stimulates the patient's immune system, which produces T-cell-mediated responses and prevents the relapse of the cancer (Coico and Eli, 2003; Janeway *et al.*, 2004; Goldsby and Orsborne, 2005).

Two examples of successful cancer vaccines licensed by the FDA comprise a prophylactic vaccine against hepatitis B virus, an infectious agent associated with liver cancer (Pagliusi, 2004) and, more recently, Gardasil, a human papillomavirus (HPV) vaccine, for the prevention and treatment of genital HPV infection and cervical cancer (Schiller and Lowy, 2006).

Cancer vaccines offer an attractive therapeutic addition, delivering treatment of high specificity, low toxicity and prolonged activity. No vaccination regimen has achieved sufficient therapeutic efficacy, necessary for clinical implementation, and a reproducible survival benefit has proved elusive. Nevertheless, several immunological advances have opened new avenues of research to decipher the biological code governing tumour immune responsiveness, and this is leading to the design of potentially more effective immunotherapeutic protocols. One hundred and five different pipeline cancer vaccines have been identified, of which 14 are in late-phase development. These existing candidates have a forecast sales potential of up to $3.1 billion in the seven major pharmaceutical markets by 2015. Cell-based vaccines use dendritic cells in combination with a specific anticancer antigen, while DNA-based vaccines utilize the properties of a wide range of genetically modified viruses to infect cells and express immunogenic genes. Some currently developed DNA cancer vaccines are listed in Table 15.2.

15.4.1 Dendritic cells

Dendritic cells from patients with cancer are deficient in number and functional activity, leading to inadequate tumour immunosurveillance, as a result of poor induction of T-cell antitumour responses. Loaded dendritic cell therapy is a vaccination strategy aimed at eliciting tumour antigen-specific, T-cell immune responses. Dendreon, in association with the Mayo Clinic, is developing the dendritic cell therapy APC-8015 (Provenge, Dendreon Corp., Seattle, WA) for the potential treatment of hormone-refractory prostate cancer. Provenge involves the use of a proprietary recombinant antigen, derived from prostatic acid phosphatase, found in approximately 95 % of prostate cancers. The target antigen is combined with the patient's own dendritic cells and reinfused into the patient to stimulate an immune response. The patents cover the composition of the prostate tumour antigen engineered by Dendreon to help stimulate the immune system. Dendreon's preregistration status uses dendritic cells for cancer vaccines. Provenge as an immunogen has shown a survival benefit in patients with metastatic hormone-refractory prostate cancer in a randomized phase III trial initiated in January 2000 (reviewed by Lin *et al.*, 2006). However, scepticism still surrounds the use of dendritic cells as a viable technology platform. Although personalized immunotherapies might offer a greater level of specificity and lower toxicity, generalized alternatives should facilitate production, help achieve economies of scale and offer broader utility in a range of tumour types.

15.4.2 Immunotherapy with synthetic tumour peptide vaccines

Cancer immunotherapies based on synthetic tumour peptide vaccines have also been developed. Tumour-specific $CD8^+$ CTLs recognize short peptide epitopes presented by MHC class I molecules that are expressed on the surface of cancer cells. Bone marrow-derived dendritic cells, grown *in vitro* in media containing combinations of GM-CSF + IL-4, when pulsed with synthetic tumour peptides (which are loaded on the surface of the dendritic cells) become potent antigen-presenting cells (APCs) capable of generating a protective antitumour immune response. Injection of these cells into naive mice protected the mice against a subsequent lethal tumour challenge; in addition, treatment of mice bearing C3 sarcoma or 3LL lung carcinoma tumours with the same type of cells resulted in sustained tumour regression in over 80 % of the animals. One of the obstacles of this method has been the difficulty in obtaining sufficient numbers of APCs; dendritic APCs have been isolated from CD34$^+$ haematopoietic progenitor cells drawn from cord blood and expanded in cell culture in the presence of GM-CSF and tumour necrosis factor-α (TNF-α); TNF-α inhibits the differentiation of dendritic cells into granulocytes. Human peripheral blood mononuclear cells or mouse bone marrow cells depleted of lymphocytes could also yield dendritic cells when cultured in the presence of GM-CSF + IL-4. One example of a synthetic peptide vaccine that is currently in clinical trials is the MUC1 vaccine, developed by Biomira and recently licensed by Merck. This vaccine is composed of a 100-amino acid peptide, corresponding to five 20-amino acid long repeats of the MUC1 variable tandem repeat region, and SB-AS2 adjuvant, and tested on patients with resected or locally advanced pancreatic cancer. This was found to be a safe vaccine that induced low but detectable mucin-specific humoral and T-cell responses in patients (Ramanathan *et al.*, 2005). A similar MUC1 peptide vaccine tried by Biomira in phase IIb clinical trials is BLP25, a liposomal formulation of the MUC1 peptide. BLP25 liposomal vaccine is a synthetic MUC1 peptide vaccine. The BLP25 liposomal vaccine incorporates a 25-amino acid sequence of the MUC1 cancer mucin, encapsulated in a liposomal delivery system. The liposome enhances recognition of the cancer antigen by the immune system and facilitates better delivery (Murray *et al.*, 2005).

15.4.3 DNA vaccines

Highly engineered vaccines have incorporated novel tumour antigens and ultimately combinations of antigens that can elicit an even more potent immune response. A list of the recently developed vaccines is presented in Table 15.2. Selective vaccines are briefly discussed below.

Gardasil is the only vaccine licensed (by Merck) until now that may help protect against HPV Types 6, 11, 16, and 18 (Wheeler, 2007). HPV types 16 and 18 cause about 70 % of HPV-related cervical cancer cases, which is one of the most common causes of death from cancer in women. Gardasil contains recombinant virus-like particles (VLPs) assembled from the L1 proteins of HPVs 6, 11, 16 and 18. The recombinant viruses lack viral DNA, so they are deficient in causing cancer. They aim is to trigger an immune response and generate specific protective antibodies for any possible infection and encounter with the HPV types mentioned above. Cervarix is an analogous vaccine developed by GlaxoSmithKine that is under phase II clinical trials.

TNFrade™ Biologic is an adenovector, expressing the gene for TNF-a, an anticancer immune agent. The vaccine is designed for the direct injection to the tumour site in combination with standard radiation and/or chemotherapy (5-fluorouracil/radiation therapy followed by gemcitabine or gemcitabine/erlotinib maintenance therapy) and the production of TNF-α. Currently in phase II/III clinical trials TNFrade™ has show promising antitumour results against a variety of solid tumours, such as pancreatic cancer.

Table 15.2   Recently developed DNA vaccines

Company	Leading product	Target	Clinical status
Merck www.merck.com	Gardasil	HPV	Clinic-licensed
Oxford BioMedica www.oxfordbio medica.co.uk	Adenovirus vaccine Trovax pox virus vector (modified vaccinia ankara, MVA)	5T4 cancer antigen delivery/cancer	Phase III: renal cell carcinoma
			Phase II: breast colorectal, prostate cancer
GenVec www.genvec.com	TNFrade™ Adenovirus vector+TNF-α gene	TNF-α anticancer agent delivery/cancer	Phase II/III: pancreatic, rectal cancer
			Phase II: metastatic melanoma
Transgene www.transgene.fr	TG4010 MVA vector (pox) + MUC1	Non-small-cell-lung Carcinoma	Phase IIb
'	TG4001 MVA vector + E6, E7 Ag of HPV16+IL2	HPV-induced precancerous cervical lesions/chronic infections	Phase II
'	TG1042 Adenovirus vector + interferon-γ	Cutaneous B- and T-cell lymphoma	Phase I/II

LentiMax™ is a highly optimized HIV-based lentiviral vector (LV) that can be engineered to express any gene or gene silencing RNA of interest. LVs are viral-based gene delivery systems that can stably deliver genes or RNAi into primary cells or cell lines with up to 100 % efficiency. LVs bind to target cells using an envelope protein, which allows for release of the LV RNA containing the gene or gene silencing sequence into the cell. The LV RNA is then converted into DNA using an enzyme called reverse transcriptase by a process called reverse transcription. The DNA preintegration complex then enters the nucleus and integrates into the target cell's chromosomal DNA. Gene delivery is stable because the target gene or gene silencing sequence is integrated in the chromosome and is copied along with the DNA of the cell every time the cell divides. One of the discriminating features of LVs is their ability to integrate into non-dividing cells, in contrast to other vectors that either don't integrate efficiently into chromosomal DNA (e.g. non-viral, adenoviral and adenoviral-associated vectors) or can only integrate upon cell division (e.g. conventional retroviral vectors).

The immune system comprises one of the most complex and powerful mechanisms that control a plethora of functions and protects the organism from factors that threat its normal healthy state including the recognition and elimination of malignant cells. This potent tool, cancer immunotherapy aims to manipulate and exploit every component towards the development of therapeutic protocols that will succeed to target cancer with the highest specificity and more importantly to limit the deleterious side effects of current chemotherapeutic treatments.

References

Albanell J, Codony J, Rovira A, Mellado B, Gascon P. Mechanism of action of anti-HER2 monoclonal antibodies: scientific update on trastuzumab and 2C4. *Adv Exp Med Biol* 2003, **532**, 253–68.

Astsaturov I, Cohen RB, Harari PM. EGFR-targeting monoclonal antibodies in head and neck cancer. *Curr Cancer Drug Targets* 2006, **6**(8), 691–710.

Avivi I, Robinson S, Goldstone A. Clinical use of rituximab in haematological malignancies. *Br J Cancer* 2003, **89**(8), 1389–94. Review.

Baselga J, Albanell J, Molina MA, Arribas J. Mechanism of action of trastuzumab and scientific update. *Semin Oncol* 2001, **28**(5 Suppl 16), 4–11.

Bell S, Kamm MA. Antibodies to tumour necrosis factor alpha as treatment for Crohn's disease. *Lancet* 2000, **355**, 858–60.

Better M, Chang CP, Robinson RR, Horwitz AH. *Escherichia coli* secretion of an active chimeric antibody fragment. *Science* 1988, **240**, 1041–3.

Binder M, Otto F, Mertelsmann R, Veelken H, Trepel M. 'The epitope recognized by rituximab. *Blood* 2006, **108**, 1975–8.

Boulianne GL, Hozumi N, Shulman MJ. Production of functional chimaeric mouse/human antibody. *Nature* 1984, **312**, 643–6.

Brekke OH, Sandlie I. Therapeutic antibodies for human diseases at the dawn of the twenty-first century. *Nat Rev Drug Discov* 2003, **2**, 52–62.

Bross PF, Beitz J, Chen G, *et al.* Approval summary: gemtuzumab ozogamicin in relapsed acute myeloid leukaemia. *Clin Cancer Res* 2001, **7**(6), 1490–6.

Carter P. Improving the efficacy of antibody-based cancer therapies. *Nat Rev Cancer* 2001, **1**, 118–29.

Cohenuram M, Saif MW. Panitumumab the first fully human monoclonal antibody: from the bench to the clinic. *Anticancer Drugs* 2007, **18**(1), 7–15.

Coico R SG, Benjamini Eli. *Immunology: a Short Course.* Wiley-Liss, New York, 2003.

Colombo MP and Piconese S. Opinion: Regulatory T-cell inhibition versus depletion: the right choice in cancer immunotherapy. *Nat Rev Cancer* 2007, **7**, 880–887.

Cragg MS, French RR, Glennie MJ. Signalling antibodies in cancer therapy. *Curr Opin Immunol* 1999, **11**, 541–7.

Cunningham D, Humblet Y, Siena S, *et al.* Cetuximab monotherapy and cetuximab plus irinotecan in irinotecan-refractory metastatic colorectal cancer. *N Engl J Med* 2004, **351**, 337–45.

Curnow RT. Clinical experience with CD64-directed immunotherapy. An overview. *Cancer Immunol Immunother* 1997, **45**, 210–11.

Di Lorenzo G, Figg WD, Fossa SD, *et al.* Combination of bevacizumab and docetaxel in docetaxel-pretreated hormone-refractory prostate cancer: a phase 2 study. *Eur Urol* 2008, [Epub ahead of print]

Faderl S, Thomas DA, O'Brien S, *et al.* Experience with alemtuzumab plus rituximab in patients with relapsed and refractory lymphoid malignancies. *Blood* 2003, **101**(9), 3413–5.

Gattinoni L, Powell Jr. DJ, Rosenberg SA, Restifo P. Adoptive immunotherapy for cancer: building on success. *Nat Rev Immunol* 2006, **6**, 383–93.

Giles FJ, Kantarjian HM, Kornblau SM, *et al.* Mylotarg (gemtuzumab ozogamicin) therapy is associated with hepatic venoocclusive disease in patients who have not received stem cell transplantation. *Cancer* 2001, **92**(2), 406–13.

Gill GN, Kawamoto T, Cochet C, *et al.* Monoclonal anti-epidermal growth factor receptor antibodies which are inhibitors of epidermal growth factor binding and antagonists of epidermal growth factor binding and antagonists of epidermal growth factor-stimulated tyrosine protein kinase activity. *J Biol Chem* 1984, **259**, 7755–60.

Goldsby RA, Orsborne KT, Kuby BA. *Immunology*, 4th edition. WH Freeman & Company, New York, 2005, p 670.

Govindan R. Cetuximab in advanced non-small cell lung cancer. *Clin Cancer Res* 2004, **10**(12 Pt 2): 4241s–4s.

Goren D, Horowitz AT, Zalipsky S, Woodle MC, Yarden Y, Gabizon A. Targeting of stealth liposomes to erbB-2 (Her/2) receptor: *in vitro* and in vivo studies. *Br J Cancer* 1996, **74**, 1749–56.

Green LL. Antibody engineering via genetic engineering of the mouse: XenoMouse strains are a vehicle for the facile generation of therapeutic human monoclonal antibodies. *J Immunol Methods* 1999, **231**, 11–23.

Guinn B, Kasahara N, Farzaneh F, Habib NA, Norris JS, Deisseroth AB. Recent advances and current challenges in tumor immunology and immunotherapy. *Mol Ther* 2007, **15**(6), 1065–71.

Hale G, Bright S, Chumbley G, Hoang T, Metcalf D, Munro AJ, Waldmann H. Removal of T cells from bone marrow for transplantation: a monoclonal antilymphocyte antibody that fixes human complement. *Blood* 1983, **62**, 873–82.

Hiddemann W, Buske C, Dreyling M, Weigert O, Lenz G, Forstpointner R, Nickenig C, Unterhalt M, Treatment strategies in follicular lymphomas: current status and future perspectives. *J Clin Oncol* 2005, **23**(26), 6394–9.

Holliger P, Prospero T, Winter G. 'Diabodies' small bivalent and bispecific antibody fragments. *Proc Natl Acad Sci U S A* 1993, **90**, 6444–8.

Hudziak RM, Lewis G D, Winget M, Fendly BM, Shepard HM, Ullrich A. p185HER2 monoclonal antibody has antiproliferative effects *in vitro* and sensitizes human breast tumor cells to tumor necrosis factor. *Mol Cell Biol* 1989, **9**, 1165–72.

Hurwitz E, Stancovski I, Sela M, Yarden Y. Suppression and promotion of tumour growth by monoclonal antibodies to ErbB-2 differentially correlate with cellular uptake. *Proc Natl Acad Sci U S A* 1995, **92**, 3353–7.

Hynes NE, Lane HA. ERBB receptors and cancer: the complexity of targeted inhibitors. *Nat Rev Cancer* 2005, **5**(5), 341–54. Review. Erratum in: *Nat Rev Cancer* 2005, **5**(7), 580.

Iliades P, Kortt AA, Hudson PJ. Triabodies: single chain Fv fragments without a linker form trivalent trimers. *FEBS Lett* 1997, **409**, 437–41.

Janeway CA TP, Walport M, Schlomchik MJ. *Immunobiology*, 4th edition. Garland Publishing, New York, 2004, pp 14–1, 5.

Jones PT, Dear PH, Foote J, Neuberger MS, Winter G. Replacing the complementarity-determining regions in a human antibody with those from a mouse. *Nature* 1986, **321**, 522–5.

June CH. Adoptive T cell therapy for cancer in the clinic. *J Clin Invest* 2007, **117**(6), 1466–76.

Kaminski MS, Zelenetz AD, Press OW, *et al*. Pivotal study of iodine I 131 tositumomab for chemotherapy-refractory low-grade or transformed low-grade B-cell non-Hodgkin's lymphomas. *J Clin Oncol* 2001, **19**(19), 3918–28.

Kazkaz H, Isenberg D. Anti B cell therapy (rituximab) in the treatment of autoimmune diseases. *Curr Opin Pharmacol* 2004, **4**(4), 398–402. Review.

Keating MJ, Flinn I, Jain V, *et al*. Therapeutic role of alemtuzumab (Campath-1H) in patients who have failed fludarabine: results of a large international study. *Blood* 2002, **99**(10), 3554–61.

Khazaeli MB, Conry RM, LoBuglio AF. Human immune response to monoclonal antibodies. *J Immunother* 1994, **15**, 42–52.

Kohler G, Milstein C. Continuous cultures of fused cells secreting antibody of predefined specificity. *Nature* 1975, **256**, 495–7.

Kontermann RE, Martineau P, Cummings CE, Karpas A, Allen D, Derbyshire E, Winter G. Enzyme immunoassays using bispecific diabodies. *Immunotechnology* 1997, **3**, 137–44.

Kortt AA, Dolezal O, Power BE, Hudson PJ. Dimeric and trimeric antibodies: high avidity scFvs for cancer targeting. *Biomol* 2001, **18**, 95–108.

Lamanna N, Kalaycio M, Maslak P, *et al*. Pentostatin, cyclophosphamide, and rituximab is an active, well-tolerated regimen for patients with previously treated chronic lymphocytic leukaemia. *J Clin Oncol* 2006, **24**(10), 1575–81.

Larrick JW, Fry KE. Recombinant antibodies. *Hum Antibodies Hybridomas* 1991, **2**, 172–89.

Lilenbaum RC. The evolving role of cetuximab in non-small cell lung cancer. *Clin Cancer Res* 2006, **12**(14 Pt 2), 4432s–5s.

Lin AM, Hershberg RM, Small EJ. Immunotherapy for prostate cancer using prostatic acid phosphatase loaded antigen presenting cells. *Urol Oncol* 2006, **24**(5), 434–41. Review.

McCafferty J, Griffiths AD, Winter G, Chiswell DJ. Phage antibodies: filamentous phage displaying antibody variable domains. *Nature* 1990, **348**, 552–4.

Mendez MJ, Green LL, Corvalan JR, *et al*. Functional transplant of megabase human immunoglobulin loci recapitulates human antibody response in mice. *Nat Genet* 1997, **15**, 146–56.

Molina MA, Codony-Servat J, Albanell J, Rojo F, Arribas J, Baselga J. Trastuzumab (herceptin), a humanized anti-Her2 receptor monoclonal antibody, inhibits basal and activated Her2 ectodomain cleavage in breast cancer cells. *Cancer Res* 2001, **61**(12), 4744–9.

Morrison SL, Johnson MJ, Herzenberg LA, Oi VT. 1984. Chimeric human antibody molecules: mouse antigen-binding domains with human constant region domains. *Proc Natl Acad Sci U S A* **81**, 6851–5.

Murray N, Butts C, Maksymiuk A, *et al*. A liposomal MUC1 vaccine for treatment of non-small cell lung cancer (NSCLC); updated survival results from patients with stage IIIB disease. *J Clin Oncol* 2005, **23**(16S), 7037.

Nahta R, Esteval FJ. HER-2-targeted therapy – lessons learned and future directions. *Clin Cancer Res* 2003, **9**, 5078–48.

Nicholson IC, Zou X, Popov AV, *et al*. Antibody repertoires of four- and five-feature translocus mice carrying human immunoglobulin heavy chain and kappa and lambda light chain yeast artificial chromosomes. *J Immunol* 1999, **163**, 6898–906.

Nielsen UB, Kirpotin DB, Pickering EM, *et al*. Therapeutic efficacy of anti-ErbB2 immunoliposomes targeted by a phage antibody selected for cellular endocytosis. *Biochim Biophys Acta* 2002, **1591**, 109–18.

Pagliusi SR. Efficacy and other milestones for human papillomavirus vaccine introduction. *Vaccine* 2004, **23**, 569–78.

Park JW, Hong K, Carter P, *et al*. Development of anti-p185HER2 immunoliposomes for cancer therapy. *Proc Natl Acad Sci U S A* 1995, **92**, 1327–31.

Peng D, Fan Z, Lu Y, *et al*. Anti-epidermal growth factor receptor monoclonal antibody 225 up-regulates p27KIP1 and induces G1 arrest in prostatic cancer cell line DU145. *Cancer Res* 1996, **56**, 3666–9.

Qian C, Prietro J. Gene therapy of cancer: induction of anti-tumour immunity. *Cell Mol Immunol* 2004, **1**(2), 105–11.

Ramanathan RK, Lee KM, McKolanis J, *et al*. Phase I study of a MUC1 vaccine composed of different doses of MUC1 peptide with SB-AS2 adjuvant in resected and locally advanced pancreatic cancer. *Cancer Immunol Immunother* 2005, **54**(3), 254–64.

Renga M, Pedrazzoli P, Siena S. Present results and perspectives of allogeneic non-myeloablative hematopoietic stem cells transplantation for treatment of human solid tumors. *Ann Oncol* 2003, **14**, 1177–84.

Romond, EH; Perez EA, Bryant J, *et al*. Trastuzumab plus adjuvant chemotherapy for operable HER2-positive breast cancer. *N Engl J Med* 2005, **353**, 1673–84.

Rosenberg SA, Restifo NP, Yang JC, Morgan RA, Dudley ME. Adoptive cell transfer: a clinical path to effective cancer immunotherapy. *Nat Rev Cancer* 2008, Apr;**8**(4), 299–308.

Saif MW, Cohenuram M. Role of panitumumab in the management of metastatic colorectal cancer. *Clin Colorectal Cancer* 2006, **6**(2), 118–24.

Saltz LB, Meropol NJ, Loehrer PJ Sr, Needle MN, Kopit J, Mayer RJ. Phase II trial of cetuximab in patients with refractory colorectal cancer that expresses the epidermal growth factor receptor. *J Clin Oncol* 2004, **22**, 1201–8.

Sandler AB, Gray R, Brahmer J, Dowlati A, Schiller JH, Perry MC, Johnson DH. Randomized phase II/III Trial of paclitaxel (P) plus carboplatin (C) with or without bevacizumab (NSC # 704865) in patients with advanced non-squamous non-small cell lung cancer (NSCLC): An Eastern Cooperative Oncology Group (ECOG) Trial-E4599. *J Clin Oncol* 2005, **23**, LBA4.

Schiller JT, Lowy DR. Prospects for cervical cancer prevention by human papillomavirus vaccination. *Cancer Res* 2006, **66**, 10229–32.

Segal DM, Weiner GJ, Weiner LM. Bispecific antibodies in cancer therapy. *Curr Opin Immunol* 1999, **11**, 558–62.

Skerra A, Pluckthun A. Assembly of a functional immunoglobulin Fv fragment in *Escherichia coli*. *Science* 1988, **240**, 1038–41.

Srinivas U, Tagliabue E, Campiglio M, Menard S, Colnaghi MI. Antibody-induced activation of p185HER2 in the human lung adenocarcinoma cell line Calu-3 requires bivalency. *Cancer Immunol Immunother* 1993, **36**, 397–402.

Slamon DJ, Leyland-Jones B, Shak S, *et al*. Use of chemotherapy plus a monoclonal antibody against HER2 for metastatic breast cancer that overexpresses HER2. *N Engl J Med* 2001, **344**(11), 783–92.

Smith I, Procter M, Gelber RD, *et al*. HERA study team. 2-year follow-up of trastuzumab after adjuvant chemotherapy in HER2-positive breast cancer: a randomised controlled trial. *Lancet* 2007, **369**(9555), 29–36.

Takahashi Y, Harashima N, Kajigaya S, Yokoyama H, Cherkasova E, McCoy JP, Hanada K, Mena O, Kurlander R, Tawab A, Srinivasan R, Lundqvist A, Malinzak E, Geller N, Lerman MI, Childs RW. Regression of human kidney cancer following allogeneic stem cell transplantation is associated with recognition of an HERV-E antigen by T cells. *J Clin Invest*. 2008 Mar;**118**(3):1099–109.

Trinchieri G. Interleukin-12 and the regulation of innate resistance and adaptive immunity. *Nat Rev Immunol* 2003, **3**(2), 133–46.

van Leeuwen F, van de Vijver MJ, Lomans J, van Deemter L, Jenster G, Akiyama T, Yamamoto T, Nusse R. Mutation of the human neu protein facilitates down-modulation by monoclonal antibodies. *Oncogene* 1990, **5**, 497–503.

van Ojik HH, Repp R, Groenewegen G, Valerius T, van de Winkel JG. Clinical evaluation of the bispecific antibody MDX-H210 (anti-Fc gamma RI x anti-HER-2/neu) in combination with granulocyte-colony-stimulating factor (filgrastim) for treatment of advanced breast cancer. *Cancer Immunol Immunother* 1997, **45**, 207–9.

van Spriel AB, van Ojik HH, van De Winkel JG. Immunotherapeutic perspective for bispecific antibodies. *Immunol Today* 2000, **21**, 391–7.

Vuist WM, Levy R, Maloney DG. Lymphoma regression induced by monoclonal anti-idiotypic antibodies correlates with their ability to induce Ig signal transduction and is not prevented by tumour expression of high levels of bcl-2 protein. *Blood* 1994, **83**, 899–906.

Wadleigh M, Richardson PG, Zahrieh D, *et al*. Prior gemtuzumab ozogamicin exposure significantly increases the risk of veno-occlusive disease in patients who undergo myeloablative allogeneic stem cell transplantation. *Blood* 2003, **102**(5), 1578–82.

Wahl RL. Tositumomab and (131)I therapy in non-Hodgkin's lymphoma. *J Nucl Med* 2005, **46**(Suppl 1), 128S–40S.

Waldmann TA. The biology of interleukin-2 and interleukin-15: implications for cancer therapy and vaccine design. *Nat Rev Immunol* 2006, **6**(8), 595–601.

Weiner LM, Holmes M, Richeson A, Godwin A, Adams GP, Hsieh-Ma ST, Ring DB, Alpaugh RK. Binding and cytotoxicity characteristics of the bispecific murine monoclonal antibody 2B1. *J Immunol* 1993, **151**, 2877–86.

Weir AN, Nesbitt A, Chapman AP, Popplewell AG, Antoniw P, Lawson AD. Formatting antibody fragments to mediate specific therapeutic functions. *Biochem Soc Trans* 2002, **30**, 512–6.

Wheeler CM. Advances in primary and secondary interventions for cervical cancer: human papillomavirus prophylactic vaccines and testing. *Nat Clin Pract Oncol* 2007, **4**(4), 224–35.

Wiseman GA, White CA, Sparks RB, *et al*. Biodistribution and dosimetry results from a phase III prospectively randomized controlled trial of Zevalin radioimmunotherapy for low-grade, follicular, or transformed B-cell non-Hodgkin's lymphoma. *Crit Rev Oncol Hematol* 2001, **39**(1–2), 181–94.

Wiseman GA, Gordon LI, Multani PS, *et al.* Ibritumomab tiuxetan radioimmunotherapy for relapsed or refractory non-Hodgkin's lymphoma patients with mild thrombocytopenia: a phase II multicenter trial. *Blood* 2002, **99**.

Witzig TE, Gordon LI, Cabanillas F, *et al.* Randomized controlled trial of yttrium-90-labeled ibritumomab tiuxetan radioimmunotherapy versus rituximab immunotherapy for patients with relapsed or refractory low-grade, follicular, or transformed B-cell non-Hodgkin's lymphoma. *J Clin Oncol* 2002, **20**(10), 2453–63.

16
Gene Therapy

Maria Belimezi, Teni Boulikas and Michael L. Roberts

16.1 The Concept of Gene Therapy

Gene therapy is a revolutionary field of biomedical research aimed at introducing therapeutically important genes into somatic cells of patients to alleviate disease symptoms.
Monumental progress in several fields including DNA replication, transcription factors
and gene expression, repair, recombination, signal transduction, oncogenes and tumour
suppressor genes, genome mapping and sequencing, and on the molecular basis of human
disease are providing the foundation of gene therapy. Disease targets include acquired
immune deficiency syndrome, cystic fibrosis, adenosine deaminase deficiency, cardiovascular
diseases (restenosis, familial hypercholesterolaemia, peripheral artery disease), Gaucher
disease, α 1-antitrypsin deficiency, rheumatoid arthritis and a few others. The main emphasis
of gene therapy, however, has been cancer.

Cancers tested in clinical trials for gene transfer include the cancers of breast, head and
neck, ovarian, prostate, brain, non-small and small cell lung, colorectal, as well as melanoma,
lymphoma, chronic myelogenous leukaemia, neuroblastoma, glioma, glioblastoma, astrocytoma, and others. A wide variety of delivery vehicles for genes have been tested including
murine retroviruses, recombinant adenoviral vectors, adeno-associated virus (AAV), herpes
simplex virus (HSV), Epstein–Barr virus (EBV), human immunodeficiency virus (HIV) vectors, and baculovirus. Nonviral gene delivery methods use cationic or neutral liposomes,
direct injection of plasmid DNA, and polymers. Various strategies to enhance efficiency of
gene transfer have been tested such as fusogenic peptides in combination with liposomes,
or polymers, to enhance the release of plasmid DNA from endosomes. Recombinant retroviruses stably integrate into the DNA and require host DNA synthesis; adenoviruses can infect
non-dividing cells but cause immune reactions leading to the elimination of therapeutically
transduced cells. AAV is not pathogenic and does not elicit immune responses but new strategies are required to obtain high AAV titres for preclinical and clinical studies. Wild type AAVs

Anticancer Therapeutics Edited by Sotiris Missailidis
© 2008 John Wiley & Sons, Ltd

integrate into chromosome 19 whereas recombinant AAVs are deprived of site-specific integration and may also persist episomally; HSV vectors can infect non-replicating cells such as neurons, have a high payload capacity for foreign DNA but inflict cytotoxic effects. It seems that each delivery system will be developed independently of the others and that each will prove its strengths for specific applications.

A number of anticancer genes are being tested in preclinical or clinical cancer trials including *p53*, *RB, BRCA1, E1A, bcl-2, MDR-1, HER2*, *p21*, *p16*, *bax, bcl-xs, E2F*, antisense *IGF-I*, antisense c-fos, antisense c-myc, antisense K-ras and the cytokine genes for granulo-cyte–macrophage colony-stimulating factor (GM-CSF), interleukin (IL)-12, IL-2, IL-4, IL-7, interferon-γ (IFN-γ), and tumour necrosis factor-α (TNF-α). A promising approach is transfer of the HSV thymidine kinase (HSV-*tk*) gene (suicide gene) and systemic treatment with the prodrug ganciclovir, which is converted by HSV-tk into a toxic drug killing dividing cells. Theoretically, expression of therapeutic genes preferentially in cancer cells could be achieved by regulatory elements from tumour-specific genes such as carcinoembryonic antigen.

Gene therapy approaches have suffered from the inadequate transduction efficiencies of replication-defective vectors. Replication-competent vectors, particularly adenoviruses that cause cytolysis as part of their natural life cycle, represent an emerging technology that shows considerable promise as a novel treatment option, particularly for locally advanced or recurrent cancer. Especially promising are adenoviruses that selectively replicate in tumour cells that have shown promising preliminary results in clinical trials, especially in combination with chemotherapy (Bernt *et al.*, 2002; reviewed in Boulikas, 1998). Liposomal formulations of genes may overcome significant hurdles in gene therapy applications in a clinical setting (reviewed in Martin and Boulikas, 1998). Many of those are in preclinical or cell culture studies and very few have surfaced to human clinical trials. Preliminary data from clinical studies using the liposomal encapsulation of replication incompetent Semliki Forest Viruses carrying therapeutic genes, such as the human IL-12 gene, showed promise in phase I/II clinical trials as a cancer immunotherapy regimen (Lundstrom and Boulikas, 2002).

An emerging concept is that combinations of gene therapy regimes with chemotherapy have synergistic antitumour effects. IFN-β inhibits cell cycle progression as an S phase block; pre-treatment of tumour cells with IFN-β could significantly potentiate the cytotoxicity of cis-platin, 5-fluorouracil, paclitaxel and gemcitabine in cell cultures (Brickelmaier *et al.*, 2002). Platinum-based chemotherapy enhances mutations in *p53* in the heterogenous cell population; transfer of the wild type *p53* gene enhanced the sensitivity of chemoresistant cells to cisplatin and cisplatin-induced apoptosis (Kigawa *et al.*, 2002).

16.2 Steps for Successful Gene Therapy

Gene therapy involves a long journey full of adventures and knowledge, like the return of Odysseus to Ithaka. Several key factors or steps appear to be involved for effective gene transfer to somatic cells in a patient: (i) the type of vehicle used for gene delivery (liposomes, adenoviruses, retroviruses, AAV, HSV, EBV, polymer, naked plasmid) which will determine not only the half-life in circulation, the biodistribution in tissues, and efficacy of delivery but also the route through the cell membrane and fate of the transgene in the nucleus; (ii) interaction of the gene-vehicle system with components in the serum or body fluids (plasma proteins, macrophages, immune response cells); (iii) targeting to the cell type, organ, or tumour, and binding to the cell surface; (iv) port and mode of entrance to the cell (poration through the cell membrane, receptor-mediated endocytosis); (v) release from cytoplasmic compartments (endosomes, lysosomes); (vi) transport across the nuclear envelope (nuclear import); (vii) type and potency of regulatory elements for driving the expression of the transferred gene in a particular cell type including DNA sequences that determine integration versus maintenance of a plasmid or recombinant virus/retrovirus as an extrachromosomal element; (viii) expression (transcription) of the transgene producing heterogeneous nuclear RNA (HnRNA) which is then (ix) spliced and processed in the nucleus to mature mRNA and is (x) exported to the cytoplasm to be (xi) translated into protein. (xii) Some therapeutic proteins need to be secreted outside the transduced cell or in the blood stream for their therapeutic effect to be seen, such as IL-12. Additional steps may include post-translational modification of the protein and addition of a signal peptide (at the gene level) for secretion.

All steps can be experimentally manipulated and improvements in each one can enormously enhance the level of expression and therapeutic index of a gene therapy approach. A mere 2-fold improvement in each one of the 12 steps enhances 2^{12} or 4096 times the overall outcome of a gene therapy protocol.

A variety of viral vectors have been developed to exploit the characteristic properties of each group to maintain persistence and viral gene expression in infected cells. Retroviral vectors and AAV integrate into target chromosomes and the transgene they carry can be inactivated from position effects from chromatin surroundings. Vectors with persistence/integration functions may not result in high levels of gene delivery *in vivo*. The first generation adenoviruses and retroviruses which were of the most frequently used vehicles for gene transfer could accommodate up to 7 kb of total foreign DNA into their genome because of packaging limitations. This precluded their use for the transfer of large genomic regions. Transfer of intact yeast artificial chromosome (YAC) into transgenic mice has enabled the analysis of large genes or multigenic loci such as human beta-globin locus. A small portion of plasmid molecules crossing the cell membrane will escape degradation from nucleases in the lysosomes and become released to the cytoplasm; even a smaller portion of these molecules will enter nuclei; finally, after successfully reaching the nucleus, plasmids with therapeutic genes are usually degraded by nuclear enzymes and transgene expression is permanently lost after about 2–7 days from animal tissues following successful gene delivery. During the peak of transgene expression (usually 7–48 h from injection) the transgene transcript can follow the normal fate of other nuclear transcripts when proper polyadenylation signals are provided; its processed mRNA will be exported to the cytoplasm and translated into the therapeutic protein.

AAV, HSV-1, EBV, baculovirus vectors, hybrid HSV-1/EBV vectors, and a number of additional viral vectors have been developed for gene therapy.

16.3 Retroviruses in Cancer Gene Therapy

The recombinant Moloney murine leukaemia virus (Mo-MLV or MLV) has been extensively used for gene transfer. Retroviral vectors derived from Mo-MLV promote the efficient transfer of genes into a variety of cell types from many animal species; up to 8 kb of foreign DNA can be packaged in a retroviral vector. Recombinant retroviruses have been the most frequently used and promising vehicles for the delivery of therapeutic genes in human gene therapy protocols. Retroviral vectors cause no detectable harm as they enter their target cells; the retroviral nucleic acid becomes integrated into chromosomal DNA, ensuring its long-term persistence and stable transmission to all future progeny of the transduced cell.

The life cycle of the retrovirus is well understood and can be effectively manipulated to generate vectors that can be efficiently and safely packaged. An important contribution to their utility has been the development of retrovirus packaging cells, which allow the production of retroviral vectors in the absence of replication-competent virus.

Recombinant retroviruses stably integrate into the DNA of actively dividing cells, requiring host DNA synthesis for this process. Although this is a disadvantage for targeting cells at G_0, such as the totipotent bone marrow stem cells, it is a great advantage for targeting tumour cells in an organ without affecting the normal cells in the surroundings. This approach has been used to kill gliomas in rat brain tumours by injection of murine fibroblasts stably transduced with a retroviral vector expressing the HSV-tk gene.

The use of retroviral vectors in human gene therapy requires a packaging cell line that is incapable of producing replication-competent virus and which produces high titres of replication-deficient vector virus. The packaging cell lines have been stably transduced with viral genes and produce constantly viral proteins needed by viruses to package their genome. Wild-type virus can be produced through recombinational events between the helper virus and a retroviral vector. Methods are also available for generating cell lines which secrete a broad host range retrovirus vectors in the absence of helper virus.

The traditional retroviral vector enters the target cell by binding of a viral envelope glyco-protein to a cell membrane viral receptor. Co-infection of cells with a retrovirus and VSV (vesicular stomatitis virus) produces progeny virions containing the genome of one virus encapsidated by the envelope protein of the other (pseudotypes of viruses); this led to the development of pseudotyped retroviral vectors where the MLV *env* gene product is replaced by the VSV-G protein able to interact with other membrane-bound receptors as well as with some components of the lipid bilayer (phosphatidylserine); because of the ubiquitous distri-bution of these membrane components pseudotyped particles display a very broad host range. Use of pseudotyped vectors has been a significant advancement for retroviral gene transfer. Because the VSV-G protein is toxic to cells when constitutively expressed, steroid-inducible and tetracycline-modulated promoter systems have been used to derive stable producer cell lines capable of substantial production of VSV-G pseudotyped MLV particles. Despite the extensive use of retroviral vectors in gene therapy, there are still problems to be solved and there is an ultimate need for the development of new, improved retroviral vectors and pack-aging systems to fuel further advances in the field of human gene therapy. Transduced cells producing retrovirus are tissue-incompatible and are, therefore, expected to be attacked by the immune system; this will lead to the elimination of therapeutic cells from the body, a phenomenon markedly associated also with adenoviral gene transfer. A privileged exception is brain tumour cells expressing recombinant retrovirus which persist without immunological rejection. Because retrovirus vectors are integrated into the genome, transcriptional repres-sion of transduced genes will often take place from position effects exerted from neighbouring chromatin domains; two matrix-attached regions (MARs), one at either flank of the transgene, are proposed here to insulating the gene in the retrovirus vector from chromatin effects at the integration site by creating an independent realm of chromatin structure harbouring the trans-gene. MAR insulators have been used and can enhance up to 2,000-fold the expression of genes in transgenic animals and plants.

16.4 Adenoviruses in Cancer Gene Therapy

Adenoviruses are also widely used as gene transfer vectors to deliver cytostatic or tumour suppressor genes into a variety of cancers. One of the most successful adenoviral vectors used to date is that which is designed to deliver p53 into tumours. This vector is currently

Table 16.1 Current gene augmentation strategies for cancer

Drug name	Mode of action	Indication	Developer	Phase
Advexin	Adenoviral p53	SCCHN[1]	Introgen Therapeutics, Interferon Sciences	III
		Bladder cancer	Introgen Therapeutics, NCI	I
		Breast cancer	Introgen Therapeutics	II
		Lung cancer	Introgen Therapeutics	II
TGDCC-E1A	Lipoplexed Ad E1A	SCCHN	Targeted Genetics	II
Cerepro	Adenoviral thymidine kinase	Brain cancer	Ark Therapeutics	II
IFN-Beta Gene	Adenoviral IFN-β	Brain cancer	Biogen/Idec	I
MVA-MUC1-IL2	Vaccinia IL-2	Breast cancer	Transgene	II
		Kidney cancer	Transgene	II
		Lung cancer	Transgene	II
		Prostate cancer	Transgene	II
INGN 225	P53 vaccine	Lung cancer	Introgen Therapeutics	I
TG1042	Adenoviral IFN-γ	Lymphoma	Transgene	II
Allovectin-7	Adenoviral HLA-B7	Melanoma	Vical	II
INGN241	Adenoviral MDA-7	Solid tumours	Introgen Therapeutics	II
TNFerade	Adenoviral TNF α	Pancreatic cancer	GenVec	II
		Solid tumours	GenVec	II
			GenVec	II

[1] Squamous Cell Carcinoma of the Head and Neck.

undergoing phase III clinical trials and has been approved for the treatment of non-small cell lung cancer (NSCLC) in China, where it is marketed as Gendicine and developed by Shenzhen-based SiBiono Gene Technology Co Ltd. Table 16.1 illustrates other gene-based therapeutics that are currently under evaluation in the clinic.

Adenoviruses posses a well-defined origin of replication which is stimulated by transcription factors NFI and NFIII. Adenoviruses replicate episomally; they need to attach to the nuclear matrix of the host cell for their replication. Two adenoviral proteins have been found attached to the nuclear matrix and presumably mediating the anchorage of the adenovirus: (i) the E1a protein (11 kDa), a transcription and replication factor sufficient to immortalize primary rodent cells and (ii) the adenovirus terminal protein (55 kDa) which is covalently attached to the 5′ end of Ad DNA and initiates DNA replication; the adenovirus terminal protein mediated adenovirus anchorage to nuclear matrix.

Expression of the adenovirus E1A protein stimulates host DNA synthesis and induces apoptosis; on the contrary E1B 19 kDa associates with Bax protein and inhibits apoptosis. The E1A oncogene of adenovirus exerts its effect *via* p53 protein. Indeed, expression of E1A increases the half-life of p53 resulting in accumulation of p53 molecules in adenovirus-infected cells leading to apoptosis. Although induction of host DNA synthesis by E1A provides a suitable environment for virus replication, the induction of apoptosis by the same protein impairs virus production since virus-infected cells are eliminated. E1A represses HER-2/neu transcription and functions as a tumour suppressor gene in HER-2/neu-overexpressing cancer cells. Transfer of the E1A gene into cancer cells that overexpress HER-2/neu is an interesting aspect of gene therapy.

Adenoviruses can achieve high levels of gene transfer. However, the duration of transgene expression is limited (i) by clearance of the infected cells because of the cellular and humoral immune response (including those mediated by cytotoxic T lymphocytes) to adenoviral antigens, and (ii) by loss of adenoviral episomes in progeny cells. In order to circumvent the elimination of adenovirus-transduced cells by immune responses and for achieving persistence of transgene expression strategies to reduce the potential for viral gene expression have been developed. The elimination of therapeutically important cells from the body after recombinant adenovirus-mediated delivery by activated T lymphocytes can be partially circumvented by daily administration of the immunosupressant ciclosporin A or with anti-T-cell receptor monoclonal antibodies. A new area of investigation is directed toward surface modification of recombinant adenoviruses to render them safer and to minimize the strong immune responses against the virus and virus-infected cells.

16.5 Gene Therapy of Cancer

The genes used for cancer gene therapy in human clinical trials include a number of tumour suppressor genes (*p53, RB, BRCA1, E1A*), antisense oncogenes (antisense c-fos, c-myc, K-ras), and suicide genes (HSV-tk, in combination with ganciclovir, cytosine deaminase in combination with 5-fluorocytosine) (for a complete list see Table 16.1).

Tumour cells have lost the function of p53 because of mutations mainly in the DNA-binding region of the molecule in over 50 % of all human malignancies. Transfer of the wild type (wt) p53 gene was able to suppress tumour cell proliferation. The long-term follow-up of heavily pre-treated patients with recurrent ovarian cancer by *p53* gene replacement using the adenoviral vector SCH 58 500 followed by multiple cycles of platinum-based chemotherapy has been evaluated by Buller and coworkers (2002a, b). The median survival of individuals who received multiple doses of the adenoviral p53 with chemotherapy was 12–13.0 months, compared to only 5 months for those treated with a single-dose of adenoviral p53; this compared favorably to the 16-month median survival for individuals treated with paclitaxel at the time of initial recurrence of this disease and was more than double the 5-month survival seen with palliative radiotherapy or paclitaxel failure. In spite of adenoviral-induced inflammatory changes, intraperitoneal adenoviral p53 treatment was safe as a 5-day regimen using 7.5×10^{13} adenoviral particles per dose per day followed by intravenous carboplatin/paclitaxel chemotherapy and resulted in a significant reduction of the serum tumour marker CA125 (Buller *et al.*, 2002a, b).

16.6 Cancer Immunotherapy with Cytokine Genes

The combination of immunotherapy with conventional treatments such as radio- and chemotherapy may be necessary to eradicate minimal residual disease. Advanced therapies involve the transfection of lymphocytes in culture with cytokine genes followed by selection of the successfully transfected cells with a selectable marker such as the bacterial neomycin-resistance gene. The development of tumour cells transduced with cytokine genes and their exploitation as tumour vaccines in patients with cancer is a very promising field. Cytokine genes used for cancer immunotherapy include those of IL-2, IL-4, IL-7, IL-12, IFNs, GM-CSF, TNF-α in combination with genes encoding co-stimulatory molecules, such as B7-I. The major goal of the use of immunostimulatory cytokines is the activation of tumour-specific T lymphocytes capable of rejecting tumour cells from patients with low tumour burden or to protect patients from a recurrence of the disease. As distant metastasis is the major cause for therapeutic failures in clinical oncology, treatment of patients having a low tumour volume with immunotherapy could protect the patient from recurrence of disease.

The transduction of the tumour cells of the patient with cytokine genes *ex vivo* and the development of tumour vaccines depends on the establishment of primary cell culture from the solid tumour. Although malignant melanomas are easy to culture, it is difficult to establish cell lines from most other primary human tumours using convenient methods; primary tumour cultures are being used (i) for the transduction of autologous cells from the cancer patient with cytokine genes to develop cancer vaccines after intradermal implantation to the patient; (ii) for characterization of tumour-specific cytotoxic T lymphocytes in order to identify specific antigens on the human primary culture; (iii) for extensive phenotypic characterization of the tumour in cell culture.

16.7 IL-12 in Cancer Immunotherapy

16.7.1 IL-12 biology

IL-12 is a heterodimeric pro-inflammatory cytokine that induces the production of INF-γ and favours the T helper1 (Th1) response creating a link between innate resistance and adaptive immunity. The phagocytes (monocytes/macrophages and neutrophils) and dendritic cells are the main physiological producers of IL-12 in response to (a) pathogens (bacteria, fungi, intracellular parasites and viruses) through Toll-like receptors (TLRs) and other receptors, to (b) membrane-bound and soluble signals from activated T cells and natural killer (NK) cells, and to c) components of the inflammatory extracellular matrix (for example, low-molecular-weight hyaluronan) through CD44 and TLRs.

The physiologically most important target cells of IL-12 are lymphocytes. IL-12, in synergy with other colony-stimulating factors, induces proliferation of NK cells and T cells, enhancement of cytotoxicity and of the expression of cytotoxic mediators. It also promotes the production of cytokines, particularly IFN-γ, as well as favouring differentiation of cells that produce type-1 cytokines (Th1, TC1 and NK1 cells) (the Th1 response). IL-12 has the ability, directly or through the effects of type-1 cytokines such as IFN-γ, to enhance the activation and production of Th1-associated classes of immunoglobulin from B cells (for example, IgG2a in the mouse).

16.7.2 Antitumour activity of IL-12

Treatment with IL-12 has been shown to have a marked antitumour effect on mouse tumours, by inhibiting establishment of tumours or by inducing regression of established tumours. The antitumour action of IL-12 is complex and uses effector mechanisms of both innate resistance and adaptive immunity. So, specific recognition of tumour antigens might not always be required for the effects of IL-12. Cytotoxic lymphocytes (CD8$^+$ T cells, NK cells and NKT cells) are often involved in the mechanism of action of IL-12, but their cytotoxic activity is not required in some cases for their anti-tumour activity. IFN-γ and a cascade of other secondary and tertiary pro-inflammatory cytokines that are induced by IL-12 have a direct toxic effect on the tumour cells and/or might activate potent anti-angiogenic mechanisms. The ability of IL-12 to induce antigen-specific immunity relies mainly on its ability to induce or augment the responses of Th1 cells and cytotoxic T lymphocytes (CTLs) (see also Chapter 15 and figures on IL-12).

However, during the acute pro-inflammatory reaction that is induced by treatment with IL-12, expression of inducible nitric oxide synthase (iNOS) and production of nitric oxide (NO) might induce a temporary immunosuppression that decreases the contribution of antigen-specific immunity to the anti-tumour effects of IL-12. In addition to the induction of cellular immune responses, IL-12, by inducing Th1 responses, also augments the production of opsonizing and complement-fixing classes of IgG antibodies that have been shown to have anti-tumour activity *in vivo*. Because of its ability to induce an inflammatory response, the administration of IL-12 results in considerable toxicity that resembles the inflammatory syndromes that are associated with the response to severe infections. The promising data that were obtained in preclinical models of antitumour immunotherapy indicated that IL-12 might be a powerful therapeutic agent against cancer. However, excessive toxicity and a modest response in clinical trials have dampened this enthusiasm greatly.

16.7.3 Cancer immunotherapy with the IL-12 gene

IL-12 gene therapy is one of the more novel and promising approaches in cancer therapy. IL-12 is a heterodimeric cytokine composed of two subunits, p40 and p35, that requires the simultaneous expression of both the p35 and p40 chain genes from the same cell for production of biologically active IL-12. Coordinate expression of the IL-12 p40 and p35 genes in several solid tumour models has been found to induce strong and specific antitumour immune responses. A variety of biological functions have been attributed to IL-12 including the induction of IFN-γ and the promotion of predominantly Th1-type immune responses to antigens.

The local secretion of IL-12 achieved by gene transduction suppressed tumour growth and promoted the acquisition of specific antitumour immunity in mice. This was shown by intradermal inoculation of mice with NIH3T3 cells transduced with expression plasmids or a retroviral vector expressing the murine IL-12 gene admixed with murine melanoma BL-6 cells; CD4$^+$ and CD8$^+$ T cells, as well as NK cells, were responsible for the observed antitumour effects resulting from IL-12 paracrine secretion.

The antitumour effect of several transgene expression plasmids encoding the cytokines IL-2, IL-4, IL-6, IL-12, IFN-γ, TNF-α, and GM-CSF was tested using the gene gun-mediated DNA delivery into the epidermis overlying an established intradermal murine tumour; this study showed that IL-12 gene therapy was much more effective than treatment with any other tested cytokine gene for induction of tumour regression as determined from the increased CD8$^+$ T-cell-mediated cytolytic activity in the draining lymph nodes of tumour-bearing mice; treated animals were able to eradicate not only the treated but also the untreated solid tumours at distant sites; elevated systemic levels of IFN-γ, were found after IL-12 gene therapy. This approach is providing a safer alternative to IL-12 protein therapy for clinical treatment of cancers.

16.7.4 Gene therapy using liposomal genes

Gene therapy approaches have suffered from the inadequate transduction efficiencies of replication-defective vectors. Replication-competent vectors, particularly adenoviruses that cause cytolysis as part of their natural life cycle, represent an emerging technology that shows considerable promise as a novel treatment option, particularly for locally advanced or recurrent cancer. Oncolytic adenovirus therapy shows the best results and achieves an enhanced tumouricidal effect when used in combination with chemotherapeutic agents such as cisplatin, leucovorin and 5-fluorouracil. Improvement of oncolytic adenoviruses is directed at molecular engineering tumour cell-specific binding tropism, selective modifications of viral early genes and incorporation of cellular promoters to achieve tumour-specific replication, augmentation of anti-tumour activity by incorporation of suicide genes, and manipulation of the immune response (Yoon *et al.*, 2001).

An important advancement has been the liposomal encapsulation of Semliki Forest virus (SFV) carrying both subunits of the human IL-12 gene and its introduction into phase I/II clinical trials in 2001 by our group. This is the first international protocol using a liposomal virus; the encapsulation procedure allows repetitive intravenous administration of the virus to humans without eliciting an immune response to the virus because it is made 'invisible' to the immune system by the liposome capsule and its coating by polyethylene glycol (PEG; (Figure 16.1)). This technology is especially important for clinical gene therapy applications using adenoviruses known to elicit immune responses. Furthermore, the virus is not destroyed, and because of its nanoparticle structure and its long circulation it is being targeted to tumours and inflammatory areas and avoids opsonization in the blood. The production of a functional IL-12 by the viral gene secreted in sera of patients can be measured by enzyme-linked immunosorbent assay, as is the concomitant increase in TNF-α and IFN-γ (Ren *et al.*, 2003). Gene therapy, in addition to its anticipated benefits in combination with tumour targeted chemotherapy could find application in reducing the side effects of chemotherapy drugs. For example, GM-CSF and erythropoietin (EPO) are used as recombinant proteins to enhance production of haematopoietic cells from bone marrow after killing of bone marrow by certain chemotherapy drugs, including cisplatin. Recombinant human interleukin-3 (rhIL-3) shortened the duration of neutropenia and thrombocytopenia induced by vincristine, ifosfamide, mesna, and carboplatin, did not enhance myelotoxicity and improved bone marrow recovery (Biesma *et al.*, 2002). Liposomal delivery of the GM-CSF, EPO and IL-3 or other IL genes could circumvent this problem, the genes can be delivered in a dose-dependent manner and their production by the cells of the patient could be continuous and at low levels circumventing the danger of cardiac episodes by overdose of the recombinant proteins.

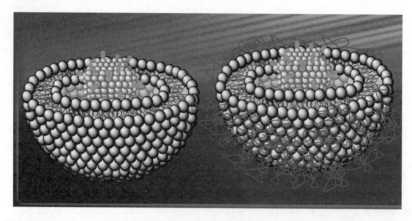

Figure 16.1 Liposomal encapsulation of viruses hides them from the immune system, prevents binding of serum protein during circulation, can lead to targeting of the nanoparticle to tumours and metastases by extravasation through their leaky vasculature and allows repetitive administration without eliciting an immune reaction

The E1A gene of adenovirus functions as a tumour inhibitor by repressing oncogene transcription; E1A also modulates gene expression resulting in cellular differentiation and induces apoptosis in cancer cells. Finally, E1A sensitizes cancer cells to chemotherapeutic drugs such as etoposide, cisplatin, and taxol. Transfer of the wild type (wt) p53 gene is known to suppress tumour cell proliferation. Besides E1A and p53 there are hundreds of genes of therapeutic potential that emerged from decades of research and the completion of the human genome project. Liposomal delivery of appropriate viral vectors expressing such genes will make possible systemic viral delivery *via* nanoparticles. One property of the liposomal particles is their ability to concentrate also in inflammatory regions including atherosclerotic sites and arthritic joins. Thus, liposomal delivery of genes combined with liposomal drugs is expected to have important applications, not only in cancer treatment, but also in the combat of cardiovascular disease and arthritis.

16.8 Viruses able to Kill Cancer Cells

A novel class of targeted anticancer agents endowed with unique mechanisms of action has emerged from the study of therapeutic oncolytic viruses creating the field of virotherapeutics (Liu *et al.*, 2007). Despite their promising preclinical data, however, corresponding clinical trials have disappointed. Hurdles may arise from low penetration of viruses into the tumour cell mass, usually the lack of a systemic delivery method, lack of targeting to primary tumours and metastases without expression in normal tissue, their destruction by the immune system within the tumour once the previous problems are solved. Understanding the field of virotherapeutics necessitates consideration of innate immune defences in human tumour cells.

The oncolytic measles virus expressing an enhanced green fluorescent protein (MV-eGFP) has been used to induce IFN production in human myeloma and ovarian cancer cells; the P gene of wild type measles virus encodes P/V/C proteins known to antagonize IFN induction and/or response; A MV was generated (MV-eGFP-Pwt) expressing the P gene from wild type IC-B strain with reduced IFN sensitivity; intravenous injection of this virus significantly enhanced oncolytic potency compared to MV-eGFP in mice bearing human myeloma xenografts (Haralambieva *et al.*, 2007).

Genetically engineered transgene-expressing 'armed' oncolytic viruses can transfer a number of therapeutic genes such as TNF-α. In previous clinical studies non-replicative adenovirus vectors have been used to deliver TNF-α directly to the tumour; Han and coworkers (2007) have used an ICP34.5 deleted, oncolytic herpes simplex virus (HSV) for delivery of TNF-α. This oncolytic HSV permits to increase expression levels and spread through the tumour; furthermore, use of the US11 true late HSV promoter was used to limit expression at tumour sites where the virus replicates in animal models.

16.8.1 Oncolytic adenoviruses

A number of oncolytic adenoviruses designed to replicate selectively in tumour cells by targeting molecular lesions inherent in cancer, or by incorporation of tissue-specific promoters driving the early genes that initiate viral replication, are currently under clinical evaluation (see Table 16.2).

Oncolytic adenovirus therapy shows the best results and achieves an enhanced tumouricidal effect when used in combination with chemotherapeutic agents such as cisplatin, leucovorin and 5-fluorouracil. Improvement of oncolytic adenoviruses is directed at molecular engineering tumour cell-specific binding tropism, selective modifications of viral early genes and incorporation of cellular promoters to achieve tumour-specific replication, augmentation of anti-tumour activity by incorporation of suicide genes, and manipulation of the immune response (Yoon *et al.*, 2001). Replication-activated adenoviral vectors have been developed to express a secreted form of β-glucuronidase and a cytosine deaminase/uracil phosphoribosyltransferase. β-glucuronidase activates the prodrug 9-aminocamptothecin glucuronide to 9-aminocamptothecin. The cytosine deaminase/uracil phosphoribosyltransferase activates the prodrug 5-fluorocytosine to 5-fluorouracil (5-FU) and further to 5-fluoro-UMP. The combination of this adenoviral vector with prodrug therapy enhanced viral replication and its spread in liver metastases derived from human colon carcinoma or cervical carcinoma in a mouse model (Bernt *et al.*, 2002).

Interestingly, the rights to the proto-typical oncolytic adenovirus, Onyx-15, were recently sold to Shanghai Sunway Co, Ltd, a Chinese company that have developed a similar construct termed H101. Shanghai Sunway are currently developing their construct for use in the clinic and are presently running phase III trials for the application of H101 in Squamous Cell Cancer

Table 16.2 Current oncolytic viral therapy clinicial trials

Drug name	Mode of action	Indication	Developer	Phase
ONYX-015	Oncolytic virus	SCCHN[2]	Onyx Pharmaceutics	III
		Lung cancer		III
G207	Oncolytic Herpes	Brain cancer	MediGene	I
Oncolytic HSV	Oncolytic virus	Liver cancer	Medigene	II
CG7870	Oncolytic virus	Prostate cancer	Cell Genesys	I

[2]Squamous Cell Carcinoma of the Head and Neck

of the Head and Neck (SCCHN) and NSCLC and as of November 2006, it has been approved by Chinese FDA for the use in these two indications.

The human adenovirus type 5 (Ad5) early region 1A (E1A) proteins have been shown to have potent antitumour effects, due to their ability to reprogram oncogenic signalling pathways in tumour cells. The E1A gene of adenovirus functions as a tumour inhibitor by repressing oncogene transcription; E1A also modulates gene expression resulting in cellular differentiation and induces apoptosis in cancer cells. Finally, E1A sensitizes cancer cells to chemotherapeutic drugs such as etoposide, cisplatin, and taxol. An adenovirus vector deleted of all viral protein coding sequences with the exception of E1A reduced the proliferative capacity of the human lung adenocarcinoma cell line A549, the ability of these cells to form colonies in soft agarose and gave a 10-fold greater sensitivity to cisplatin (Hubberstey *et al.*, 2002).

References

Bernt KM, Steinwaerder DS, Ni S, Li ZY, Roffler SR, Lieber A. Enzyme-activated prodrug therapy enhances tumor-specific replication of adenovirus vectors. *Cancer Res* 2002, **62**, 6089–98.

Biesma B, van Kralingen KW, van Leen RW, Koster MC, Postmus PE. Recombinant human interleukin-3 administered concomitantly with chemotherapy in patients with relapsed small cell lung cancer. *J Exp Ther Oncol* 2002, **2**, 47–52.

Boulikas T. Status of gene therapy in 1997: molecular mechanisms, disease targets, and clinical applications. *Gene Ther Mol Biol* 1998, **1**, 1–172.

Brickelmaier M, Carmillo A, Goelz S, Barsoum J, Qin XQ. Cytotoxicity of combinations of IFN-beta and chemotherapeutic drugs. *J Interferon Cytokine Res* 2002, **22**, 873–80.

Buller RE, Runnebaum IB, Karlan BY, *et al*. A phase I/II trial of rAd/p53 (SCH 58500) gene replacement in recurrent ovarian cancer. *Cancer Gene Ther* 2002a, **9**, 553–66.

Buller RE, Shahin MS, Horowitz JA, *et al*. Long term follow-up of patients with recurrent ovarian cancer after Ad p53 gene replacement with SCH 58500. *Cancer Gene Ther* 2002b, **9**, 567–72.

Han ZQ, Assenberg M, Liu BL, Wang YB, Simpson G, Thomas S, Coffin RS. Development of a second-generation oncolytic Herpes simplex virus expressing TNFalpha for cancer therapy. *J Gene Med* 2007, **9**, 99–106.

Haralambieva I, Iankov I, Hasegawa K, Harvey M, Russell SJ, Peng KW. Engineering oncolytic measles virus to circumvent the intracellular innate immune response. *Mol Ther* 2007, **15**, 588–97.

Hubberstey AV, Pavliv M, Parks RJ. Cancer therapy utilizing an adenoviral vector expressing only E1A. *Cancer Gene Ther* 2002, **9**, 321–9.

Kigawa J, Sato S, Shimada M, Takahashi M, Itamochi H, Kanamori Y, Terakawa N. p53 gene status and chemosensitivity in ovarian cancer. *Hum Cell* 2002, **14**, 165–71.

Liu TC, Galanis E, Kirn D. Clinical trial results with oncolytic virotherapy: a century of promise, a decade of progress. *Nat Clin Pract Oncol* 2007, **4**(2), 101–17.

Lundstrom K, Boulikas T. Breakthrough in cancer therapy: encapsulation of drugs and viruses. *Curr Drug Discov* 2002, **11**, 19–23.

Martin F, Boulikas T. The challenge of liposomes in gene therapy. *Gene Ther Mol Biol* 1998, **1**, 173–214.

Ren H, Boulikas T, Lundstrom K, Söling A, Warnke PC, Rainov NG. Immunogene therapy of recurrent glioblastoma multiforme with a liposomally encapsulated replication-incompetent Semliki forest virus vector carrying the human interleukin-12 gene–a phase I/II clinical protocol. *J Neurooncol* 2003, **64**, 147–54.

Yoon TK, Shichinohe T, Laquerre S, Kasahara N. Selectively replicating adenoviruses for oncolytic therapy. *Curr Cancer Drug Targets* 2001, **1**, 85–107.

17
Antisense Agents

Huma Khan and Sotiris Missailidis

17.1 Introduction

In this era of functional genomics, on which completion of the human genome project has made a significant impact, targeting the genetic cause of cancer has become an extremely viable therapeutic strategy. Sequence analysis of the human genome, in collaboration with gene silencing methods, has allowed researchers to take a step further into identifying gene(s) involved in the progression of cancer. Consequently, direct targeting of such oncogenes, using gene silencing methods, has attracted immense interest within the drug design and discovery sector. This, in turn, has led to an explosion in oligonucleotide-based drug design, preceded by antisense-based technologies. Discovered in the 1970s by Zamecnik and Stephenson (Zamecnik and Stephenson, 1978), antisense technology has remained to date the most successful method for gene suppression, and therefore proven to be a powerful research tool in identifying gene function as an attractive therapeutic mechanism. The core of antisense therapy is based on blocking the expression of a target mRNA, which encodes for a disease-related protein. Acting at the genetic level, antisense therapy offers higher specificity than many protein-targeting therapies, which can often lead to the inhibition of other related proteins.

Once a target mRNA and its genetic sequence has been identified, researchers working in the antisense field can rapidly and effectively design short oligonucleotides which are complementary, hence antisense, to the target. Many biotechnology companies have taken advantage of this and specialize in the design of antisense oligonucleotides (ASOs) against targets of interest. Antisense molecules are therefore designed to bind tightly, in a sequence-dependent manner, by Watson–Crick base pairing, to their complementary mRNA and, consequently, prevent translation of the gene target and its eventual expression into a protein product. Once bound, the antisense molecule can interfere with translation of its target by different methods, based upon the type of antisense molecule employed. As such, advances in antisense strategies have generated three types of antisense-based molecules: (i) traditional single-stranded

Anticancer Therapeutics   Edited by Sotiris Missailidis
© 2008 John Wiley & Sons, Ltd

antisense oligonucleotides (ASOs), which predominantly block mRNA expression by steric blockage or by recruiting a nuclease to degrade its target; (ii) catalytically active oligonucleotides, termed ribozymes or DNAzymes, which bind and cleave the target themselves and; (iii) small interfering RNA (siRNA), molecules which exploit Nature's RNA interference (RNAi) machinery to degrade their target. Each of these antisense strategies will be briefly discussed below, with their current drug candidates in clinical trials for cancer related diseases.

17.2 Traditional Antisense Oligonucleotides (ASOs)

Traditional antisense molecules are single stranded oligonucleotides, typically 15–20 nucleotides in length, the sequence of which is designed to be complementary to their target (mRNA). Hence, these molecules are characterised by high specificity and affinity for their target, an attribute central to the design of therapeutic drugs. Following binding, inhibition of mRNA translation can be achieved in various ways. First, the duplex formed between the antisense agent and its target acts to sterically block the binding of ribosomal machinery, necessary for translation. In addition, or alternatively, many ASOs recruit RNase H, a cellular endogenous enzyme that specifically degrades RNA in a RNA–DNA heteroduplex, and therefore degrade the target. While these two mechanisms are the most prominent ways that antisense agents instigate their action, other mechanisms have been proposed. Such methods, whereby the antisense oligonucleotide enters the nucleus, involves preventing mRNA transport, inhibition of mRNA splicing and inhibition of 5′ cap formation of mRNA, all of which prevent maturation of mRNA (Chan et al., 2006).

Most often, hybridization of the antisense molecule to its RNA target leads to permanent degradation of the mRNA (by recruitment of RNase H). This, in turn, frees the ASO to bind to another target molecule, thereby allowing the antisense agent to be recycled; an attractive asset that many other drug molecules are unable to offer.

Despite their encouraging potential, in the past two decades, antisense agents have faced major problems which have limited their clinical applicability. Predominantly, their low stability in vivo hindered their initial success (see below for other limitations associated with ASOs). This low stability in biological samples stems from the susceptibility of oligonucleotides to degradation by endogenous nucleases, hence their low half-life in vivo impedes their effective therapeutic action. Nonetheless, the ability to easily modify oligonucleotides has greatly facilitated overcoming this shortfall of antisense molecules. In general, modifications of nucleotides can be categorized into three main types, termed first, second and third generation chemistries (Figure 17.1).

First-generation ASOs contain phosphorothioate (PS)-modified backbones, where one of the non-bridging oxygen atoms in the phosphodiester bond is substituted with a sulfur atom. This modification often confers increased resistance to nuclease degradation and therefore prolongs the half-life of ASOs. In addition, first-generation ASOs also tend to support RNase H mediated cleavage of mRNA. Conversely, PS modifications tend to reduce the affinity of ASOs to their target. However, this shortcoming is overcome in second-generation ASOs, which contain 2′ alkyl modifications (2′-O-methyl or 2′-O-methoxyethyl) to the ribose of PS-modified ASOs. Although such modifications further increase the stability and enhance the affinity of ASOs for RNA, they are unable to activate RNase H, limiting their main mode of action. This has, however, been resolved by generating 'gapmers', where 2′-modified nucleotides are only positioned at the ends of a chimeric oligonucleotide. This leaves the central region (i.e. a gap of around 10 PS-nucleotides) free for RNase H to bind to, while

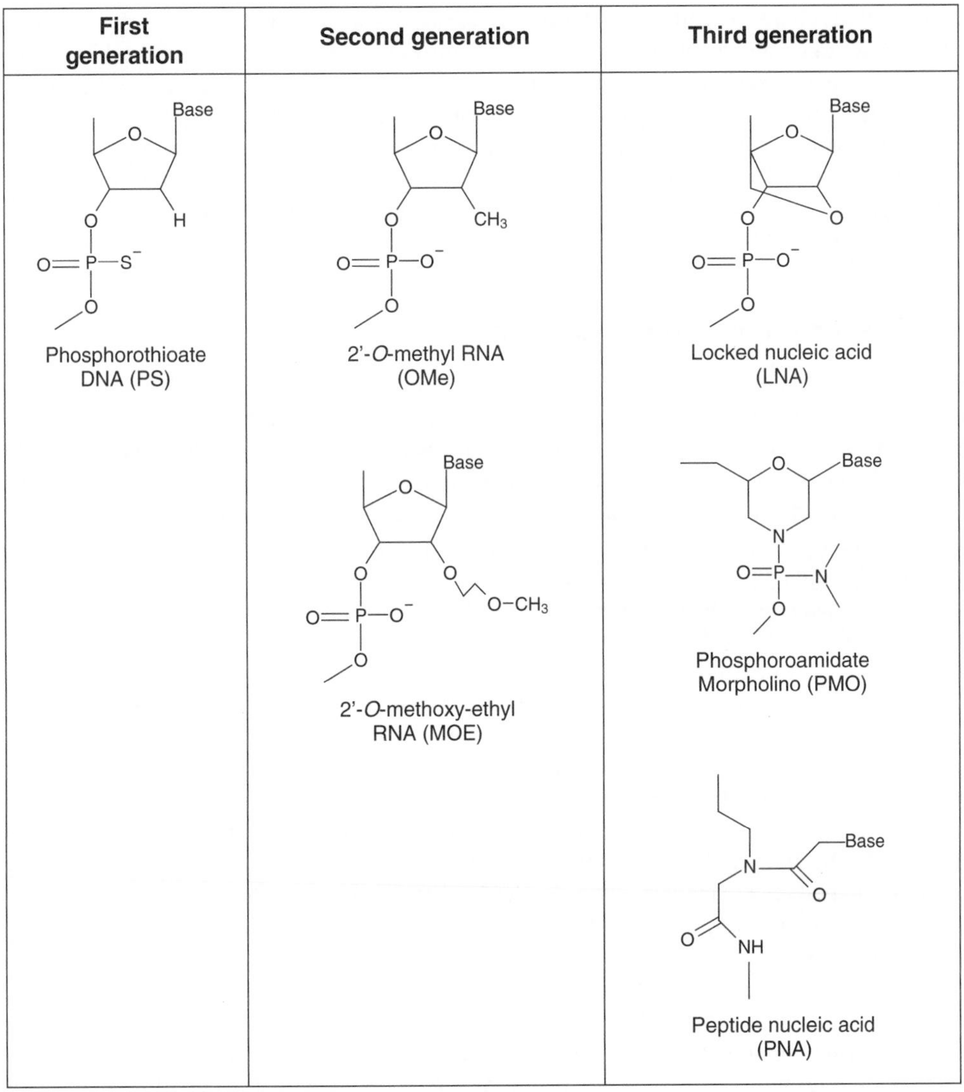

First generation	Second generation	Third generation
Phosphorothioate DNA (PS)	2'-*O*-methyl RNA (OMe)	Locked nucleic acid (LNA)
	2'-*O*-methoxy-ethyl RNA (MOE)	Phosphoroamidate Morpholino (PMO)
		Peptide nucleic acid (PNA)

Figure 17.1 Chemical modification of antisense oligonucleotides. Such modifications to oligonucleotides are typically made to protect them from nuclease degradation in biological samples

the flanking ends (containing the 2'-alkyl-modified nucleotides) confer resistance to nuclease degradation. Further improvements in affinity, stability and enhanced pharmacokinetic properties are envisaged in third generation ASOs. The main third-generation ASOs include peptide nucleic acids (PNA), locked nucleic acids (LNA) and phosphoroamidate morpholino oligomers (PMO), which consist of modifications to the furanose ring. In PNAs, the phosphodiester backbone is replaced by polyamide linkages, LNAs contain a 2'-*O*,4'-C-methylene bridge on the ribose unit and in PMOs the ribose unit is substituted with a morpholino moiety and the phosphodiester bond replaced with a phosphoroamidate linkage. While each of these modifications confers excellent resistance to nuclease degradation and enhances hybridization of the ASO to complementary DNA or RNA, activation of RNase H by these ASOs is hindered. Hence, the mode of action of most third generation ASOs is mainly elicited by steric blockage of ribosomal scanning (Gleave and Monia, 2005; Chan *et al.*, 2006).

Whilst many ASOs degrade their target by recruiting an endogenous enzyme, other antisense agents have been engineered so that they themselves fulfil the catalytic function of RNase H. As such, ribozymes have been immensely useful in exploiting this catalytic function in the antisense field of therapeutics.

17.3 Ribozymes and DNAzymes

Ribozymes (RNA enzymes) are RNA molecules capable of catalysing the cleavage of either their own RNA or other RNA substrates. Many naturally occurring ribozymes exist, of these the hammerhead ribozyme has been extensively studied (with its structure being widely used as a foundation for engineering other ribozymes). The principal structure of ribozymes consists of a catalytic motif flanked by a substrate-binding domain. The substrate-binding domain entails a sequence antisense to the target mRNA, allowing the enzyme to hybridize specifically to its substrate by Watson–Crick base pairing. This brings the RNA substrate close to the catalytic domain, which cleaves the substrate at a specific site recognized by the ribozyme (Famulok and Verma, 2002). The mechanism of catalysis undertaken by these RNA enzymes is a 2′ oxygen nucleophile attack of the adjacent phosphate in the RNA backbone, leading to the formation of 2′, 3′-cyclic phosphate and 5′ hydroxyl terminus (Khan, 2006). The resultant products are subsequently degraded by ribonucleases, guaranteeing permanent inactivation of the target. Following cleavage, the ribozyme is able to dissociate itself from the RNA products and bind to another RNA molecule to be cleaved.

Although hammerhead ribozymes are intrinsically self-cleaving enzymes (Famulok, and Verma, 2002), they can be engineered into target-specific, *trans*-cleaving enzymes, offering magnificent potential in medical research. Hence, therapeutic ribozymes have been potentially synthesized to target almost any RNA sequence of interest. By incorporating the catalytic domain of ribozymes into short oligonucleotides antisense to the target mRNA (i.e. changing the sequence of the substrate recognition domains), highly specific drug molecules have been engineered to bind and cleave targets. Consequently, the application of RNA cleaving enzymes in suppressing the expression of a range of therapeutically relevant genes has shown considerable promise. However, one of the major problems hindering the development of ribozyme therapeutics, (and other oligonucleotide-based drugs; see below) is the sensitivity of RNA molecules to degradation by endogenous RNA nucleases. Although advances in nucleotide chemistry (as mentioned for ASOs) have immensely helped in solving this problem, researchers have developed an alternative and more 'natural' approach in solving this problem, by the engineering of DNAzymes. DNAzymes are DNA backbone based enzymes capable of specifically cleaving RNA targets. Although their structure (catalytic domain and two substrate-binding domains) and mechanism of catalysis is analogous to ribozymes, naturally, DNA molecules are known to be less sensitive to nucleases than RNA, hence DNAzymes offer higher stability in biological media. Given that DNAzymes are not known to exist naturally, they are developed by *in vitro* selection methods. Thus, ultimately, DNAzymes (and engineered ribozymes) can be considered as catalytic DNA aptamers (see chapter 18). Owing to their catalytic nature, ribozymes and DNAzymes have a significant advantage over other drugs by having a recyclable mechanism of action (as with ASOs).

17.4 RNA Interference and siRNAs

First discovered in the late 1990s by Fire and Mello (Fire *et al.*, 1998), RNA interference (RNAi) has since received considerable interest in its application as a powerful research tool and more importantly as a potent therapeutic. Their research in the nematode *Caenorhabditis elegans*, showed that naturally occurring dsRNA molecules effectively led to the down-regulation of gene expression, for which they coined the term RNAi. Since then, this natural process for gene silencing has been observed in many organisms, ranging from plants to humans (Cullen, 2002; Hannon, 2002).

Acting at the post-transcriptional level, RNAi is a process whereby dsRNA molecules trigger the degradation of target mRNA by utilizing two proteins conserved among all multicellular organisms. The mechanism of RNAi can be briefly summarized as:

1. The RNAi mechanism is endogenously triggered by the presence of long dsRNA molecules. In response, the RNAi process begins by utilizing a cytoplasmic enzyme, Dicer, to cleave the dsRNA into smaller fragments. These short interfering RNAs (siRNA) are typically 20–23 base pairs in length and contain two nucleotide overhangs at both 3′ ends.

2. The siRNAs are then recognized by another enzyme complex, RNA induced silencing complex (RISC), which unwinds the two strands of the dsRNA using its helicase activity. Subsequently, one of the siRNA strands (the antisense strand) becomes incorporated into RISC and functions as guide for RISC to seek out target mRNA with sequences complementary to the incorporated siRNA strand. The remaining free RNA strand gets degraded by endonucleases.

3. Target mRNA hybridizes to the RISC-associated RNA strand by sequence-specific base pairing. In turn, this activates the endonuclease activity of RISC, which leads to cleavage of the target mRNA. The cleaved mRNA fragments are then fully degraded by endogenous nucleases, which ultimately leads to post-transcriptional gene silencing.

Hence, whilst the traditional antisense mechanism exploits the use of RNase H, RNAi differs by utilizing Dicer, and primarily RISC, as a means to regulate gene expression.

As mentioned, RNAi is a naturally occurring phenomenon; hence multicellular organisms contain innate dsRNA molecules to trigger RNAi. This highly conserved method of gene regulation is widely believed to be an evolutionary mechanism to protect the genome of organisms from invading pathogens such as viruses and transposans and defective mRNAs (Cullen, 2002; Hannon, 2002). Scientists, however, have exploited this natural defence mechanism as a powerful therapeutic tool, by rationally designing short dsRNA molecules, i.e. siRNAs, and introducing them into cells to silence targeted genes. Transfection of larger dsRNAs (analogous to the endogenous ones that trigger RNAi) in mammalian cells, often leads to activation of the innate immune response (interferon response) and cell death. siRNAs are typically 21–25 nucleotides in length, with the antisense strand designed to be complementary to any target mRNA. Hence, these drugs can be used to treat a variety of diseases which involve alterations in gene expression patterns. Several papers have reported the successful transfection of chemically synthesized siRNAs into mammalian cells, which lead to the down-regulation of target genes by RNAi. Although RNAi and siRNA drugs are still in their infancy, they are considered to be one of the strongest candidates, among antisense and other types of drug molecules, for the future treatment of a range of diseases including infection, neurodegenerative diseases and cancer. Consequently, for the discovery of RNAi and given the significant impact it has made within the scientific community, Fire and Mello received the 2006 Nobel Prize in Physiology and Medicine.

17.5 Shortcomings of Antisense Therapeutics

Acting at the genetic level, antisense agents offer several advantages over other drugs which confer their mode of action at the proteomic level. Primarily, antisense agents offer high specificity by binding in a sequence dependent manner to the corresponding mRNA of target proteins. In turn, this avoids the potential of inhibiting the action of proteins related to the target (such as protein isoforms) or even non-related proteins, a drawback often associated with many protein targeting drugs. Other desirable properties that antisense molecules offer include low cost of synthesis, broad applicability, rapid development and their potential in being easy to modify for appropriate applications (see above). As such, antisense technologies have dominated the market in genetic research and are also rapidly expanding into the therapeutic field of medicine. However, despite the encouraging pre-clinical results observed with antisense drugs, their progress into clinical trials has been hindered by their low stability, toxicity, target site identification and accessibility, and low cellular uptake. Being the oldest of antisense agents, research with ASOs and knowledge of their associated problems has significantly benefited the development of the younger antisense molecules. Consequently, the modifications made to ASOs to overcome their low stability, such as the use of phosphothiorate backbones and ribose modifications (see ASO section above), has been equally applied to ribozymes, DNAzymes and siRNAs. Such modifications have produced compounds with enhanced resistance to nucleases and good tissue distribution. Nevertheless, an important consideration that needs to be taken when modifying the oligonucleotides is that these modifications should not hinder their binding or mode of action. The introduction of modifications often produces conformational changes in oligonucleotides, which, in the case of ribozymes and DNAzymes in particular, can have a knock on effect in abolishing their catalytic activity. Hence, following modification, the binding and kinetic properties of each antisense agent needs to be assessed. Another issue encountered in the design of antisense agents is the problem of target site identification and its accessibility to hybridization. Simply knowing the gene sequence of the target mRNA is not enough for the design of effective antisense agents, as more than often, long RNA molecules adopt secondary structures that significantly influence the binding of antisense agents. Therefore, sites chosen for targeting, have to be easily accessible for hybridization to their complementary drug. A number of computer software programs exist for the prediction of secondary structures of oligonucleotides (m-fold, RNAstructure), which have provided an economical mean into the design of antisense agents. However, these are theoretical predictions and therefore other methods facilitating the development of antisense molecules, such as *in vitro* combinatorial selection techniques, have been used. The use of computational programs can also aid in reducing the number of antisense molecules for screening, hence a combination of computational predictions and *in vitro* selection methods may be the best approach into assisting the design of effective antisense agents.

In general, the toxicity profiles associated with antisense agents are transient and mild to moderate. While the toxic effects of antisense agents are largely sequence independent, they are generally attributed to backbone modifications, such as phosphorothioate-oligonucleotides which bind non-specifically to plasma proteins (Chan *et al.*, 2006). By far, the most significant challenge that each of the antisense agents face is their systemic delivery to targeted cells. For antisense molecules to elicit their effect, they must be able to efficiently enter cells to reach their target mRNA. However, internalization of oligonucleotides is usually limited owing to their inherent negative charge, making it difficult for them to efficiently cross the hydrophobic cell membrane. As a result, considerable efforts have been made in designing efficient *in vivo* and *in vitro* delivery vehicles, with many antisense technology companies focusing intensely on solving this problem. The most common, and so far the most successful, delivery system has entailed the use of liposomes and charged lipids. Most of these lipid vehicles consist of cationic lipids that can either encapsulate the antisense molecule in its centre or, facilitated by the opposing charges, form complexes with it. Using liposomes as their delivery vehicle, Alnylam and Protiva were the first group to demonstrate the effective targeted delivery of a siRNA, targeting the apoB gene in the liver of monkeys. Liposomes are typically taken up by cells in the form of endosomes *via* endocytosis. Following cellular uptake, the antisense molecule needs to be released from the endosomal compartment to exert its action; hence, delivery vehicles have to be designed to act in such a way. Sirna, a leading pharmaceutical company in the design of antisense agents, also work on the design of lipid nanoparticles that change under certain biological conditions, such as a change in pH, which leads to disruption of the endosomal membrane, setting the antisense molecule free (Arnaud, 2006). Other liposomes contain helper liquids that disrupt the endosomal membrane and aid cytosolic release of the antisense molecule. While liposomes have shown good potential, a number of other macromolecular delivery systems have also been designed, such as cyclodextrin based nanoparticles, dendrimers (highly branched 3D polymers), amino acids and sugars. Nastech, for example, use peptide carriers as a delivery method, whereby a siRNA is directly conjugated to a peptide. The siRNA is synthesized as a slightly longer version (around 25–30 nucleotides) so that it can act as substrate for Dicer. As a result, following delivery, the peptide moiety is cleaved by Dicer, which concomitantly frees the siRNA molecule, allowing it to be incorporated into RISC for gene silencing (Arnaud, 2006). Often, the various delivery systems are conjugated to polyethylene glycol (PEG) to prolong circulating/tissue half life, and/or ligands which allow for tissue specific targeting of the antisense molecules. For effective targeting, along with efficient cellular uptake, ligands such as receptors, antibodies and aptamers have been conjugated to the delivery vehicle-antisense complex or even the antisense molecule directly. Receptor mediated endocytosis is a process whereby the antisense agent is coupled to a ligand or antibody that is specific for a certain receptor, mediating its uptake into targeted cells. Shi and co workers (Shi *et al.*, 2000) conjugated a peptide nucleic acid (PNA) antisense molecule to an antibody for the receptor transferrin, which allowed for the antisense agent to cross the blood brain barrier. In another study, the antibody for transferrin was conjugated to a ribozyme, which led to a three-fold increase in cellular uptake compared to the free ribozyme (Hudson *et al.*, 1999). The use of such targeting ligands is becoming increasingly popular, with Nastech using peptides as their targeting molecule and Calando Pharmaceuticals using transferrin complexed with a cyclodextrin nanoparticle to target cancer cells. Various other types of targeting ligands are also being investigated, such as the use of fusion proteins (e.g. antibody–protamine fusion protein complexed to siRNA, for delivery and targeting of siRNAs; Song *et al.*, 2005) and aptamers. A recent study showing promising results in xenograft models demonstrated the use of an aptamer conjugated to siRNA *via* a double stranded RNA linker, for the targeting of prostate cancer cells. The aptamer used was specific for PMSA, a receptor that is overexpressed and internalized in prostate can-cer cells, while the siRNA was used to silence survival genes overexpressed in cancer cells. Incubation of the aptamer–siRNA complex in cells expressing PMSA, led to internalization of the complex and subsequent incorporation of the siRNA into the RNAi pathway, ultimately leading to targeted gene silencing and cell death. The aptamer–siRNA complex, however, did not bind or function in control cells that didn't express PSMA. Furthermore, the anti-sense complex also inhibited prostate tumour growth in xenograft models (McNamara *et al.*, 2006). Such studies, in combination with exploring the use of other delivery/targeting vehi-cles, show promising potential in overcoming the major delivery challenge associated with antisense therapeutics. Overall, advances in nucleotide chemistry and biochemical engineer-ing have significantly aided in upholding the initial therapeutic potential anticipated from antisense molecules. As such, we are now seeing the emergence of antisense agents entering clinical trials for cancer.

17.6 Antisense Agents in Clinical Trials

While many antisense agents have proven their efficacy in the lab, they have yet to show their full potential in the clinic. For the fight against cancer, only ASOs have shown signif-icant progress, with many of them now entering phase I/II clinical trials. Although siRNAs have entered clinical trials against various other diseases, such as human immunodeficiency virus (HIV), asthma and age-related macular degeneration, currently no clinical trials for can-cer have been established. In the case of ribozymes, only two candidates have so far shown encouraging results. Hence, this section will only briefly mention ribozyme and siRNAs ther-apeutics in pre-clinical/clinical stages against cancer and focus more on ASOs drugs.

17.6.1 Antisense oligonucleotides (ASOs)

Being the oldest of the antisense molecules, ASOs are slowly proving their value for the treat-ment of a range of diseases including cancer. As such, many companies have advanced on the product pipeline with some ASOs now entering phase II and III trials for cancer (Table 17.1), although only a few of them will be discussed in detail below.

Table 17.1 Antisense oligonucleotides in clinical trials, with details on their target, trial phase and the companies that have developed them

Drug	Target	Phase	Company
OGX-011	Clusterin	II	ISIS and OncoGenex
LY2181308	Survivin	I	ISIS and Lilly
LY2275796	eIF-4E	I	ISIS and Lilly
OGX-427	Hsp27	I	ISIS and OncoGenex
Oblimersen (genasense/G-3139)	Bcl-2	III	Genta Inc.
GTI-2040	Ribonucleotide reductase	II	Lorus
GTI-2501	Ribonucleotide reductase	II	Lorus
LErafAON-ETU	c-raf	I	Neopharm
AEG35156	XIAP	II	Aegera Therapeutics
AP12009	TGF-β2	II	Antisense Pharma
ISIS-2503	H-ras	II	NCI

GTI-2040 (first-generation ASO)

Lorus Therapeutics developed two antisense drugs, GTI-2040 and GTI-2501, as part of their anticancer therapeutic regime. These drugs target ribonucleotide reductase (RNR), an enzyme that is essential for DNA synthesis and repair. RNR catalyses the synthesis of $2'$-deoxyribonnucleotides by removing an oxygen molecule from the corresponding ribonucleotide. Formation of $2'$-deoxyribonnucleotides is essential for DNA replication, and therefore vital for the survival and proliferation of cancer cells. Hence, RNR makes an attractive target for many cancer therapeutics, as synthesis of $2'$-deoxyribonnucleotides depends solely on the activity of RNR. The RNR is composed of two subunits, R1 and R2, with the dimerization of both subunits being necessary for enzyme activity. While the R1 subunit is stably expressed, expression of the R2 component is regulated differentially throughout the cell cycle (Nocentini, 1996). The R2 subunit has shown to be overexpressed in cancer cells and can alter the malignant potential of cancer cells by acting as a signalling molecule. Overexpression of R2 also appears to increase the drug resistance of cancer cells to various chemotherapy drugs (Huang *et al.*, 1997). GTI-2501 and GTI-2040 have been shown to act as antisense agents to the R1 and R2 subunits, respectively. GTI-2040 is a 20 bp phopshorothioate oligonucleotide targeting the coding region of the R2 subunit of RNR. Pre-clinical studies have shown that GTI-2040 exhibits marked antitumour activity against a wide range of human cancers in xenograft models (Lee *et al.*, 2003). Toxicology and pharmacokinetic studies indicated that GTI-2040 was safe to use in humans at concentrations expected to inhibit RNR. GTI-2040 completed its phase I clinical trials, showing that the drug was well tolerated in humans as a single agent (Desai *et al.*, 2005). Recently, results of its phase I/II trials in combination with capecitabine have been reported in patients with renal cell carcinoma. Although the administration of both drugs was well tolerated, the response rate did not look promising. No marked changes were observed in tumour size, time to progression or survival. Given that the results did not meet the predefined criteria, the combination of GTI-2040 with capecitabine at the tested doses has not been recommended for further evaluation in metastatic renal cancer (Stadler *et al.*, 2007). However, Lorus are collaborating with the US National Cancer Institute (NCI) in multiple phase II clinical programmes to assess the effect of GTI-2040 in a variety of other cancer types (Lorus Therapeutics).

OGX-011 (second-generation ASO)

Another antisense agent to enter clinical trials for cancer is OGX-011, developed by OncoGenex Technologies and ISIS Pharmaceuticals. OGX-011 is a 21-mer phosphorothioate gapmer, containing $2'$ MOE modifications to the four bases at each of its ends. This second generation antisense agent specifically targets the site of translation initiation on the mRNA encoding for human clusterin. Clusterin acts as a cell survival protein and plays an important role in tumourigenesis and cancer progression. Upregulation of clusterin levels are typical in various cancers such as prostate, breast, renal, non-small cell lung (NSCL) and ovarian. The protein is also overexpressed in response to anticancer agents such as chemotherapy, radiation and hormone ablation therapy (Chi *et al.*, 2005; Biroccio *et al.*, 2005). Hence, as well as facilitating cancer progression, increased levels of clusterin also confer resistance to anticancer treatments. Pre-clinical studies with OGX-011 showed that this antisense oligonucleotide enhanced the efficacy of anticancer drugs by inhibiting the expression of clusterin. Phase I results evaluating OGX-011, alone or in combination with docetaxel, in patients with high risk prostate cancer, demonstrated that the antisense drug was well tolerated. In a single agent study, candidates for prostatectomy were administered OGX-011 by IV at intervals over 29 days. On days 30–36, prostatectomy was performed, which provided the opportunity to correlate dose-dependent clusterin expression and tissue concentration. Results of this trial indicated that OGX-011 achieved excellent tissue drug concentration and strongly inhibited clusterin expression in prostate cancer (Chi *et al.*, 2005). Currently, OGX-011 is in five phase II trials in prostate, lung and breast cancer. Preliminary results have been reported in June 2007 for three of these phase II trials (ISIS Pharmaceuticals), with two of them showing encouraging results. In the prostate cancer study, patients received OGX-011 with docetaxel and prednisone or docetaxel and prednisone alone. Early indication showed prolonged durations of progression free survival and increased disease stabilization. In patients with advanced non-small cell lung cancer (NSCLC), administration of OGX-011 with gemcitabine plus cisplatin or carboplatin led to increased overall survival (14 months) compared to the expected effects of gemcitabine plus cisplatin or carboplatin administration (8–11 months). The percentage of patients surviving at one year also increased to 54 % compared to the reported 33–43 % for the chemotherapy regime alone. Despite the fact that the results for the prostate and NSCL cancer trials look encouraging, preliminary data presented on the breast cancer trial seem disappointing. Although clinical activity was observed in patients given OGX-011 and docetaxel, the results of the trial did not meet the pre-determined criteria of tumour size reductions to progress into a second stage of accrual. OncoGenex is expected to announce the results of the remaining phase II trials, from which they will determine whether to proceed into phase III trials and for which cancer types.

LY2275796

ISIS Pharmaceuticals has also developed another antisense therapeutic for cancer, in collaboration with Eli Lilly and company. LY2275796 has been developed to target the eukaryotic translation initiation factor 4E (eIF4E), which is commonly involved in tumour progression, angiogenesis and metastasis, including breast, prostate, lung, colon and other cancers. Pre-clinical data, published in September 2007 (Graff *et al.*, 2007), described the antitumour effects of various antisense oligonucleotides developed to suppress translation of eIF4E. Mammalian cultured cell experiments demonstrated that the antisense agents inhibited expression of eIF4E and eIF4E regulated proteins, induced apoptosis and prevented formation of vessel-like structures in endothelial cells. Significant reduction of eIF4E expression by I.V administration of the antisense oligonucleotides was also observed in human tumours grown in mouse hosts, which led to significant suppression of tumour growth. Although administration of the oligonucleotides targeted and suppressed mouse eIF4E in the liver (by 80 %), no adverse effects were observed in body weight or liver function. Consequently, this provided evidence that cancers may be more susceptible to eIF4E inhibition than normal tissue, concurrently prompting the development of LY2275796. Currently, LY2275796 is in phase I clinical trials for the treatment of human cancers (ISIS International).

Oblimersen Sodium (also referred to as Genasense or G3139)

It is also noteworthy to mention Oblimersen Sodium, which is the most advanced ASO to be developed for cancer. Developed by Genta, this ASO is an 18-mer phosphorothioate DNA oligonucleotide designed to target the first six codons of the human mRNA for Bcl-2. The Bcl-2 protein is known to play a vital role in inhibiting apoptosis and is overexpressed in many cancers. Overexpression of the protein is also associated with conferring drug resistance in cancer cells. Hence, downregulation of this protein is expected to induce apoptosis and/or enhance the effectiveness of anticancer drugs when used in combination studies (Chanan-Khan, 2005). Following encouraging results in preclinical models, Oblimersen progressed into a number of clinical trials, with it being evaluated in patients with various types of cancers. Generally, the results of such studies indicate the Oblimersen is moderately tolerated. Randomized phase III trials of oblimersen in patients with malignant melanoma, chronic lymphocytic leukaemia (CLL), multiple myeloma and acute myeloid leukaemia (AML) have been conducted and phase I and II trials are also underway for other cancer types. The effect of Oblimersen has also been evaluated with a number of chemotherapy agents. Results of these trials generally indicate that Oblimersen accentuates the effects of a number of chemotherapy agents and is associated with increased response rates in patients (O'Brien *et al.*, 2007; Rheingold *et al.*, 2007). However, these effects of Oblimersen in the individual trials have not been deemed as clinically significant by the FDA (with Genta appealing against some of the decisions made; Drugs RD, 2007). Despite this, clinical trials evaluating the efficacy of Oblimersen alone and in combination with a number of chemotherapy agents are still ongoing (see Gleave and Monia, 2005 and Drugs RD, 2007 for a detailed overview of the clinical trials for Oblimersen).

17.6.2 Ribozymes

Only two ribozyme-based therapeutics for cancer have entered clinical trials. One of these candidates is the chemically synthesized ribozyme, Angiozyme, developed by Ribozyme Pharmaceuticals Inc (RPI; now known as Sirna Therapeutics). The enzyme is a hammerhead ribozyme which has been successfully modified to show improved stability in human serum, without perturbing catalytic activity (Beigelman *et al.*, 1995). Angiozyme was selected and designed to inhibit angiogenesis, the formation of new blood vessels, a prerequisite for sustained tumour growth. Its action is exhibited by blocking the expression of the receptor tyrosine kinase Flt-1 (vascular endothelial growth factor, VEGF-R1), a high affinity receptor for VEGF. These growth factors are secreted by tumour cells and interact with their receptors located on nearby blood vessels. This results in the formation of new blood vessels which branch into closely located tumours, supplying them with oxygen and allowing them to proliferate further. Hence, by inhibiting the expression of the receptors for VEGF, angiogenesis is inhibited, leading to restricted tumour growth. In clinical trials, Angiozyme was shown to be well tolerated and maintained in plasma for several hours following intravenous administration. The drug has been entered in multiple phase I/II trials, being evaluated for its effects alone and in combination with other chemotherapy drugs. Although each study showed that Angiozyme was well tolerated, the company reported that its effects could not be separated from the chemotherapeutic drugs, but it may result in the downregulation of the VEGF receptor and improve clinical outcome (Schubert and Kurreck, 2004). Other clinical studies have also demonstrated that Angiozyme may lead to disease stabilization (Kobayashi *et al*, 2005; Weng *et al*, 2005).

Herzyme is another ribozyme, developed by RPI, that entered clinical trials for breast and ovarian cancer. This enzyme was used to target human epidermal growth factor receptor 2 (HER-2), which is commonly overexpressed in breast cancer. Like Angiozyme, Herzyme too was well tolerated in clinical studies. However, further data for these ribozymes have not been reported and current studies involving Angiozyme and Herzyme seem to have ceased, with the company now focusing on RNAi as a therapeutic strategy for the treatment of a variety of diseases.

17.6.3 siRNAs

Whilst only five siRNAs have entered clinical trials, none of them have been against cancer. However, this is about to change with Calando Pharmaceuticals (US California) announcing earlier this year that they hope to enter the first targeted cancer siRNA into clinical trials. Calando have developed CALAA-01, a nanoparticle containing siRNA and a protein targeting agent, transferrin, formulated with Calandos RONDEL (RNA/oligonucleotide nanoparticle delivery) technology. The siRNA was developed to target the M2 subunit of ribonucleotide reductase. The siRNA has demonstrated potent anti-proliferative activity in a range of cancer models (Heidel *et al.*, 2007a). The use of a transferrin protein targeting agent, on the surface of the nanoparticle, is to specifically target the siRNA to cancer cells. Since this protein is found in a variety of tumours, binding of the transferrin protein to its receptor will facilitate the specific uptake of the nanoparticle primarily in cancer cells, followed by subsequent release of the siRNA. Incorporation of the siRNA into the RNAi machinery is then anticipated to silence the expression of the mRNA encoding for the R2 subunit of ribonucleotide reductase. In September 2007, Calando published encouraging results demonstrating the targeted uptake of CALAA-01 by tumour cells in mice (Heidel *et al.*, 2007a). Currently, the company has commenced toxicity studies of the drug in rats and monkeys, with results showing the targeted delivery of the drug and that it is well tolerated (Heidel *et al.*, 2007b). Calando hope these encouraging preclinical results will approve their investigational new drug (IND) application with the US food and drug administration. Early next year, Calando plan to conduct dose-escalation phase I clinical trails in patients with metastatic solid tumours.

Another siRNA, targeting the R2 mRNA of ribonucleotide reductase, is siRNA 1284, developed by Lorus Therapeutics. Currently in its preclinical phase, this drug has shown sequence-specific down-regulation of R2, with concomitant antiproliferative effects in cancer cells *in vitro*. In xenograft models, administration of siRNA 1284 led to suppression of tumour growth (Avolia *et al.*, 2007). A lead company in the field of RNAi, Silence Therapeutics, has also developed four siRNAs, for various cancer types. Currently in late stage pre-clinical, Atn027 and Atn093 have been developed for gastrointestinal cancer and non-small cell lung cancer (NSCLC) respectively, with the company hoping to initiate phase I clinical studies of these compounds in 2008. For prostate and liver cancer, Silence Therapeutics has developed Atn111 and Atn150 siRNAs, respectively, but these are still in the early preclinical stage of development (Silence Therapeutics). Working on liver cancers, Alnylam have produced ALNVSP01, which targets VEGF and kinesin spindle protein (KSP), both of which are involved in the growth and development of tumours. They use their novel liposome technology to deliver the drug to the liver. Results of preclinical studies have demonstrated the ability of ALNVSP01 to silence VEGF and KSP expression in the liver and prevent the growth of cancer cells. Intradigm Corporation's lead siRNA candidate, ICS-283, has been formulated as their peptide targeted nanoparticle; RNAi Nanoplex technology. Intravenous administration of ICS-283 has demonstrated *in vivo* knockdown of VEGF pathway genes in mouse tumour models, leading to inhibition of tumour growth. The company is hoping to start clinical studies with this drug in 2008. Overall, the future of siRNA therapeutics looks promising, with many companies hoping to initiate cancer related clinical trials of their drugs, in combination with their targeted systemic delivery technology.

17.7 Concluding Remarks

Exploiting nature's ingenuity has become a trademark for scientists searching for possible therapeutic novelties. Antisense technology is one such strategy that initially held out high therapeutic potential, but, despite being in over twenty years of development, the hurdles in its medical application have still not been overcome. Nonetheless, the lessons learnt with ASOs and the advances made in this area have significantly aided in the development of ribozymes and more so siRNAs. Hence, with several companies and research groups focussing on overcoming the problems associated with antisense agents, particularly the engineering of efficient delivery vehicles, the hope for antisense molecules to precede the market in cancer therapy still remains an exciting and viable strategy. Despite having to yet prove its therapeutic efficacy, antisense technology has been and remains undoubtedly one of the most powerful research tools in aiding our scientific understanding into the genetic basis of many diseases.

References

Arnaud CH. Delivering RNA interference. Developing siRNA therapeutics depends on synthetic delivery systems. *Chem Eng News* 2006, **84**(46), 16–23.

Avolio TM, Lee Y, Feng N, *et al*. RNA interference targeting the R2 subunit of ribonucleotide reductase inhibits growth of tumour cells *in vitro* and *in vivo*. *Anticancer Drugs* 2007, **18**(4), 377–88.

Beigelman L, McSwiggen JA, Draper KG, *et al*. Chemical Modification of Hammerhead Ribozymes. *J Biol Chem* 1995, **270**, 25702–8.

Biroccio A, D'Angelo C, Jansen B, Gleave ME, Zupi G. Antisense clusterin oligodeoxynucleotides increase the response of HER-2 gene amplified breast cancer cells to Trastuzumab. *J Cell Physiol* 2005, **204**, 463–9.

Chan JHP, Lim S, Wong WSF. Antisense oligonucleotides: from design to therapeutic application. *Clin Exp Pharmacol Physiol* 2006, **33**, 533–40.

Chanan-Khan A. Bcl-2 antisense therapy in B-cell malignancies. *I* 2005, **19**, 213–21.

Chi KN, Eisenhauer E, Fazli L, Jones EC, Goldenberg SL, Powers J, Tu D, Gleave ME. A phase I pharmacokinetic and pharmacodynamic study of OGX-011, a 2'-methoxyethyl antisense oligonucleotide to clusterin, in patients with localized prostate cancer. *J Natl Cancer Inst* 2005, **95**, 1287–96.

Cullen BR. RNA interference: antiviral defence and genetic tool. *Nat Immunol* 2002, **3**, 597–99.

Desai AA, Schisky RL, Young A, *et al*. A phase I study of antisense oligonucleotide GTI-2040 given by continuous intravenous infusion in patients with advanced solid tumours. *Ann Oncol* 2005, **16**, 958–65.

Drugs RD. Oblimersen: Augmerosen, Bcl-2 antisense oligonucleotide – Genta, G 3139, GC 3139, Oblimersen Sodium. *Drugs RD* 2007, **8**, 321–34.

Famulok M, Verma S. In vivo-applied functional RNAs as tools in proteomics and genomics research. *Trends Biotechnol* 2002, **20**(11), 462–6.

Fire A, Xu S, Montgomery MK, Kostas SA, Driver SE, Mello CC. Potent and specific genetic interference by double-stranded RNA in Caenorhabditis elegans. *Nature* 1998, **391**, 806–11.

Gleave ME, Monia BP. Antisense therapy for cancer. *Nat Rev Cancer* 2005, **5**, 468–79.

Graff JR, Konicek BW, Vincent TM, *et al*. Therapeutic suppression of translation initiation factor eIF4E expression reduces tumour growth without toxicity. *J Clin Invest* 2007, **117**(9), 2638–48.

Hannon GJ. RNA interference. *Nature* 2002, **418**, 244–51.

Heidel JD, Liu JY-C, Yen Y, Zhou B, Heale BSE, Rossi JJ, Bartlett DW, Davis ME. Potent siRNA inhibitors of ribonucleotide reductase subunit RRM2 reduce cell proliferation *in vitro* and *in vivo*. *Clin Cancer Res* 2007a, **13**(7), 2207–15.

Heidel JD, Yu Z, Liu JY-C, Rele SM, Liang Y, Zeidan RK, Kornbrust DJ, Davis ME. Administration in non-human primates of escalating intravenous doses of targeted nanoparticles containing RRM2 siRNA. *Proc Natl Acad Sci U S A* 2007b, 10.1073/pnas.0701458104.

Huang A, Fan H, Taylor WR, Wright JA. Ribonucleotide reductase R2 gene expression and changes in drug sensitivity and genome stability. *Cancer Res* 1997, **57**, 4876–81.

Hudson AJ, Normand N, Ackroyd J, Akhtar S. Cellular delivery of hammerhead ribozymes conjugated to a transferrin receptor antibody. *Int J Pharm* 1999, **182**(1), 49–58.

Khan AU. Ribozyme: A clinical tool. *Clin Chim Acta* 2006, **367**(1–2), 20–7.

Kobayashi H, Gail ES, Lockridge JA, *et al*. Safety and pharmacokinetic study of RPI.4610 (ANGIOZYME), an anti-VEGFR-1 ribozyme, in combination with carboplatin and paclitaxel in patients with advanced solid tumours. *Cancer Chemother Pharmacol* 2005, **56**, 329–36.

Lee Y, Vassilakos A, Feng N, *et al*. GTI-2040, an antisense agent targeting the small subunit component (R2) of human ribonucleotide reductase, shows potent antitumour activity against a variety of tumours. *Cancer Res* 2003, **63**, 2802–11.

McNamara JO, Andrechek ER, Wang Y, Viles KD, Rempel RE, Gilboa E, Sullenger BA, Giangrande PH. Cell type-specific delivery of siRNAs with aptamer-siRNA chimeras. *Nat Biotechnol* 2006, **24**, 1005–15.

Nocentini G. Ribonucleotide reductase inhibitors: new strategies for cancer chemotherapy. *Crit Rev Oncol Hematol 1996*, **22**, 89–126.

O'Brien S, Moore JO, Boyd TE, *et al*. Randomised Phase III trial of fludarabine plus cyclophosphamide with or without oblimersen sodium (Bcl-2 antisense) in patients with relapsed or refractory chronic lymphocytic leukaemia. *J Clin Oncol* 2007, **25**, 1114–20.

Rheingold SR, Hogarty MD, Blaney SM, Zwiebel JA, Sauk-Schubert C, Chandula R, Krailo MD, Adamson PC and Children's Oncology Group Study. Phase I trial of G3139, a Bcl-2 antisense oligonucleotide, combined with doxorubicin and cyclophosphamide in children with relapsed solid tumours: a Children's Oncology Group Study. *J Clin Oncol* 2007, **25**, 1512–18.

Schubert S, Kurreck J. Ribozyme- and deoxyribozyme strategies for medical applications. *Curr Drug Targets* 2004, **5**, 667–81.

Shi N, Boado RJ, Pardridge WM. Antisense imaging of gene expression in the brain *in vivo*. *Proc. Natl. Acad. Sci. U S A* 2000, **97**, 14709–14.

Song E, Zhu P Lee S-K, *et al*. Antibody mediated *in vivo* delivery of small interfering RNAs via cell-surface receptors. *Nat Biotechnol* 2005, **23**, 709–17.

Stadler WM, Desai AA, Quinn DI, *et al*. A Phase I/II study of GTI-2040 and capecitabine in patients with renal cell carcinoma. *Cancer Chemo Pharm* 2008, **61**, 689–94.

Weng DE, Masci PA, Radka SF, *et al*. A phase I clinical trial of a ribozyme-based angiogenesis inhibitor targeting. vascular endothelial growth factor receptor-1 for patients with refractory solid tumors. *Mol Cancer Ther* 2005, **4**, 948–55.

Zamecnik PC, Stephenson ML. Inhibition of Rous sarcoma virus replication and cell transformation by a specific oligodeoxynucleotide. *Proc. Natl. Acad. Sci. U S A* 1978, **75**, 280–84.

Zellweger T, Miyake H, Cooper S, Chi K, Conklin BS, Monia BP, Gleave ME. Antitumour activity of antisense clusterin oligonucleotides is improved in vitro and in vivo by incorporation of 2'-O-(2-methoxy)ethyl chemistry. *J Pharmacol Exp Ther* 2001, **298**(3), 934–40.

18
Aptamers as Anticancer Agents

Vaidehi Makwana, Suzanne Simmons and Sotiris Missailidis

18.1 Introduction

The continuous demand and rapid pace in pharmaceutical development, led by the need to deliver better anticancer agents faster, has provided a stimulus for the discovery of methodologies which promote recognition of disease-specific molecular targets, based on biological molecules. Until 10 years ago, amongst chemical compounds and biological entities, antibodies were the prevalent genus molecules used for molecular recognition. However, in the early 1990s, comparable molecules, referred to as aptamers, found a niche of their own in the area of molecular target validation, diagnostics, biotechnology applications in biosensors and as therapeutics. Aptamers are now at the forefront of molecular recognition, with their increasing popularity ensuing from their numerous advantages not only over antibodies and peptides, but also over small organic compounds.

Aptamers are short, synthetic, single-stranded RNA or DNA oligonucleotides that possess the ability to bind to their molecular targets with great specificity and high affinities, in the nanomolar to picomolar range. Aptamers typically range from 15 to 40 nucleotides in length, which are adequate to form stable secondary structures, allowing them to interact with their selected target *via* hydrogen bonding, electrostatic interactions and van der Waals forces. Oligonucleotides exceeding 30 nucleotides in length essentially exist in the initial stages, during the *in vitro* selection process. However, these molecules usually also contain primer sequences, which are not necessarily involved in the binding process. Thus, the selected aptamer is usually truncated to obtain the sequence region specifically involved in binding to the preferred target, predominantly to reduce the cost of production when utilized in diagnostic or therapeutic applications, but also the fidelity of production and flexibility of use. High affinity and specificity aptamers are, primarily, generated *via* an *in vitro* selection process referred to as SELEX (Systematic Evolution of Ligands by EXponential enrichment; see also Chapter 2) introduced autonomously by two groups, led by Ellington and Gold

Anticancer Therapeutics   Edited by Sotiris Missailidis
© 2008 John Wiley & Sons, Ltd

respectively (Tuerk and Gold, 1990; Ellington and Szostak, 1990). However, recently, different selection processes have also been developed to allow aptamer selection without the SELEX methodology (Drabovich *et al.*, 2005; Berezovski *et al.*, 2005).

The basic SELEX protocol is an evolutionary, iterant stringent process involving a combinatorial library of randomized nucleic acid sequences, with structural variations of more than 10^{15} different molecules, flanked by primers that allow polymerase chain reaction amplification. This is subjected to the selected molecular target for a series of events of binding, partitioning of unbound aptamers from the bound, followed by amplification of the bound aptamers, which are subsequently further subjected to the target. This process is repeated for several rounds, typically from 8 to 12, to obtain, through competitive binding, one or few aptamer sequences with high specificity and affinity for the chosen molecular target. At this stage, selected aptamers are cloned and sequenced, to reveal the binding sequences. Many alterations can be made to the basic SELEX protocol, such as the approach employed to present the target or the manner in which the aptamer-target complexes can be partitioned. Alterations in the presentation of the target can entail a method of counter selection or ToggleSELEX technique, involving cross-reactivity, whereas the variations in partitioning of aptamer-target complexes include photocrosslinking and capillary electrophoresis (CE) reviewed by Hamula *et al.* (2006).

Aptamers can function by site specifically manipulating their target molecules, such as gene products or epitopes directly, with the purpose of preventing the original protein–ligand interaction from taking place or by causing an alteration in the normal signal transduction cascade. However, the function of any non-targeted molecular product remains intact. As aptamers are versatile molecules and can be modified relatively easily in many ways, including the attachment of radiolabels, fluorescent labels, polymers or organic drug molecules, they can be equally used in diagnostic, imaging or therapeutic applications. These characteristics have favoured aptamers for diagnostic and therapeutic purposes in many areas, including anti-infectives, anticoagulants, anti-inflammatory, anti-angiogenesis, antiproliferation and immunotherapeutics. There have been a number of aptamers under development against many different oncological disease-related targets and are used for tumour diagnosis or therapy by specific inhibition of malignant cell proliferation.

18.2 Aptamers in Cancer

18.2.1 Aptamers against the vascular endothelial growth factor (VEGF)

One of the leading aptamers in drug development is the anti-VEGF aptamer, also commonly referred to as Macugen, Pegaptanib and NX1838. The anti-VEGF aptamer has been specifically generated against VEGF, which has a key role in angiogenesis and vasculogenesis. VEGF is a glycoprotein, disulfide-linked in a homodimeric form with differing molecular weights, each corresponding to the various isoforms formed as a product of alternative splicing. The prime splice variants are $VEGF_{165}$, $VEGF_{121}$, $VEGF_{185}$, and $VEGF_{206}$, where each isoform is identified by the number of amino acids present to compose the dimeric protein; for instance $VEGF_{165}$ is a dimeric isoform composed of 165 amino acids (Gatto and Cavalli, 2006). The over expression of $VEGF_{165}$ (in general referred to as VEGF) has been strongly allied with tumour proliferation in the most frequently diagnosed cancers and malignancies resulting in fatality. Furthermore, raised levels of VEGF have also been associated with other diseases such as ocular neovascularization. Progression of this disease can lead to severe loss of vision, due to conditions such as retinopathy of prematurity, diabetic retinopathy and age-related macular degeneration (AMD). Acute growth of blood vessels promoted by VEGF is characteristic of these diseases and, as a consequence, if these blood vessels begin to leak, then complete blindness is observed in the wet form of AMD. Other areas in which VEGF has a significant role involve inflammatory disorders, including rheumatoid arthritis and psoriasis (Gatto and Cavalli, 2006; Ruckman *et al.*, 1998).

Despite the fact that many other drug molecules are available in clinical use for the targeting of VEGF, the anti-VEGF aptamer is advancing incredibly well into the later stages of clinical trials. The anti-VEGF aptamer is an RNA oligomer that was selected against the VEGF$_{165}$ protein *via* the SELEX process utilizing a pre-SELEX 2′ fluoropyrimidine modified RNA oligonucleotide library, which consists of either 30 or 40 nucleotide randomized regions. Following the isolation of aptamers, which possess particularly high affinities, in the picomolar range, to the VEGF, the ligands underwent a post-SELEX modification process of exchanging the 2′OH with 2′OMe on the purines that are not involved in binding to the VEGF. The pre- and post-SELEX modification allows superior stabilization of the RNA oligonucleotides against nuclease degradation and the selected anti-VEGF aptamer has been further modified at the 5′ end with its conjugation to a 40 kDa polyethylene glycol (PEG), hence the name Pegaptanib (Ruckman *et al.*, 1998). In 2004 Pegaptanib was the first and is the only aptamer to date to have received an US Food and Drug Administration (FDA) approval for the use in treatment of AMD, as it inactivates the function of VEGF upon binding to it. The results obtained in preclinical studies permitted the aptamer to enter clinical trials, in which phase II/III trials have demonstrated encouraging results for the treatment of diabetic retinopathy and AMD (Zhou and Wang, 2006; Adamis *et al.*, 2006). Pegaptanib is suggested to have a direct effect on retinal neovascularization, as significant regression of neovascularization has been observed. The two year safety results demonstrated that all the doses of Pegaptanib given to patients were well tolerated and any adverse events were momentary owing to the injection preparation and procedures rather than the Pegaptanib drug (D'Amico *et al.*, 2006). Eyetech Pharmaceuticals and Pfizer have started the phase IV clinical trials for the safety and efficacy of injections of Pegaptanib when administered to patients with AMD every 6 weeks (www.ClinicalTrials.gov).

In addition to the anti-VEGF aptamer's role in retinal neovascularization, NX-1838 has also been shown to have significant consequences as an anti-cancer agent. Huang *et al.* demonstrated that tumour growth and metastasis is effectively suppressed by the aptamer. The experimental animal models with cultured Wilms' tumour cells were continuously treated on a daily basis for a period of 5 weeks and found a reduction in tumour weight by 84 % without any adverse effects, compared to the untreated control animal models (Huang *et al.*, 2001). Following the continuous encouraging results from *in vivo* and preclinical studies, the anti-VEGF aptamer may enter clinical studies for the treatment of numerous different cancers in the near future.

18.2.2 Aptamers against nucleolin

An aptamer prominent in the research field of specific targeted oligonucleotide therapy, AS-1411 has proceeded the farthest and is the first aptamer to enter clinical trials for cancer therapy. This nucleolin-binding aptamer is composed of only 26 guanosine and thymidine nucleotides, which has evolved from earlier studies of single-stranded guanosine-rich oligonucleotides (GROs). AS-1411, previously known as AGRO-100, was initially developed from the 29mer 3' amino alkyl modified phosphodiester oligonucleotide GRO29A. The GRO29A analogue, GRO29A-OH is the unmodified phosphodiester oligonucleotide with the removal of three nucleotides (5' thymidines) making the AS-1411 aptamer a 26-mer in length. The AS-1411 has not been selected like most aptamers *via* the SELEX methodology, but rather credited to the discovery that oligonucleotides can bring about a biological response without acquiring a relationship of sequence specific interactions with its target. The GROs have the potential to form G-quartet structures and their ability to arrange into these three-dimensional constructs allows them to bind to surface or intracellular proteins. Results from early investigations carried out by Bates *et al.* suggested that the GROs bind specifically to the cellular protein nucleolin and consequently demonstrated anti-proliferative activity, accompanied by inducement of apoptosis in many cancer cell lines (Bates *et al.*, 1999). The GRO-binding nucleolin is a phosphoprotein, primarily found in the cytoplasm and nucleolus of proliferating cells, where increased amount of the protein is detected in malignant cells, suggesting nucleolin is essential for cell division and survival of cancer cells. Other than the role of nucleolin involved in cell proliferation, it has been suggested that it acts as a shuttling protein between the cytoplasm and the nucleus in a cell, transporting viral and cellular proteins. Nucleolin has also been found to be expressed on the surface of cancer cells, hence creating an avenue to utilize the protein as a marker for early diagnosis of cancer.

The initial outlining concept of AS-1411 binding to nucleolin at the surface of the cell leads onto its anti-proliferative effects upon internalization of the aptamer. However, this is due to its unique ability to form a G-quadruplex structure, which has an additional benefit as the aptamer presents stability without any form of modification in serum and is not susceptible to nuclease degradation. Establishment of AS-1411 specifically binding to nucleolin was carried out using biotinylated GROs and employing monoclonal and polyclonal antibodies to nucleolin for competition based assays. Further research carried out with the GRO29A by Xu *et al.* (2001) revealed that the several different malignant cell lines, when treated with the aptamer, all demonstrated implication on the cell cycle. It was determined that the antiproliferative effects were due to the GRO29A preventing the progressions of the cell cycle from the S phase, hence accumulation of the cells was found in the S phase with the inhibition of DNA replication. However, RNA and protein synthesis remained unaffected, hence suggesting that hindering of DNA replication is the initial cause of inhibition of cell division that ultimately leads to induction of cell apoptosis (Xu *et al.*, 2001). Numerous analogues of GRO29A, with varying modifications to the sugar–phosphate backbone, were explored to determine their extent of antiproliferative properties. The outcome of this examination strongly indicates that, although the GRO29A with a phosphodiester backbone is the most unstable *in vivo*, it is the most appropriate form of backbone which allows the aptamer to be taken further as a therapeutic agent. The analogue with a phosphothioate backbone is more stable *in vivo*, however, poses more non-specific toxic side effects due to its polyanionic characteristics and is identified in prompting the immune system (Dapic *et al.*, 2002). AS-1411 has demonstrated encouraging results in the phase I clinical trials (Ireson and Kelland, 2006) and has now entered phase II trials for treatment for acute myeloid leukemia with plans in place for renal cancer.

18.2.3 Aptamers against the E2F group of proteins

The E2F proteins are a family of transcriptional factors consisting of eight members known to date, found in the mammalian cell, which are strongly associated with progression of the cell cycle, cellular proliferation and apoptosis. The progression of the cell cycle is regulated by the retinoblastoma (Rb) pathway, which comprises of the Rb susceptibility gene encoding for the tumour suppressor Rb protein (pRb) that is responsible for maintaining a stringent control of the cell cycle. The ability of pRb to interact with transcriptional factor E2F will determine the cellular proliferative effects. Any form of inactivation of pRb will most often lead to apoptosis. However, malignant cells have been found to possess mutations in the Rb and E2F pathway. Phosphorylation of the pRb induces the release of the E2F transcriptional factor, a protein that serves the purpose of an activator for vital genes involved in the progression of the cell cycle. Hence, pRb conveys signals in order to allow the cell to advance further in the cell cycle from the G_1 phase into the S phase, thus contributing to cell proliferation. However, there has also been evidence suggesting the E2F is involved in the initiation of apoptosis, hence signalling to associated genes to halt the cell cycle progression at the G1 phase, causing inhibition of cellular growth and eventually resulting in apoptosis. Various studies carried out by different research groups have suggested that over-expression of E2F leads to tumour growth *via* cellular proliferation and is correlated to numerous malignancies such as epithelial ovarian cancer (Suh *et al.*, 2007), non-small cell lung carcinoma (Gorgoulis *et al.*, 2002; Zacharatos *et al.*, 2004), and oesophageal squamous cell carcinoma (Yamazaki *et al.*, 2005; Ebihara *et al.*, 2004).

Besides E2F proteins' potential role in the development of malignancies, E2F has also been associated with cardiovascular disease. The main methods employed for the treatment of cardiovascular diseases are coronary bypass surgery and angioplasty. However, upon the required post-therapeutic application, intimal hyperplasia develops as a consequence of vascular smooth muscle cell (VSMC) proliferation. The transcriptional factor E2F is fundamental for controlling cellular proliferation and is the principal activator of VSMC proliferation. E2F binds tightly, displaying great specificity and affinity for a double stranded oligonucleotide containing an 8 bp consensus sequence (Hiebert *et al.*, 1989). Morishita *et al.* (1995) generated a 14-mer and a 30-mer oligonucleotide, both containing the E2F binding 8 bp consensus sequence. Results obtained from the investigation using these two decoy oligonucleotides demonstrated that they both entirely inhibited proliferation of VSMC *in vitro*, in comparison to the mismatched decoy, which had no effect. It is suggested that the E2F decoy binds to the E2F transcriptional factor, preventing the protein from binding to the tumour suppressor protein pRb. Hence, activation of the necessary genes are prevented from promoting the cell to enter the S phase of the cell cycle, thereby inhibiting tumour growth (Morishita *et al.*, 1995). The 14-mer decoy oligonucleotide has provided encouraging results *in vivo* and in preclinical studies that has allowed the E2F decoy to enter clinical trials in which the decoy has completed phase II and phase III trials for coronary artery and vascular diseases. To improve the efficacy of the decoy oligonucleotide upon binding to the transcriptional factor, an investigation was carried out using the E2F decoy oligonucleotide along with another two oligonucleotides which targeted nuclear factor-κB and Stat3 (signal transducer and activator of transcription-3). The study involved forming a complex decoy oligonucleotide (heterogeneous) which consisted of all three decoy oligonucleotides that contain three different binding sites for the three individual transcriptional factors. Another complex, which only consisted of the decoy oligonucleotide (homogeneous) for each individual transcriptional factor and the single decoy, were used as a control for comparison. The outcome of the study demonstrated that the heterogeneous complex decoy increased efficacy and potency compared to the homogeneous complex and the single decoy oligonucleotide (Gao *et al.*, 2006). There are also many other groups that are developing aptamers against specific E2F transcriptional factor members *via* the means of SELEX, such as E2F-1 and E2F-3, for the inhibition of excessive proliferation. However, more research is required before the E2F decoy can be entered into clinical trails for the application of the decoy oligonucleotide as a cancer therapeutic agent.

18.2.4 Aptamers against platelet-derived growth factor (PDGF)

There are potentially numerous growth factors that may possess a significant role in cell proliferation, many of which remain unidentified. However, one growth factor that is well recognized is PDGF. PDGF is a 30 kDa dimeric protein, which comprises two different polypeptide chains, both connected to each other by disulfide bonds. The polypeptide chains were named PDGF-A chain and PDGF-B chain, according to their order of elution in chromatography using a high resolution reverse phase HPLC separating system,which was necessary due to their similarity in size and homology (Heldin, 1992). PDGF is an essential and effective mitogen with a significant function in embryonic development, initiating healing of wounds and formation of connective tissue in a number of organs. However, overexpression of PDGF associates the growth factor with various different proliferative diseases, all of which are not malignant, such as atherosclerosis and glomerulonephritis. The levels of expression of both PDGF chains A and B found in the stroma of tumour cells is far greater in malignant tumours compared to the expression of PDGF in non-malignant tumours. The majority of solid tumours are known to exhibit excessive interstitial fluid pressure (IFP), which possess a crucial problem in the uptake of chemotherapeutic agents, thus resulting in diminution of efficacy for the anticancer drug employed in treatment. The fact that the regulation of IFP appears to be controlled by PDGF and the over activity of the growth factor demonstrated in numerous tumour cells, has led many researchers towards the development of agents as potential PDGF inhibitors for their application in cancer therapeutics. Selective inhibition of PDGF signalling in the stroma of tumour cells is expected to decrease tumour IFP, which subsequently stimulates the transport of chemotherapy drugs into tumours, hence enhancing therapeutic effectiveness as substantiated by Pietras and colleagues (Pietras *et al.*, 2001, 2002). Preliminary studies showed that the small molecule STI-571, a PDGF tyrosine kinase inhibitor, and the anti-PDGF-B aptamer were both found to reduce IFP in tumours using experimental rat colonic carcinoma models. The results from initial studies prompted further investigations aimed at determining the efficacy of the chemotherapeutic drugs, Taxol and 5-fluorouracil upon inhibition of PDGF using the antagonists STI-571 and the anti-PDGF aptamer. The KAT-4 tumours in severe combined immunodeficiency (SCID) mice were utilized and co-treated with PDGF inhibitors and chemotherapeutic agents. The results obtained supported the hypothesis that reduction in IFP elevates the uptake of cytotoxic drugs, as both inhibitors improved the effect of Taxol and 5-fluorouracil on tumour growth. The inhibitors alone, however, suggested no evidence of any antiproliferative outcomes in the absence of cytotoxic agents, hence co-treatment is necessary to provide an enhanced therapeutic effect. The PDGF-B aptamer has many advantages over the small drug STI-571, of which one critical benefit is the specificity the aptamer possesses for PDGF. Although STI-571 inhibits PDGF, the drug also inhibits the customary effects of other tyrosine kinases. This problem is not encountered with the aptamer, due to its high specificity for PDGF-B.

The PDGF-B chain aptamer was isolated from a single-stranded DNA library containing a randomized region of 40 nucleotides, *via* the SELEX procedure. The selected aptamer has a high affinity (100 pM) for PDGF-B, with a much reduced affinity for the PDGF-A, hence displaying great specificity for the B-chain growth factor (Green *et al.*, 1996). The PDGF-B aptamer has been studied in other diseases that have been associated with the overexpression of PDGF, such as in renal and cardiovascular diseases, and the results have been promising. The advanced renal diseases are often linked with proliferation of glomerular mesangial cells and accumulation of mesangial matrix, which appears to be promoted by PDGF. Floege *et al.*'s investigation demonstrated the PDGF-B aptamer, coupled to a 40 kDa PEG, had caused a reduction in proliferating mesangial cells, considerably reducing the over production of glomerular extracellular matrix, and resulted in declining expression of PDGF-B chain in a rat mesangioproliferative glomerulonephritis model. The specificity of the aptamer was further reassured, as no beneficial effect was observed using a pegylated scrambled aptamer sequence (Floege *et al.*, 1999; Ostendorf *et al.*, 2001). Further studies would be necessary to determine the potential of the PDGF-B aptamer in the diseases mentioned above, before the aptamer can make an entry into the clinical trials. However, other than VEGF-A, PDGF also appears to play an important role in new blood vessel growth in ocular pathologies. Evidence suggests inhibiting signalling initiated by VEGF-A and PDGF-B simultaneously is more effective in regression of neovascular growth, than inhibiting VEGF-A alone (Jo *et al.*, 2006). The anti-PDGF pegylated aptamer has thus entered phase I clinical trial for neovascular AMD.

18.2.5 Aptamers against the prostate-specific membrane antigen (PSMA)

There are numerous distinctive aptamers that have been developed for different cancer associated targets. These aptamers are advancing effectively in their *in vivo* and preclinical studies, of which many are also progressing successfully in their application to various types of chemical modification for delivery, diagnostics and therapeutics. An epitome of this is the anti-PSMA aptamer. Lupold *et al.* generated two aptamers with nanomolar affinity to the extracellular domain of PSMA, a type II membrane-bound glycoprotein. This protein is found to be excessively expressed in prostate cancer, especially in significantly advanced malignancies, where approximately a 100-fold increase of the membrane-bound form is observed, compared to that found in the normal prostate cells where a cytosolic form is mainly detected. The two aptamers A9 and A10 developed for binding specifically to PSMA are distinct as they share no consensus sequence and the aptamer A9 possess a non-competitive enzyme inhibition constant (K_i) of 2.1 nM whereas the A10 aptamer competitively inhibits enzyme activity with a K_i of 11.9 nM (Lupold *et al.*, 2002; Liu *et al.*, 2002). Farokhzad *et al* (2004, 2006) employed the anti-PSMA aptamer A10 to form a bioconjugate with a copolymer poly (lactic-acid)-block PEG with a carboxylic acid terminal group (PLA-PEG-COOH). The PLA-PEG-COOH copolymer encapsulated rhodamine-labelled dextran, which was used as a drug model, and these nano-particle aptamer bioconjugates were investigated for their use as delivery vehicles to prostate cancer cells. Results from the investigation revealed that the nano-particle aptamer bioconjugates were internalized successfully at 2 h, specifically by prostate cancer cells that express the PSMA protein compared to the control cell line, which does not express the PSMA protein (Farokhzad *et al.*, 2004). This success led the group to further develop the nano-particles for their use in therapeutics as they employed a chemotherapy drug Docetaxel in place of the rhodamine-labelled dextran. The aim of encapsulating the drug within the copolymer and forming a bioconjugate with the aptamer was to increase cytotoxicity and antitumour efficacy of specifically targeted prostate cancer cells. The *in vitro* and *in vivo* data demonstrated a significant outcome with the Docetaxel encapsulated aptamer bioconjugate when compared to the encapsulated drug without the aptamer, the drug without the nano-particle and the nano-particle alone. The Docetaxel encapsulated aptamer nano-particle improved specific cellular toxicity, hence complete tumour reduction in xenograft nude mice was observed along with a reduction in overall toxicity (Farokhzad *et al.*, 2006). The coupling of the A9 to siRNAs without any further modification i.e. attachment to any form of polymer, can be used to target PSMA expressing cells and was found to be taken up by the cells within 30 minutes (Chu *et al.*, 2006c). Other forms of conjugation explored include coupling the anti-PSMA to the toxin gelonin for specifically annihilating prostate cancer cells in therapeutic applications (Chu *et al.*, 2006a) and conjugating the aptamer to quantum dots for specific PSMA expressing cellular imaging (Chu *et al.*, 2006b).

18.2.6 Other aptamers in preclinical cancer studies

The aptamers currently in clinical studies have advanced significantly, based on their particularly successful results. However, a large number of other aptamers are currently entering the pipeline with more and more promising preclinical data becoming available.

Basic fibroblast growth factor (bFGF) is also a potent angiogenic factor (Basilico and Moscatelli, 1992; Jellinek *et al.*, 1995). Jellinek *et al.* (1995) produced a $2'$-aminopyrimidine modified RNA aptamer to bFGF with a K_D of 35 nM. The modification ensured that the aptamer was 1000-fold more stable in 90 % human serum than an unmodified version. Binding studies carried out on the aptamer to denatured bFGF and other members of the FGF family; acidic FGF, FGF-4, FGF-5, FGF-6 and FGF-7 plus four other heparin-binding proteins, showed considerably lower affinity, ranging from 4.1×10^{-2} to $>10^{-6}$ (Jellinek *et al*, 1995). Furthermore, aptamers showed promising data in tissue culture experiments, where the aptamer inhibited bFGF binding to the FGF receptor 1.

Raf-1, a serine threonine kinase, is the first in the mitogen-activated protein (MAP) kinase cascade (called a MAP kinase), which is activated by a chain of events from the receptor tyrosine kinase (RTK) (Cowley *et al.*, 2004), leading to cell growth and proliferation. Raf-1 is activated upon binding of Ras to the Ras binding domain (RBD); amino acids 51–131 of Raf-1. Ras binds to Raf-1 with an affinity of 18 nM (Herrmann *et al.*, 1995; Vojtek *et al.*, 1993; Chuang *et al.*, 1994). Using only the RBD of Raf-1 for selection, RNA aptamers were produced, which were all similar to one of two different sequences; RNA 9A and 9B. The two sequences showed binding affinities of 152 and 361 nM respectively (Kimoto *et al.*, 2002) for the RBD of Raf-1. In a competition experiment, the addition of the two aptamers significantly reduced Ras binding to the RBD of Raf-1. The RNA aptamer 9A also bound to a GST fusion form of Raf-1 RBD; B-Raf RBD, with an affinity of 285 nM, but in the competition experiment it showed specificity for RBD of Raf-1 and did not inhibit the interaction between Ras and B-Raf RBD (Kimoto *et al.*, 2002). It was found that RNA 9A and B inhibited the Raf-1 activation reaction in an experiment using full-length Raf-1, but that addition of the aptamers after the activation reaction had occurred provoked no inhibition by the aptamers, suggesting that the aptamers function as inhibitors of the activation reaction and not of Raf-1 itself. These experiments showed that the most potent aptamer, RNA 9A, could be a useful regulatory drug for mediating Ras and Raf-1 interactions in cells, therefore influencing cell growth and proliferation (Kimoto *et al.*, 2002).

Angiogenin is another molecule involved in angiogenesis and cell proliferation (Bussolino *et al.*, 1997). It is a ribonuclease thought to become active when it translocates to the nucleolus of the cell (Moroianu and Riordan, 1994a, b). DNA aptamers were produced (Nobile *et al.*, 1998), which showed inhibition of angiogenin's ribonucleolytic activity by >50 when present in a threefold excess. The aptamer was then truncated and tested for specificity by observing its effect on RNase A (another ribonuclease which is 33 % homologous to angiogenin) for its substrate poly(C), to which there was no significant inhibition, showing that the aptamer was specific for angiogenin. In an assay with human endothelial cells, it was found that the aptamer inhibited angiogenin-induced cell proliferation by 65 % at a concentration of 1 µM and by 100 at 5 µM, although not inhibiting, rather co-translocating with angiogenin to the nucleus (Nobile *et al.*, 1998).

Tenascin-C (TN-C), an extracellular matrix glycoprotein which is overexpressed during tissue remodelling, including tumour growth was a target for Hicke *et al.* (2001). Recent evidence has shown that the fibrinogen-like domain of TN-C may bind to a critical integrin involved in angiogenesis, $\alpha_V\beta_3$ (Yokoyama *et al.*, 2000), and that expression of TN-C in tumours correlates with poor prognosis. Hicke *et al.* (2001) used a method of 'target switching' between purified protein and cell SELEX on TN-C expressing U251 glioblastoma cells, to generate aptamers to TN-C (Hicke *et al.*, 2001). This was subsequently used in an *in vivo* preclinical targeting study, investigating the biodistribution of a radiolabelled anti-Tenascin-C aptamer in nude mice bearing either U251 glioblastoma or MDA-MB-435 breast tumour xenografts (Hicke *et al.*, 2006). The Tenascin-C targeting aptamer was radiolabelled with ^{99m}Tc using mercaptoacetyl glycene (MAG$_2$) and DTPA. Imaging studies of the complex demonstrated rapid blood clearance with a biological half-life of less than 10 min and rapid tumour penetration (6 % injected dose per gram) at 10 min. A tumour to blood ratio of 50:1 was achieved after 3 h.

A similar approach has been undertaken by our own group in targeting the MUC1 tumour marker for diagnostic, imaging and therapeutic applications (Ferreira *et al.*, 2006, 2008; Borbas *et al.*, 2007). MUC1 is an epithelial mucin coded for by the *MUC1* gene and is a transmembrane molecule, expressed by most glandular epithelial cells (Shimizu and Yamauchi, 1982). MUC1 mucin is restricted to the apical cell surface by interactions with the microfilament network. Although MUC1 is widely expressed by normal glandular epithelial cells (Zotter *et al.*, 1988), the expression is dramatically increased when the cells become malignant. This has been well documented for breast and ovarian cancer (Girling *et al.*, 1989), as well as some lung, pancreatic and prostate cancers (Girling *et al.*, 1989; Burdick *et al.*, 1997; Zhang *et al.*, 1998). Recently, it has also been shown that MUC1 is a valuable marker for bladder and has been used for the diagnosis of bladder cancer in a number of studies (Simms *et al.*, 2001; Hughes *et al.*, 2000). Antibody studies also showed that MUC1 is not only overexpressed in carcinomas, but the pattern of its glycosylation is altered. Thus, in the breast cancer mucin, glycosylation changes result in certain epitopes in the core protein being exposed, which are masked in the mucin produced by the lactating mammary gland (Zhang *et al.*, 1998). These characteristics of MUC1 have been explored over the years in a number of immunotherapeutic approaches, mainly involving radiolabelled antibodies against breast cancer (Maraveyas *et al.*, 1994; Riethmuller *et al.*, 1994; DeNardo *et al.*, 1997; Biassoni *et al.*, 1998) and bladder cancer (Simms *et al.*, 2001; Hughes *et al.*, 2000). Other attempts on active specific immunotherapies based on MUC1 have also taken place and have recently resulted in a MUC1 vaccine in clinical trials (see Chapter 16), which was originally produced by BioMira and is now licensed by Merck.

We have used immunogenic MUC1 peptides to generate DNA aptamers, which are much more stable than their RNA counterparts, with subnanomolar dissociation constants that bind selectively to MUC1 molecules in the surface of tumour cells (Ferreira *et al.*, 2006). Furthermore, we have used these aptamers in diagnostic immunoassays for the identification of circulating MUC1 in serum, to generate novel aptamer-based diagnostic assays for the prognosis of primary and metastatic disease, on the basis that the selected aptamers demonstrate much lower dissociations constants than their antibody counterparts utilized in currently commercial assays such as the CA15–3 (Ferreira *et al.*, 2008). Finally, we have labelled the aptamers with novel and commercially available chelators to generate monomeric and multimeric aptamer constructs for diagnostic imaging and targeted radiotherapy of breast and other epithelial cancers (Borbas *et al.*, 2007). Our aptamer-based radiopharmaceuticals have demonstrated exceptional tumour penetration compared to previously used antibodies and are currently at the preclinical and toxicity assessment stage.

Another way in which aptamers have been used as therapeutic or diagnostic agents is represented by the aptamer produced against neutrophil elastase (Smith *et al.*, 1995; Charlton *et al.*, 1997a, b) to provide a useful therapeutic and imaging tool. Neutrophil elastase is a serine protease that can degrade a variety of proteins, including connective tissue, and is found in the granules of neutrophils. When there is infection and inflammation in the body, neutrophils are recruited to the site (Repine 1992; Varani and Ward, 1994; Weis, 1989), whereupon they release their granules, and hence, neutrophil elastase (Wright, 1988). Aptamers were generated that showed a rate constant of $1-3 \times 10^8 \, \text{M}^{-1} \, \text{min}^{-1}$ for neutrophil elastase (Smith *et al.*, 1995; Charlton *et al.*, 1997b) and were shown by flow cytometry to bind preferentially to activated neutrophils (Charlton *et al.*, 1997b). The aptamers were covalently attached to known neutrophil elastase inhibitors, resulting in highly specific and highly potent irreversible inhibitors of neutrophil elastase (Charlton *et al*, 1997b). Furthermore, the aptamers were tested for their imaging properties *in vivo* in a rat model and it was found that, compared with an anti-elastase IgG (which is the method clinically used), not only was the overall intensity higher, but the background signal was lower, all over a shorter time period. This is due to the rapid clearance that the aptamer has from the blood because of its smaller size, and minimal perfusion into the surrounding tissues. The results showed that aptamers could be a viable alternative in imaging compared to antibodies (Charlton *et al.*, 1997b).

The human epidermal growth factor receptors (HER family) have been at the centre of pharmaceutical development in the past few years, after the massive success of Herceptin in the treatment of breast cancer. Aptamers were soon to follow this route and have now been described for the targeting of the human epidermal growth factor receptor-3 (HER3) (Chen and Landgraf, 2003). HER2, the target of Herceptin has mostly been targeted by a different oligonucleotide therapeutic, a ribozyme. Ribozymes (RNA enzymes) are RNA molecules capable of catalysing the cleavage of either their own RNA or other RNA substrates, and have sometimes been used in close combination with aptamers (see Chapter 17). A ribozyme that is currently in clinical trials to treat breast and ovarian cancer is Herzyme, used to target human epidermal growth factor 2 (HER2) (Zaffaroni and Folini, 2004).

A molecule called a spiegelmer has also been developed as a therapeutic reagent in cancer (Wlotzka *et al.*, 2002) and is designed to overcome the problem of *in vivo* stability that has been associated with aptamers. Spiegelmers are comprised of L-ribose or L-2′-deoxyribose, creating mirror image oligonucleotides with enhanced *in vivo* stability compared with aptamers. The spiegelmer in question inhibits the cascade for release of gonadotropin-releasing hormone (GnRH) (Wlotzka *et al.*, 2002), the hormone responsible for releasing gonadotropins, luteinizing hormone (LH) and follicle-stimulating hormone (FSH) in mammalian reproduction (Conn and Crowley, 1994). This is important as prostate and breast cancers are caused by these hormones (Schally, 1999; Kettel and Hummel, 1997), as well as endometriosis. The spiegelmer showed a K_D of 20 nM for GnRH, and high specificity, as, when GnRH was substituted with an analogue of GnRH (Buserelin), showing a closely related structure, the spiegelmer showed no antagonistic activity towards it (Wlotzka *et al.*, 2002). Animal studies showed that the coupling of the spiegelmer with PEG, not only had no effect on its specificity and biological activity, but caused the molecule to be retained for a longer period of time, causing its activity to be comparable with the drug Cetrorelix, which is a marketed GnRH antagonist. It also exhibited no immune response, which is important in assessing the safety of the molecule for long-term use (Wlotzka *et al.*, 2002).

18.3 Final comments

The continuous advances in technology and method development have made progress of aptamer development possible in the field of target recognition, diagnostics and therapeutics for many different diseases. Further advancements in aptamer research will surely allow scientists to develop tailor-made aptamers for specific targets and diseases with great efficacy. Aptamers have shown no immunogenicity, little or no side effects, offer great selectivity and specificity and great tumour penetration. Furthermore, they clear quickly from the system, through first-pass from the kidneys and are excreted in urine, resulting in little or no organ toxicity. Original stability problems with aptamers have now been widely overcome and PEGylation is becoming standard in therapeutic applications, as it improves the pharmacokinetic properties of the molecules. One of the major issues is that aptamers are still considered expensive for therapeutic applications. This is mostly a result of technological development, or lack thereof. Only a handful of companies are currently equipped to produce aptamers on a truly large scale, necessary for clinical applications. However, as more aptamers reach the clinic, demand is expected to reduce production costs, thus overcoming this obstacle. A final issue, currently standing, is the complex intellectual property field associated with aptamer development. Aptamers are mostly selected using the SELEX methodology, which is a proprietary technology. Thus, Archemix is the company that currently controls aptamer therapeutic applications whereas Somalogic are focused mostly on diagnostics. Yet, more and more companies sign agreements on therapeutic developments in a field that will open significantly in the next 5 to 10 years.

References

Adamis AP, Altaweel M, Bressler NM, *et al.*, Changes in retinal neovascularization after pegaptanib (Macugen) therapy in diabetic individuals. *Ophthalmology* 2006, **113**(1), 23–8.

Basilico C, Moscatelli D. The FGF family of growth factors and oncogenes. *Adv Cancer Res* 1992, **59**, 115–65.

Bates PJ, Kahlon JB, Thomas SD, Trent JO, Miller DM. Antiproliferative activity of G-rich oligonucleotides correlates with protein binding. *J Biol Chem* 1999, **274**(37), 26369–77.

Berezovski M, Drabovich A, Krylova SM, Musheev M, Okhonin V, Petrov A, Krylov SN. Nonequilibrium capillary electrophoresis of equilibrium mixtures: a universal tool for development of aptamers. *J Am Chem Soc* 2005, **127**(9), 3165–71.

Biassoni L, Granowska M, Carroll MJ, *et al.* 99mTc-labelled SM3 in the preoperative evaluation of axillary lymph nodes and primary breast cancer with change detection statistical processing as an aid to tumour detection. *Br J Cancer* 1998, **77**, 131–8.

Borbas KE, Ferreira CSM, Perkins A, Bruce JI, Missailidis S. Design and synthesis of mono- and multimeric targeted radiopharmaceuticals based on novel cyclen ligands coupled to anti-MUC1 aptamers for the diagnostic imaging and targeted radiotherapy of cancer. *J Bioconj Chem* 2007, **18** (4), 1205–12.

Burdick MD, Harris A, Reid CJ, Iwamura T, Hollinqsworth MA. Oligosuccharides expressed on MUC1 produced by pancreatic and colon tumor cell lines. *J Biol Chem* 1997, **272**, 24198–24202.

Bussolino F, Mantovani A, Persico G. Molecular mechanisms of blood vessel formation. *Trends Biochem Sci* 1997, **22**(7), 251–6.

Charlton J, Kirschenheuter GP, Smith D. Highly potent irreversible inhibitors of neutrophil elastase generated by selection from a randomized DNA-valine phosphonate library. *Biochemistry* 1997, **36**(10), 3018–26.

Charlton J, Sennello J, Smith D. *In vivo* imaging of inflammation using an aptamer inhibitor of human neutrophil elastase. *Chem Biol* 1997, **4**(11), 809–16.

Chen C-HB, Landgraf R. Aptamers to human epidermal growth factor receptor-3. http://www.freshpatents.com/Aptamers-to-human-epidermal-growth-factor-receptor-3-dt20070201ptan20070027096.php 2003 (accessed January 2008).

Chu TC, Marks JW 3rd, Lavery LA, Faulkner S, Rosenblum MG, Ellington AD, Levy M. Aptamer: toxin conjugates that specifically target prostate tumor cells. *Cancer Res* 2006a **66**(12), 5989–92.

Chu TC, Shieh F, Lavery LA, Levy M, Richards-Kortum R, Korgel BA, Ellington AD. Labeling tumor cells with fluorescent nanocrystal-aptamer bioconjugates. *Biosensors Bioelectronics* 2006b. **21**(10), 1859–66.

Chu TC, Twu KY, Ellington AD, Levy M. Aptamer mediated siRNA delivery. *Nucl Acids Res* 2006c, **34**(10), e73.

Chuang E, Barnard D, Hettich L. Critical binding and regulatory interactions between Ras and Raf occur through a small, stable N-terminal domain of Raf and specific Ras effector residues. *Mol Cell Biol* 1994, **14**(8), 5318–25.

Conn PM, Crowley Jr WF. Gonadotropin-releasing hormone and its analogs. *Annu Rev Med* 1994, **45**, 391–405.

Cowley S, *et al. The Dynamic Cell*. S377 Molecular and Cell Biology, edited by G. Bearman *et al.* Vol. 2. Open University Press, Milton Keynes, 2004, pp 117–19.

D'Amico DJ, Masonson HN, Patel M. Pegaptanib sodium for neovascular age-related macular degeneration: two-year safety results of the two prospective, multicenter, controlled clinical trials. *Ophthalmology* 2006, **113**(6), 992–1001 e6.

Dapic V, Bates PJ, Trent JO. Antiproliferative activity of G-quartet-forming oligonucleotides with backbone and sugar modifications. *Biochemistry* 2002, **41**(11), 3676–85.

DeNardo SJ, Kramer EL, O'Donnell J, *et al.* Radioimmunotherapy for breast cancer using indium-111/yttrium-90 BrE-3: result of a phase I clinical trial. *J Nucl Med* 1997, **38**, 1180–5.

Drabovich A, Berezovski M, Krylov SN. Selection of smart aptamers by equilibrium capillary electrophoresis of equilibrium mixtures (ECEEM). *J Am Chem Soc* 2005, **127**(32), 11224–5.

Ebihara Y, Miyamoto M, Shichinohe T, *et al.* Over-expression of E2F-1 in esophageal squamous cell carcinoma correlates with tumor progression. *Dis Esophagus* 2004, **17**(2), 150–4.

Ellington AD, Szostak JW. *In vitro* selection of RNA molecules that bind specific ligands. *Nature* 1990, **346**(6287), 818–22.

Farokhzad OC, Jon S, Khademhosseini A, Tran TN, Lavan DA, Langer R. Nanopartide-aptamer bioconjugates: A new approach for targeting prostate cancer cells. *Cancer Res* 2004, **64**(21), 7668–72.

Farokhzad OC, Cheng J, Teply BA, Sherifi I, Jon S, Kantoff PW, Richie JP, Langer R. Targeted nanoparticle-aptamer bioconjugates for cancer chemotherapy *in vivo. Proc Natl Acad Sci U S A* 2006, **103**(16), 6315–20.

Ferreira CSM, Matthews CS, Missailidis S. DNA aptamers that bind to MUC1 tumour marker: Design and characterization of MUC1-binding single stranded DNA aptamers. *Tumor Biol* 2006, **27**, 289–301.

Ferreira CSM, Papamichael K, Guilbault G, Schwarzacher T, Gariepy J, Missailidis S. DNA aptamers against MUC1: Design of aptamer-antibody sandwich ELISA for early tumour diagnosis. *Anal Bioanal Chem* 2008, in press (published online).

Floege J, Ostendorf T, Janssen U, *et al.* Novel approach to specific growth factor inhibition *in vivo*: antagonism of platelet-derived growth factor in glomerulonephritis by aptamers. *Am J Pathol* 1999, **154**(1), 169–79.

Gao H, Xiao J, Sun Q, *et al.* A single decoy oligodeoxynucleotides targeting multiple oncoproteins produces strong anticancer effects. *Mol Pharmacol* 2006, **70**(5), 1621–9.

Gatto B, Cavalli M. From proteins to nucleic acid-based drugs: the role of biotech in anti-VEGF therapy. *Anticancer Agents Med Chem* 2006, **6**(4), 287–301.

Girling A. Bartkova J., Burchell J, Gendler S, Gillett C, Taylor-Papadimitriou J. A core protein epitope of the polymorphic epithelial mucin detected by the monoclonal antibody SM-3 is selectively exposed in a range of primary carcinomas. *Int. J. Cancer* 1989, **43**, 1072–6.

Gorgoulis VG, Zacharatos P, Mariatos G. Transcription factor E2F-1 acts as a growth-promoting factor and is associated with adverse prognosis in non-small cell lung carcinomas. *J Pathol* 2002, **198**(2), 142–56.

Green LS, Jellinek D, Jenison R. Inhibitory DNA ligands to platelet-derived growth factor B-chain. *Biochemistry* 1996, **35**(45), 14413–24.

Hamula CLA, Guthrie JW, Zhang H. Selection and analytical applications of aptamers. *Trac-Trends Anal Chem* 2006, **25**(7), 681–91.

Heldin CH. Structural and functional studies on platelet-derived growth factor. *EMBO J* 1992, **11**(12), 4251–9.

Herrmann C, Martin GA, Wittinghofer A. Quantitative analysis of the complex between p21ras and the Ras-binding domain of the human Raf-1 protein kinase. *J Biol Chem* 1995, **270**(7), 2901–5.

Hicke BJ, Stephens AW, Gould T, *et al.* Tumour targeting by an aptamer. *J Nucl Med* 2006, **47**, 668–78.

Hicke BJ, Marion C, Chang YF, Tenascin-C aptamers are generated using tumor cells and purified protein. *J Biol Chem* 2001, **276**(52), 48644–54.

Hiebert SW, Lipp M, Nevins JR. E1A-dependent trans-activation of the human MYC promoter is mediated by the E2F factor. *Proc Natl Acad Sci U S A* 1989, **86**(10), 3594–8.

Huang J, Moore J, Soffer S, *et al.* Highly specific antiangiogenic therapy is effective in suppressing growth of experimental Wilms tumors. *J Pediatr Surg* 2001, **36**(2), 357–61.

Hughes ODM, Bishop MC, Perkins AC, *et al.* Targeting superficial bladder cancer by the intravesical administration of copper-67-labeled anti-MUC1 mucin monoclonal antibody C595. *J Clin Oncol* 2000, **18**, 363–70.

Ireson, C. R. and L. R. Kelland, Discovery and development of anticancer aptamers. *Molecular Cancer Therapeutics*, 2006. **5**(12), 2957–2962.

Jellinek D, Green LS, Bell C, *et al.* Potent 2′-amino-2′-deoxypyrimidine RNA inhibitors of basic fibroblast growth factor. *Biochemistry* 1995, **34**(36), 11363–72.

Jo N, Mailhos C, Ju M, *et al.* Inhibition of platelet-derived growth factor B signaling enhances the efficacy of anti-vascular endothelial growth factor therapy in multiple models of ocular neovascularization. *Am J Pathol* 2006, **168**(6), 2036–53.

Kettel LM, Hummel WP. Modern medical management of endometriosis. *Obstet Gynecol Clin North Am* 1997, **24**(2), 361–73.

Kimoto M, Shirouzu M, Mizutani S, Koide H, Kaziro Y, Hirao I, Yokoyama S. Anti-(Raf-1) RNA aptamers that inhibit Ras-induced Raf-1 activation. *Eur J Biochem* 2002, **269**(2), 697–704.

Leppanen O, Janjic N, Carlsson MA, *et al.* Intimal hyperplasia recurs after removal of PDGF-AB and -BB inhibition in the rat carotid artery injury model. *Arterioscler Thromb Vasc Biol* 2000, **20**(11), E89–95.

Liu C, Huang H, Donate F. Prostate-specific membrane antigen directed selective thrombotic infarction of tumors. *Cancer Res* 2002, **62**(19), 5470–75.

Lupold SE, Hicke BJ, Lin Y, Coffey DS. Identification and characterization of nuclease-stabilized RNA molecules that bind human prostate cancer cells via the prostate-specific membrane antigen. *Cancer Res* 2002, **62**(14), 4029–33.

Maraveyas A, Snook D, Hird V, Kosmas C, Meares CF, Lambert HE, Epenetos AA. Pharmacokinetics and toxicity of an yttrium-90-CITC-DTPA-HMFG1 radioimmunoconjugate for intraperitoneal radioimmunotherapy of ovarian cancer *Cancer* 1994, **73**, 1067–1075.

Morishita R, Gibbons GH, Horiuchi M, *et al.* A gene therapy strategy using a transcription factor decoy of the E2F binding site inhibits smooth muscle proliferation *in vivo. Proc Natl Acad Sci U S A* 1995, **92**(13), 5855–9.

Moroianu J, Riordan JF. Identification of the nucleolar targeting signal of human angiogenin. *Biochem Biophys Res Commun* 1994a, **203**(3), 1765–72.

Moroianu J, Riordan JF. Nuclear translocation of angiogenin in proliferating endothelial cells is essential to its angiogenic activity. *Proc Natl Acad Sci U S A* 1994b, **91**(5), 1677–81.

Nobile V, Russo N, Hu G, Riordan JF. Inhibition of human angiogenin by DNA aptamers: nuclear colocalization of an angiogenin-inhibitor complex. *Biochemistry* 1998, **37**(19), 6857–63.

Ostendorf T, Kunter U, Grone HJ, *et al.* Specific antagonism of PDGF prevents renal scarring in experimental glomerulonephritis. *J Am Soc Nephrol* 2001, **12**(5), 909–18.

Pietras K, Rubin K, Sjoblom T. Inhibition of PDGF receptor signaling in tumor stroma enhances antitumor effect of chemotherapy. *Cancer Res* 2002, **62**(19), 5476–84.

Pietras K, Ostman A, Sjoquist MInhibition of platelet-derived growth factor receptors reduces interstitial hypertension and increases transcapillary transport in tumors. *Cancer Res* 2001, **61**(7), 2929–34.

Rabbani F, Richon VM, Orlow I. Prognostic significance of transcription factor E2F-1 in bladder cancer: genotypic and phenotypic characterization. *J Natl Cancer Inst* 1999, **91**(10), 874–81.

Repine JE. Scientific perspectives on adult respiratory distress syndrome. *Lancet* 1992, **339**(8791), 466–9.

Riethmuller G, Schneider-Gadicke E, Schlimok G, *et al.* Randomised trial of monoclonal antibody for adjuvant therapy of resected Dukes' C colorectal carcinoma. German Cancer Aid 17-1A Study Group. *Lancet* 1994, **343**, 1177–83.

Ruckman J, Green LS, Beeson J. 2′-fluoropyrimidine RNA-based aptamers to the 165-amino acid form of vascular endothelial growth factor (VEGF(165)) – inhibition of receptor binding and VEGF-induced vascular permeability through interactions requiring the exon 7-encoded domain. *J Biol Chem* 1998, **273**(32), 20556–67.

Sampson T. Aptamers and SELEX: the technology. *World Patent Information*, 2003, **25**(2), 123–9.

Schally AV. Luteinizing hormone-releasing hormone analogs: their impact on the control of tumorigenesis. *Peptides* 1999, **20**(10), 1247–62.

Shimizu M, Yamauchi K. Isolation and characterization of mucin-like glycoprotein in human milk fat globule membrane. *J.Biochem* 1982, **91**, 515–24.

Simms MS, Murray A, Denton G, Scholfield DP, Price MR, Perkins AC, Bishop MC. Production and characterisation of a C595 antibody-99mTc conjugate for immunoscintigraphy of bladder cancer. *Urol Res* 2001, **29**, 13–19.

Smith D, Kirschenheuter GP, Charlton J, Guidot DM, Repine JE. *In vitro* selection of RNA-based irreversible inhibitors of human neutrophil elastase. *Chem Biol* 1995, **2**(11), 741–50.

Suh DS, Yoon MS, Choi KU, Kim JY. Significance of E2F-1 overexpression in epithelial ovarian cancer. *Int J Gynecol Cancer* 2008, **18**, 492–8.

Tuerk C, Gold L. Systematic evolution of ligands by exponential enrichment – RNA ligands to bacteriophage-T4 DNA-polymerase. *Science* 1990, **249**(4968), 505–10.

Varani J, Ward PA. Mechanisms of neutrophil-dependent and neutrophil-independent endothelial cell injury. *Biol Signals* 1994, **3**(1), 1–14.

Vojtek AB, Hollenberg SM, Cooper JA. Mammalian Ras interacts directly with the serine/threonine kinase Raf. *Cell* 1993, **74**(1), 205–14.

Weiss SJ. Tissue destruction by neutrophils. *N Engl J Med* 1989, **320**(6), 365–76.

Wlotzka B, Leva S, Eschgfaller B, *et al*. *In vivo* properties of an anti-GnRH Spiegelmer: an example of an oligonucleotide-based therapeutic substance class. *Proc Natl Acad Sci U S A* 2002, **99**(13), 8898–902.

Wright DG. Human neutrophil degranulation. *Methods Enzymol* 1988, **162**, 538–51.

Xiao Q, Li L, Xie Y, *et al*. Transcription factor E2F-1 is upregulated in human gastric cancer tissues and its overexpression suppresses gastric tumor cell proliferation. *Cell Oncol* 2007, **29**(4), 335–49.

Xu XH, Hamhouvia F, Thomas SD, *et al*. Inhibition of DNA replication and induction of S phase cell cycle arrest by G-rich oligonucleotides. *J Biol Chem* 2001, **276**(46), 43221–30.

Yamazaki K, Hasegawa M, Ohoka I, Hanami K, Asoh A, Nagao T, Sugano I, Ishida Y. Increased E2F-1 expression via tumour cell proliferation and decreased apoptosis are correlated with adverse prognosis in patients with squamous cell carcinoma of the oesophagus. *J Clin Pathol* 2005, **58**(9), 904–10.

Yokoyama K, Erickson HP, Ikeda Y, Takada Y. Identification of amino acid sequences in fibrinogen gamma -chain and tenascin C C-terminal domains critical for binding to integrin alpha vbeta 3. *J Biol Chem* 2000, **275**(22), 16891–8.

Zacharatos P, Kotsinas A, Evangelou K, *et al*. Distinct expression patterns of the transcription factor E2F-1 in relation to tumour growth parameters in common human carcinomas. *J Pathol* 2004, **203**(3), 744–53.

Zaffaroni N, Folini M. Use of ribozymes in validation of targets involved in tumor progression. *Drug Discov Today: Technol* 2004, **1**(2), 119–24.

Zhang S, Zhang HS, Reuter VE, Slovin SF, Scher HI, Livingston PO. Expression of potential target antigens for immunotherapy on primary and metastatic prostate cancer. *Clin Cancer Res* 1998, **4**, 295–302.

Zhou B, Wang B. Pegaptanib for the treatment of age-related macular degeneration. *Exp Eye Res* 2006, **83**(3), 615–9.

Zotter S, Hageman PC, Lossnitzer A, Mooi WJ, Hilgers J. Tissue and tumor distribution of human polymorphic epithelial mucin. *Cancer Rev* 1988, **11–12**, 55–101.

SECTION III

Other Aspects in Anticancer Therapeutic Development

19
Treatment of Cancer in Conjunction with Other Agents

Gary Robert Smith

19.1 Introduction

Inflammation has strong links with cancer, and this relationship can usefully be simplified into two phases of the disease; in the earlier phase, inflammation and other risks associated with infection or injury can lead to increased risk of genetic damage and carcinogenesis; and in the later phase, as the growing cancer exerts stress (hypoxic, oxidative and physical) on the surrounding tissue, induced chronic inflammation provides vital processes for malignancy. This cancer-induced wounding in the stroma provides many benefits to cancer progression, including destruction and remodelling of competing healthy tissue, angiogenesis and immune suppression.

If genetic damage is the 'match that lights the fire' of cancer, some types of inflammation may provide the 'fuel that feeds the flames' (Balkwill and Mantovani, 2001).

This chapter will focus on the use of agents, some of which are well established, some more novel, that can not only be used to reduce the risk of carcinogenesis, yet perhaps more excitingly offer significantly increased prospects for the regression of tumours in combination with other therapies.

19.1.1 Anti-inflammatory and anti-infective agents as a means to protect against carcinogenesis

In the 'The hallmarks of cancer', the authoritative work by Hanahan and Weinberg (2000), the evolutionary acquired capabilities necessary for cancer cells to become life-threatening

Anticancer Therapeutics   Edited by Sotiris Missailidis
© 2008 John Wiley & Sons, Ltd

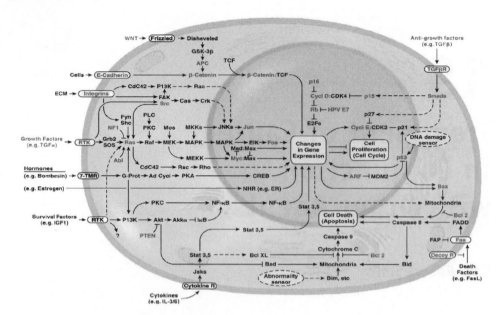

Figure 19.1   The emergent cell circuitry utilized to govern cell behaviour such as growth, anti-growth, controlled cell death through apoptosis and reaction to stress. The figure illustrates that there are many checks and balances designed to keep cells in homeostasis with their environment and that a series of failures are required for cancers to occur. From Hanahan D, Weinberg RA. The hallmarks of cancer. *Cell* 2000, **100**(1), 57–70., with permission from Elsevier Ltd

tumours are described. Figure 19.1, an extract from Hanahan and Weinberg (2000) highlights some of those common defects in growth and death controls that allow normal cells to become cancerous.

There is a wide array of cytokines, chemokines and other mediators produced in the inflammatory environment that can increase the risk for tumour genesis. Persistent production of nitrogen and oxygen intermediates by phagocytes causes tissue stress and damage to the DNA of cells. Damage to the DNA can lead to mutations, as well as altering cellular repair mechanisms. This may also lead to the inhibition of cell death by apoptosis, allowing the continued growth of mutated cells, a key hallmark in tumour development. This is one rationale for considering the use of anti-inflammatory drugs and anti-infective agents as a means of protection against cancers.

Viruses, in particular, offer an additional risk to carcinogenesis. All (even those considered benign) have to hijack and promote host cell growth in order to replicate. Numerous mechanisms are employed, DNA and RNA sequences are inserted into the host, growth factors are stimulated and anti-growth factors suppressed. Viral infections thus increase the risk of cell mutation and population growth, and for these reasons provide another target of interest for cancer prevention, some well-known examples being shown in Table 19.1.

19.1.2 Anti-inflammatory and anti-infective agents as a means to suppress malignancy

The Hallmarks of Cancer also suggested that cancer researchers should look not just at the cancer cells, but also at the environment in which they interact, with cancers eliciting the aid

Table 19.1 The marked association between a number of specific cancers and viral infections

Infection	Cancer	% virus-positive	No of cases worldwide/year	Vaccine developed
HPV	Cervical cancer	100	490 000	Yes
HBV	Liver cancer	50	340 000	Yes
HCV	Liver cancer	25	195 000	No
EBV	Burkitt's lymphoma	>90		No
	Hodgkin's lymphoma	>50		
	Post-transplant lymphoma	>80	113 000	
	Nasopharyngeal carcinoma	100		
KSHV	Kaposi's sarcoma	100		
	Primary effusion lymphoma	100	66 000	
	Multi Castleman's disease	>50		
HTLV1	Adult T-cell leukaemia	100	3000	No
H. pylori	Gastric carcinoma	30	603 000	No
	MALT lymphoma	100		

EBV, Epstein–Barr virus; HBV, hepatitis B virus; HCV, hepatitis C virus; HPV, human papillomavirus; HTLV1, human T-cell lymphotropic virus-1; KSHV, Kaposi's sarcoma-associated herpes virus.
Source: Cancer Research UK http://info.cancerresearchuk.org/cancerstats/causes/infectiousagents/virusesandcancer/?a=5441#basicmech

of fibroblasts, endothelial cells and immune cells. The immune system can be thought of as a surveillance system, which ensures that tissues of the body are free of invading organisms and pathogens. Another function is to remove cells that have been damaged through injury to a tissue, or by infection. In the case of malignancy, the immune system lacks effectiveness against cancerous cells, as they have evolved resistance to death signals and apoptosis. Thus, the influx of immune cells caused by the stress signals induced by the environmental pressures of growing cancer exert instead their effects on the surrounding healthy tissue, (Figure 19.2), thus leading to apoptosis and remodelling of healthy tissue, angiogenesis, invasion and eventual metastasis. (Smith and Missailidis, 2004)

The injury response induced by cancers involves the production of a host of cytokines, chemokines and other factors. Many of these factors, including transforming growth factor-β (TGF-β) and vascular endothelial growth factor (VEGF) are well known for their immunosuppressive effects. Surprisingly, immune suppression activities for tumour necrosis factor-α (TNF-α) and interleukin (IL)-1β have also been reviewed. These factors induce immature myeloid cells, tumour-associated macrophages, fibroblasts and regulatory T cells, and inhibit the process of antigen presentation and processing, resulting in the inhibition of dendritic cell maturation and T-cell activation in a tumour-specific immune response. In addition, where chronic inflammation predominates, IL-2, a vital growth factor in the progression of adaptive immunity, is significantly down regulated. This effect is put to good use by the pathogen *Helicobacter pylori* in its evasion of the immune system and progression in peptic ulcer disease (Figure 19.3), for which Marshall and Warren were awarded the 2005 Nobel Prize (http://nobelprize.org/nobel_prizes/medicine/laureates/2005/press.html).

In addition to viral strategies that directly stimulate factors for immune suppression, viruses also cause inflammation when their host cells die and release viral products. This can be

The selfish growth of Cancer causes progressive stress in its environment. The result is an ever expanding wound that never heals.

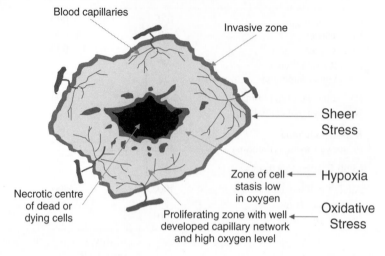

Figure 19.2 Irresolvable inflammation is often observed when a cancerous growth extends beyond 2 mm in size and this is not necessarily caused by a genetic change. A more compelling explanation is that this is simply due to the stress that the cancer and surrounding area is under. Adapted from http://www.abpi.org.uk/publications/publication_details/targetBreastCancer/pharm4.asp

Chronic Inflammation can be positively unhelpful in fighting invaders

The injury response diverts the attention of the immune cells – reducing the prospects for a specific trained adaptive response. Having one disease significantly opens up the prospects of co-infection - In the case of Helicobacter Pylori its relationship with Gastric Cancer.

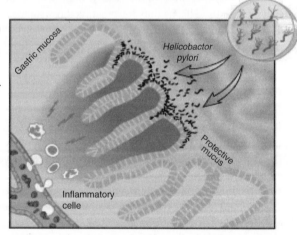

Figure 19.3 The invasion of *Helicobacter pylori* is associated with chronic inflammation. This type of inflammation actually suppresses the adaptive immune system and speeds the progression of the disease. Similar causative pathogens are being sought and investigated in an effort to better understand and treat what we currently term perhaps incorrectly 'autoimmune' diseases. Adapted from http://nobelprize.org/nobel_prizes/medicine/laureates/2005/press.html

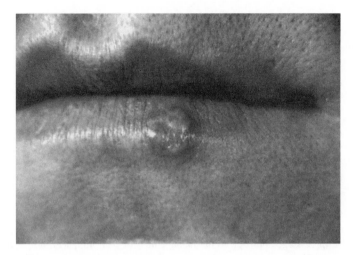

Figure 19.4 The typical burn like bubbles experienced during a cold sore attack. Herpes simplex is a 'benign' virus, which manifests in this active state when the immune system is suppressed by co-infections, sun burn or other stresses

visibly seen in measles and cold sores (Figure 19.4), and can be felt in the symptoms of influenza.

Just like cancer-induced inflammation, this also hinders antigen presentation, and the maturation of dendritic cells, thus providing protection for future generations of the virus. The toxicity of this inflammatory response, triggered by some viruses, is deadly and is responsible for the millions of deaths associated with the influenza pandemics of the past and concerns for the future. Virus-induced chronic inflammation is another gateway to co-infection and cancer due to cell damage and immune suppression.

The invasiveness and immune suppression of many cancers appears dependent on induced chronic inflammation. In addition, work such as that by Slaviero and colleagues (2003) suggests the effectiveness of conventional drug approaches is impeded by the inflammatory response. Given the key importance of immune suppression and injury responses for tumour protection and progression, strategies to resolve cancer induced inflammation and wounding must form a vital component in therapy if we are to conquer this disease.

19.2 Non-steroidal Anti-inflammatory Drugs

19.2.1 Pain relief in cancer; are we missing an opportunity?

The World Health Organization describes a three-tier approach to control the progression of pain associated with cancer (Figure 19.5).

Stage one recommends the prompt oral administration of non-opioids (aspirin and paracetamol); a progression to mild opioids (codeine); and then strong opioids such as morphine, until the patient is free of pain (http://www.who.int/cancer/palliative/painladder/en/) Aspirin is a prominent example of a non steroidal anti-inflammatory drug (NSAID), drugs that have analgesic, antipyretic and anti-inflammatory effects. Most NSAIDs act as non-selective inhibitors of the enzyme cyclo-oxygenase, inhibiting both the cyclo-oxygenase-1 (COX-1)

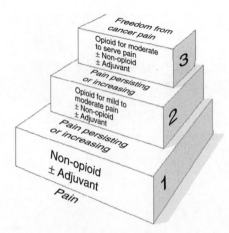

Figure 19.5 The three stage approach to the management of pain in cancer, non-opioids playing an important role as an adjuvant. Source: http://www.who.int/cancer/palliative/painladder/en/

and cyclo-oxygenase-2 (COX-2), isoenzymes that catalyse the formation of prostaglandins. Aspirin has a slight preference towards inhibition of COX-1 over COX-2.

Paracetamol is not generally considered an NSAID as, although its mechanism is similar, inhibiting COX-3, its effects are analgesic and antipyretic but not anti-inflammatory. Some suspect that this lack of anti-inflammatory action may be due to the paracetamol inhibiting cyclo-oxygenase predominantly in the central nervous system. Simmons and colleagues (2004) provide a good review of cyclo-oxygenase enzymes.

NSAIDs play a key role in the first step of the WHO guidelines for management of cancer pain. Nearly 90 % of patients with bone metastasis that present with pain are treated with NSAIDs as the most effective agents for treatment of patients with this condition, as prostaglandins appear to play an important role. NSAIDs treatment is often continued into the second step of pain management, although, given the role of chronic inflammation in the progression of cancer, this should perhaps be extended further and at higher doses. The evidence to support this is examined later in this chapter.

NSAIDs are commonly used, often in combination with an opioid, for treatment of cancer pain. Short-term studies have shown that NSAIDs alone are effective in managing cancer pain, with side effects similar to placebo and in about 50 % of studies, increasing the dose of NSAID can increase efficacy without increasing the incidence of side effects. Similar studies have not demonstrated a large clinical difference when combining an opioid with an NSAID versus either medication alone. Insufficient long-term studies have been conducted to provide information on chronic safety and effectiveness of NSAIDs alone or with opioids in treating cancer pain. (McNicol *et al.*, 2005)

19.2.2 NSAIDs as a preventative, balancing the pros and cons

The use of NSAIDs in cancer is not simply limited to the control of pain; epidemiological evidence is accumulating that aspirin or NSAID use is protective against oesophageal and gastric cancer, and possibly also against cancers of the prostate, ovary and lung. An inverse association has also been reported between aspirin use and breast cancer, indicating

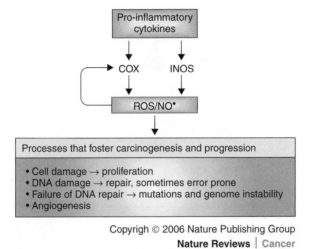

Nature Reviews | Cancer

Ulrich *et al. Nature Reviews Cancer* **6**, 130–140 (February 2006) | doi:10.1038/nrc1801

Figure 19.6 There is a mutually supportive relationship between reactive oxygen species and cyclo-oxygenase. The use of COX inhibitors can help to reduce this cycle and may provide some benefit in cancer and other diseases. From Ulrich CM, Bigler J, Potter JD, *et al.* Non-steroidal anti-inflammatory drugs for cancer prevention: promise, perils and pharmacogenetics. *Nat Rev Cancer* 2006, **6**(2), 130–40, with permission from Nature Publishing Group

that this type of cancer might also be amenable to chemoprevention by NSAIDs. One likely rationale for NSAIDs having this effect is their potential to inhibit the production of reactive oxygen species in inflammation, which would otherwise lead to cell and DNA damage (Figure 19.6; (Ulrich *et al.*, 2006).

A major drawback to the long-term use of NSAIDs as a preventative would appear to be the development of side effects: stomach irritation and gastrointestinal bleeding, associated with the older NSAIDs, or increased risk of heart attack and stroke in the newly developed molecules that more specifically target COX-2.

In 1988, when two specific COX-2 inhibitors Vioxx and Celebrex became available, huge interest was raised in these among cancer researchers, as it was thought that these drugs could be used without the detrimental effects of inhibiting COX-1 that was seen in earlier NSAIDs. Numerous clinical trials were initiated to investigate the use of these drugs in the prevention of cancer. In 2004 both drugs were withdrawn from the market, when an increased risk of developing heart attacks and strokes had been identified.

The side effects of NSAIDs can be explained on examination of the differing roles of COX-1 and COX-2 enzymes in tissue maintenance, repair and healing (Figure 19.7).

The most compelling case for the use of NSAIDs as a preventative can be found in colorectal cancers, although even in this case NSAID use is controversial. A recent large cohort study (Jacobs *et al.*, 2007) of cancer incidence populations, among whom colorectal, prostate, and breast cancers are common, indicates that long-term daily use of adult-strength aspirin is associated with modestly reduced overall cancer incidence. Dosage is undoubtedly important,

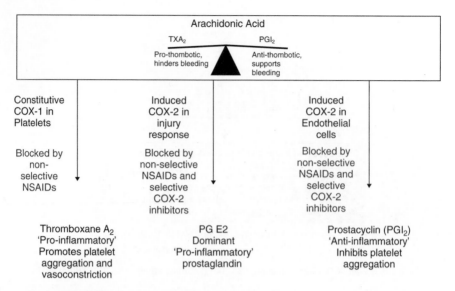

Figure 19.7   An overview of the sources and relationships between COX and some of its important products. COX-1 is constitutively expressed in essentially every organ, having a role in the daily maintenance of tissue repair and regeneration. Platelet-derived COX-1 is particularly important for the production of thromboxane A_2 (TXA_2) and its action in the aggregation of platelets. This, not only accounts for why low-dose aspirin is effective against thrombotic cardiovascular disease, but also why inhibition of COX-1 can also cause gastric ulceration and bleeding. Unlike COX-1, COX-2 is induced in inflammatory conditions and of all of the arachidonic products catalysed by COX-2, PGE_2 is the most pro-inflammatory. Prostacyclin (PGI_2) is another COX-2-catalysed product, induced in the endothelium in injured tissue. PGI_2 has an important role in wound resolution, being anti-inflammatory, and works in opposition to TXA_2 by inhibiting platelet aggregation. It is currently believed that the cardiovascular and stroke risk associated with selective inhibition of COX-2 is due to platelet accumulation, particularly in damaged tissue arising from an imbalance of PGI_2 and TXA_2

generally benefits being attributed to higher doses. However, this is not universally the case, with some studies indicating that subadult doses provide superior benefits.

Weighing up the current evidence both for and against, the US Preventative Services Task Force, convened by the US Public Health Service in 1984 and sponsored by the Agency for Healthcare Research and Quality (AHRQ) since 1998, concluded that harms outweigh the benefits of aspirin and NSAID use for the prevention of colorectal cancer. In their recommendation statement (2007) regarding routine aspirin or NSAIDs for the primary prevention of colorectal cancer, they stated that aspirin appears to be effective at reducing the incidence of colonic adenoma and colorectal cancer, especially if used in high doses for more than 10 years. However, the possible harms of such a practice require careful consideration. Further evaluation of the cost-effectiveness of chemoprevention compared with, and in combination with, a screening strategy is required.

19.2.3 Mechanisms of suppression of cancer progression

In oncology, cardiovascular and other risks are not unusual. Side-effect profiles of drugs used to treat cancer are often more severe than seen for drugs used to treat other non-life threatening conditions. If NSAIDs were found to have the ability to suppress cancer progression, or even to have the potential to support regression, then the risks of side effects might be outweighed by the benefits.

A recent study of aspirin-associated reduction in colorectal cancer risk, too late for consideration in the USPSTF review, concluded that protection appeared to be limited to COX-2–expressing cancers (which is in keeping with the induced nature of COX-2 in inflammation). This study identifies a potential subset of patients who would benefit from NSAIDs (Chan *et al.*, 2007).

COX-2 over expression is seen in many malignancies, including lung, breast, prostate, colorectal, oesophageal and pancreatic cancer. Elevated levels of tumour prostaglandin E2 (PGE_2), a major COX-2 metabolite, have been implicated in angiogenesis, tumour growth and invasion, apoptosis resistance and suppression of anti-tumour immunity and this has led to growing interest in the therapeutic potential of NSAIDs (and more recently specific PGE_2 inhibitors) as an adjunct to existing radiotherapy and chemotherapy (Figure 19.8; Mann *et al.*, 2005).

Ferrari *et al.* (2006a) report that gemcitabine, in combination with celecoxib, during a phase II trial showed low toxicity, good clinical benefit rate and good disease control. Forty-two consecutive patients with histological or cytologically-confirmed pancreatic adenocarcinoma entered the trial. Twenty-six patients were metastatic and 16 had locally advanced disease. The schedule consisted of Gemcitabine $1000\,mg/m^2$ (as a 30 min intravenous infusion) on days 1, 8 every 3 weeks and celecoxib 400 mg bid. Median survival was 9.1 months (95 % CI, 7.5–10.6 months).

Reckamp *et al.* (2006) report that erlotinib (an epidermal growth factor receptor (EGFR) tyrosine kinase inhibitor) in combination with celecoxib, during a phase I trial, showed objective responses with an acceptable toxicity profile. The celecoxib had been added to the trial in order to overcome COX-2 induced resistance to the effects of Erlotinib. Twenty-two subjects with stage IIIB and/or IV non-small cell lung cancer (NSCLC) were enrolled. Seven patients showed partial responses (33 %), and five patients developed stable disease (24 %).

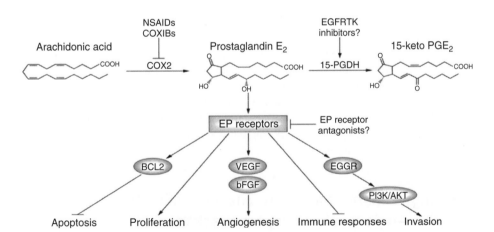

Figure 19.8   The involvement of the PGE_2 receptors in cancer. From: Mann JR, Backlund, MG, DuBois R, *et al.* (2005). Mechanisms of disease: inflammatory mediators and cancer prevention. *Nat Clin Pract Oncol* 2(4): 202–10, with permission from Nature Publishing Group

In Altorki *et al.*'s (2003) phase II trial report of 29 patients with stages IB to IIIA NSCLC, the patients were treated with two preoperative cycles of paclitaxel and carboplatin, as well as daily celecoxib, followed by surgical resection. The addition of celecoxib abrogated the marked increase in levels of PGE_2 detected in primary tumours after treatment with paclitaxel and carboplatin alone. The overall clinical response rate was 65 % (48 % with partial response; 17 % with complete response). Grade 3 or 4 neutropenia was observed in 18 patients (62 %). Twenty-eight patients were explored and underwent complete resection of their tumours. There were no complete pathological responses, but seven patients (24 %) had minimal residual microscopic disease. The results suggested that the addition of a selective COX-2 inhibitor enhanced the response to preoperative paclitaxel and carboplatin in patients with NSCLC.

In contrast, Lilenbaum *et al.* (2006) report that the addition of celecoxib failed to deliver any additional benefits when combined with Docetaxel/irinotecan or Gem-citabine/irinotecan during a phase II trial in the second-line treatment of NSCLC. Correspondence from Gradilone *et al.* (2007) comments on the disappointing results from the above trial, discussing a possible role for celecoxib in the induction of multidrug resistance-associated protein-4, whilst others have discussed the adequacy of the dose of celecoxib, as well as upregulation of PGE_2.

In a study of 586 patients with prostate cancer who have had radiotherapy, Khor *et al.* (2007) have reported an association of COX-2 expression with patient outcome. Increasing COX-2 expression was significantly associated with biochemical failure, distant metastasis, and any failure in treatment. The report further suggests that COX-2 inhibitors might improve patient response to radiotherapy in those treated with or without androgen deprivation.

19.2.4 Concluding remarks

The subject of NSAIDs and their potential for treatment is extremely complex with huge amounts of conflicting and therefore confusing data. Classically, when thinking about cancer prevention, it is easy to forget that in the life of our billions of cells, carcinogenesis is an every-day occurrence and that generally, when cancers are discovered, it is only because the cancer has progressed significantly. With this in mind, a distillation of the evidence indicates that protective effects, through the premise of reduced inflammatory reactive oxygen species and cell damage, against carcinogenesis are not that noticeable. Protective effects against malig-nancy, when chronic inflammation is being induced, may be more significant, malignancy being marked by the expression of COX-2. In conclusion, whist NSAIDs may not necessar-ily significantly reduce carcinogenesis, a large body of laboratory and early clinical evidence suggests that NSAIDs treatment could offer significant benefits in inhibiting the progress of invasion, metastasis and immune suppression. Further clinical trials for NSAIDs as an adjunct to other therapies are warranted.

19.3 Angiotensin-converting Enzyme (ACE) Inhibitors and Angiotensin Receptor Blockade

Although significant progress has been made in cancer drug therapy, our established biochem-ical methods are still highly limited in the battle against tumours that have progressed to their invasive and metastasis stages. Manipulation of the Angiotensin system, using Angiotensin

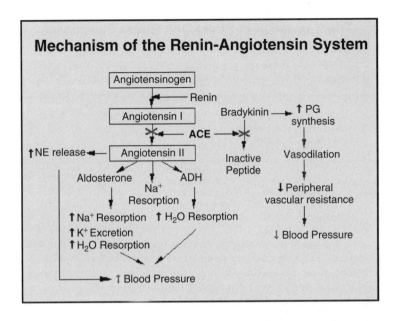

Figure 19.9   An overview of the classically defined renin angiotensin system

Receptor Blockade and ACE Inhibitors is, however, promising vastly increased prospects in this area.

Angiotensin II (Ang II) is a peptide hormone within the renin–angiotensin system (RAS), generated from the precursor protein angiotensinogen, by the actions of renin, angiotensin converting enzyme, chymases and various carboxy- and amino-peptidases. Ang II is the main effector of the RAS; an overview of the classically defined RAS is shown in Figure 19.9, which has been shown to play an important role in the regulation of vascular homeostasis, with growing interest as key mediator for vascular and chronic inflammation.

Drugs specifically targeting the RAS began with ACE inhibitors that not only blocked the production of Ang II, but also inhibited the inactivation of bradykinin. This additional action on bradykinin led to mild side effects such as cough and the need for the development of angiotensin receptor blockers that specifically antagonize the AT1 receptor.

The opposing roles of its two mutually antagonistic receptors AT1 and AT2 in maintaining blood pressure, water and electrolyte homeostasis are well established. It is, however, becoming recognised that the RAS is a key mediator of inflammation Figure 19.10; Smith and Missailidis, 2004) with the expression and activation of Ang II receptors governing the transcription of pro-inflammatory mediators both in resident tissue, tumour cells and in inflammatory cells such as macrophages.

In addition to the mediators, including COX-2, reviewed by (Suzuki *et al.*, 2003) a number of vital molecules in wound response processes are induced by the AT1 receptor (Figure 19.11). These include IL-1b in activated monocytes, TNF-α, plasminogen activator inhibitor type 1 (PAI-1), adrenomedullin and notably TGF-β, all of which have been shown to have active participation in various aspects of cancer development and immune suppression.

Through its ability to activate these pathways and the production of these mediators, AT1 would appear an attractive target in inhibiting the chronic inflammatory response required

The translation of Stress into Stress Response.

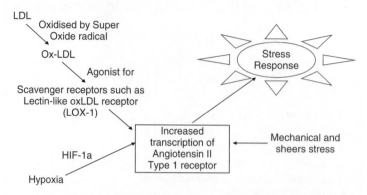

Figure 19.10   AT1 expression is upregulated in tissue stress and injury by the action of oxidized low-density lipoprotein (LDL) on scavenger receptors, such as the lectin-like oxidized LDL receptor, intracellular hypoxia sensing mechanisms, such as HIF-alpha, and mechanical and physical cellular stress. It is gradually becoming accepted that the AT1 receptor plays a pivotal role in the translation of these stresses into 'stress response'

Stress Response/Wound signalling role of AT1

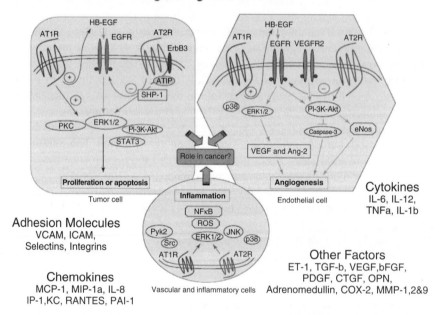

Figure 19.11   The involvement of not just cancer cells, but also those normal cells co-opted to support cancer progression in producing and releasing a full spectrum of 'stress and wound response' mediators. ADD Adapted from Deshayes, F. and C. Nahmias (2005). Angiotensin receptors: a new role in cancer? *Trends Endocrinol Metab* **16**(7): 293–9, with permission from Elsevier Ltd. Reproduced from Deshayes, F. and C. Nahmias (2005). "Angiotensin receptors: a new role in cancer?" Trends Endocrinol Metab 16(7): 293–9. with kind permission from Elsevier Ltd

for malignancy (Deshayes and Nahmias, 2005). A growing body of laboratory evidence in animal studies confirms that blockade of the AT1 receptor with commercially available drugs, at three to five times the cardiovascular dosage, has proven effective in halting the progression of tumour with no toxicity reported. Clinical studies of angiotensin receptor blockers of up to five times normal doses have proven effective in the treatment of chronic kidney disease with no reported toxicity, (Weinberg *et al.*, 2006; Schmieder *et al.*, 2005).

In humans, insertion/deletion polymorphisms in the ACE gene, which affect the efficiency of the enzyme in cleaving Ang I, have been noted to affect the progression of cancers. The DD phenotype, which is the most efficient in producing Ang II, being noted to increase invasion, metastasis and decrease survival in a variety of solid tumours in comparison to ID and II phenotypes: gastric (Rocken *et al.*, 2007a), oral (Vairaktaris *et al.*, 2007), prostrate (Yigit *et al.*, 2007), colorectal (Rocken *et al.*, 2007b) and breast cancer (Yaren *et al.*, 2007).

AT1 receptor expression has also been found to correlate with increased invasiveness and poorer patient outcome in a number of cancers. In their further investigation of the influence of ACE polymorphism in the progression of gastric cancer, (Rocken *et al.*, 2007a) aimed to substantiate the putative significance of the AT1 and AT2 receptors in gastric cancer biology by investigating the correlation of their expression with various clinicopathological variables and patient survival. In the 100 patients examined, the study found that combination of AT1 expression and ACE I/D gene polymorphism directly correlated with nodal spread and decreased patient survival. In ovarian cancer (Ino *et al.*, 2006), assessed whether AT1 expression is correlated with clinicopathological parameters, angiogenic factors and patient survival. The analysis of AT1 and VEGF through immunohistochemical staining in 67 ovarian cancer tissues demonstrated that AT1 receptors were expressed in 85 % of the cases examined, and that 55 % were strongly positive. AT1 expression was positively correlated with VEGF expression intensity, tumour angiogenesis and poor patient outcome. In cervical carcinoma, Kikkawa *et al.* (2004) examined the expression of AT1 receptor by immunohistochemistry in normal and neoplastic cervical tissues. Mean staining intensity level was stronger in invasive carcinoma cells than in normal dysplasia, and carcinoma *in situ* tissues. Angiotensin II induced the secretion of VEGF from Siha cells and promoted their invasive potential. A further study by Shibata *et al.* (2005) found that in the examination of 94 cases of primary endometrial carcinomas, AT1 expression predicted a significantly poorer prognosis.

Uemura *et al.* (2005) offer the first results published regarding the use of angiotensin receptor blockade in their pilot study in hormone-refractory prostate cancer. Twenty-three patients with advanced hormone-refractory prostate cancer who had already received secondary hormonal therapy using dexamethasone, and who were no longer receiving conventional therapy, were enrolled. All of the patients received candesartan 8 mg (half the usual maximum cardiovascular dose) once daily per os and, simultaneously, androgen ablation. Eight patients (34.8 %) showed responsive prostate-specific antigen (PSA) changes; six showed a decrease immediately after starting administration and two showed a stable level of PSA. Six men with a PSA decline of more than 50 % showed an improvement in patient survival. The mean time to PSA progression (TTPP) in responders was 8.3 months (range, 1–24 months) and one half of the patients showed stable or improved patient survival during treatment. With regard to toxic effects, only one patient showed hypotension during treatment. The reverse transcription–polymerase chain reaction showed that AT1 receptor expression in well-differentiated adenocarcinoma was higher than that in poorly differentiated adenocarcinoma.

More recently, Brown *et al.* (2006) describe two patients with pheochromocytoma in whom treatment with angiotensin receptor blocker was associated with cessation of tumour growth. Dosage of 300 mg irbesartan per day was utilised in one patient and 16 mg candesartan per day in the other. Both approaches appeared well tolerated with no apparent downsides to use. The paper also noted an observation that in a patient with carcinoid syndrome and hypertension, introduction of candesartan at 16–32 mg for his hypertension has been associated with an arrest of tumour growth and 5HIAA excretion over a 3-year period.

19.3.1 Concluding remarks

A number of studies now show that availability of Ang II and AT1 receptor expression are important factors in the degree of invasiveness of solid tumours. Early clinical evidence suggests that angiotensin receptor blockade alone may prove an exceptionally useful tool in slowing or even halting the progression of solid tumours. Prospects for AT1 receptor blockade do not end there though. In their ability to resolve chronic inflammation and injury responses, angiotensin receptor blockers pose a logical adjunct to radiotherapy (Zhao *et al.*, 2007), surgery, and established chemotherapy that target growth and resistance to apoptosis. Smith and Missailidis (2004) also propose that AT1 receptor blockade will greatly enhance the effectiveness of immune therapy such as dendritic cell vaccines.

19.4 Partners in Crime – Dealing with Co-infections

Secondary infections of bacteria, viruses, protozoa/parasites, or fungi are common in cancer. Systemic suppression of the immune system is not only caused by cancer itself, but it can also be an unfortunate side of current treatment options: surgery, radiotherapy and chemotherapy.

In keeping with the logical relationship between chronic inflammation, immune suppression and the progression of disease, it is not surprising that co-infections in cancer patients are common and pose a serious further threat to patient survival. An additional factor, however, and one that is only recently being investigated is the influence of the co-infection on the progression of their cancer.

19.4.1 The accelerators of malignancy

Bacterial toxins, including staphylococcal enterotoxins (SEs), have been implicated in the pathogenesis of cutaneous T-cell lymphomas (CTCLs). Recent research by Woetmann *et al.* (2007) in lymphatic cancer has demonstrated that these bacteria can cause the cancer to be more aggressive and that patients with skin lymphoma may benefit from antibiotic treatments used for bacterial infections.

Ferreri *et al.* (2006b) explore the association between ocular adnexal MALT lymphoma (OAL) and *Chlamydia psittaci* (Cp) infections, aiming to confirm reports suggesting that doxycycline treatment causes tumour regression in patients with Cp-related OAL. In a prospective trial, 27 OAL patients (15 newly diagnosed and 12 having experienced relapse) were given a 3-week course of doxycycline therapy. At a median follow-up of 14 months, lymphoma regression was complete in six patients, and a partial response≥50 % reduction of all measurable lesions) was observed in seven patients (overall response rate [complete and partial responses] = 48 %). Lymphoma regression was observed in both Cp DNA-positive patients (seven of 11 experienced regression) and Cp DNA-negative patients (six of 16 experienced regression) (64 % versus 38 %; $P = 0.25$, Fisher's exact test). The three patients with regional lymphadenopathies and three of the five patients with bilateral disease achieved objective response. In relapsed patients, response was observed both in previously irradiated and non-irradiated patients. The 2-year failure-free survival rate among the doxycycline-treated patients was 66 % (95 % CI = 54 to 78), and 20 of the 27 patients were progression free. Doxycycline proved a fast, safe, and active therapy for Cp DNA-positive OAL that was effective even in patients with multiple failures involving previously irradiated areas or regional lymphadenopathies.

19.5 Discussion

The aim of this chapter is to highlight the importance of cancer and infection induced inflammation in carcinogenesis and malignancy and to examine the emerging evidence for the broader use of conventional drugs such as NSAIDs as adjuncts to existing therapy. Furthermore, it aims to bring to the attention of readers the significant potential for angiotensin receptor blockers in treating malignant tumours and other diseases of 'wounds that will not heal'.

References

US Preventive Services Task Force recommendation statement. Routine aspirin or nonsteroidal anti-inflammatory drugs for the primary prevention of colorectal cancer. *Ann Intern Med* 2007, **146**(5), 361–4.

Altorki NK, Keresztes RS, Port JL, *et al*. Celecoxib, a selective cyclo-oxygenase-2 inhibitor, enhances the response to preoperative paclitaxel and carboplatin in early-stage non-small-cell lung cancer. *J Clin Oncol* 2003, **21**(14), 2645–50.

Balkwill F, Mantovani A. Inflammation and cancer: back to Virchow? *Lancet* 2001, **357**(9255), 539–45.

Brown MJ, Mackenzie IS, Ashby MJ, Balan KK, Appleton DS. AT2 receptor stimulation may halt progression of pheochromocytoma. *Ann N Y Acad Sci* 2006, **1073**, 436–43.

Chan AT, Ogino S, Fuchs CS, *et al*. Aspirin and the risk of colorectal cancer in relation to the expression of COX-2. *N Engl J Med* 2007, **356**(21), 2131–42.

Deshayes F, Nahmias C. Angiotensin receptors: a new role in cancer? *Trends Endocrinol Metab* 2005, **16**(7), 293–9.

Ferrari V, Valcamonico F, Amoroso V, *et al*. Gemcitabine plus celecoxib (GECO) in advanced pancreatic cancer: a phase II trial.' *Cancer Chemother Pharmacol* 2006a, **57**(2), 185–90.

Ferreri AJ, Ponzoni M, Guioboni M, *et al*. Bacteria-eradicating therapy with doxycycline in ocular adnexal MALT lymphoma: a multicenter prospective trial. *J Natl Cancer Inst* 2006b, **98**(19), 1375–82.

Gradilone A, Pulcinelli FM, Lotti LV, Martino S, Mattiello T, Frati L, Agliano AM, Gazzaniga P. Celecoxib induces MRP-4 in lung cancer cells: therapeutic implications. *J Clin Oncol* 2007, **25**(27), 4318–20; author reply 4320.

Hanahan D, Weinberg RA. The hallmarks of cancer. *Cell* 2000, **100**(1), 57–70.

Ino K Shibata K, Kajiyama, Yamamoto E, Nagasaka T, Nawa A, Nomura S, Kikkawa F. Angiotensin II type 1 receptor expression in ovarian cancer and its correlation with tumour angiogenesis and patient survival. *Br J Cancer* 2006, **94**(4): 552–60.

Jacobs EJ, Thun MJ, Bain EB, Rodriguez C, Henley SJ, Calle EE. A large cohort study of long-term daily use of adult-strength aspirin and cancer incidence. *J Natl Cancer Inst* 2007, **99**(8), 608–15.

Khor LY, Bae K, Pollack A, *et al*. COX-2 expression predicts prostate-cancer outcome: analysis of data from the RTOG 92-02 trial. *Lancet Oncol* 2007, **8**(10), 912–20.

Kikkawa F, Mizuno M, Shibata K, Kajiyama H, Morita T, Ino K, Nomura S, Mizutani S. Activation of invasiveness of cervical carcinoma cells by angiotensin II. *Am J Obstet Gynecol* 2004, **190**(5), 1258–63.

Lilenbaum R, Socinski MA, Altorki LK, *et al*. Randomized phase II trial of docetaxel/irinotecan and gemcitabine/irinotecan with or without celecoxib in the second-line treatment of non-small-cell lung cancer. *J Clin Oncol* 2006, **24**(30), 4825–32.

Mann JR, Backlund MG, DuBois RN, *et al*. Mechanisms of disease: Inflammatory mediators and cancer prevention. *Nat Clin Pract Oncol* 2005, **2**(4), 202–10.

McNicol E, Strassels SA, Goudas L, Lau J, Carr DB. NSAIDS or paracetamol, alone or combined with opioids, for cancer pain. *Cochrane Database Syst Rev* 2005, (1), CD005180.

Reckamp KL, Krysan K, Morrow JD, *et al*. A phase I trial to determine the optimal biological dose of celecoxib when combined with erlotinib in advanced non-small cell lung cancer. *Clin Cancer Res* 2006, **12**(11 Pt 1), 3381–8.

Rocken C, Rohl FW, Diebler E, Lendeckel U, Pross M, Carl-McGrath S, Ebert MP. The angiotensin II/angiotensin II receptor system correlates with nodal spread in intestinal type gastric cancer. *Cancer Epidemiol Biomarkers Prev* 2007, **16**(6), 1206–12.

Rocken C, Neumann K, Carl-McGrath S, *et al*. The gene polymorphism of the angiotensin I-converting enzyme correlates with tumor size and patient survival in colorectal cancer patients. *Neoplasia* 2007b, **9**(9), 716–22.

Schmieder RE, Klingbeil AU, Fleischmann EH, Veelken R, Delles C. Additional antiproteinuric effect of ultrahigh dose candesartan: a double-blind, randomized, prospective study. *J Am Soc Nephrol* 2005, **16**(10): 3038–45.

Shibata K, Kikkawa F, Mizokami Y, Kajiyama H, Ino K, Nomura S, Mizutani S. Possible involvement of adipocyte-derived leucine aminopeptidase via angiotensin II in endometrial carcinoma. *Tumour Biol* 2005, **26**(1), 9–16.

Simmons DL, Botting RM, Hla T, *et al*. Cyclooxygenase isozymes: the biology of prostaglandin synthesis and inhibition. *Pharmacol Rev* 2004, **56**(3), 387–437.

Slaviero KA, Clarke SJ, Rivory LP, *et al*. Inflammatory response: an unrecognised source of variability in the pharmacokinetics and pharmacodynamics of cancer chemotherapy.' *Lancet Oncol* 2003, **4**(4), 224–32.

Smith GR, Missailidis S. Cancer, inflammation and the AT1 and AT2 receptors. *J Inflamm (Lond)* 2004, **1**(1), 3.

Suzuki Y, Ruiz-Ortega M, Lorenzo O, Ruperez M, Esteban V, Egido J. Inflammation and angiotensin II. *Int J Biochem Cell Biol* 2003, **35**(6), 881–900.

Uemura H, Hasumi H, Kawahara T, *et al*. Pilot study of angiotensin II receptor blocker in advanced hormone-refractory prostate cancer. *Int J Clin Oncol* 2005, **10**(6), 405–10.

Ulrich CM, Bigler J, Potter JD, *et al*. Non-steroidal anti-inflammatory drugs for cancer prevention: promise, perils and pharmacogenetics. *Nat Rev Cancer* 2006, **6**(2), 130–40.

Vairaktaris E, Yapijakis C, Tsigris C, *et al*. Association of angiotensin-converting enzyme gene insertion/deletion polymorphism with increased risk for oral cancer. *Acta Oncol* 2007, **46**(8), 1097–102.

Weinberg AJ, Zappe DH, Ramadugu R, Weinberg MS. Long-term safety of high-dose angiotensin receptor blocker therapy in hypertensive patients with chronic kidney disease. *J Hypertens Suppl* 2006, **24**(1), S95–9.

Woetmann A, Lovato P, Eriksen KW, *et al*. Nonmalignant T cells stimulate growth of T-cell lymphoma cells in the presence of bacterial toxins. *Blood* 2007, **109**(8), 3325–32.

Yaren A, Turgut S, Kursunluoglu R, Oztop I, Turgut G, Degirmencioglu S, Kelten C, and Erdem E. Insertion/deletion polymorphism of the Angiotensin I-converting enzyme gene in patients with breast cancer and effects on prognostic factors. *J Invest Med* 2007, **55**(5), 255–61.

Yigit B, Bozkurt N, Narter F, Yilmaz H, Yucebas E, Isbir T. Effects of ACE I/D polymorphism on prostate cancer risk, tumor grade and metastatis.' *Anticancer Res* 2007, **27**(2), 933–6.

Zhao W, Diz DI, Robbins ME, *et al*. Oxidative damage pathways in relation to normal tissue injury.' *Br J Radiol* 2007, **80 Spec No 1**, S23–31.

20

Clinical Trials in Oncology

Tim Friede, Janet Dunn and Nigel Stallard

20.1 Clinical Trials

Altman defines a clinical trial as 'a planned experiment on human beings which is designed to evaluate the effectiveness of one or more forms of treatment' (Altman, 1991). Here treatment is understood in the widest sense. It includes drugs or other forms of medical interventions like surgeries, radiotherapies, diets, and ways of delivering health care (e.g. telemedicine). As clinical trials methodology has been developed only in part in the setting of oncology, we start this chapter with a general section introducing key concepts in clinical trials. In particular, we describe a classification of clinical trials into different phases. We then discuss clinical trials in oncology, explaining how and why these may differ from clinical trials in other indications, before describing phase I, phase II and phase III oncology clinical trials in detail. Inevitably a chapter such as this cannot cover all aspects of cancer clinical trials. Further material is given by, for example Girling *et al.* (2003) and Green *et al.* (2003).

Clinical trials are defined as *planned* experiments. This means that prior to the study the research plan for the trial is laid out in a formal document, the so-called *study protocol*. The main features of the study protocol include study objectives, patient population, treatment schedules, methods of patient evaluation, trial design, statistical analysis, patient consent, and administrative responsibilities (Pocock, 1983).

Different forms of clinical trial classification exist. For instance, trials are characterized by their purpose. We distinguish therapeutic trials from prophylactic trials in disease prevention (e.g. stroke) or evaluation studies of screening programmes (e.g. breast cancer screening). In drug development a classification of trials by development phase is common. The phases can be broadly described as follows:

- *Phase I* comprises studies to investigate clinical pharmacology (pharmacokinetics and pharmacodynamics) and drug safety. In most disease areas, these studies are usually carried out

Anticancer Therapeutics Edited by Sotiris Missailidis
© 2008 John Wiley & Sons, Ltd

in healthy volunteers. An exception to this, as described in detail below, is the testing of anti-cancer drugs which are often too toxic to be tested in healthy volunteers.

- In *phase II* the experimental treatment is first studied in patients with a view of an initial assessment of efficacy. Phase II is also often used to identify a safe and effective dose level for further development.

- *Phase III* is the confirmatory phase in which the experimental drug is compared to placebo or standard therapies in large scale trials.

- Trials carried out after marketing authorization are usually called *phase IV* studies. They can help to discover uncommon side effects or to study long-term effects.

Unless the disease history is entirely understood a control group is required in order to assess the efficacy and safety of a new treatment. Concurrent controls are preferred to historical ones because circumstances might have changed over time or the assessment might be different. The 'gold standard' for the evaluation of a new therapy is the *randomized controlled trial*. The purpose of the randomization is to avoid bias due to differences in clinical and demographic patient characteristics. A variety of approaches for random treatment allocation exists. They all have in common that the chance to receive a certain treatment is known, but that it is not predictable which treatment each patient will receive. The most basic form of randomization is the so-called *simple randomization*. The treatments are allocated, for instance, by tossing a coin. This means the treatment allocations are independent and with fixed probabilities. As an alternative to tossing a coin, random number tables in books or computer random number generators can be used. The problem with simple randomization is that it is very likely to lead to unequal numbers in treatment groups. *Block randomization* (also known as *restricted randomization*) is designed to avoid this problem. For example, there are six possible ways to arrange two treatments A and B in blocks of size four with equal treatment allocation. These are AABB, ABAB, ABBA, BBAA, BABA, and BAAB. In the block randomization process, blocks are chosen at random and thereby a series of A's and B's is created. If the trial is stopped at any time the difference in treatment group size can never be more than two in this example.

When important prognostic factors exist it is desirable to balance the treatment groups with regard to these factors. One way to achieve this is *stratified randomization*. In stratified randomization, separate block randomizations are carried out for each stratum. For example, in multi-centre studies the randomization is normally stratified by centre, to achieve treatment balance within centres, since outcomes might differ by study centre. Other randomization techniques exist and the reader is referred to Rosenberger and Lachin (2002) for more information.

In order to avoid bias due to differences in treatment or outcome assessment, patients and investigators often are not told during the ongoing study which treatment the patient receives. This is called *double-blinding*. If only the patient does not know which treatment is being received, the study is called *single-blind*. In addition to double-blinding it is good practice that the trial team, including the trial statistician, do not know who receives which treatment during the ongoing study. Sometimes this is referred to as *triple-blinding*. In a drug trial, treatment blinding, which is also sometimes called *masking*, can be achieved for example by including a placebo treatment in the trial. A placebo is a treatment that has the same appearance, taste, and smell but is lacking the effective component. When two active treatments are compared and it is impossible for them to have the same appearance (e.g. comparison of a capsule and a tablet) blinding can be achieved by the *double-dummy technique*. One would use two placebos, one similar to treatment A and one similar to treatment B. Then each patient would get either treatment A and the placebo to B, or the placebo to A and treatment B.

Blinding is an important design feature in preventing bias, but it is not always easy to achieve and may even be impossible. For instance, surgeons carrying out an operation cannot be blinded. In these circumstances, however, it is important to conceal the treatment allocation until the patient is entered into the trial to avoid selection bias. Furthermore, the assessment of the outcome should be blind to treatment allocation. It should be noted that even in studies that are set up to be double-blind unblinding might occur, for example through treatment specific side effects.

As clinical trials are 'experiments' with human beings, there are many ethical issues involved. For instance, randomization takes the treatment decision away from the doctor and patient and makes it subject to a randomization procedure. Ethically this can only be justified if it can be assumed at the outset of the study that the treatments under investigation are equally beneficial or harmful. This assumption is referred to as the *equipoise* assumption, and has been the subject of long-standing debate. An important ethical aspect of clinical trials is that patients consent to participation in the study after having had the trial procedures and associated risks clearly explained to them on a patient information sheet. This is referred to as *informed consent*. Furthermore, all clinical trials must undergo ethical review before initiation by so-called ethics committees or institutional review boards. The World Medical Association (WMA) has published guidance on ethical issues, which is known as the Declaration of Helsinki. This is regularly updated and can be found at the WMA's website (www.wma.net).

From consideration of clinical trials in a general setting, we now turn to focus specifically on oncology. Here the chronic and serious nature of the disease and the toxicity of many treatments lead to a different emphasis in the conduct of clinical trials. As in other areas, cancer clinical trials are divided into phases I to IV as described above. A major difference between cancer clinical trials and those in other areas, however, is that new drugs, or new drug combinations, are tested on patients with advanced disease rather than on healthy volunteers. Patients entering phase I cancer trials have usually been treated with standard treatments before they are considered for inclusion in these experimental trials. The entry criteria for a trial may allow different primary cancers to be included, thus providing a mixed disease site study population. The reason for entering phase I trials is often one of pure altruism on the part of the patient as there is usually only a remote chance of a 'cure' and potentially a very good chance of experiencing harmful side-effects from the experimental drugs. This is a very different situation to the development of non-cancer treatments where toxicity is limited and new drugs are tested on healthy volunteers.

New cytotoxic anticancer agents have the ability to cause serious harm to patients from the associated toxicities, many of which could be life threatening. Phase I trials must therefore include a high level of monitoring and investigation for the patients of pharmacological endpoints. Phase I trials in oncology also generally include dose escalation. This is predicated on the assumption that both increased drug activity and increased toxicity are associated with increased dose of the drug. For safety reasons, a low dose of the drug is initially used, this being increased through the course of the trial until a level with acceptable toxicity is reached, as described in more detail in the next section. In the oncology setting, therefore, selection of an appropriate dose for further assessment is usually made in a phase I trial on the basis of safety data rather than in a phase II trial on the basis of efficacy data. Exceptions to this type of phase I cancer trial are the new cytostatic compounds being tested, which may not incur severe toxicities and for which it is then difficult to determine the optimal therapeutic dose.

Once the optimal dose has been achieved within the phase I trial, the treatment can be tested within a phase II trial where the primary endpoint is efficacy as determined by some measure of response. Phase II trials in oncology usually consider one primary disease site where again patients may have relapsed from conventional standard therapy but for which the disease is either less advanced or the patient is fitter that the phase I trial patients. Phase II patients tend to be heavily preselected and as a consequence may not be a representative sample. Often phase II trials are conducted in a single centre and high response rates are reported, but this may be due to the selection criteria and prescreening of the patients as opposed to the activity of the drug itself. For this reason several attempts have been made to optimize the phase II design, including the use of randomized phase II trials comparing the new drug against a control group. After a detailed safety review, randomized phase II trials can be easily rolled into randomized phase III trials to assess survival or recurrence endpoints. These types of trials are common when the study drug has been shown to be advantageous in another disease sites and thus the safety profile is known.

Phase I and II trials and phase III trials in oncology are discussed in detail in the next two sections.

20.2 Early-Phase (Phase I and Phase II) Clinical Trials in Oncology

As described above, prior to evaluation in large-scale definitive phase III clinical trials, anti-cancer drugs are assessed in humans in a series of early-phase clinical trials, divided into phase I and phase II clinical trials. The aim of these trials is to ensure, as far as possible, that treatments evaluated in the time-consuming and expensive phase III trials are those that are the most effective, whilst also being safe and administered at the correct dosage. The emphasis in these early-phase trials is thus on the relatively rapid screening of drugs to determine which are suitable for evaluation in phase III trials, and on the choice of an appropriate dose for further testing and hopefully, following phase III, for actual clinical use.

The aims of phase I and phase II trials are rather different. Phase I trials are sometimes called '*first in man*' trials, since often the drug is being used in humans for the first time. Because of this, the emphasis is necessarily on the safety of the novel drug therapy, and in particular on the identification of an acceptably safe dose for continued testing. As explained above, in oncology, the highly toxic nature of many drug therapies makes testing on healthy volunteers unethical, so that phase I trials usually involve patients for whom previous therapy has failed. These patients therefore may represent a very different patient population than that which would be treated with the drug in general clinical use following successful testing. The assessment of efficacy is thus difficult, emphasizing the main role of the phase I trial in assessing drug safety.

The first primary assessment of efficacy is the role of the phase II clinical trial, although of course the question of safety assessment is also present in these trials. The main role of the phase II clinical trials is thus to provide a screening based on efficacy and safety of drug prior to phase III. The phase II trial may involve comparison of several doses of the drug under evaluation to further inform the choice of dose for phase III testing. The desire to obtain as much information on the new drug as possible from a small number of patients enrolled in the phase I or phase II clinical trials means that sometimes no control group is used and, when a control group is used for blinding purposes it is often much smaller in size than the experimental treatment group. In order to evaluate drugs as quickly as possible, both phase I and phase II are based on the observation of short-term or medium-term endpoints for toxicity and efficacy, such as dose-limiting toxicity, disease progression or remission. This is in contrast to the phase III clinical trial in which double-blind randomization to a concurrent control group of similar size to the experimental therapy group is standard and long-term endpoints, often survival times or progression-free survival times, are used.

As explained above, the primary aim of the phase I clinical trial in oncology is evaluation of drug safety, and in particular the identification of a safe dose of the drug for further testing. Because to maximize the chance of efficacy it is desirable to give as large a dose as possible, this usually means identification of the *maximum tolerated dose* (MTD). A number of definitions exist for the MTD. It is sometimes defined to be the highest dose tested that leads to no dose-limiting toxicity, or the highest dose that does not lead to a significant increase in dose-limiting toxicity relative to a placebo control. A problem with such definitions, however, is that they depend in part on the sample size used to test the toxicity of the new drug, with larger sample sizes leading in general to a lower MTD. An alternative definition, which does not depend on the sample size of the clinical trial, and often leads to a similar MTD in practice, is that dose that would lead to dose-limiting toxicity in 20 % of the population under investigation. This definition reflects the fact that a small level of toxicity is acceptable in an effective anticancer treatment, and often leads to an MTD close to that found not to lead to toxicity in a small phase I clinical trial.

The fact that the phase I clinical trial often comprises the first use in humans of the new drug means that testing usually starts at a low dose with patients treated one at a time or in small groups, the dose level being escalated only after observation of the safe use of the drug at lower dose levels. A standard approach is the '3 + 3' design. Using this approach, patients are treated in cohorts of size three at one of a series of predefined dose levels and assessed. The initial cohort receives the lowest dose, which from laboratory investigation and animal studies is believed to be safe. If no toxicity is observed at this dose a cohort of three patients is treated at the next higher dose, with dose escalation continuing if these patients do not show any toxicity. If toxicity is seen in a single patient at any dose level, the next cohort of three patients is treated at the same dose, with dose escalation continuing if none of these three patients experience toxicity. The trial continues until two or more patients experience toxicity at some dose level. The dose immediately below that leading to toxicity in two patients is then chosen as the MTD.

A feature of the '3 + 3' rule is that many patients are treated with doses well below the MTD. Whilst this reduces the risk of dose-limiting toxicity, it may also lead to the use of ineffective doses for these patients. An alternative approach, in which patients are treated one at a time to allow more rapid dose escalation when this is safe, is the *continual reassessment method* (CRM) proposed by O'Quigley *et al.* (1990). Using this method, the MTD is estimated after treatment of each patient, and the next patient is given the dose closest to this estimate. Modifications of the CRM have also been proposed (Faries, 1994; Goodman *et al.*, 1995) in which the first patient always receives the lowest dose level, and in which the dose is never increased by more than one step at a time.

As explained above, in contrast to the phase I clinical trial, the focus of the phase II trial is treatment effectiveness, usually as indicated by some rapidly observable binary (success/fail) response. Based on the observed efficacy data, the trial will lead to a decision either to continue with treatment development into phase III or to terminate clinical development of the new treatment. The random nature of the observed data means that there is a risk of making an incorrect decision. An incorrect decision can be made in one of two ways; either concluding that the novel therapy is worthy of further development when in fact it is not, that is a *type I error*, or concluding that development should be abandoned when the treatment is actually effective, a *type II error*. The majority of phase II clinical trials in oncology are conducted using what is called the *frequentist* approach so as to control the type I error rate and the power, that is the converse of the type II error rate, to detect some specified treatment effect. The control of the probabilities of each type of error leads to specification of the number of patients required and of the number of treatment successes that need to be observed in order to conclude that the treatment is effective and worthy of investigation in further clinical trials. Requiring stricter control of the type I and type II error probabilities means that a larger number of patients are needed. In order to keep the number of patients required to a reasonable level, it is common to set the type I error rate to be 10 % or even 20 % whereas in larger phase III studies 5 % would be the standard.

The frequentist approach can be illustrated using the example of a phase II clinical trial in poor prognosis acute myelogenous leukaemia patients described by Thall and Simon (1994). The standard treatment for these patients was fludarabine + ara-C, for which the rate of complete remission (CR) of the leukaemia was known to be 50 %. The aim of the single-arm phase II trial was to evaluate a novel therapy, comprising the use of granulocyte colony-stimulating factor (GCSF) in addition to the standard therapy. This new therapy would be considered beneficial if the CR rate was increased to 70 %. The clinical trial could be conducted so as to control the probability of erroneously concluding that the new therapy is effective if the CR rate was really 50 % or of erroneously concluding that the new therapy is ineffective if the CR rate was really 70 %, for example to each be 10 %.

The use of rapidly observable endpoints in phase II clinical trials, together with the desire, particularly for ethical reasons, to monitor the results from patients already in the trial, mean that sequential methods, in which the data are analysed at a series of interim analysis conducted through the course of the clinical trial, are particularly desirable. Many phase II clinical trials are thus conducted in two or more stages, with stopping possibly at the end of each stage if the results are not sufficiently promising. The chance to stop the study early also leads to a reduction in the average number of patients required relative to a single-stage study.

The most commonly used method is that proposed by Simon (1989). This is a two-stage design, with the analysis at each stage conducted so as to control the probabilities of errors of types I and II as described above. In addition to the number of patients required for the whole study, a decision must be made regarding the number of patients recruited in the first stage. This choice is usually made in such a way as to minimize the total expected sample size under the scenario that the experimental treatment is ineffective. As an example, for the GCSF trial described above, controlling type I and type II error rates at the 10 % level, the design proposed by Simon would first treat 21 patients. If the treatment was successful for fewer than twelve of these patients, the study would stop at this point. Otherwise a further 24 patients would be treated, giving 45 in total. If the total number of successes was 27 or more, the new treatment would be considered sufficiently promising to proceed to phase III evaluation, otherwise development would be terminated. A similar method for trials with three has been proposed by Chen (1997). An alternative approach is to use a sequential approach in which any number of interim analyses may be conducted, possibly also with the chance to stop early if the experimental therapy proves sufficiently effective. One such method has been proposed by Fleming (1982), while Stallard and Todd (2000) give a similar approach particularly suitable for small sample sizes or rare events.

A slightly different method, also based on the frequentist approach, was proposed by Gehan (1961). This approach considers, in addition to error rates, estimation of the effectiveness of the novel therapy. In this approach, the trial is also conducted in two stages. At the end of the first stage, a decision is made to abandon the trial if no treatment successes are observed, but otherwise to continue to a second stage. The size of the first stage is chosen so as to control the type I error rate and the size of the second stage is chosen to give specified precision for the estimate of the success rate for the novel therapy. For a type I error rate of 5 % and a standard success rate of 20 %, the first stage includes 14 patients.

20.3 Confirmatory (Phase III) Trials in Oncology

Before they can be accepted as the new gold standard, potential new drugs or drug combinations need to be evaluated within a definitive prospective multicentre randomized phase III controlled clinical trial. There are many pitfalls to overcome when designing phase III trials and the International Conference on Harmonization of Technical Requirements for Registration of Pharmaceuticals for Human Use (ICH) clinical trial guidelines E7 to E11 (http://www.ich.org) attempt to address issues relating to design considerations in a general setting. The main points to consider when designing and assessing phase III trials include a literature review, choice of a control treatment, choice of the study population, choice of endpoints, the sample size and power of the trial and the duration of follow-up and timing of assessments. These points are considered in turn in the following paragraphs.

A *systematic literature review* is essential before conducting a clinical trial. There may be information about the treatment or treatment combinations in another study population, or inconclusive evidence in a series of smaller trials which can be assessed using meta-analysis. There are several registers of ongoing trials or completed trials that make the searching for reported trials easier; in particular most trial protocols now register for an International Standard Randomized Controlled Trial Number (ISRCTN), which is included on the front of the protocol. Conducting proper systematic reviews and meta-analysis is a skill in itself and there are several centres that specialize in providing training, e.g. Cochrane Collaboration at Oxford or Centre for Systematic Reviews at York. For the common disease sites such as early breast or colorectal cancers, there are ongoing international overview collaborations that are regularly updated. These overviews do not only provide information about the possible benefit that may be detected from certain treatments or combination of treatments but they also provide information about what may be expected from the control group.

The Early Breast Cancer Trialists' Collaborative Group (EBCTCG) is a good example of ongoing meta-analysis collaboration. Since 1983 information on over 264 000 women recruited into around 400 trials has been collected, analysed and followed up for recurrences and deaths. Results are presented as per their randomized comparison across a variety of baseline prognostic factors such as nodal status, age, menopausal status and receptor status. This is an excellent source of information when planning a new trial. Publications are listed on the website (http://www.ctsu.ox.ac.uk/projects/ebctcg).

Careful *choice of control group* is essential for the evaluation of a new treatment. A suboptimal choice of control may inflate the treatment effect, if one exists, or may fail to detect a difference if patients and clinicians feel disadvantaged and subsequently elect for more active therapy. When evaluating non-cancer treatments on healthy volunteers, a placebo or dummy treatment is often the optimal choice of control. However, in cancer trials, it is often not possible, or indeed ethical, to use a non-active control treatment. It is surprising how often there is lack of consensus about the optimal control arm in a cancer trial, especially for trials of advanced disease where the standard treatment is 'best supportive care'. For trials of early breast or colorectal cancer, there may be several treatments which a patient may have to undergo, such as surgery, chemotherapy, radiotherapy and maintenance treatment. The control arm in this case may be a 'treatment package' which maps onto standard care.

Choice of the patient population is also important. Phase III trial participants are typically newly diagnosed, have not had cancer before in another site (except basal cell carcinoma or cervical intraepithelial neoplasia) and are fit enough to receive treatment. Other eligibility criteria include willingness to be randomized and to give informed consent. There are inevitably differences between study populations across centres; populations served by inner city hospitals will be different to seaside or rural centres. There are also differences in prognostic factors across patients. The study population will have a dramatic effect on the ability to determine a treatment benefit, if it exists. The most influential factor to affect outcome is the spread of the disease, followed by the fitness of the patient. Stratification within the randomization can stop any imbalances across the treatment allocation. Early breast cancer trial patients tend to be females who are relatively young (median age 48 years) and fit. For example, in early breast chemotherapy cancer trials we typically stratify by centre, nodal involvement (negative, positive), age ($\leqslant 50$, >50 years) or menopausal status (pre-, post-) and other tumour characteristics such as oestrogen receptor status. In comparison, advanced pancreatic cancer patients tend to be older (median age 65 years) and frail; stratification factors may include surgeon, resection margin involvement (clear, involved) and performance status (0, 1, 2; indicating restricted activity but fit enough to receive the treatment). It is important to decide carefully the exact population under consideration and to continuously monitor the type of patients entering the study, as this will affect the number of events occurring within the study period and the ability of the trial to detect treatment differences. It is essential to collect baseline information about the study participants and to present a patient characteristic table alongside the treatment results.

The *primary endpoint* in phase III cancer clinical trials is often survival. The advantage of using such an unambiguous endpoint is obvious; it is non-subjective and even patients who are deemed 'lost to follow-up' for trial purposes can be included in future survival analysis as a copy of their death certificate can usually be obtained. However, trials in early disease, for which the majority of patients are expected to remain alive for many years, may use recurrence or relapse-free survival as the primary endpoint. The most important consideration is the ability to be able to detect the outcome equally well across the study treatment arms. When comparing six cycles of a new drug to best supportive care, the ability to detect response or relapse may be higher in the new drug arm unless follow-up patterns are clearly set out in the protocol.

Quality of life and healthcare economics are secondary endpoints within phase III trials. If several drugs are equally effective, then the cost and effect on patient's quality of life will be the deciding factor as to which drug is chosen to be the new standard treatment.

The *sample size* of a phase III clinical trial should be chosen to ensure adequate *power* to be able to detect clinically worthwhile differences. Before calculating the sample size, it is necessary to determine the population under consideration and to know what to expect of the control arm. Academic pragmatic trials often have loose entry criteria, e.g. anyone diagnosed with early breast cancer may be entered into the trial, but a more detailed estimation of the population needs to be determined. In an early breast cancer trial, an estimate of the proportion of node-positive and node-negative cancers must be made in order to determine the event rate expected. The best source of information about what to expect on the control group is a previous trial using the same population of patients.

The next consideration in sample size estimation is the size of the difference that might reasonably be detected between the treatments. It is rare in cancer trials to expect differences in excess of 10 %. Past trials may have failed to detect small but clinically worthwhile differences because they were underpowered. In early breast cancer or colorectal cancer there are several thousand patients diagnosed each year and it is not unreasonable to carry out a two-arm randomized trial of around 5000 patients in order to detect 3–5 % differences between treatments. These small differences are clinically worthwhile for common cancers. However, in advanced pancreatic cancer a trial of 500 patients would be sufficient to detect differences in excess of 10 % between treatment groups, since this is a poor prognostic population with a high event rate. Sample size calculations are often called 'the numbers game' as we need to adequately power a trial to determine clinically worthwhile differences but we also have to consider resources and ability to recruit into the trial. Machin and Campbell (1987) explain sample size calculations in detail for a variety of different situations and trials.

All trials have fixed *assessments* and *follow-up* set into the protocol. A typical trial of advanced disease may require patients to be treated monthly for six cycles and then followed up every 3 months until death. In contrast, a trial of early disease may require 6 months of treatment and then 6 monthly follow-up for the first 2 years post-treatment and annually thereafter up to 5 or 10 years. The important consideration, as discussed in the collection of endpoints, is to have equal assessments and follow-up patterns across the treatment arms. If this is not the case then one group may be disadvantaged since recurrences may be detected earlier, not because they occur earlier but because the follow up is more frequent. Ineligibles, withdrawals and protocol violations need to be recorded and followed-up as near as possible to the trial protocol so that they can be included in an Intention-To-Treat (ITT) analysis. This is possible if the trial patient assessment and follow-up patterns match onto normal clinical practice.

For all multicentre phase III trials it is essential to have an *independent data and safety monitoring board* (DSMB) established, whose remit it is to provide an unbiased review of the safety and conduct of the trial. It is usual to devise a charter when setting up a DSMB, which clearly sets out the roles and responsibilities of the DSMB members, the frequency of the meetings and reporting structure for their recommendations. A standard charter template can be downloaded from the UK Medical Research Council website (http://www.ctu.mrc.ac.uk/files/DMCcharter_general.pdf) or found in the book by Ellenberg *et al.* (2002). The trials office should also supply DSMB members with a protocol, set of blank case report forms and statistical analyses plan for the trial. The extension of monitoring of safety data is the use of a formal sequential analysis method to allow the clinical trial to be stopped as soon as sufficient evidence of a treatment effect is observed (see e.g. Jennison and Turnbull (2000) and Whitehead (1997)).

Pharmacovigilence monitoring is a legal requirement within the UK for an Investigational Medicinal Product (IMP) trial. Each individually reported Serious Adverse Event (SAEs) and Suspected Unexpected Serious Adverse Reaction (SUSARs) is reviewed by the Clinical Chief Investigator of the trial, reported to the regulatory authority and followed-up until resolution by the clinical trials unit running the trial. However, the overall safety review must be carried out by an experienced independent committee to determine if the amount of reported toxicities is as expected when comparing the trial treatments. The DSMB have the power to make recommendations about stopping the trial if the level of reported toxicity is in excess of the expected toxicity, especially if there is an excess of treatment related deaths.

Other considerations made by the DSMB include recruitment targets, event rates, follow-up and treatment compliance. Power calculations may have to be revisited if the population recruited differs from the planned population, or if external evidence becomes available suggesting a bigger treatment effect. Membership on a DSMB is a serious commitment and should not be undertaken lightly.

20.4 Further Issues in Clinical Trials in Oncology

In this section we briefly discuss two important issues in clinical trials in oncology. These are subgroup analysis and the conduct of translational substudies.

All experienced trialists know the dangers of *subgroup analysis* and this is well illustrated by the ISIS-2 collaborative group where they demonstrated a significant treatment benefit within a subgroup, which was determined by the patients' zodiac sign (ISIS-2 Collaborative, 1988). However, there may be genuine reasons why exploratory data analyses of subgroups are important within a trial. Preplanned subgroup analyses based on scientific rationale should be stated in the protocol with appropriate adjustment for multiple testing so that any differences seen are more plausible (Follmann, 2004). Rather than explore treatment effects within each subgroup separately, it is better to use forest plots to explore treatment effects within and between subgroups.

The aim for commercial development is to maximize the effect of the drug in a large population of patients. Interestingly, the benefit of Herceptin may have been lost when exploring the treatment effect in a trial of early breast cancers as the benefit is confined to the Her2+ve subgroup ($<30\%$ of the population). Subgroup analysis does have a place in trial reporting but must be handled with caution and confirmed in further independent trials.

Translational substudies are often added onto the back of the main clinical trial. These studies could be in the form of a collection of baseline blood and tissue samples to predict prognosis, or repeat blood samples to assess potential biomarkers. Ethical approval and patient consent must be obtained specifically for each substudy as well as the main trial. Retrospective collection of samples is problematic, and if possible prospective sample collection should be undertaken to specific standard operational procedures. There is an increase in the regulatory requirements for tissue collection, transportation and storage which has led to an increase in quality standards but also an increase in the resources required when carrying out these studies. All translational sub-studies must have a separate protocol and be adequately powered to answer the question.

References

Altman DG. *Practical Statistics for Medical Research*. Chapman & Hall, London, 1991.

Chen TT. Optimal three-stage designs for phase II cancer clinical trials. *Statistics in Medicine* 1997, **16**, 2701–11.

Ellenberg SS, Fleming TR, DeMets DL. *Data Monitoring Committees in Clinical Trials: A Practical Perspective*. Wiley, Chichester, 2002.

Faries D. Practical modification of the continual reassessment method for phase I clinical trials. *J Bio-pharm Stat* 1994, **4**, 147–64.

Fleming TR. One-sample multiple testing procedure for phase II clinical trials. *Biometrics* 1982, **38**, 143–51.

Follmann D. Subgroups and interactions. In: *Advances in Clinical Trial Biostatistics*, edited by NL Geller. Marcel Dekker, New York, 2004, p 121–40.

Gehan EA. The determination of the number of patients required in a preliminary and follow-up trial of a new chemotherapeutic agent. *J Chronic Dis* 1961, **13**, 346–53.

Girling D, Parmar M, Stenning S, Stephens R, Stewart L. *Clinical Trials in Cancer: principles and practice*. Oxford University Press, Oxford, 2003.

Goodman SN, Zahurak ML, Piantadosi S. Some practical improvements in the continual reassessment method for phase I studies. *Statistics in Medicine* 1995, **14**, 1149–61.

Green S, Benedetti J, Crowley J. *Clinical Trials in Oncology*. Chapman & Hall, London, 2003.

ISIS-2 COLLABORATIVE GROUP. Randomised trial of intravenous streptokinase, oral aspirin, both, or neither among 17 187 cases of suspected acute myocardial infarction: ISIS-2. *Lancet* 1988, **2**, 349–60.

Jennison C, Turnbull BW. *Group Sequential Methods With Applications to Clinical Trials*. Chapman & Hall, London, 2000.

Machin D, Campbell MJ. *Statistical Tables for the Design of Clinical Trials*. Blackwell Scientific Publications, Oxford, 1987.

O'Quigley J, Pepe M, Fisher L. Continual reassessment method: a practical design for phase I clinical trials in cancer. *Biometrics* 1990, **46**, 33–48.

Pocock SJ. *Clinical Trials: A Practical Approach*. John Wiley, Chichester, 1983.

Rosenberger WF, Lachin JM. *Randomization in Clinical Trials*. John Wiley, Chichester, 2002.

Simon R. Optimal Two-Stage Designs for Phase II Clinical Trials. *Controlled Clinical Trials* 1989, **10**, 1–10.

Stallard N, Todd S. Exact sequential tests for single samples of discrete responses using spending functions. *Statistics in Medicine* 2000, **19**, 3051–64.

Thall PF, Simon R. Practical Bayesian guidelines for phase IIB clinical trials. *Biometrics* 1994, **50**, 337–49.

Whitehead J. *The Design and Analysis of Sequential Clinical Trials*. Wiley, Chichester, 1997.

21

Representative Cancers, Treatment and Market

Teni Boulikas and Nassos Alevizopoulos

21.1 Lung Cancer

Lung cancer includes small cell (SCLC) and non-small cell (NSCLC) lung cancers, with NSCLC accounting for 75 % of all cases. In 2006, more than 338 000 cases of the disease were diagnosed in the seven major pharmaceutical markets. In early-stage NSCLC, surgical resection is the treatment of choice; treatment may include surgery followed by a combination of chemotherapy and radiation. In locally advanced NSCLC, chemoradiation is superior to radiation alone. First-line chemotherapy consists of cisplatin or carboplatin in combination with any of the following agents: paclitaxel, docetaxel, gemcitabine, vinorelbine, irinotecan, etoposide, vinblastine and, more recently, bevacizumab (Avastin). Single agent docetaxel or pemetrexed, or the tyrosine kinase inhibitor erlotinib are established second-line agents. Erlotinib is also indicated in third line. On the other hand, SCLC is rarely amenable to curative treatment by resection. SCLC is almost universally treated with a combination of cisplatin/carboplatin with etoposide.

High unmet needs still persist for this tumour type and despite two decades of extensive R&D and chemotherapy, the average 1-year overall survival of NSCLC patients is below 25 %. Amgen's pantitumumab targets epidermal growth factor receptor (EGFR) and has a favourable toxicity profile. With pantitumumab seeking to gain initial approval in the first-line setting, the agent might rapidly gain ground on BMS/Merck KgGA's Erbitux, despite the first-to-market advantage of Erbitux.

Over the forecast period to 2015 Avastin might be the sole drug to achieve blockbuster status within the NSCLC market. Avastin's 2015 sales of $1285 m are forecast to eclipse those of Tarceva ($737 m) and Erbitux ($696 m). This is largely attributed to the size of Avastin's target population within the first-line setting. In attempting to avoid the fate

Anticancer Therapeutics   Edited by Sotiris Missailidis
© 2008 John Wiley & Sons, Ltd

of AstraZeneca's recently withdrawn drug Iressa, developers must allow for adequate maturation of trial data in order to explicitly demonstrate the clinical benefit of novel pipeline agents.

21.2 Breast Cancer

Breast cancer is the most common form of cancer among women in North America and almost all of Europe. It is estimated that each year the disease is diagnosed in over 1 million women worldwide and is the cause of death in over 400 000 women. Recent data identifies a substantial reduction in breast cancer mortality, which is probably due to improvements in diagnosis and treatment. Early detection, *via* breast cancer screening is a potentially important strategy for reducing mortality. The clinical manifestations of breast cancer can range from a localized tumour to a widely metastatic neoplasm.

Breast cancer mainly occurs as ductal carcinoma *in situ* (DCIS) and lobular carcinoma *in situ* (LCIS), which can be easily treated if detected, invasive breast carcinoma, inflammatory breast cancer, as well as other more rare forms such as Paget's disease. The type of treatment administered against invasive breast carcinoma depends on the stage of the disease. In the early stages, successful treatment can be accomplished without chemotherapy through a combination of surgery (lumpectomy or partial mastectomy) and radiotherapy. At later stages of disease, the chances of curative treatment are substantially reduced.

Treatment options that may be considered are surgery, radiotherapy, cytotoxic chemotherapy and hormonal manipulation or endocrine therapy. Treatment options for breast cancer are initially identified according to the stage of the disease. Treatment for advanced stages of breast cancer may involve more aggressive surgery (total or radical mastectomy) in combination with adjuvant chemotherapy and radiotherapy. Anthracyclin (doxorubicin or epirubicin) or taxane (paclitaxel or docetaxel)-based regimens are used in the neo-adjuvant (preoperative) setting for locally advanced invasive breast cancer. In the more advanced recurrent or metastatic (stage IV) disease, menstrual status as well as existence of oestrogen receptors (ER) are important determinants of pharmacological therapy. Tamoxifen, letrozole, anastrozole and exemestane are key drugs used in the adjuvant setting of ER-positive patients of this class, while ER-negative adjuvant patients are treated with trastuzumab and several chemotherapeutic cocktails consisting of doxorubicin, epirubicin, liposomal doxorubicin, paclitaxel, docetaxel, capecitabine, vinorelbine, gemcitabine and albumin-bound paclitaxel.

21.3 Prostate Cancer

The prostate is a gland in a male's reproductive system that makes and stores seminal fluid. Prostate cancer has a high cure rate in the early stages (stage I/II), where active treatment will usually consist of surgery or radiotherapy. Many cases of early-stage prostate cancer may be managed with observation only, if the disease is causing no side effects and treatment would have a greater impact on quality of life than living with the disease would. Drug therapy may be incorporated in the advanced stages of the disease. Treatment options include hormonal therapy aimed at reducing testosterone levels by using luteinizing hormone-releasing hormones (LHRHs) or LHRH agonists, or by blocking the effect of androgens with antiandrogens. These two drug classes are often combined to form a total androgen blockade

(TAB), which is effectively chemical castration. Typical LHRH agonists are leuprolide and goserelin, with bicalutamide and flutamide typically used as antiandrogens. The cytotoxics mitoxantrone and estramustine are commonly employed in hormone-refractory disease, while docetaxel-based regimens are now the standard of care in metastatic prostate cancer. Finally, biphosphonates may be used in the management of bone disease generated by prostate cancer metastasis, while several new therapies may hold promise for the management of the disease.

21.4 Colorectal Cancer

Colorectal cancer (CRC) is the third most common cancer worldwide. The incidence of CRC is higher in countries in the developed world, where it is the second most common cancer. The World Health Organization estimates that there are over 940 000 cases annually worldwide, with almost 500 000 deaths. CRC, if detected early, is associated with a high survival rate. Surgery is usually the prime treatment option. However, many patients are still diagnosed with advanced disease, and in this setting recent advances in chemotherapy and novel targeted agents, particularly drugs that target growth factor receptors, have changed the treatment paradigm.

Colorectal cancer includes the development of malignant tumours within the colon and rectal segment of the gastrointestinal tract. As with other cancers, the optimal treatment depends on the stage of the disease, and early detection increases the patient's chance of survival. Surgical resection is the principal treatment for both colon and rectal cancer, particularly in the early stages of the disease. Surgical resection of early-stage polyps (polypectomy) corresponds to a survival rate of 90 %. More extensive surgical procedures may be employed in more advanced stages of the disease, involving partial or complete removal of the colon and rectum. Radiation and pharmacological therapy play an increasingly important role at the later stages of the disease both as adjuvants to surgery and in the reduction of symptoms.

The chemotherapeutic treatment of colorectal cancer (CRC) has undergone somewhat of a revolution in the past 5–10 years, with a variety of new drugs and regimens being either approved or under investigation. Worldwide, there is no consensus on the optimal treatment of CRC; this is particularly apparent with respect to advanced or metastatic disease. 5-fluorouracil (5-FU) has been used to treat CRC for about 50 years. In the 1980s, studies showed that the addition of leucovorin (LV; also known as folinic acid) improved the efficacy of 5-FU without a large increase in toxicity, thus 5-FU/LV became the standard treatment for CRC. Several 5-FU/LV regimens have been developed, with infusional regimes favoured in European countries over bolus administration, which was common practice in the USA. Today, however, the chemotherapy for CRC is more likely to consist of 5-FU/LV in combination with oxaliplatin or irinotecan (i.e. FOLFOX or FOLFIRI). Another drug, capecitabine, may also be included. This is an orally administered fluoropyrimidine, and a prodrug of 5-FU. Capecitabine is at least as effective as bolus 5-FU/LV for the treatment of colorectal cancer, and is an alternative to infusional 5-FU/LV for the treatment of advanced disease. In advanced or metastatic disease pharmacological treatment options include 5FU-leucovorin, capecitabine, irinotecan, oxaliplatin, bevacizumab and panitumumab in various combinations. Repeat hepatectomy for patients with colorectal cancer metastases is safe and provides survival benefit equal to that of a first liver resection.

21.5 Ovarian Cancer

Ovarian cancer can arise from a number of different cell types; these can be grouped into three main classes: epithelial cancers occurring in cells lining the ovaries; germ cell cancers derived from germ cells within the ovaries and cancers of the sex cord and stromal cells. The epithelial ovarian cancers constitute about 95 % of all ovarian cancers. The most common treatment of ovarian cancer is surgery to remove the uterus, fallopian tubes, ovaries and any large nodules of cancer. This procedure is known as a total abdominal hysterectomy/bilateral salpingo-oophorectomy (TAH-BSO). Removal of segments of the bowel or a colostomy may also be required. In young, early-stage patients who wish to have children after treatment, removal of only one ovary may be considered (unilateral oophorectomy). Postoperative chemotherapy or radiotherapy (internal or external) may also be employed to prevent recurrence of the disease. First-line drug therapy for ovarian cancer has changed little in the last 5 years, with the carboplatin/cisplatin and paclitaxel regimen firmly established as the gold standard. The combination of paclitaxel with carboplatin/cisplatin has demonstrated improved efficacy over the older standard treatment of cyclophosphamide with cisplatin, while vinca alkaloids (vinblastine, vincristine) and antibiotics (dactinomycin) may occasionally be used as well.

21.6 Pancreatic Cancer

Pancreatic cancer, the malignant tumour of the pancreatic gland, continues to constitute a major unresolved health problem, affecting over 230 000 people worldwide each year. Pancreatic cancers are very hard to diagnose because they grow deep in the abdomen in the absence of alarming symptoms. About 85 % of the patients are usually diagnosed at an advanced stage of the disease resulting in overall high mortality rates even though its incidence is roughly ten times lower than that of lung cancer. Indeed, only 1–4 % of the patients are expected to have survived pancreatic cancer five years post diagnosis (as compared to a staggering 90 % of melanoma patients, for example). Regional incidence rates vary in the global community. For example, developed countries of the western World have five- to sevenfold higher incidence rates than developing countries, suggesting that lifestyle-dependent factors might contribute to the pathogenesis of the disease. As the mortality rates are essentially equal to incidence rates, the death toll is high. Indeed, being the 10th most common cancer, pancreatic adenocarcinoma is the sixth leading cause of cancer-related deaths in Europe, where over 50 000 individuals are expected to die every year of pancreatic cancer.

The causes of pancreatic cancer are poorly understood, and only a limited number of demographic, environmental, risk and genetic factors are associated with the disease. For instance, advanced age, smoking, long-standing chronic pancreatitis, diabetes and obesity may confer increased risk, albeit of different magnitude. Cigarette smoking seems to be the most firmly established risk factor, as long-term administration of tobacco-specific N-nitrosamines induces pancreatic malignancies in laboratory animals. Furthermore, multiple case–control or cohort studies in humans suggest that 30 % of the pancreatic cancers are caused by cigarette smoking. Similarly, the long-term risk of pancreatic cancer in patients with chronic pancreatitis may be attributed to tobacco and alcohol consumption rather than to a causatively related transformation of pancreatitis into pancreatic adenocarcinoma. Occupational exposure to chemicals, increased body mass index and animal fat uptake may lead to an increased risk of pancreatic cancer, as may high fat-cholesterol diets, in contrast to diets rich in fruit and vegetables, which may be protective (reviewed by Evans *et al.*, 2004.)

An increased incidence of pancreatic cancers has been associated with various genetic syndromes or mutations. For example, individuals with germline mutations in the gene encoding the cyclin-dependent-kinase inhibitor p16 may a have a risk of pancreatic cancer up to 20 times higher than the general population. Moreover, families with hereditary nonpolyposis colon cancer caused by mutations in several DNA mismatch repair genes or individuals with Peutz–Jeghers syndrome caused by mutations in the 19p13 tumour suppressor locus also have an increased risk to develop pancreatic cancer. One commonly identified point mutation in pancreatic adenocarcinoma pathology specimens pinpoints at the codon 12 of the K-ras proto-oncogene. In addition, the breast cancer susceptibility (*BRCA2*) gene is mutated in up to 10 % of patients with familial (two first-degree relatives affected) pancreatic cancer, while the hereditary non-polyposis colorectal cancer (*hMSH2, hMLH1*) hereditary pancreatitis (*PRSS1*) and Ataxia–telangectasia (*ATM*) and Fanconi anaemia genes may also be implicated in the pathogenesis of the disease.

Diagnosis of pancreatic cancer is usually delayed, due to the absence of characteristic early symptoms and signs. Abdominal discomfort is the primary symptom in 70 % of patients with pancreatic cancer (probably resulting from its location). Discomfort refers to a characteristic dull pain, which begins at the periumbilical area, radiates to the middle of the back, and improves by bending forward. The pancreas may produce a significant amount of insulin, causing symptoms ranging from weakness and chills to diarrhoea and muscle spasms. About 50 % of patients present at the time of diagnosis with jaundice, which causes yellowing of the eyes and skin. It is caused by a build-up of bilirubin, produced in the liver. Painless jaundice is the commonest first symptom in a patient with respectable pancreatic tumour. Other symptoms include weight loss, fatigue, loss of appetite, pruritus, steatorrhoea and glucose intolerance. An enlarged gallbladder or having trouble with breathing, due to blood clots formed in the veins from substances released by cancer cells, which travel to the lungs, may indicate pancreatic cancer. The patient might not be fully aware of the progression of these relatively nonspecific symptoms and the doctor may interpret them as a manifestation of another condition. Altogether, the above analysis underscores the gloomy prospects of this devastating disease.

Surgery, chemotherapy and radiotherapy are the standard therapeutic options for the management of resectable disease, while locally advanced or metastatic patients are mainly treated with gemcitabine, 5-FU in first-line and erlotinib in second line.

21.7 Gastric Cancer

Gastric cancer is the fourth most commonly diagnosed cancer and is the second leading cause of cancer death worldwide. The highest incidences of gastric carcinoma are seen in east Asia with the lowest typically observed in North America. While early diagnosis and encouraging survival rates are common in Japan, the disease is associated with greater mortality in the West, thus representing a significant health burden.

Although the overall incidence of gastric cancer has declined over the past several decades, the site of origin within the stomach has changed. Moreover, patients with gastric cancer typically present with locally advanced or unresectable disease. As a result of these factors, only 30 % of the patients are expected to be alive after 5 years. 5-FU in combination with cisplatin and epirubicin are also commonly used against gastric cancer. Very recently, docetaxel also was formally approved for the systemic treatment of the disease, while there is mounting evidence for a possible role of oxaliplatin and capetibanine in disease management.

Gastric cancer has never been the most attractive indication for drug developers, which is evidenced by the current lack of approved agents for its treatment. This situation has changed over the past few years and numerous targeted therapies are now in phase II trials, which are expected to eventually revolutionize treatment of gastric cancer. Positive results from the recent MAGIC trial showed the feasibility of neoadjuvant therapy and its advantages over adjuvant therapy. The ECF regimen constitutes standard first-line chemotherapy for advanced gastric cancer. Several large-scale randomized clinical trials are currently assessing the newer cytotoxics for this treatment setting but, despite seemingly positive results, only the REAL-2 study is expected to cause a great impact.

21.8 Combination Chemotherapy

Combination chemotherapy is important in most cancer treatment regimens to exploit synergy between chemotherapy drugs. The molecular mechanisms of chemotherapeutic drugs may differ; thus their combination will combat tumour cells more effectively. For example, the advantage using combinations of gemcitabine, topotecan, liposomal doxorubicin, and prolonged oral etoposide with platinum has been attributed to inhibition of DNA synthetic pathways involved in the repair of platinum–DNA adducts. Gemcitabine and cisplatin act synergistically, increase platinum–DNA adduct formation and induce concentration and combination dependent changes in ribonucleotide and deoxyribonucleotide pools in ovarian cancer cell lines (Van Moorsel *et al.*, 2003). The combination of nedaplatin and irinotecan, a topoisomerase I inhibitor, showed synergistic interaction in cell cultures by concurrent exposure to both drugs; on the other hand, sequential exposure to the two drugs led only to additivity (Kanzawa *et al.*, 2001). Neither cisplatin nor carboplatin co-administration affected significantly the pharmacokinetics of etoposide on a randomized crossover clinical trial involving 15 patients. Thus, in this case, the interaction between etoposide and platinum drugs is small and the clinical impact is unlikely to be significant (Thomas *et al.*, 2002). Many drug combinations involving platinum complexes have been explored, but those with taxanes are particularly noteworthy. Paclitaxel in combination with a platinum agent is now accepted as a standard component of first-line treatment for ovarian cancer, and produces improved survival (reviewed by Judson and Kelland, 2000). Recent clinical trials comparing concurrent chemotherapy and radiation with radiation alone in cervical cancer have shown that chemoradiation reduces the risk of death by 30–50 % (Koivusalo *et al.*, 2002). In addition, recently randomized trials show an overall survival advantage of 30 % for cisplatin-based chemotherapy given concurrently with radiation therapy (Berclaz *et al.*, 2002). The molecular mechanism lies in the induction of strand breaks by the ionizing radiation that adds to platinum crosslinks and adds a formidable task to the DNA repair machinery.

21.8.1 Abbreviations of chemotherapy regimens

ABV	doxorubicin, bleomycin and vinblastine
ABVD	doxorubicin, bleomycin, vinblastine, and dacarbazine
ACVBP	doxorubicin, cyclophosphamide, vindesine, bleomycin, prednisone
ADE	cytosine arabinoside,daunorubicin, and etoposide
BAP	bleomycin, actinomycin-D and cisplatin
BEACOPP	bleomycin, etoposide, doxorubicin, cyclophosphamide, vincristine, procarbazine, prednisolone
BEP	bleomycin, etoposide and cisplatin,
CAV	cisplatin, adriamycin, vindesine
CEV	carboplatin, etoposide phosphate and vincristine
CHOP-Bleo	cyclophosphamide, doxorubicin, vincristine, prednisone, bleomycin
CHVP	cyclophosphamide, doxorubicin, teniposide, and prednisone
DC	docetaxel plus cisplatin
EMA-CO	etoposide, high-dose methotrexate with folinic acid, actinomycin D, cyclophosphamide and vincristine
FOLFIRI.3	a new regimen combining 5-fluorouracil, folinic acid and irinotecan
G-FLIP	gemcitabine, 5-fluorouracil, leucovorin and cisplatin
MOPP	mechlorethamine, vincristine, procarbazine and prednisone
MOPPEBVCAD	mechlorethamine, vincristine, procarbazine, prednisone, epidoxirubicin, bleomycin, vinblastine, lomustine, doxorubicin and vindesine
MVP	mitomycin C, vindesine, and cisplatin
OPEC	vincristine, cisplatin, teniposide, and cyclophosphamide
OPEC	vincristine, cisplatin, teniposide, and cyclophosphamide
PCV	procarbazine, lomustine, and vincristine
ProMACE-MOPP	methotrexate, doxorubicin, cyclophosphamide, etoposide, mechlorethamide, vincristine, procarbazine and prednisone
VAB-6	vinblastine, dactinomycin, bleomycin, cyclophosphamide, and cisplatin
VAC	vincristine, dactinomycin, and cyclophosphamide
VAD	vincristine, doxorubicin, and dexamethasone
VC	vinorelbine and cisplatin
VCAP	vindesine. doxorubicin, cyclophosphamide and prednisone
VECP-Bleo	vindesine, epirubicin, cyclophosphamide, prednisone, and bleomycin
VIP	vindesine–ifosfamide–cisplatin

21.9 The Pharmaceutical World of Anticancer Drugs

21.9.1 Annual cost of some treatments – pharmacoeconomics of cancer

Usually it takes 500 million to 1 billion dollars in a 10–15-year time frame for the development of a drug from inception to registration. The success rate is extremely low; only 0.1 % of

Table 21.1   Sales of top selling drugs

Company	Commercial name	Generic name or INN	2006 sales worldwide $million
Roche/Genentech	Mabthera	rituximab	2 769
Novartis	Glivec	imatinib	2500
Roche/Genentech	Avastin	bevacizumab	2243
Roche/Genentech	Herceptin	trastuzumab	2178
Sanofi-Aventis	Taxotere	paclitaxel	2011
Sanofi-Aventis	Eloxatin	oxaliplatin	1955
Eli Lilly & Co	Gemzar	gemcitabine	1335
AstraZeneca	Arimidex	anastrozole	1181
AstraZeneca	Zoladex	goserelin	1004
BMS	Taxol	paclitaxel	747
Novartis	Femara	letrozole	719
Schering AG	Betaferon	interferon beta 1a	694
Merck KgGA	Erbitux	cetuximab	686
Roche/Genentech	Tarceva	erlotinib	533
Roche/Genentech	Xeloda	capecitabine	531
Pierre Fabre	Navelbine	vinorelbine	501
Pfizer	Camptosar	irinotecan	235
Schering Plough	Caelyx (Doxil)	doxorubicin, liposomal	200* (excludes US sales)

the drugs conceived will make it to the market explaining the high cost of the newly developed drugs.

Avastin costs $4400 per treatment taken once or twice per month and is taken as a long-term treatment (one year or longer). ImClone's Erbitux costs $30 000 for 8 weeks of treatment. Genentech's Herceptin costs $3200 per month for a 1–2-year treatment regime. A full treatment of colorectal cancer by Eloxatin (Sanofi) costs $32 000. Sales of top-selling drugs are listed in Table 21.1 and their annual costs, increase or decrease in sales, monthly cost per treatment and indication is shown in Table 21.2. From the data in the following tables it is obvious that the newest, biological treatments are particularly expensive compared to their small molecule counterparts, a fact that has often generated problems as to the inclusion of such molecules to national health systems and their free provision to cancer patients. However, where a biological marker has been associated with the disease indication, it has been apparent that such therapeutics can be of great value to the treatment of individuals and have been particularly profitable for the companies producing them. Although until recently there has been some hesitation as to the use of novel biological treatments by big pharmaceutical industries worldwide, this is increasingly overcome by the success of some of these molecules in the international market.

Table 21.2 Market data for top selling cancer drugs and their indications

Company	Drug	Drug INN	2004 sales worldwide ($ million)	2005 sales worldwide ($ million)	Growth 05–04 (%)	US monthly treatment cost per cycle (2 sqm person)	Indication
Roche/Genentech	Mabthera/Rituxan	rituximab	4298	5154	19.9	15 048	NHL
Roche/Genentech	Avastin	bevacizumab	1490	2465	65.5	4 40 0/8 80 0/8 800	Colorectal/breast/lung
Roche/Genentech	Herceptin	trastuzumab	1639	2464	50.3	3672	Breast
Novartis	Glivec	imatinib mesylate	1667	2200	32.0	2290	CML
Sanofi-Aventis	Taxotere	paclitaxel	1783	2011	12.8	3580	Breast
Sanofi-Aventis	Eloxatin	oxaliplatin	1497	1955	30.6	6424	Colorectal
Eli Lilly & Co	Gemzar	gemcitabine	1214	1335	9.9	3591/1850	Lung/pancreatic
AstraZeneca	Arimidex	anastrozole	811	1181	45.6	250	Breast
AstraZeneca	Zoladex	goserelin	917	1004	9.5		Prostate/breast
Pfizer	Camptosar	irinotecan	554	910	64.3		Colon/rectum
BMS	Taxol	docetaxel	991	747	−24.6	1945	Lung
Merck KgGA	Erbitux	cetuximab	502	686	36.5	7800	Colorectal
Roche/Genentech	Xeloda	capecitabine	433	637	47.0	1500	Colorectal
Roche/Genentech	Tarceva	erlotinib	27	585	2071.5		NSCLC/pancreatic
Novartis	Femara	letrozole	388	536	38.0		Breast
Pierre Fabre	Navelbine	vinorelbine	464	501	8.0		Breast/NSCLC
Pfizer	Aromasin	exemestane	143	247	72.7	5100	Breast
Schering Plough	Caelyx (Doxyl)	doxorubicin, liposomal	182 (excl.US)	206 (excl.US)	13.0	4200	Breast

CML, chromic myelogenous leukaemia; NHL, non-Hodgkin's lymphoma.

Further Reading

Ajani JA. Evolving chemotherapy for advanced gastric cancer. *Oncologist* 2005, **10**(Suppl 3), 49–58.

Antoniou A, Lovegrove RE, Tilney HS, Heriot AG, John TG, Rees M, Tekkis PP, Welsh FK Meta-analysis of clinical outcome after first and second liver resection for colorectal metastases. *Surgery* 2007, **141**(1), 9–18.

Bhoola S, Hoskins WJ. Diagnosis and management of epithelial ovarian cancer. *Obstet Gynecol* 2006, **107**(6), 1399–410.

Brand TC, Tolcher AW. Management of high risk metastatic prostate cancer: the case for novel therapies. *J Urol* 2006, **176**(6 Pt 2), S76–80; discussion S81–2.

Colucci G, Guiliani F, Gebbia V, *et al.* Gemcitabine alone or with cisplatin for the treatment of patients with locally advanced and/or metastatic pancreatic carcinoma: a prospective, randomized phase III study of the Gruppo Oncologia dell'Italia Meridionale. *Cancer* 2002, **94**(4), 902–10.

Globocan 2002, Cancer Incidence, Mortality & Prevalence, International Agency for Research on Cancer (IARC): http://www-dep.iarc.fr/

Nahta R, Hortobagyi GN, Esteva FJ. Novel pharmacological approaches in the treatment of breast cancer. *Exp Opin Investig Drugs* 2003, **12**(6), 909–21.

Saif MW. Pancreatic cancer: highlights from the 42nd annual meeting of the American Society of Clinical Oncology, 2006. *JOP* 2006, **7**(4), 337–48.

Saltz LB. Metastatic colorectal cancer: is there one standard approach? *Oncology* 2005, **19**(9), 1147–54.

Saunders M, Iveson T. Management of advanced colorectal cancer: state of the art. *Br J Cancer* 2006, **95**(2), 131–8.

Stinchcombe TE, Fried D, Morris DE, Socinski MA. Combined modality therapy for stage III non-small cell lung cancer. *Oncologist* 2006, **11**(7), 809–23. Erratum in *Oncologist*. 2006, **11**(8), 958.

Terstriep S, Grothey A. First- and second-line therapy of metastatic colorectal cancer. *Exp Rev Anticancer Ther* 2006, **6**(6), 921–30.

Thomas HD, Porter DJ, Bartelini I, *et al.* Randomised cross-over clinical trial to study potential pharmacokinetic interactions between cisplatin or carboplatin and etoposide. *Br J Clin Pharmacol* 2002, **53**(1), 83–91.

Thomson BN, Banting SW, Gibbs P. Pancreatic cancer – current management. *Aust Fam Physician* 2006, **35**(4), 212–7.

References

Berclaz G, Gerber E, Beer K, *et al.* Long-term follow-up of concurrent radiotherapy and chemotherapy for locally advanced cervical cancer: 12-year survival after radiochemotherapy. *Int J Oncol* 2002, **20**(6), 1313–18.

Evans DB, Wolff RA, Crane CH, Pisters PW. In: *UICC (International Union Against Cancer) Manual of Clinical Oncology*, 8th edition, edited by RE Pollock, JH Doroshow, D Khayat, A Nakao and B O'Sullivan. Wiley-Liss, NJ, 2004.

Judson I, Kelland LR. New developments and approaches in the platinum arena. *Drugs* 2000, **59**, 29–36.

Kanzawa F, Koizumi F, Koh Y, *et al. In vitro* synergistic interactions between the cisplatin analogue nedaplatin and the DNA topoisomerase I inhibitor irinotecan and the mechanism of this interaction. *Clin Cancer Res* 2001, **7**(1), 202–9.

Koivusalo R, Krausz E, Ruotsalainen P, *et al.* Chemoradiation of cervical cancer cells: Targeting human papillomavirus E6 and p53 leads to either augmented or attenuated apoptosis depending on the platinum carrier ligand. *Cancer Res* 2002, **62**(24), 7364–71.

Van Moorsel CJA, Smid K, Voorn DA, *et al.* Effect of gemcitabine and cis-platinum combinations on ribonucleotide and deoxyribonucleotide pools in ovarian cancer cell lines. *Int J Oncol* 2003, **22**, 201–207.

22

Future Trends in Cancer Therapeutics

Sotiris Missailidis

22.1 Introduction

Our knowledge about cancer, or rather cancers, as they represent a variety of different diseases, has changed dramatically over the last few years and, with it, our approach to diagnosis and therapy. A lot has become known about the causes, onset and spread of the disease, the different features of individual cancers and their genetic origin. With the rapid development of genomics, a number of genes have been identified that increase the predisposition for cancer, such as *BRCA1* and *BRCA2* for breast cancer. Furthermore, a number of genes have been identified as responsible for the promotion and development of cancer, such as *HER2/neu*, responsible for the overproduction of the HER2 protein, believed to lead to continuous growth-promoting signals being transmitted to affected cells, as a key part of the development of cancer. Whilst the former genes have led to the prediction of risk of cancer, the later has been responsible for the development of Herceptin, one of the most successful drugs of recent years.

The best therapy against cancer is prevention. Our increasing knowledge on the aetiology of cancer, the links between cancer and diet, smoking or alcohol, as exemplified by the 2007 WCRF/AICR report on 'Food, Nutrition, Physical Activity and the Prevention of Cancer', environmental and socioeconomic factors, inflammation and infection, can offer us significant potential to delay or postpone indefinitely the onset of the disease, or to avoid it altogether. With improved diet and exercise, our possibilities of avoiding the disease are significantly greater. The link between tobacco smoking and cancer has now become clear and reduction in smoking through public education could lead to reduction in lung cancer incidence. Similarly, the link between infection and cancer has led to better screening levels, such as those attained with the use of the smear (PAP) test for human papillomavirus (HPV) and cervical cancer.

Anticancer Therapeutics   Edited by Sotiris Missailidis
© 2008 John Wiley & Sons, Ltd

HPV is sexually transmitted with a prevalence of about 7 % in developed and 15 per cent in developing countries and has been clearly associated with cervical cancer. Supported by the success of the screening program, a vaccine for the prevention of HPV infection has been developed and FDA approved. There is now the hope that HPV vaccination may provide an opportunity to profoundly affect cervical cancer incidence worldwide.

Another significant change in our fight against cancer that is continuously moving in our favour is the technological advancement in medical imaging. In the past a tumour could only be imaged at 3 mm size and above, at a stage that the cancer had often already become metastatic; current technological developments, such as spiral computerized tomography, can offer significant advancements in the diagnosis of tumours, such as those of the lungs, at an earlier stage and smaller size. Whilst this would not prevent cancer, it would allow an earlier therapeutic intervention that has significantly higher chances of success even with today's standard treatments.

Yet, the topic of this book has been 'Anticancer Therapeutics' and as such, it has been focusing on the actual treatment of cancer, based on molecular therapeutics. As such, this chapter will close the book with some future trends that appear probably clearly throughout the book, where the various authors have referred to molecules in clinical and preclinical development stage. There is, clearly, a continuous effort to develop novel therapeutic molecules, based on the various approaches encountered at the early chapters of this book. These may attempt to improve on previous molecules, find new ways to target previous targets, or generate novel agents that are targeting new markers identified from large proteomics efforts, in a manner often not dissimilar from that used for a number of years. However, new approaches have also been developing in parallel, delivering agents both at the clinic and at early experimental stages that may form the therapeutics of the next decades to come. The two trends in cancer therapeutic development that I will mostly refer to, apart from the above mentioned developments in prophylactic, diagnostic or imaging techniques, are the areas of personalized medicine and delivery systems. These two areas will be considered briefly in the sections below.

22.2 Personalized Medicines

What in the 2005 report of the Royal Society 'Personalised Medicines: Hopes and Realities' had been described as a secondary alternative for the pharmaceutical industry, only 3 years down the line it is already becoming the standard in consideration of novel drugs for development. In this report, it was speculated that industry will continue to favour drug development that avoids the effect of genetic variation, whilst, where this is not possible, drugs will be developed with associated diagnostic genetic tests to identify the subgroup where such drugs would be effective. Thus, such molecules would be strictly associated with the identification and validation of a biomarker that could be subsequently tested for. It would, of course, be ideal if the concept of the traditional 'one disease–one drug' approach were a viable one. However, there is no such thing as an average person and the statistical analysis of percentages is really not adequate in selecting and treating people, often resulting in overmedication or treatment of patients with completely ineffective medicines. Thus, with the realization that generic drugs, even though they are still the most commonly used in the clinic, only work on perhaps 50 % of patients, and the recent successes of molecules like Glivec and Herceptin, the pharmaceutical industry is already changing its focus. The increasing knowledge and development in genetics, coupled to pharmacological advancements, has given rise to the disciplines of pharmacogenetics and pharmacogenomics, and they are expected to play an important role in future pharmaceutical developments in the area of cancer.

The promise of these scientific and pharmaceutical areas of development is to deliver more specific drugs, targeted to specific groups of people, with the knowledge and certainty that these drugs will be effective against these groups of people, offering higher efficacy and reduced side effects. Thus, each patient will be genetically profiled to detect which variant of cancer they are suffering from and which therapies would work best for them. They will then be treated by carefully selecting the appropriate drugs, targeted at a very specific subset of disease. The efficacy of treatment is expected to be dramatically improved, and the risk of undesirable side effects greatly reduced. It is in fact becoming a trait in pharmaceutical industry to consider in-licensing or development of new drugs only when these are associated with a biomarker test that will automatically identify the subset where the particular drug would be most effective. Furthermore, there are extensive clinical studies to identify the genetic population subsets that some of the currently available medicines may be more effective, thus increasing a more targeted use of these medicines. Thus, in childhood acute lymphocytic leukaemia, the children with *AML1* mutations are treated differently from those with cells bearing the Philadelphia translocation, significantly increasing the success in curing this disease. Still, one drug alone would almost certainly not be effective against even the particular population subset for which the drug has been developed. This has become apparent by the fact that Glivec has often been very successful in treating patients initially, but then resistance arises in the tumour, making the drug ineffective. Thus, a combination of molecules would have to be used at each instance to ensure that successful treatment is effected, avoiding resistance through mutations or genetic variations in tumour cells.

The development of personalized medicines has a number of clear advantages and disadvantages in terms of worldwide application. Though they are seemingly more costly in terms of development, needing assay developments associated with the drug and appropriate testing, they are actually poised to reduce costs by reduction of failures of drugs in clinical trials, as only carefully selected population subgroups would participate, and would ensure a higher number of molecules coming to the market. On the other hand, such drugs, often coming from biologics such as antibodies or oligonucleotide therapeutics, are expensive for developing countries that are still basing treatments on cheap, widely available generic drugs. This is set to further increase the gap between treatments available in the developed from the developing world.

The other issue associated with the focus of the industry to personalized medicines is the fact that only subgroups of cancers that are considered economically viable would be specifically targeted, leaving cancers that affect smaller numbers of the population out of consideration. In fact, some forms of leukaemia have already passed into the category of 'orphan diseases', which large pharmaceutical industries regard as non-viable targets for commercial success since so few people suffer from them. Although this appears to be a dilemma that may affect the positive outlook on personalized medicines, it has been shown that this is not necessarily the case, as these diseases are often associated with specific disease markers and they can offer smaller *niche* markets that could be viable for and appeal to small or medium biotech companies and there have already been successful examples of such companies specializing on the development of 'orphan' drugs.

22.3 Delivery Systems

Small molecular therapeutics, such as cisplatin, have been at the forefront of the fight against cancer for the past many years. Many variations and optimizations have been made to the original cisplatin, resulting as we saw in Chapter 5 in several second- and third-generation compounds. Yet, these compounds have only marginally improved on the original molecule and further development seems to have reached a limit. Thus, very little is now expected with regards to the development of new platinum based compounds. The major efforts for these and other potent generic compounds is now shifted from the attempt to improve on their structure, therapeutic index or pharmacokinetic properties, to the attempt to deliver them specifically to cancer cells, thus exerting their cytotoxic action at the site of need, increasing their potency and minimizing their side effects. One such example we have already seen in Chapter 5, with the development of Lipoplatin and Lipoxal, liposomal formulations of known platinum compounds, and the liposomal formulations of a peptide vaccine (Chapter 15) and gene therapy (Chapters 15 and 16). Liposomes have the ability to encapsulate a hydrophilic molecule inside a hydrophobic membrane, or have hydrophobic molecules dissolved into the membrane. The lipid bilayer of the liposome can subsequently fuse with other bilayers, such as cell membranes, thus delivering the liposomal content. A number of variations on the liposome design can effect other beneficial delivery characteristics.

Liposomes are but one of the newly developed delivery systems that are in various clinical trials and are ready to enter the clinic. Other such formulations that have been designed for encapsulation of toxic molecules, or compounds with undesired pharmacokinetic properties include dentrimers, cucurbit[n]urils, cyclodextrins or even carbon nanotubes.

Already in the market, but with a great potential for further future development are more traditional delivery approaches of potential therapeutic agents, based on the targeting and recognition ability of molecules such as antibodies (see Chapter 15) and aptamers (see Chapter 19).

Radiopharmaceuticals, although they have not been covered in this book, have traditionally only been used when there is a particularly high uptake by a particular part of the body, such as ^{131}I by the thyroid. Otherwise, the use of a radiopharmaceutical would be dangerous, often causing more damage than therapeutic effect, and the majority of radiotherapy approaches are focused on beam therapy that can now be focused quite accurately to the cancer site. However, with the use of appropriate targeting agents, such as antibodies and aptamers, a new generation of targeted radiopharmaceuticals has emerged, with molecules already in the market, such as Zevalin, and others following up in clinical development. Aptamers have also been used as delivery agents for radiotherapy to the cancer site and they are currently at the preclinical stage. With the development of coupling techniques to such delivery agents has come the use of new chelators and different metals that could emit alpha or beta particles for cancer radiotherapy and can now be directed specifically at the tumour site. Similarly, toxins or chemotherapy agents that have demonstrated high cytotoxicity but little specificity can be coupled directly to such molecules and delivered more specifically to the tumour site, creating targeted chemotherapy agents.

Aptamers and antibodies can equally be coupled to the liposomal and other delivery formulations described above, to improve even further their delivery potential, making them deliver their therapeutic load specifically to cancer cells.

Another such use of antibodies' targeting capabilities have been utilized in the development of prodrugs. Prodrugs, molecules that are inactive themselves but get activated in the body, either through liver metabolism or by differences between the tumour site and the normal tissue, such as hypoxia and differences in pH, are already available in the market (see Chapter 9 for some examples). However, more complex approaches have been adopted, such as ADEPT (antibody-directed enzyme prodrug therapy), where an antibody against a specific tumour marker is linked to an enzyme and injected to the blood, resulting in selective binding of the enzyme to the tumour. Subsequently, a prodrug is administered into the circulation, which is converted to the active cytotoxic drug by the enzyme, only at the tumour site. Selectivity is achieved by the tumour specificity of the antibody and by delaying the prodrug administration until there is a large differential between tumour and normal tissue enzyme levels. The approach has shown some beneficial activity in preclinical and early clinical studies. Several other strategies have followed the antibody directed enzyme prodrug therapy, including gene-directed enzyme prodrug therapy (GDEPT), virus-directed enzyme prodrug therapy (VDEPT), polymer-directed enzyme prodrug therapy (PDEPT), lectin-directed enzyme-activated prodrug therapy (LEAPT) and clostridial-directed enzyme prodrug therapy (CDEPT). All these approaches aim to deliver and activate drugs specifically at the tumour site, so as to confer higher specificity and improved properties to previously identified potent agents.

22.4 Closing Remarks

I would like to close this book with a positive outlook on the international efforts against cancer. Thousands of scientists and billions of dollars are devoted annually to the battle against cancer. Education of the public, improved lifestyles, actions of charities, governments, scientists and medical professionals as well as the general public, constantly improve our chances to prevent or treat cancer. Whilst this group of diseases has been elusive and the outcome is still not as positive as one would like to believe, great improvements in diagnosis, imaging and therapy are happening daily. A number of traditional and novel approaches are continuously being employed in the development of antitumour agents and their mode of action is becoming better understood, assisting scientific and medical professionals to make better choices for their use. A number of cancers are already treatable, and many others are now treated as chronic diseases, so much so that patients will succumb to old age or other conditions before cancer can claim them. Furthermore, new directions in chemotherapeutic approaches, coupled with technological advancements in imaging, radiotherapy and surgery, constantly improve treatment, making it possible to take a positive outlook on these diseases, which are conquered slowly but steadily, one by one, or delayed enough to offer previously untreatable patients the opportunity to live out their lives. Whilst we are a long way from eradicating this group of diseases, we are getting closer every day, one step at a time.

Index

17AAG (tanespimycin)   241, 273
3-aminobenzamide   275, 276
3-aminobenzamide   33
5-(aryltio)quinazolinone   43
9-aminocamptothecin   315
9-aminocamptothecin glucuronide   315
Abdominal pain   94
ABT-888   275
acetaminophen (paracetamol)   291
Acolbifene   168
Acquired immune deficiency syndrome (AIDS)   305
Acridine orange   211
Acromegaly   179
Actinic keratosis   210
Actinomyces antibioticus   111
Actinomycetes   9
Actinomycin (dactinomycin)   53, 54, 111, 112, 380, 383
Adenine deaminase (ADA)   106, 107
Adenocarcinoma   6, 100, 113, 251, 357, 361, 380, 381
Adenoma   209
Adenosine deaminase deficiency   305
Adenosine deaminase   53
Adenovirus(es)   305, 306, 307, 308, 309, 310, 313, 314, 315, 316
ADPM01   211
ADPM06   211
Adrenal cortex (cancer of)   159
Adriamycin   134, 292, 383
Advexin   309
AEG35156   324
AG14361   275, 276
AG14447   275, 276
Alkaloid   6, 9, 18

Alkane sulfonates / Alkylsulfonates   54, 143
Alkylating agents   54, 97, 133, 136, 137, 138, 140, 143, 144, 146, 149, 150, 151, 287
Allopurinol   105
Allovectin-7   309
ALNVSP01   328
Alopecia   63, 84, 85, 106, 112, 116, 120, 124, 138, 139, 150
Altretamine   140
Alzheimer's disease   163
AMD473 (also picoplatin)   66
Amenorrhoea   142
Amifostine   57
Aminopterin   91
Amnesia   146
Amyloidosis   136, 137
Anaemia   84, 105, 113, 143, 245, 246, 254, 255
Anal (cancer)   97, 99, 113
Anaphylaxis   150
Anastrozole (Arimidex)   169, 171, 173, 378, 384, 385
Androgen   32, 161, 172, 378
Androst-4-ene-3,17-dione   32
Angina pectoris   291
Angioedema   291
Angiogenesis   58, 225, 226, 227, 231, 232, 233, 235, 237, 239, 246, 247, 249, 250, 253, 254, 255, 285, 288, 289, 291, 293, 326, 327, 332, 339, 349, 351, 355, 357, 360, 361
Angiogenin   339
Angiotensin-converting Enzyme (ACE) (inhibitors)   358, 359, 361
Angiozyme   327, 328
ANI (4-amino-1,8-naphthalimide)   275, 276

Anorexia (loss of appetite)   68, 113, 116, 119,
    150, 159, 381
Anthracycline(s)   53, 54, 82, 85, 100, 113, 118,
    120, 121, 268, 269, 292, 378
Antiangiogenesis (drug)   58
Antiapoptotic agents   57
Antiemetics   57, 62, 63
Antifolates   53
antihormone therapy   162
Antimetabolites   53, 54, 91
Antimicrotubule agents   53, 54, 79, 82, 83
Antimitotic agents   79, 83
Anti-oestrogen(s)   162, 163, 166, 168, 169, 172
Antioxidant   11, 13, 57
Antisense (ASO)   51, 73, 194, 252, 317, 318,
    319, 320, 321, 322, 323, 324, 325, 326, 329
α1-antitrypsin deficiency   305
Antitumour antibiotics   53, 54, 111
AP12009   324
Apaziquone (EO9)   113, 115
APC-8015   297
Aphasia   146
Aplastic anaemia   137
Aptamer(s)   23, 24, 25, 26, 27, 51, 331, 332, 333,
    334, 335, 336, 337, 338, 339, 340, 341, 390
Aromatase inhibitors   159, 160, 169, 170, 171,
    172, 173, 175, 176
Aromatase   32
Aroplatin   69, 70
Arrhythmia   291
Arthralgia   172
Arthritis   172
Arzoxifene   167, 168
AS-1411   333, 334
L-asparaginase   105
Aspirin   353, 354, 355, 356, 357
Asthenia   55, 60, 63, 152, 168
Asthma   324
Astrocytoma   61, 151, 305
AT7519   236
Ataxia   142, 146, 147
Ataxia−telangectasia (ATM)   381
Atherosclerosis   166, 336
Atypical hyperplasia   176
Ayurvedic medicine   4, 10
Azacitidine   53, 102
Azinomycin   37
Azoospermia   142
Bacteriochlorins   193, 204, 208
Baculovirus IAP repeat domain (BIR3)   41, 43
Barrett's oesophagus   200, 207, 226
Basal cell carcinoma   97, 209, 372

BBR3464   66, 67
Beh,cet's syndrome   137
Bendamustine   136, 137
Benzamide(s)   275
Benzimidazole-4-carboxamide   33
Benzvix   188, 197, 198
Bevacizumab (Avastin)   53, 233, 234, 239, 288,
    289, 377, 379, 384, 385
BI   2536 236
Bicalutamide   379
Biliary tract (cancer of)   99, 100
Bilirubin   63, 381
Biphosphonates   379
Biricodar   270, 271
Bladder cancer   50, 55, 64, 92, 99, 100, 113, 115,
    119, 137, 140, 197, 198, 309, 339
Bleeding   84, 167
Blenoxane   115
Bleomycin   9, 53, 54, 84, 115, 116, 117, 118,
    134, 150, 383
Bloating   180
BLP25   298
Bone (cancer/metastasis)   63, 64
Bone marrow toxicity   63, 97, 112, 135, 145, 146,
    172
Bowel cancer   49, 97, 178
Bowen's disease   198, 210
Brain cancer   92, 125, 137, 139, 141, 145, 146,
    147, 152, 197, 207, 305, 308, 309, 315
Brain swelling   146
Breast cancer   7, 13, 22, 49, 50, 55, 82, 83, 84,
    85, 86, 87, 92, 97, 99, 100, 101, 103, 106,
    113, 115, 116, 119, 120, 121, 122, 125,
    134, 136, 137, 138, 139, 146, 159, 160,
    161, 162, 163, 164, 165, 166, 167, 168,
    169, 170, 171, 172, 173, 174, 175, 176,
    197, 199, 203, 205, 207, 223, 230, 231,
    233, 235, 236, 237, 238, 239, 241, 245,
    249, 250, 251, 253, 264, 268, 275, 284,
    288, 289, 291, 299, 305, 309, 325, 326,
    327, 339, 340, 341, 355, 357, 361, 365,
    371, 372, 373, 374, 378, 381, 385, 387
Bronchospasm   291
Bruising   84
Busulfan   54, 143, 144
O6-(4-bromothenyl) guanine
    (lomeguatrib/PaTrin-2)   276
O6-benzylguanine (O6-BG)   276, 277

2′-cyano-2′-deoxyarabinofuranosylcytosine
    (CNDAC)   104

2-Chloroethyl-3-sarcosinamide-1-nitrosourea (SarCNU)   149
5-chloro-2,4-dihydroxypyridine (CDHP)   102
Caffeine   274
CALAA-01   328
Campath (alemtuzumab)   287, 288
cAMP-dependent protein kinase   72
cAMP-specific phosphodiesterase   72
Camptosar   7
Camptotheca acuminata   6, 12, 123
Camptothecin(s)   6, 7, 11, 12, 53, 54, 68, 123, 124, 275
Candesartan   361, 362
Capecitabine   53, 92, 99, 100, 101, 113, 236, 239, 325, 378, 379, 382
Carbamazepine   62
Carboplatin   53, 54, 56, 57, 61, 62, 65, 71, 141, 147, 236, 239, 313, 326, 358, 377, 380, 382, 383
Carcinoembryonic antigen (CEA)   284, 306
Carcinogenesis   13
Carcinoid syndrome   176, 179, 362
Carcinomatous meningitis   92, 101
Cardiac dysfunction   138, 172
Cardiomyopathy   119
Cardiotoxicity   41, 60, 63, 119, 120, 121
Cardiovascular disease   13, 160, 174
Carmustine   54, 145, 146, 147, 276
Castleman's disease   351
Catharanthus roseus   10, 11, 82
CB3717   42
CD20   284, 288, 289, 290, 291
CD33   288, 289
CD44   311
CD52   287, 288
cDDP   64
Celebrex   355
Celecoxib   255, 357, 358
CEP-6800   275, 276
Cerebral oedema   146
Cerebrovascular accident   164, 172
Cerepro   309
Cervarix   298
Cervix (cancer of)/cervical carcinoma   50, 55, 116, 124, 137, 139, 194, 197, 199, 245, 255, 315, 351, 361, 372, 382, 387, 388
Cetrorelix   341
Cetuximab (Erbitux)   227, 239, 284, 288, 289, 290
Cetuximab (see also Erbitux)   53, 384
CG7870   315
CGS 16949A   32

Chemical ecology   9
Chemoenzymatic   12
Chemoprevention   13
Chemopreventive   13
Chills   94
Chlamydia psittaci   362
Chlorambucil (Leukeran)   135
Chloramphenicol   93
Chlorin e6   188, 191, 192, 193, 194, 196, 205, 208
Chlorin   188, 191, 192, 193, 194, 196, 197, 204, 205, 206, 208, 209
Chlormethine   133, 134, 135
Cholesterol   32
CHOP therapy (cyclophosphamide, doxorubicin, vincristine and prednisone)   85
Chorioadenoma destrues   92
Choriocarcinoma (gestational)   92
Churg–Strauss syndrome   137
Cimetidine   98
Cirrhosis   94
Cisplatin   37, 53, 54, 55, 56, 57, 58, 59, 60, 61, 62, 64, 65, 66, 67, 68, 70, 71, 72, 73, 82, 95, 101, 113, 140, 147, 268, 275, 292, 306, 313, 314, 315, 316, 326, 377, 380, 382, 383, 389
Cladribine   53, 107
Clofarabine   108
Cloretazine   148, 149
Codeine   353
Colon cancer   7, 50, 55, 59, 63, 71, 86, 100, 116, 145, 162, 177, 237, 239, 248, 288, 289, 315, 326, 379
Colonic adenoma   356
Colorectal cancer   62, 63, 69, 70, 85, 95, 98, 99, 100, 102, 103, 115, 122, 124, 147, 149, 160, 223, 233, 235, 238, 239, 288, 289, 290, 295, 299, 305, 355, 356, 357, 361, 371, 372, 373, 379, 381, 384, 385
    Non-polyposis   381
Congenital abnormalities   94
Constipation   84, 152
Convulsions   146
Cordycepin   107
Cordyceps militaris   107
Cortisone   6
Cough   94
CP-4055   101, 102
CPT-11   7
Creatinine   57
Crohn's disease   162
Cryptogenic fibrosing alveolitis   137
Curcumin   13

Cyanophenyl 33
Cyclin-dependent kinases 235, 236
Cyclodextrin 323, 390
Cyclophosphamide 54, 82, 99, 134, 135, 136, 137, 138, 139, 140, 143, 144, 231, 380, 383
Cyclosporine 269, 270, 310
Cystemustine 149
cystic fibrosis 305
Cytarabine 53, 101, 102
Cytochrome P-450 32
Cytoprotective 57
Cytosine arabinose 54, 383

10-deacetylbaccatin III 9
17DMAG 241
Dacarbazine 84, 134, 149, 150, 151, 152, 276, 383
Dactinomycin (see actinomycin)
Dasatinib 230
Daunorubicin 53, 101, 118, 119, 120, 383
Decitabine 103
Deep venous thrombosis 164
Dendritic cells 292, 293, 296, 297, 298, 311, 351, 353, 362
Dentrimer(s) 191, 323, 390
Deoxythymidylate (dTMP) 41, 42
Deoxyuridylate (dUMP) 41, 42, 43
Depression 142
Depressive psychosis 140
Dexamethasone 101, 136, 361, 383
Dexniguldipine 270, 271
Dexverapamil 270
Dezaguanine 109
DHFR inhibitors 53
Diabetes mellitus 10
Diarrhoea 60, 63, 84, 94, 95, 97, 98, 113, 119, 124, 142, 143, 150, 152, 179
Diazoxid 179
Diethylstilboestrol (DES) 172
Difluorodeoxycytidine 100
Dihydrofolate (FH2) 41, 92
Dihydrofolate reductase (DHFR) 92, 95, 96
Dihydropyrimidine dehydrogenase enzyme (DPD) 98, 102, 103
Dipalmitoyl phosphatidyl glycerol (DPPG) 57, 58
Diphenhydramine 291
Diterpenoids 9
Diuretics 57
Dizziness 94

Docetaxel 53, 61, 79, 80, 81, 82, 86, 87, 95, 100, 101, 236, 239, 289, 325, 326, 337, 338, 358, 377, 378, 379, 382, 383, 385
Dose-limiting toxicity (DLT) 62, 63, 69
Doxifluridine 103
Doxorubicin 9, 53, 68, 70, 72, 82, 84, 96, 99, 113, 118, 119, 120, 121, 126, 134, 150, 151, 231, 275, 287, 378, 382, 383, 384, 385
Doxycycline 362, 363
Droloxifene 165
Drowsiness 140
Dukes' C carcinoma 98, 100
Duodenum (cancer of) 176, 177
Dysarthria 147
Dyspareunia 164
Dyspnoea 94, 291
Dysuria 139

Ecteinascidia turbinata 5, 121
Ecteinascidin-743 5
Elacridar 269, 271, 272
EM-800 168
Emetogenesis 65
Endobronchial (cancer) 197
Endocrine therapy 162
Endometriosis 22, 160, 174
Endometrium (cancer of)/Endometrial carcinoma 137, 159, 160, 163, 164, 165, 167, 172, 176
Endothelial growth factor receptor (EGFR) 51, 53, 284, 288, 289, 290
Enediyne 9
Enterocolitis 98
Eosin 187
Epidemiological 13
Epidermal growth factor (receptor) (EGF/EGFR) 226, 227, 228, 229, 239, 247, 288, 289, 290, 377
Epidoxirubicin 383
Epigallocatechin-3-gallate 13
Epipodophyllotoxin 8
Epirubicin 118, 121, 231, 269, 272, 378, 382, 383
Epothilone(s) 53, 82
ERA-923 167, 168, 241
ErbB-2 22
Erbitux 53, 377, 384, 385
ERCC-1 57, 62
Erlotinib (Tarceva) 227, 228, 229, 357, 377, 381, 384, 385
Erythema multiforme 94, 205
Erythropoietin (EPO) 57, 246, 247, 254, 255, 313

Escherichia coli 22
Estramustine phosphate 61, 82, 289, 379
Estrogen 32
Estrogen receptor 32
Ethinyloestradiol (EE2) 172
Ethnobotany 7
Ethnomedicine 7, 8, 9
Ethnopharmacological 8
Ethnopharmacology 7
Ethylenimines 140
Etioporphyrin 209
Etopophos 126
Etoposide 8, 53, 57, 61, 72, 101, 125, 126, 134, 275, 314, 316, 377, 382, 383
Everolimus 181
Ewing's sarcoma 111, 125
Exemestane 171, 173, 378, 385
exfoliative dermatitis 94
Extravasation 58

5-fluoro-2′-deoxyuridine-5′-phosphate (FUDRMP) 103
5-Fluoro-2-deoxycytidine 99
5-fluorocytosine 315
5-fluorodeoxyuridine monophosphate (5-F-dUMP) 41, 42, 97, 98
5-fluorouracil (5-FU) 41, 42, 53, 59, 62, 72, 82, 91, 95, 97, 98, 99, 100, 102, 103, 113, 233, 290, 298, 306, 313, 315, 336, 379, 383
5-fluoroxyuridine monophosphate (F-UMP) 97, 98, 315
Factor inhibiting HIF-1 (FIH-1) 247, 248
Familial hypercholesterolaemia 305
Fanconi anaemia 381
Fatigue 94, 152, 168, 291, 381
Fetal death 94
Fever 94, 143, 152, 291
Fibroblast growth factor (FGF) 338
Fibrosis 22, 94
Flavonoids 18
Flavopiridol 235
Floxuridine 103
Fludarabine 53, 92, 106, 287
Fluoropyrimidines 54, 99
Fluorouridine (FUR) 98
Fluorouridine diphosphate (FDP) 98
Fluorouridine triphosphate (FUTP) 98
Flutamide 379
Folate antagonists 92
Folate(s) 91, 92, 93
Folic acid 91, 92

Folinic acid (see also Leucovorin) 91, 95, 98, 99, 100, 102, 383
Formestane 171
Fotemustine 145, 147
Fullerenes 211
Fulvestrant 169, 170, 173
Furosemide 57

G207 315
Gabapentin 62
Ganciclovir 306, 310
Gapmers 318
Gardasil 297, 298, 299
Gastric cancer 59, 62, 63, 82, 97, 99, 100, 103, 113, 115, 119, 197, 351, 352, 354, 361, 381, 382
Gastrin 176, 179
Gastrointestinal bleeding 355
Gastrointestinal cancer/malignancy 103, 123, 125, 197, 328
Gastrointestinal toxicity 55, 60, 62, 63, 112, 124, 126, 140
Gaucher disease 305
Gefitinib (Iressa) 227, 228, 229, 239
Geldanamycin 241, 253
Gelonin 338
Gemcitabine 53, 54, 57, 59, 60, 61, 62, 100, 101, 306, 326, 357, 358, 377, 378, 381, 382, 383, 384, 385
Gemtuzumab ozogamicin (mylotarg) 288, 289
Gendicine 309
Gene therapy 51, 73, 252, 283, 292, 294, 305, 306, 307, 308, 310, 312, 313, 389
Genistein 13
Gestational trophoblastic neoplasia 112, 125, 137
GL-331 126
Glioblastoma 55, 61, 147, 151, 233, 305, 339
Glioma 107, 146, 147, 149, 150, 152, 254, 255, 276, 305, 308
Glivec (Gleevec, Imatinib) 230, 235, 238, 388, 389
Glomerular kidney disease/glomerulonephritis 137, 336, 337
Glucagonoma syndrome 179
Glutathione 55, 62, 72, 84
Glycinamide ribonucleotide formyltransferase (GARFT) 109
Golgi vesicular membrane-golvesin 72
Gonadal suppression 138
Goserelin 22, 379
GPI 15427 275

Granulocytopenia   10
GTI-2040   324, 325
GTI-2501   324, 325
GW5638   170
Gynaecomastia   143

4-hydroxytamoxifen   241
Haemangioma   197, 208
Haematological toxicity   62, 121, 147, 152
Haematopoiesis   94
Haematoporphyrin   192, 193, 198
Haematuria   139
Haemorrhage   105, 150
Haemorrhagic cystitis   123, 137, 138, 139
Hallucinations   142
Hammerhead ribozymes   71
Headache   142, 146, 150, 152, 291
Head-and-neck cancer   55, 82, 92, 95, 97, 99,
        100, 103, 115, 197, 236, 253, 254, 288,
        289, 305, 309, 315, 316
Heat-shock protein 90 (HSP90)   253
Helicobacter pylori   351, 352
Hematoporphyrin   188
Hepatic dysfunction   81
Hepatocellular cancer/adenocarcinoma   59, 96,
        99, 100, 103, 236, 253
Hepatoma   125
Hepatotoxicity   60, 63, 94, 100, 105, 146, 147,
        150
Her2/neu   53, 284, 288, 291, 327, 374, 387
Herceptin (see also trastuzumab)   27, 53, 231,
        239, 284, 285, 288, 291, 292, 374, 387, 388
Herpes simplex   353
Herzyme   327, 328
Hexamethylmelamine   140
Hexvix   188, 197, 198
High mobility group (HMG) non-histone proteins
        57
Hodgkin's disease/lymphoma   84, 85, 101, 104,
        115, 119, 125, 134, 136, 137, 141, 145,
        146, 150, 237, 351
Hormone therapies   159, 170, 174, 175, 378
Hormone(s)   159
HPPH   188, 197, 208, 239, 241
Human epidermal growth factor receptor 2
        −(HER2)   231, 232, 247, 249, 340
human immunodeficiency virus (HIV)   324
Hycamtin   7
Hydatidiform mole   92
Hydroporphyrin   190, 204, 208
Hydroxyurea   143

Hyperglycaemia   148
Hyperpigmentation   106, 143, 198, 209
Hypertension   362
Hyperuricaemia   105
Hyperuricosuria   105
Hypoglycaemia   179
Hypotension   126, 146, 291
Hypoxanthine-guanine phosphoribosyltransferase
        (HGPRT)   104, 107
Hypoxia   94, 232, 245, 246, 247, 248, 249, 250,
        251, 252, 253, 254, 255
Hypoxia-inducible factor 1 (HIF-1)   232, 241,
        246, 247, 248, 249, 250, 251, 252, 253,
        254, 255

Ibritumomab tiuxetan (Zevalin)   27, 288, 289,
        290, 291, 390
ICS-283   328
Idarubicin   118, 121
Idiopathic thrombocytopenic purpura   137
Idoxifene   165, 166
IFN-Beta Gene   309
Ifosfamide   57, 135, 136, 138, 139, 140, 313, 383
Imatinib (Glivec)   181, 289, 384
Immunosuppression   84, 112
Indole alkaloids   9
Indole-3-carbinol   273
Infertility   142
INGN   225 309
INGN241   309
Inositol polyphosphates   73
Insomnia   142
Insulin resistance   160
Insulinoma(s)   176
Interferon beta 1a (Betaferon)   384
Interferon-alpha (INF-α)   143, 177, 180, 181, 292
Interferon-beta (INF-β)   147, 292, 306, 309
Interferon-gamma (INF-γ)   292, 306, 311, 312,
        313
Interleukin (IL)   51, 249, 292, 293, 294, 298, 306,
        307, 309, 311, 312, 313, 351, 359
Intestinal ulceration   106
Intestine (cancer of)   284
Intrahepatic cholestasis   105
Iproplatin   66
Iressa   378
Irinotecan   7, 53, 113, 123, 124, 233, 239, 254,
        255, 290, 358, 377, 379, 382, 383,
        384, 385
Irriversible steroidal activators   171
ISIS-2503   324

Ispinesib   235, 236
Ixabepilone (BMS-247   550) 82

Jaundice   172
Jejunum/ileum   177
JM-216 (also satraplatin)   66, 73

Keratose(s)   97
Kidney (cancer)   55, 70, 233, 293, 309
Klinefelter's syndrome   159
KU59436   275

Laniquidar   271, 272
Lanreotide   178, 179
Lapatinib (Tyverb)   239, 240
LentiMax™   299
LErafAON-ETU   324
Lethargy   147
Letrozole (Femara)   171, 173, 378, 384, 385
Leucopenia   94, 105, 113, 135, 141, 143, 150, 295
Leucovorin   94, 98, 233, 313, 315, 379, 383
Leukaemia   8, 82, 85, 91, 92, 96, 101, 104, 105, 121, 133, 137, 143, 223, 229, 236, 264, 289, 307, 389
    Acute lymphoblastic (ALL)   92, 101, 104, 105, 106, 107, 108, 119, 120, 137
    Acute lymphocytic   125, 126, 389
    Acute myelocytic   121
    Acute myelogenous   125, 370
    Acute myelogenous   137
    Acute myeloid (AML)   101, 104, 105, 106, 107, 108, 119, 120, 124, 235, 289, 327, 334
    Acute non-lymphocytic (ANLL)   101, 106, 142
    Acute promyelocytic (APL)   92, 101, 104
    Adult T-cell   351
    Chronic lymphoblastic   137
    Chronic Lymphocytic (CLL)   85, 106, 107, 135, 136, 287, 288, 327
    Chronic myelogenous/myeloid (CML)   101, 106, 108, 121, 125, 137, 143, 229, 230, 236, 238, 305
    Hairy cell   107
Leuprolide   22, 379
Levulan (ALA)   188, 197, 198, 201, 208
Lipoplatin   54, 57, 58, 59, 60, 64, 71, 390
Liposome(s)   57, 58, 191, 305, 306, 323, 390
Lipoxal   54, 62, 63, 64, 390
Liver cancer   50, 55, 63, 100, 177, 209, 351

L-NDDP   69, 70
Lobaplatin   65
Lomustine   145, 146, 276, 383
Lung cancer   49, 50, 55, 83, 85, 92, 100, 119, 121, 124, 125, 134, 137, 139, 145, 146, 147, 162, 205, 223, 224, 226, 228, 236, 251, 254, 284, 289, 290, 298, 299, 309, 315, 316, 339, 354, 357, 377, 380, 385, 387, 388
    Non-Small Cell (NSCLC)   61, 82, 85, 86, 94, 95, 100, 115, 162, 226, 228, 233, 235, 236, 237, 239, 251, 290, 299, 305, 309, 325, 326, 328, 335, 357, 358, 377, 385
    Small-cell (SCLC)   8, 61, 69, 113, 124, 126, 141, 176, 236, 305, 377
Lung fibrosis   116
Lupron   22
Lupus erythematosus   160, 174, 189, 290
Lutetium texaphyrin   188, 197, 202, 203
Lutrin   197, 202
LY2181308   324
LY2275796   324, 326
Lymphadenopathies   363
Lymphoma   8, 83, 85, 92, 94, 101, 104, 119, 121, 133, 134, 136, 147, 276, 305, 309, 351, 362
    Burkitt's   92, 137, 351
    Central nervous system (CNS)   92
    Cutaneous T-cell   107, 134, 137, 198, 362
    Follicular   135, 237
    High grade   92
    Histiocytic   84
    Lymphocytic   84
    MALT   351, 362
    Mantle-cell   237
    Non-Hodgkin's (NHL)   85, 92, 100, 101, 106, 107, 115, 119, 121, 125, 126, 134, 136, 137, 141, 145, 146, 147, 236, 284, 287, 288, 289, 290
    T-cell lymphoblastic   107
Lyn tyrosine kinase   72

2-(3-methoxyphenyl)benzimidazole-4-carboxamide   34
2-methoxyoestradiol (2ME2)   253
Macroglobulinaemia   135
Macugen (Pegaptanib/NX1838)   332, 333
Macular degeneration (AMD)   324, 333, 337
Malaise   94, 291
Malignant pleural effusions   115
Malignant pleural mesothelioma (MPM)   94
Mannitol   57
MAP-kinase (MAPK)   70, 248, 249

Materia medica   9

Maximum tolerated dose (MTD)   62, 63

Mechlorethamine   54, 134, 383

Medroxyprogesterone   172

Medulloblastoma   146, 147

Megace   269, 270

Megesterol acetate (Megace)   172, 173

Melanoma   50, 55, 83, 102, 106, 112, 116, 123, 136, 145, 146, 147, 149, 150, 152, 197, 236, 253, 272, 276, 277, 293, 294, 295, 299, 305, 309, 311, 312, 327

Melatonin   57

Meningitis lymphomatous   101

Mephalan   54, 61, 85, 112, 135, 136

Mercaptan   139

Mercaptopurine   53, 91, 104, 105, 106

Mesalazine   105

Mesna   139, 313

Mesothelioma   55, 69, 70

Metallothionein   72

Methotrexate   53, 54, 72, 91, 92, 93, 94, 96, 99, 383

Methylene blue   211

Methylene tetrahydrofolate (FH4)   41

Methylhydrazine   141

Methylmelamine(s)   140

Methylprednisolone   101

Metvix   188, 197, 198

MGMT (O^6-methylguanine-DNA-methyltransferase) (inhibitors)   274, 275, 276, 277

Mismatch repair proteins   57

Mitogen-activated protein kinase phosphatase-1 (MKP-1)   70, 338

Mitomycin   53, 54, 112, 113, 114, 115, 383

Mitoxantrone   379

MKC-1   236

Morphine   353

Motexafin lutetium   197

mTHPBC   188, 209

mTHPC   188, 193, 197, 206, 207, 209

Mucin 1 (MUC1)   22, 284, 298, 299, 339, 340

Mucositis   95, 138, 143

Multidrug resistance (MDR) gene/pump   71, 268, 272, 273

Multidrug resistance inhibitors (P-gp inhibitors)   269, 270, 271, 272

Multidrug resistance protein-1 (MDR-1; P-gp, ABCB1)   268, 269, 270, 271, 272, 273

Multidrug resistance   86

Multidrug resistance-associated protein (MRP)   71

Mustine   134

MVA-MUC1-IL2   309

Myalgia   172, 291

Mycosis fungoides   125, 134, 137, 145, 210

Myelodysplasia   101, 102, 104, 106, 108

Myelodysplastic syndrome(s)   237

Myelofibrosis   237

Myeloma (multiple)   85, 106, 136, 137, 145, 146, 253

Myelosuppression   60, 69, 113, 121, 123, 124, 126, 136, 139, 140, 143, 146, 152, 291

Myelotoxicity   59, 60, 62, 63, 313

Nanoparticle(s)   58, 191, 193, 212, 323, 328

Nasopharyngeal   351

Natural Products   3, 4

Nausea   59, 60, 61, 62, 63, 68, 94, 95, 97, 106, 113, 116, 119, 120, 138, 139, 142, 143, 146, 147, 150, 152, 168, 172, 291

Necrotic skin lesions   179

Nedaplatin   56, 65, 382

Nelarabine   107

Neoangiogenesis   58

Nephroblastoma   61

Nephrotoxicity   57, 59, 60, 62, 65

Neuroblastoma   119, 121, 125, 126, 136, 137, 139, 305

Neurodegenerative disease   13, 160, 162, 174

Neuroendocrine tumours   176, 177, 178, 179, 180, 181

Neuropathy   55, 57, 62, 63, 82, 142

Neuroprotection   162

Neuroretinitis   146

Neurotoxicity   41, 57, 59, 60, 62, 63, 65, 98

Neutropenia   60, 62, 63, 68, 82, 98, 124, 147, 152, 313, 358

Neutrophil elastase   340

Neutrophil gelatinase-associated lipocalin (NGAL)   57

NF-κB   10

Nightmares   142

Nilotinib   230

Nimustine   145, 147

Nitric oxide synthase inhibitors   57

Nitric oxide   57, 250, 253, 312

Nitrogen mustard(s)   54, 133, 136, 137, 140

Nitrosoureas   54, 144, 145, 148

NK611   126

Nolatrexed   96

Non-steroidal anti-inflammatory drugs (NSAIDs)   272, 353, 354, 355, 356, 357, 358, 363

Non-steroidal inhibitors   32
NPe6   188, 192, 197, 205
NU1025   275, 276
NU1085   275, 276
Nucleolin   333, 334

Obatoclax   237
Obesity   160
Oblimersen (genasense/G-3139)   324, 326, 327
Octreotide   178, 179
Ocular neovascularization   332
Oedema   172, 198, 205
Oesophagus (cancer of) / oesophageal cancer   50,
   55, 92, 97, 99, 100, 103, 113, 197, 226,
   335, 354, 357
Oestrogen deprivation therapy   170
Oestrogen receptor downregulators (ERDs)   160,
   169, 170
Oestrogen   32, 160, 161, 162, 163, 166, 169, 172,
   173, 174, 175, 176, 378
OGX-011   324, 325, 326
OGX-427   324
Oligodendroglioma   146
Oligospermia   94, 106
Oncolytic HSV   315
Ondansetron   63
ONT-093   271, 272
Onyx-15   315
Oral cavity (cancer of)   50
Oropharyngeal (cancer)   255
Orotate phosphoribosyltransferase   103
Osteoporosis   13, 160, 166
Osteosarcoma   92, 137
Ototoxicity   55, 57, 59, 60, 65
Ovarian cancer   5, 7, 55, 61, 64, 70, 72, 82, 86,
   97, 100, 101, 119, 120, 121, 122, 124, 125,
   136, 137, 139, 140, 160, 162, 191, 233,
   235, 236, 254, 287, 295, 305, 310, 314,
   325, 327, 335, 339, 340, 361, 380, 382
Oxaliplatin   53, 54, 56, 61, 62, 63, 65, 69, 71,
   379, 382, 384, 385
Oxidation   11

10-propargyl-5,8-dideazafolate   42
2-phenyl-1H-benzimidazole   33, 34
L-phenylalanine mustard-based agents   54
P276-00   236
Pacific Yew tree   7
Paclitaxel   53, 61, 79, 80, 81, 82, 86, 87, 100,
   101, 113, 231, 239, 306, 310, 311, 358,
   377, 378, 380, 382, 384, 385

Palmar–plantar erythrodysesthesia syndrome   97
Pancreas (cancer of) / Pancreatic cancer   55, 60,
   63, 86, 95, 97, 99, 100, 103, 113, 115, 148,
   176, 177, 179, 233, 237, 298, 299, 309,
   339, 357, 372, 373, 380, 381, 385
Pancreatitis   381
Pancytopenia   94
Panitumumab (Vectibix)   288, 290, 377, 379
Paracetamol   353, 354
Paraesthesia   150
Parenchymal cell necrosis   105
Parkinson's disease   163
PARP (poly(ADP-ribosylation)) (inhibitors)   274,
   275, 276
Pasireotide   178
Pc4   189, 210
Pelitrexol   109
Pentostatin   53, 106
Peptide receptor radionuclide therapy (PRRT)
   181
Peripheral artery disease   305
Permetrexed   53, 92, 94, 95, 377
Peroxisome proliferator-activated receptor ligand
   176
Pertuzumab (Omnitarg)   284
P-glycoprotein   86, 96
Phagocytosis   58
Phenytoin   93
Pheochromocytoma   176, 361
Phorbol 12,13-dibutyrate (PDBU)   36
Photochlor   197, 208, 209
Photodynamic therapy (PDT)   187, 189, 191,
   192, 193, 194, 196, 199, 200, 201, 202, 203,
   204, 205, 206, 207, 208, 209, 210, 211, 212
Photolyase   73
Photosensitizer(s) (PS)/dye   187, 188, 189, 190,
   191, 192, 193, 194, 195, 196, 197, 198,
   199, 200, 201, 202, 204, 205, 206, 207,
   208, 210, 211, 212
Phototoxicity   150, 191, 194, 198, 206
Phthalocyanine   190, 191, 193, 196, 197, 209,
   210
Phytochemical   9, 10, 11
Picoplatin   66, 68, 69
Platelet toxicity   61
Platelet-derived growth factor (receptor)
   (PDGF/PDGFR)   230, 235, 239, 336, 337
Platinum   53, 54, 100, 229, 251, 289, 306, 310,
   382
Pleiotropy   13
Pleural effusion   63
Podophyllotoxin(s)   8, 53, 54, 124, 125, 126

Podophyllum emodii 8
Podophyllum peltatum 8, 124
Poly (ADPribose) polymerase-1 (PARP-1) 33
Polyarteritis nodosa 137
Polycystic ovary syndrome 174
Polycythaemia vera 136
Polyethylene glycol (PEG) 22, 313, 323, 333, 337, 341
Polymyositis 137
Polyphenolics 11
Porphyria 189
Porphyrin 33, 190, 191, 192, 193, 194, 196, 197, 198, 201, 202, 204, 206, 208, 209, 210, 211
Porphyromycin 113, 115
Potassium oxonate 102
Pralatrexate 95, 96
Precocious puberty 22
Prednisolone, 105, 134, 135, 136, 326, 383
Procaine HCl 57
Procarbazine 134, 141, 142, 147, 383
Progesterone 174, 175, 176
Progestin(s) 161, 172, 173, 176
Prostate cancer 22, 49, 50, 55, 61, 69, 82, 106, 122, 148, 159, 160, 162, 197, 223, 233, 236, 249, 252, 253, 289, 297, 299, 305, 309, 315, 323, 324, 325, 326, 328, 337, 338, 339, 341, 354, 355, 357, 358, 361, 378, 379, 385
Prostate specific membrane antigen (PSMA) 324, 337, 338
Prostate-specific antigen (PSA) 361
Protein kinase 35, 226, 229, 230, 237
Proton pump inhibitors 179
Protoporphyrin 188, 191, 194, 197, 198
Pruritus 152, 381
PSC833 270
Pt-ACRAMTU 68
Pulmonary embolus 164
Pulmonary fibrosis 143, 146, 147
Pulmonary toxicity 100, 116, 147
Purine antimetabolites 53, 104
Purpurins 193, 204, 209
Puryltin 197
Pyrimidine antimetabolites 53, 96

Quinone PK-C agonist 36
Quinine 269, 270

Raloxifene 13, 162, 166, 167, 168, 170, 241
Raltitrexed 53, 95
Rapamycin (Sirolimus) 181, 237, 241, 249, 252
Rash 143, 152

Rectum (cancer of) 50, 177, 379
Renal cell cancer/carcinoma (RCC) 85, 100, 106, 115, 233, 234, 237, 238, 239, 290, 291, 295, 299, 325, 334, 337
Renal damage 57
Renal toxicity 55, 63, 100, 146
Resistance 41
Restenosis 305
Resveratrol 13
Retinoblastoma 101, 137, 335
Retinoic acid 13
Retinoid 176
Retinopathy 333
Retrovirus(es) 305, 306, 307, 308
Reversible nonsteroidal imidazole-based inhibitors 171
Rhabdomyosarcoma 111, 125
Rheumatoid arthritis 162, 290, 305
Rhinitis 291
Ribonucleotide reductase 53
Ribozyme 273, 318, 320, 322, 323, 324, 327, 328, 329, 340
Rituximab (Rituxan MabThera) 101, 136, 137, 288, 289, 290, 291, 384
RNA interference (RNAi/siRNA) 51, 252, 273, 318, 321, 322, 323, 324, 328, 329, 338
Rose Bengal 191, 193
RU 58 668 170

S-1 102, 103
Salicylates 93
Sapacitabine 104
Sarcoma 5, 112, 115, 119, 122, 136, 137, 139, 146, 150, 211
    Ewing's 137, 139
    Kaposi's 8, 50, 115, 119, 120, 121, 125, 197, 209, 351
Satraplatin 66
Scleroderma 137
Seizures 150
Selective oestrogen receptor modulators (SERMs) 13, 160, 162, 163, 165, 166, 167, 168, 169, 170, 172, 174, 175
Selenium 57
Serotonin 176
Sex steroid therapies 161, 172
Skin cancer 116, 197, 205, 209, 210
Skin necrosis 94
Skin rashes 106, 112, 116, 150
SLC22A16, 176
SN38 123

SnET2   189, 197, 209
SNS-032   236
SOM-230   179
Somatostatin / somatostatin analogues (SSA)
    176, 177, 178, 179, 180, 181
Somatostatinoma(s)   176
Sorafenib   238
Sphingosine-1-phosphate   72, 73
Spiegelmer   340, 341
Spironolactone   57
Squamous cell cancer/carcinoma   115, 116, 315
SR   16 234 170
Staphylococcal enterotoxins   362
Steatorrhoea   180, 381
Stem cells   266, 268
Steroidal analogues   32
Stevens–Johnson syndrome   94
STI-571   336
Stilboestrol   172
Stomach cancer   49, 50, 55, 113, 121, 176, 177
Stomatitis   98, 120, 142, 143
Streptomyces achromogenes   147
Streptomyces antibioticus   106
Streptomyces caespitosus   112
Streptomyces parvulus   111
Streptomyces peucetius   118
Streptomyces verticellus   115
Streptomyceshygroscopicus   237
Streptozocin   145, 147, 148
Sulfonamides   93
Sulindac   272, 273
Sulphasalazine   105
Sunitinib (Sutent)   181, 233, 234, 235, 238
Systematic Evolution of Ligands by Exponential
    Enrichment (SELEX)   23, 24, 331, 332,
    333, 334, 336, 339, 341
Systemic lupus eryhtematosus   137

(19R)-10-thiiranylestr-4-ene-3,17-dione   32
6-thioguanosine-5'-phosphate (6-thioGMP)   104
6-thioinosine monophosphate (TIMP)   104
Tachycardia   146
Tachyphylaxis   179, 180
Talotrexin   96
Tamoxifen   13, 159, 162, 163, 164, 165, 166, 167,
    168, 169, 170, 171, 172, 173, 174, 175, 378
Tannins   11
Tariquidar   271, 272
TAS-108   170
TAT-59   168
Taxane(s)   9, 53, 54, 62, 70, 79, 81, 82, 85, 86,
    87, 100, 264, 292, 378, 382

Taxol   7, 9, 12, 68, 79, 316, 336
Taxonomic   9
Taxotere (also docetaxel)   7, 9, 79
Taxus baccata   79
Taxus brevifolia   7, 9, 79
Taxus   9
Tegafur   102
Temozolomide   149, 150, 151, 152, 275, 276
Temsirolimus (Torisel)   237, 238, 241
Tenascin   339
Teniposide   8, 53, 125, 126, 383
Terpenoids   18
Testicular cancer   8, 55, 64, 84, 111, 115, 125,
    126, 137, 139, 223
Testosterone   32, 172, 378
Tetracyclines   93
Tetrahydrofolate   92, 93
Tetraplatin   66
TG1042   299, 309
TG4001   299
TG4010   299
TGDCC-E1A   309
Thalidomide   136
Thioguanine   107
Thiotepa   140, 141
Thiouric acid   105
Thrombocytopenia.   60, 68, 105, 113, 135, 141,
    143, 147, 150, 152, 313
Thromboembolic event(s)   172
Thymidine phosphorylase   99
Thymidine triphosphate (TTP)   97
Thymidylate synthase (TS)   41, 53, 92, 98, 103
Thymoma   137
Thyroid cancer   119, 176, 239
Tiredness   97
TNFrade^{TM} Biologic   298, 299, 309
Tookad   188, 204
TOP   53 126
Topoisomerase inhibitors   62
Topotecan   7, 53, 123, 124, 254, 382
Toremifene   165
Tositumomab (Bexxar)   288, 291
Toxic epidermal necrolysis   94
Trabectedin   5, 111, 121, 122
transforming growth factor-β (TGF-β)   351, 359,
    360
Transplatin   66
Trastuzumab (Herceptin)   53, 283, 288, 291, 292,
    378, 384, 385
Tretamine   140
Triazine(s)   149
Triciribine phosphate   108

Triethylenemelamine   140
Trimethoprimsulfamethoxazole   105
Triphenylchlorethylene   172
Triphenylmethylethylene   172
tris-tetrahydroisoquiloline   5
Trophoblastic neoplasms   111
Tubular necrosis   62
Tubulin   8
Tumour lysis syndrome   94
Tumour necrosis factor (TNF)   192, 298, 306,
    309, 311, 313, 315, 351, 359, 360
Turner's syndrome   159
Tyrosine kinase inhibitor(s)   226, 233, 234
Tyrosine kinase(s)   22, 226, 227, 228, 229, 230,
    232, 233, 234, 235, 238, 239, 247, 248,
    249, 252

Ulcerative stomatitis   94
Uracil (UFT)   102
Urticaria   291
Uterus (cancer of)   50, 55

Vaginal bleeding   172
Vaginal discharge   164, 167, 172
Vaginal dryness   164
Vandetanib (Zactima)   239
Vascular endothelial growth factor (receptor)
    (VEGF/VEGFR)   22, 53, 232, 233, 235,
    239, 247, 248, 284, 288, 327, 328, 332,
    333, 337, 351, 357, 360, 361
Vasculitic syndromes   137
Veno-occlusive disease   143
Verapamil   86, 269, 270, 271
Verteporfin   188, 197, 207, 208, 249, 250, 251,
    252, 253, 254, 255
Vidarabine   106
Vinblastine (Velbe)   10, 11, 53, 82, 83, 84, 85, 86,
    134, 150, 377, 380, 383
Vinca alkaloids   53, 54, 82, 83, 84, 85, 86, 87,
    380

Vinca rosa   10
Vincristine (Oncovin)   10, 11, 82, 83, 84, 86, 105,
    134, 147, 313, 380, 383
Vindesine (Eldesine)   10, 82, 83, 85, 383
Vinflunine (Javlor)   82, 83, 84, 85, 86
Vinorelbine (Navelbine)   10, 53, 61, 82, 83, 84,
    85, 100, 377, 378, 383, 384, 385
Vioxx   355
Virilization   172
Vitamin E   57
Vomiting   59, 60, 61, 62, 63, 68, 94, 95, 106, 113,
    116, 119, 120, 138, 139, 142, 143, 146,
    147, 150, 152, 291
Vorozole   171

Waldenstrom's macroglobulinaemia   107
Warfarin   106
Wegener's granulomatosis   137
Wilm's tumour   61, 111, 119, 125, 137, 333
Withania somnifera   4
Wortmannin   274

Xanthine oxidase inhibitor   105
Xeroderma pigmentosum   62
XIAP   43

YC-1   253
Yondelis   5

ZD   164 834 170
ZD0473   71, 73
Zevalin (see Ibritumomab tiuxetan)
Ziconotide   5
Zinc histidine complex   57
ZK   191 703 170
Zoladex   22
Zollinger–Ellison syndrome   176, 179
Zosuquidar   271, 272